Decoding the Aryabhatiya

Aryabhata's Mathematical Universe

Harshad Rane

Contents

To the seekers of cosmic truth, past and present, whose curiosity illuminates the heavens and whose courage charts the unknown.

And to my family and all souls whose sacrifices, great or small, have paved the way for me to be here.

Disclosures

Use of Digital and AI Technologies in This Work

In pursuit of the goal to bring ancient knowledge, such as the wisdom contained in the Aryabhatiya, to the modern world as effectively and rapidly as possible, this book, *Decoding the Aryabhatiya: Aryabhata's Mathematical Universe*, was developed with the strategic assistance of digital and artificial intelligence (AI) technologies.

Digitization and Initial Research: Optical Character Recognition (OCR) sped up access to primary sources by digitizing ancient Sanskrit transcripts. Early research also leveraged Large Language Models (LLMs) for preliminary translations and semantic exploration of key terms. These AI insights served as a useful reference to broaden initial inquiries and build a conceptual foundation efficiently.

Research and Drafting Support: AI tools streamlined literature searches and helped organize complex data. AI also generated initial drafts for contextual sections (e.g., Gupta period history, summaries of astronomical concepts). These drafts were starting points only, undergoing thorough human review, critical revision, and integration of original analysis to ensure accuracy.

Author's Original Contribution: While AI enhanced the speed of preliminary tasks, the core scholarly work—including the detailed interpretation of Sanskrit, final translations, and nuanced analysis—remained a meticulous manual endeavor. The synthesis of information, the intellectual arguments, the book's conceptual structure, and the final text are the author's original contributions, based on close study of primary and secondary sources and a commitment to fresh perspectives.

Visual Elements: Most diagrams were created by the author using Python libraries or custom scripts. A few illustrative images were produced with graphic design tools like Canva, which may use AI features. All visuals aim to clarify Aryabhata's ideas.

Conclusion: The use of digital tools and AI supported the efficient research and drafting necessary for this project, aiding the larger mission of disseminating ancient texts. Nevertheless, the author takes full responsibility for the scholarly integrity, critical analysis, and final articulation of *Decoding the Aryabhatiya*.

About the author

Harshad Rane

Harshad Rane, an engineer with a deep reverence for timeless knowledge, quietly seeks to bridge ancient wisdom with the modern world through his work in artificial intelligence. Holding a Engineering degree in Electronics and Telecommunications and MBA, he brings careful precision and a storyteller's curiosity to Aryabhata's 5th-century text, the *Aryabhatiya*.

His journey began with childhood tales about the origins of zero, which sparked a lifelong fascination with Aryabhata's mathematical insights, particularly his value for π. Drawn to the Gupta Empire's era of discovery, Harshad uses digital tools to uncover the *Aryabhatiya*'s enduring lessons.

In *Decoding the Aryabhatiya: Aryabhata's Mathematical Universe*, he aims to make this 1,500-year-old text accessible to today's readers, revealing its quiet relevance to fields like AI and space exploration. Inspired to share universal knowledge from any culture or era, Harshad aspires to bring other ancient works into contemporary light with clarity and care. For him, timeless insights are a gentle conversation across centuries, guiding our shared quest to understand the cosmos.

Preface

My journey to Aryabhata began with a familiar tale from my school days in India: the story of his connection to the invention of zero. That simple idea ignited a spark of wonder, but it was years later, when I stumbled upon a verse from the *Aryabhatiya* revealing his astonishingly precise approximation of π, that I was truly captivated. The contrast between the popular narrative and the sophistication of his mathematics drew me irresistibly to the source—a concise yet profound text of 121 Sanskrit verses, crafted around 499 CE by a young scholar in his early twenties, thriving in the intellectual brilliance of the Gupta Empire.

The *Aryabhatiya* unveiled a universe expressed through mathematical poetry, encompassing innovative numerical systems (*Daśagītikā*), essential computations (*Gaṇitapāda*), cosmic timekeeping (*Kālakriyāpāda*), and a groundbreaking vision of the celestial sphere (*Golapāda*). Aryabhata wove together precise calculations and a deep reverence for the cosmic order, reflecting a worldview that saw the divine in the measurable. This interplay of science, philosophy, and cosmology, so emblematic of ancient India, demanded a careful approach—one that honors the historical context while delving into the rigor of his mathematics and astronomy.

This book stems from a personal quest to understand not just Aryabhata's technical achievements but also the vibrant Gupta era that fostered such genius. What conditions enabled this golden age of innovation? How did Aryabhata build upon and advance existing knowledge? And how did his ideas echo through time, shaping Indian scholarship and the global path of science?

Through dedicated study, I've sought to bridge Aryabhata's world with our own, offering my own interpretations and synthesis in these pages. More than a technical analysis or translation, this book aims to recapture the intellectual excitement of Aryabhata's time and convey the awe I felt in uncovering the depth behind the myth. My hope is that readers will share in this sense of discovery, glimpsing the universe through Aryabhata's remarkable intellect—a universe that still speaks to us across millennia, inviting new exploration.

Acknowledgements

This book has been a journey of learning and exploration, and it would not have been possible without the inspiration, research, and support of many.

I owe an immense debt of gratitude to the long lineage of scholars—historians, mathematicians, and astronomers from Bhaskara I to our own time. Their meticulous efforts to preserve, translate, and interpret Aryabhata's genius and the rich intellectual world of ancient India provided the very foundation upon which this work stands. It is their dedication that has kept these profound insights alive across the centuries.

To my family, teachers, friends, colleagues, and mentors, I offer my heartfelt thanks. Your consistent belief, remarkable patience, and insightful feedback have been invaluable. Your encouragement often served as an anchor as I navigated the complexities of ancient Indian science, and your thoughtful critiques helped to sharpen the focus and clarity of this work. Your support was a grounding force while my thoughts journeyed through the cosmos of 5th-century India.

Finally, I wish to express my deep respect for the intellectual giants, Aryabhata and his esteemed commentator Bhaskara I, whose brilliance first sparked this endeavor. While any errors or shortcomings found within these pages are mine alone, it is my sincere hope that this book might, in some small way, reflect the profound insights of these masters and the wisdom of all those who have generously guided my understanding along this path.

This Book Is For

Decoding the Aryabhatiya: Aryabhata's Mathematical Universe is intended for a diverse audience united by a curiosity about the history of science, mathematics, and ancient civilizations. It is written for:

- **Students** of history, mathematics, astronomy, physics, and South Asian studies seeking a deeper understanding of a pivotal figure and a transformative era in scientific thought.
- **Scholars and Researchers** looking for a comprehensive synthesis of Aryabhata's work, meticulously placed within its historical and scientific context, and bridging primary textual analysis with broader comparative perspectives.
- **Educators** seeking accessible, enriching material to broaden curricula on the history of science by exploring foundational scientific achievements from civilizations beyond the European tradition.
- **Enthusiasts** of astronomy and mathematics eager to delve into the origins of key concepts and witness the profound ingenuity of ancient thinkers.
- **General Readers** with an interest in India's rich intellectual heritage and the universal human quest to comprehend the cosmos.

While firmly grounded in scholarly research, the aim of this book is to transform the dense elegance of Aryabhata's Sanskrit stanzas into an engaging and accessible narrative. My hope is that these pages awaken in you the same sense of awe that drew me to his work, allowing you to glimpse a cosmos illuminated by an ancient, yet remarkably modern, intellect—a universe waiting to be explored anew.

Introduction

Long before Copernicus challenged the cosmic order of the West, a young scholar in 5th-century India named Aryabhata unveiled his own revolutionary vision of the heavens. Working around 499 CE near the vibrant city of Kusumapura, at the zenith of the Gupta Empire's "Golden Age," he composed a masterpiece that would echo through a millennium.

Encapsulated within 121 concise Sanskrit verses, his seminal work, the *Aryabhatiya*, presented a sophisticated system that seamlessly integrated mathematics, astronomy, and a profound sense of cosmic order. It stands as a landmark achievement in the global history of science, and this book will transport you into the fertile intellectual environment that produced it.

Our journey will follow the four key sections, or *pādas*, of Aryabhata's great work:

- **The *Gītikāpāda* (The Tenfold Echoes):** The opening overture, this section lays down the foundational constants of his universe. It introduces a unique alphabetical number system to encode vast cosmic cycles and the precise rhythms of the planets.

- **The *Gaṇitapāda* (The Mathematical Engine):** Here, Aryabhata forges the essential mathematical tools for his system, including arithmetic, algebra, geometry, a remarkably accurate value for π, and the seeds of trigonometry.

- **The *Kālakriyāpāda* (The Reckoning of Time):** This section defines the very fabric of time, from fleeting breaths to vast cosmic ages. It details the methods for calculating planetary positions using a sophisticated system of "wheels upon wheels," or epicycles.

- **The *Golapāda* (The Celestial Sphere):** In his final and most revolutionary section, Aryabhata maps the heavens, explains the true, shadow-based causes of eclipses, and boldly proposes the radical idea of the Earth's daily rotation on its axis.

Throughout this exploration, we will rely on the indispensable 7th-century commentary of Bhāskara I, whose detailed explanations unlock Aryabhata's often cryptic verses. We will also delve into the instruments he likely used, the spirited debates his work ignited, and the eventual journey of his ideas across continents.

Why does Aryabhata's legacy endure? Because he represents more than a historical figure; he embodies the power of mathematical reasoning to unlock the secrets of the universe. His work contains the seeds of modern trigonometry, calculations of astonishing precision, and a holistic vision that unites scientific inquiry with a philosophical appreciation for cosmic harmony.

This book aims to make that vision accessible. Join us as we decode the universe that Aryabhata, with youthful brilliance, first expressed in verse.

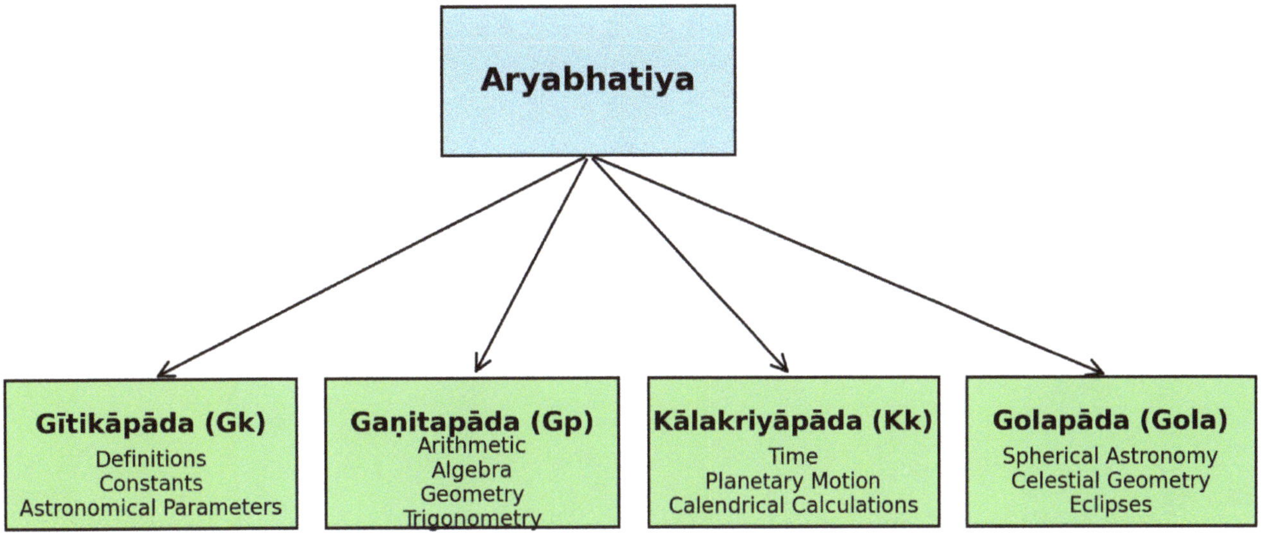

Figure Introduction.1: Structure of Aryabhata's Aryabhatiya, illustrating its division into four main sections (pādas): the Gītikāpāda (foundations and constants), the Gaṇitapāda (mathematics), the Kālakriyāpāda (time and planetary motion), and the Golapāda (spherical astronomy). This structure reflects Aryabhata's integrated approach to mapping the cosmos through calculation.

Chapter One

The Aryabhata Spark: Indian Astronomy, Bhaskara I, and the Need for Commentary

From Foundational Text to Living Tradition: Bhaskara I and the Interpretation of Aryabhata's Astronomy

1.1 The Aryabhata Departure

In the vast intellectual landscape of classical India, astronomy (*jyotiḥśāstra*) stands as a discipline intertwining meticulous observation with sophisticated cosmology, and rigorous mathematics with deep philosophical underpinnings. This tradition, enriched by centuries of inquiry and potential cross-cultural exchanges, saw astronomy serve practical needs such as calendrics and ritual timing, while also probing fundamental questions about the cosmos and humanity's place within it. Pre-Aryabhata astronomy was often characterized by sprawling treatises known as *Siddhāntas*, which typically blended technical astronomical parameters and rules with extensive cosmological narratives drawn from Purāṇic lore.

At a pivotal moment in this tradition stands Aryabhata (born 476 CE), whose remarkably concise and innovative *Aryabhatiya*, composed around 499 CE, heralded a new direction in Indian scientific thought. Composed in dense Sanskrit verse, the *Aryabhatiya* diverges significantly from the earlier *Siddhānta* style. Instead of lengthy prose, Aryabhata presented a compact work of just 121 verses (in most recensions), divided into four distinct chapters (*pādas*): the Gītikāpāda (containing the *Daśagītikā* verses with Time Reckoning and Astronomical Constants), the Gaṇitapāda (Mathematics), the Kālakriyāpāda (Planetary Motion Calculation), and the Golapāda (The Celestial Sphere). This structure prioritizes mathematical precision and poetic economy, compressing complex technical details into metrical stanzas that demand careful interpretation and commentary for their full explication.

Aryabhata's distinction lies in his dual mastery: he was both a brilliant innovator in *gaṇita* (computational mathematics) and a bold reformer of cosmological understanding. He consistently favored elegant mathematical models and demonstrated a willingness to challenge prevailing astronomical parameters and concepts based on his calculations. Most notably, he proposed that the apparent westward motion of the celestial sphere is caused by the Earth rotating eastward on its axis (found in Golapāda 9), a revolutionary hypothesis for its time that directly contradicted the static-Earth model common in authoritative texts like the *Sūrya Siddhānta* and the assumptions underpinning much of the Greco-Roman tradition exemplified by Ptolemy.

Two elements introduced early in the *Aryabhatiya* exemplify Aryabhata's innovative approach and signal his departure from convention:

- **Reformed Cosmic Time:** Verses Gītikāpāda 3-5 define a cosmic time cycle (*mahāyuga*) of 4,320,000 years, but critically imply its division into four equal subdivisions (*yugapādas*). This reformed the traditional, unequal cosmological timescales found in Purāṇic accounts, establishing a framework optimized for mathematical consistency and astronomical calculation.

- **A Novel Cipher:** Gītikāpāda 2 describes a unique alphabetic numeral system. This sophisticated cipher, using specific Sanskrit consonants for numerical values and vowels for powers of 100, allowed Aryabhata to encode the vast numbers required for astronomical constants within the tight constraints of Sanskrit meter. This demonstrated both mathematical ingenuity and a profound concern for efficient knowledge transmission.

Together, these innovations signal a significant shift: an astronomy grounded not solely in inherited tradition or mythological narrative, but increasingly in empirical calculation, mathematical abstraction, and reasoned innovation (*yukti*). However, the very conciseness and novelty of Aryabhata's work meant its enduring influence would depend heavily on interpreters capable of unlocking its meaning and championing its methods.

1.2 Bhaskara I: Illuminating the Verses

Aryabhata's verses, though mathematically precise, are often cryptic without elucidation. Unlocking their meaning required not only mathematical acumen but also a deep appreciation for Sanskrit's poetic compression and the nuances of the astronomical tradition. This crucial interpretative role fell primarily to **Bhaskara I** (fl. c. 629 CE). His commentaries (***bhāṣya***), particularly the comprehensive *Aryabhatiya-bhāṣya*, transformed Aryabhata's terse treatise into a workable, teachable system, ensuring its transmission and profound influence on subsequent Indian astronomy.

Bhaskara I was more than an expositor; he was a vital proponent and passionate defender of Aryabhata's system. He translated Aryabhata's elliptical statements into concrete computational procedures, provided invaluable examples, and championed his innovations against widespread skepticism. His *bhāṣya* is the earliest comprehensive Sanskrit commentary on the *Aryabhatiya* that survives, serving as our primary lens for understanding much of Aryabhata's original intent and mathematical methods today. In his introduction, Bhaskara I extols the profound knowledge embedded within Aryabhata's brevity, suggesting that Aryabhata had distilled the essence of astronomical science, comparing the *Aryabhatiya* metaphorically to potent nectar extracted from vast scriptures.

1.2.1 The Cipher of Numbers

One of Bhaskara I's essential contributions was explicating Aryabhata's unique alphabetic numeration system, detailed in Gītikāpāda 2. This system employed specific Sanskrit consonants (varga and avarga letters) to represent digits 1-25 and 30-100, combined with vowels to denote place values based on powers of 100 (100^0, 100^1, 100^2, etc.). This allowed large astronomical constants, like the number of planetary revolutions in a **mahāyuga** (a vast cosmic cycle of 4,320,000 years used by Aryabhata), to be encoded efficiently within the **gīti meter** (a specific Sanskrit poetic meter characterized by its syllabic structure) of the verse. Without Bhaskara I's meticulous explanation, complete with examples, decoding this sophisticated system might have remained obscure or indecipherable. He demonstrated how decoding required careful attention to both phonetic structure and the place value rules, a technique likely designed for mnemonic efficiency and accuracy in a tradition where oral transmission remained significant alongside written manuscripts.

1.2.2 Championing a Rotating Earth

Even more radical was Aryabhata's assertion (**Golapāda 9**) that the diurnal (daily/circadian) motion of the stars results from the Earth's own eastward rotation. This directly challenged traditional Indian cosmology (like the *pravaha* wind theory presented in Gola.10), the prevailing geocentric views influenced by Hellenistic (relating to greek history) astronomy, and intuitive sensory experience. Bhaskara I boldly defended this concept, employing Aryabhata's striking analogy (found in **Golapāda 9**) based on relative motion:

"Just as a man in a boat moving forward sees the stationary objects [on the bank] as moving backward, so are the stationary stars seen by people on earth as moving exactly westward."

This analogy reveals a sophisticated understanding of relative motion. For Bhaskara I, Aryabhata's rotating Earth was not mere speculation but a rational model (***yukti***) capable of explaining perception through logical reasoning. He saw it as a kinematic simplification superior to the traditional *pravaha* wind hypothesis, even though it wasn't a heliocentric proposal. This defense marks a significant moment, privileging mathematical reasoning and explanatory power in cosmology, a stance that would be fiercely contested by later traditionalists like Brahmagupta.

1.2.3 Defending Reformed Cosmic Time

Bhaskara I's commentary was also crucial in defending Aryabhata's division of the *mahāyuga* (4,320,000 years) into four **equal sub-yugas** (*yugapādas*) of 1,080,000 years each, as implied in **Gītikāpāda 5**. This symmetrical division contrasted sharply with the Purāṇic tradition and some earlier *Siddhāntas* which used unequal *yugas* (Kṛta, Tretā, Dvāpara, Kali in a 4:3:2:1 duration ratio). Aryabhata's choice greatly simplified astronomical calculations, ensuring that the mean positions of the planets calculated over these vast periods remained consistent and computationally manageable. Bhaskara I staunchly upheld this computational rationale, arguing that the time divisions within *jyotiḥśāstra* must prioritize mathematical coherence and predictive accuracy, even if it meant diverging from established mythological chronologies when necessary for scientific consistency.

Through Bhaskara I's tireless efforts, Aryabhata's dense work became an accessible, practical tool and simultaneously a philosophical statement: mathematical reasoning could refine, and sometimes supersede, tradition for the sake of scientific advancement. In championing Aryabhata, Bhaskara I became an indispensable bridge—connecting Aryabhata to later generations of Indian astronomers and facilitating an era where mathematical models increasingly illuminated the workings of the cosmos. His commentary ensured that the spark ignited by Aryabhata would continue to burn brightly, even amidst scholarly debate. Yet, before this spark could fully illuminate the path forward, it needed the practical tools of observation to ground its theoretical brilliance.

Figure 1.1: Map showing the Indian subcontinent during the classical period. This empire provided the political and cultural backdrop for Aryabhata's work in Kusumapura (modern Patna), located within this region.

***

Chapter Two

Bridging Sky and Scroll: Observational Instruments of Gupta India

The Practical Tools Enabling Aryabhata's Astronomical Vision

2.0 Introduction: Grounding Theory in Observation

The sophisticated mathematical and cosmological framework developed by Aryabhata and interpreted by Bhaskara I did not exist in a vacuum. It was intimately connected to the practical art of observing the heavens. While the *Aryabhatiya* itself is primarily a theoretical and computational text, its parameters and methods imply the use of specific instruments available to astronomers in Gupta-era India (roughly 4th-6th centuries CE) and the subsequent classical period. These relatively simple yet ingenious devices enabled the collection of empirical data—tracking the passage of time, mapping the Sun's path, determining location, and timing celestial events—that informed and validated the mathematical models. This chapter delves into the key tools likely employed by astronomers like Aryabhata, revealing the interplay between theoretical insight and practical observation.

2.1 The Gnomon's Versatility: Shadow as Celestial Clock and Compass

Long before the advent of telescopes, astronomers worldwide relied on the fundamental interplay of sunlight and shadow. Among the most ancient and essential tools was the gnomon—in India, typically a vertical rod or pillar of standardized height, known as the ***śaṅku***. Meticulous observation of its shadow (*chāyā*) allowed tracking the Sun's diurnal and annual motion. For Aryabhata and his contemporaries, the *śaṅku* was far more than a simple sundial; it functioned as a versatile instrument for timekeeping, latitude determination, and establishing cardinal directions, deeply integrated into their mathematical astronomy.

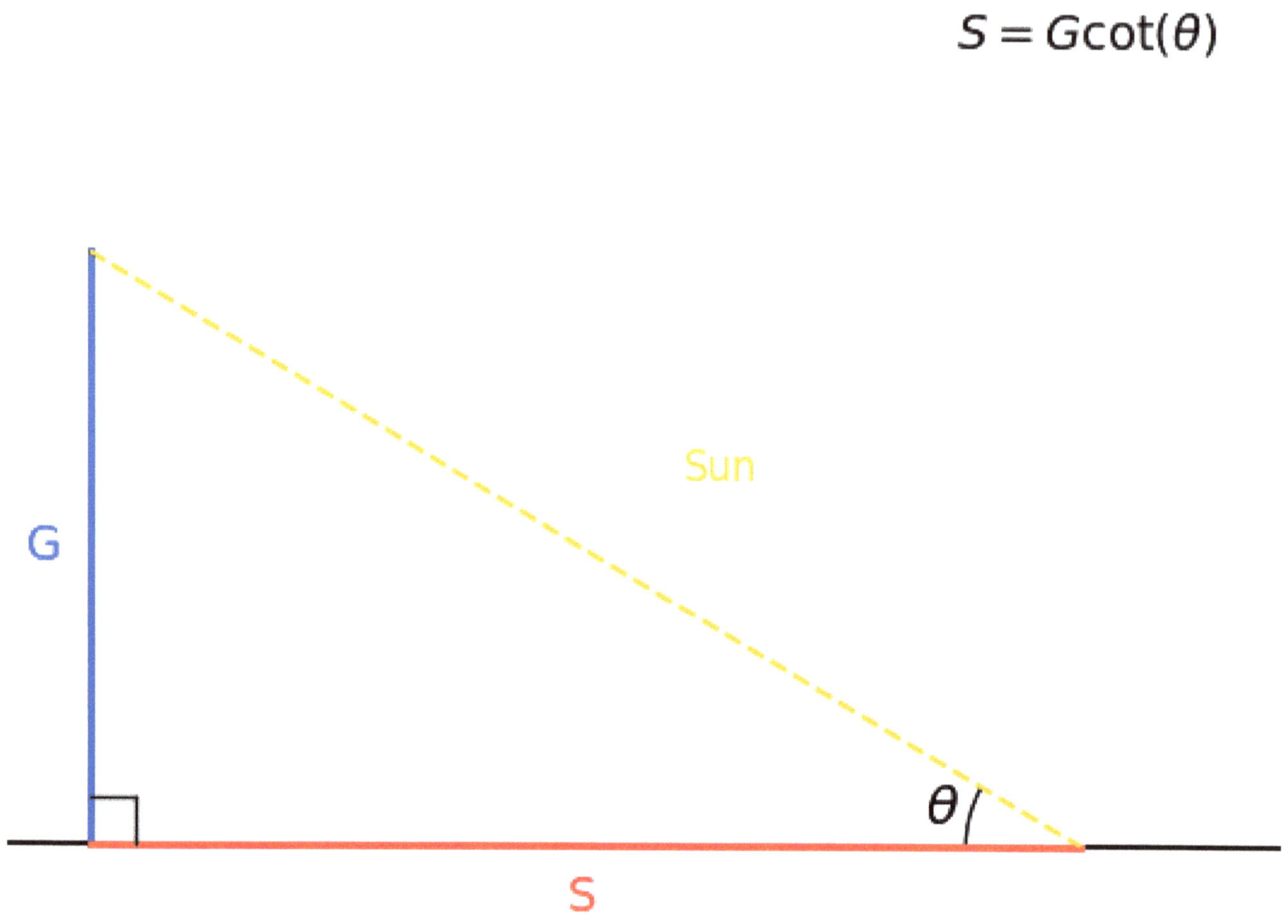

Figure 2.1: Diagram illustrating the fundamental geometry of a Gnomon (Śaṅku) casting a shadow (Chāyā). The gnomon's height (G), shadow length (S), and the Sun's ray form a right triangle. The angle θ represents the Sun's altitude above the horizon. This basic geometric relationship is used in various astronomical calculations, including determining time from shadow lengths or calculating latitude.

2.1.1 Timekeeping and Trigonometry by Shadows

The *Aryabhatiya* explicitly details methods for deriving time from shadow measurements (found in the **Gaṇitapāda**, specifically **Gp.14–16**). The process, elaborated upon by commentators like Bhaskara I, involved several key steps:

1. **Measurement:** Carefully measuring the shadow length (*chāyā*) cast by the *śaṅku* (stick) at a given moment.

2. **Calculation:** Knowing the gnomon's standardized height (often 12 *aṅgulas* or "digits"), astronomers could use geometric relationships to determine the Sun's altitude (angle above the horizon) or zenith distance (angle from the point directly overhead). This wasn't just simple proportion; it implicitly involved trigonometry, often expressed through derived lengths like the *bhuja* (base, related to shadow) and *koṭi* (perpendicular, related to gnomon height) in the right triangle formed by the gnomon, shadow, and the hypotenuse (*karṇa*, representing the sunbeam). Aryabhata's groundbreaking sine (*jyā*) table (foundation laid in **Gk.10, Gk.12**) provided the necessary trigonometric values (R·sin(altitude), R·cos(altitude)) for these precise calculations.

3. **Temporal Mapping:** By comparing shadow lengths or derived solar altitudes at different times (e.g., sunrise, noon, sunset, or intermediate points), astronomers could accurately calculate time elapsed or remaining within

the day. Determining the precise moment of local noon (when the shadow is shortest and points directly north or south) was particularly important for establishing the **meridian** (the imaginary north-south line passing directly overhead) and anchoring time calculations.

While gnomons were ubiquitous in ancient civilizations (Egypt, Babylon, Greece), the distinctive Indian contribution, exemplified by Aryabhata, lay in integrating these observational measurements directly into a sophisticated computational framework using sine-based trigonometry. This moved beyond simple empirical rules or graphical methods, allowing for greater precision and predictive power.

2.1.2 Latitude from Noon Shadows

A key use of the *śaṅku* (gnomon) was to find an observer's geographical latitude (*akṣa*). As described in the *Golapāda* (Gola.31), this involved measuring the shadow's length exactly at local noon on an equinox day.

Here's why this works:

1. **Equinox Day:** On an equinox, the Sun is directly above the Earth's equator (its declination is zero).
2. **Local Noon:** At this time, the Sun is at its highest point on the observer's north-south line (meridian).
3. **Geometric Link:** Because of this specific alignment, the Sun's angle from the overhead point (its zenith distance) is equal to the observer's latitude (φ).
4. **Calculation:**
    - The shadow length (S) and the gnomon's height (G) form a right triangle. The ratio S/G gives the tangent of the latitude ($\tan(\varphi) = S/G$).
    - Aryabhata's system, using R-sines, would calculate *akṣajyā* ($R\cdot\sin\varphi$). This was typically done by first finding the hypotenuse (*karṇa* = $\sqrt{(G^2 + S^2)}$) and then using the formula $R\cdot\sin(\varphi) = (R \times S)$ / *karṇa*.
    - **R** is Aryabhata's chosen radius (3438).

Measuring the noon shadow on solstices (when the shadow is longest or shortest) offered additional checks and helped calculate the ecliptic's tilt (the Sun's maximum north-south deviation, given as 24° in Gk.9). Knowing the latitude was vital for accurate planetary calculations, a form of early astronomical mapping. For Aryabhata, likely based in Kusumapura (modern Patna, around 25.6° N latitude), such measurements would have been standard, contributing to his system's accuracy.

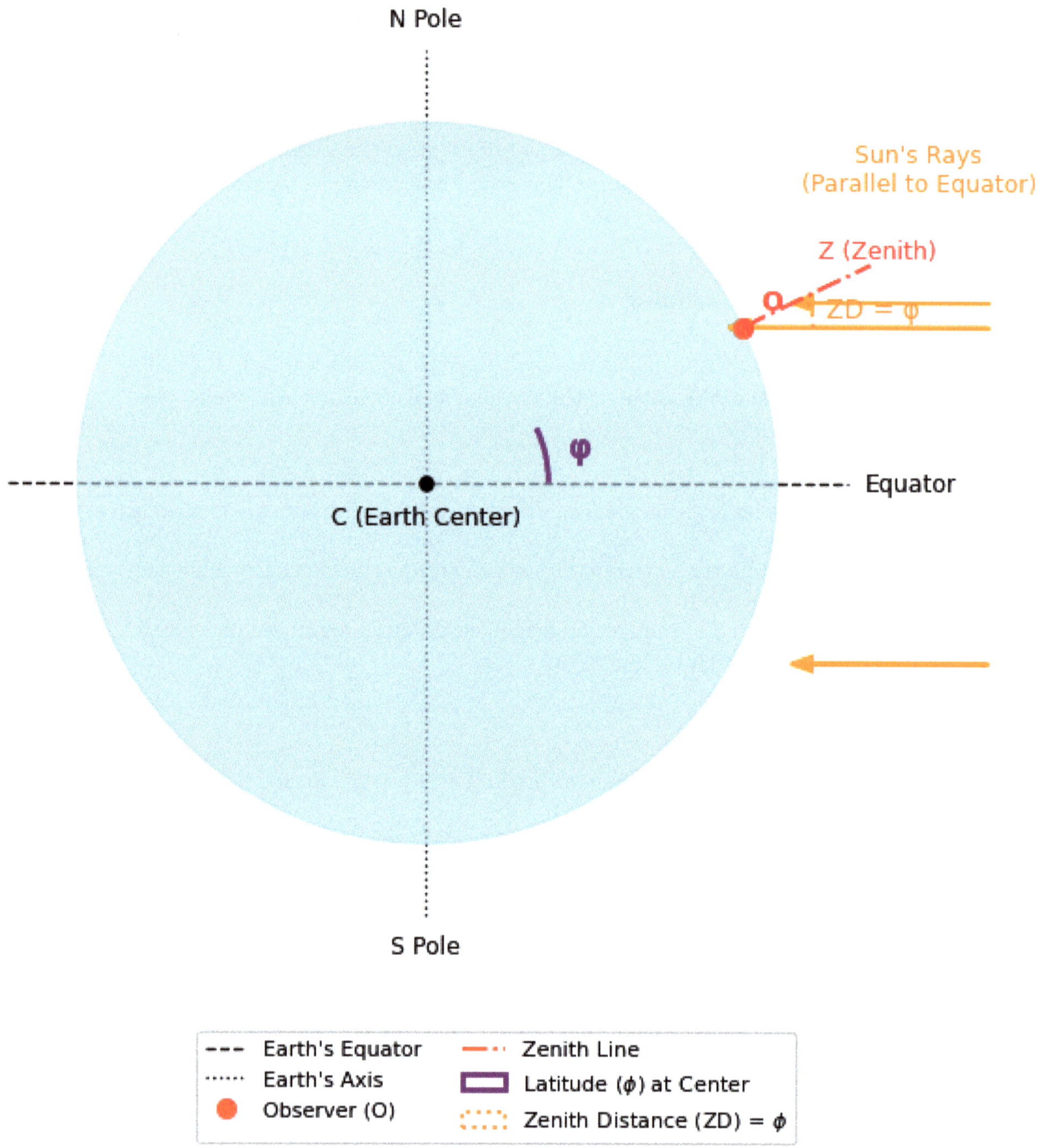

Figure 2.2a: The key takeaway is that due to the Earth's spherical shape and the Sun's parallel rays at equinox, the angle between the observer's local vertical (Zenith) and the direction of the Sun (Zenith Distance) is exactly equal to their latitude (φ).

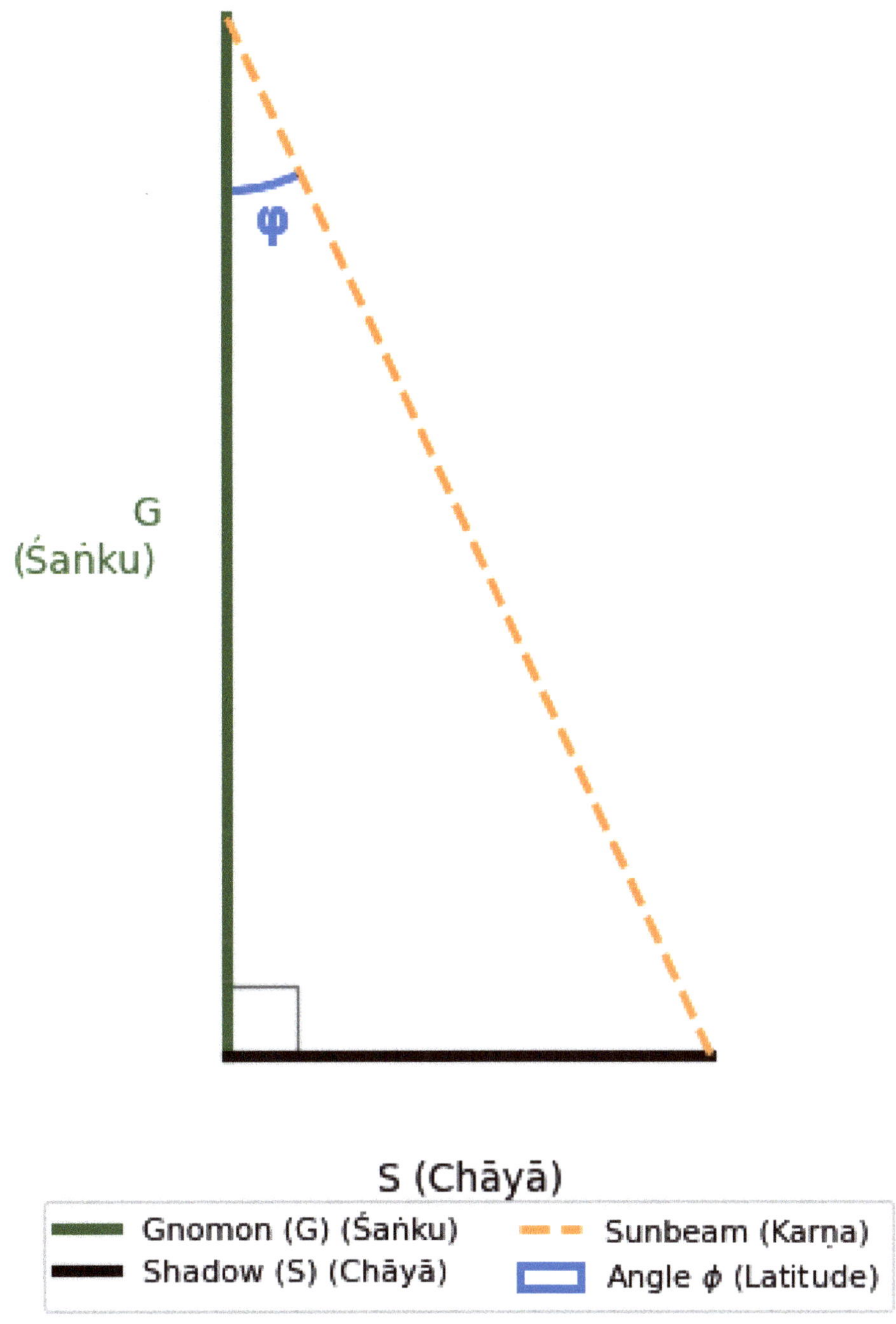

Figure 2.2b (Local Gnomon Setup): The gnomon and its shadow form a right triangle with the Sun's ray. The angle at the top of the gnomon is also φ. This is because the gnomon is aligned with the Zenith line, and the shadow direction is determined by the incoming Sun's ray. Thus, the geometry of the local gnomon triangle directly mirrors the larger global geometry. By measuring G and S, one can calculate tan(φ) = S/G, and thereby find φ.

2.1.3 The Gnomon as Practical Tool and Cosmic Link

The *śaṅku* served not only for measurement but also symbolically connected the terrestrial observer to the celestial order. It established the **cardinal directions** (east-west line determined from the line connecting shadow tips equidistant from noon; north-south from the noon shadow). In the context of Aryabhata's rotating Earth (Gola.9), the shadow's relentless movement across the ground could be interpreted not just as tracking the Sun, but as visually tracing the Earth's own spin relative to the Sun. This simple, elegant instrument grounded complex mathematical models in tangible, repeatable observation, allowing the astronomer to mathematically locate themselves within the cosmos and chart its rhythms.

2.2 Water Clocks & Celestial Models: Measuring and Visualizing Time and Space

Complementing the sun-dependent gnomon, ancient Indian astronomers required instruments for continuous time-keeping, especially at night or during cloudy weather. The primary device for this was the water clock, known generally as ***ghaṭikā yantra*** or *jala yantra*. References to time units like the ***nāḍī*** or *ghaṭikā* in the *Aryabhatiya* (***Kālakriyāpāda 1***) directly imply the use of such instruments to provide the uniform temporal framework essential for astronomical calculations.

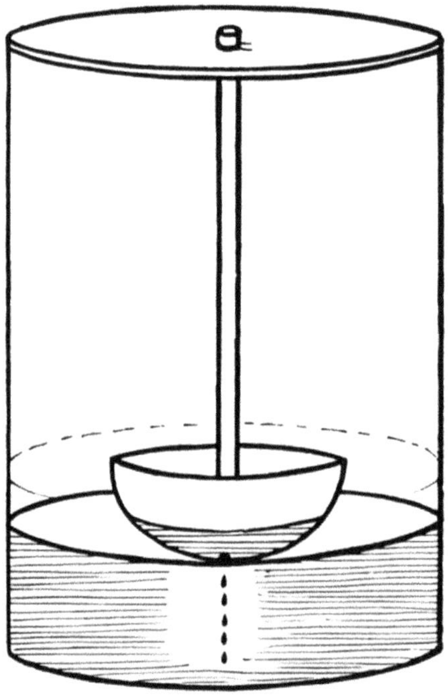

Figure 2.3: Illustration of a Ghaṭikā Yantra (water clock), typically of the inflow type. A small bowl with a calibrated hole floats in a larger basin and sinks after a specific interval, defining the standard time unit of a nāḍī or ghaṭikā (1/60th of a day), crucial for continuous timekeeping in ancient Indian astronomy.

2.2.1 The Gradual Flow of Time

While outflow clocks existed, the inflow type is more frequently described in classical Indian texts: a small copper bowl (*ghaṭikā*), carefully crafted with a tiny perforation at its base, was floated in a larger basin of clean water. Water would slowly seep into the bowl, causing it to sink after a standardized interval. This interval typically defined one ***nāḍī*** (also called *ghaṭikā*), the fundamental unit of time in classical Indian astronomy, equal to 1/60th of a full day-night cycle (*ahorātra*), corresponding precisely to 24 modern minutes. This elegant device enabled:

- **Continuous Measurement:** It provided a way to measure time continuously, day or night, independent of sunlight, crucial for timing observations of stars, planets, and especially lunar phenomena like eclipses throughout the night.
- **Standardized Intervals:** It offered a reliable basis for calculating celestial bodies' rates of motion (e.g., degrees per *nāḍī*), fundamental to Aryabhata's dynamic models and the calculation of planetary positions (*graha gaṇita*).
- **Operational Consistency:** It offered greater reliability across varying weather conditions compared to the gnomon, though careful maintenance (clean water, consistent temperature, ensuring the orifice remained clear and unworn) was necessary for accuracy. Regional variations in exact *nāḍī* calibration likely existed, but the theoretical framework aimed for standardization

Aryabhata's system formalized the division of the day into 60 *nāḍīs*, each comprising 60 *vināḍikās* (***Kālakriyāpāda 1-2***), embedding a sexagesimal structure well-suited for mathematical calculation into the fabric of time measurement itself. This structure, likely influenced by Mesopotamian precedents but integrated uniquely within the Indian framework, facilitated seamless calculations involving fractions of a day.

2.2.2 Celestial Spheres: Modeling the Cosmos

In the **Golapāda** ("On the Celestial Sphere"), Aryabhata describes the cosmos using the concept of the ***gola***—the celestial sphere. This was not merely an abstract geometric construct; Indian astronomers almost certainly used physical, three-dimensional models (***gola yantras***) for teaching, visualization, and potentially even computation. While complex geared mechanisms like the Antikythera device have not been found from Gupta India, and direct archaeological evidence for *gola yantras* from Aryabhata's precise time is scarce, textual descriptions like Aryabhata's own (**Golapāda 22**) and the later commentary tradition strongly suggest their existence and importance:

- **Armillary Spheres:** These models were likely constructed from interlocking rings (*valaya*) representing key celestial circles: the horizon (*kṣitija*), meridian (*yāmyottara*), celestial equator (*viṣuvadvṛtta*), ecliptic (*apamaṇḍala*), and perhaps circles of declination or altitude. These could be rotated to demonstrate the apparent daily motion of stars due to Earth's rotation (or the traditional *pravaha* wind), the Sun's annual path along the ecliptic causing seasons (as described in **Gola.1**), planetary movements (often incorporating representations of epicycles), and phenomena like the rising and setting points of celestial bodies at different latitudes. Bhaskara I's commentary occasionally refers to demonstrating concepts "on the surface of the *gola*," hinting at the use of such physical models. The *Aryabhatiya* itself (**Gola.22**) describes constructing a light, balanced wooden sphere that could be rotated smoothly (perhaps using mercury, oil, or water for bearings) in sync with time (*kālasamam*), strongly suggesting a functional model sphere.

- **Educational Tools:** Such *gola yantras* would have been essential for helping students grasp the complex spatial relationships involved in spherical astronomy – understanding eclipses (which Aryabhata correctly described as caused by shadows, Golapāda 37-46), planetary latitudes, conjunctions, parallax, and the geometrical basis of varying daylight hours across different locations (e.g., **Gola.19**).

- **Conceptual Framework:** The emphasis on the spherical *gola* fostered a dynamic, three-dimensional mental model of the cosmos. This contrasts somewhat with the planar, projective geometry emphasized by the astrolabe, which became a dominant observational and computational tool in the later Islamic and European traditions. The Indian *gola* arguably provided a more intuitive framework for visualizing orbits in space and conceptualizing phenomena related to Aryabhata's rotating Earth.

Fig 2.4: Representations of armillary spheres and planetary system diagrams. While these specific illustrations are likely European, they visually convey the concept of the gola yantra (model celestial sphere) described by Aryabhata in Gola.22. Such models, constructed with key celestial circles, were essential tools for visualizing, teaching, and potentially calculating celestial motions in various astronomical traditions.

2.3 Instrumental Thinking: The Synthesis of Observation and Calculation

The *śaṅku*, *ghaṭikā yantra*, and *gola yantra*, though perhaps simple by modern standards, reveal a sophisticated instrumental approach to astronomy in Aryabhata's era. They embody a mindset where time and space were not just philosophical concepts but quantities to be actively measured, quantified, and modeled. The *nāḍī*, measured by the water clock, standardized time's flow; the *gola yantra* mapped cosmic structure; and the *śaṅku* linked terrestrial observation to celestial geometry through the precise language of trigonometry applied to sunlight and shadow. These tools were not mere curiosities; they were the indispensable empirical counterparts to the sophisticated mathematical framework Aryabhata constructed in the *Aryabhatiya*. They provided the observational data and the conceptual grounding upon which the intricate calculations, soon to be unveiled, were built.

***

Chapter Three

The Tenfold Echoes – Gītikāpāda

The Poetic Code of Creation: Aryabhata's Numerals, Cosmic Ages, and the Rhythms of the Planets

3.0 Introduction: The Cosmic Overture

Having set the stage with the history, interpreters, and tools of Aryabhata's era, we now turn to the *Aryabhatiya* itself. We begin with its opening section, the **Gītikāpāda**—the foundational overture to Aryabhata's mathematical cosmos. Composed around 499 CE in the vibrant intellectual hub of Pataliputra, these thirteen verses are the genetic code from which his entire system unfolds.

This section is far more than a simple preface. It is a dense and powerful symphony of ideas, including:

- An **invocation** to the divine principles of the cosmos (Verse 1).
- A unique **alphabetical number system**, an ingenious code for capturing immense numbers (Verse 2).
- A set of nine core **parameter verses** that define the rhythms of the planets, the scale of the universe, the data for the sine function, and the crucial link between time and space (Verses 3-11).
- The elegant **rule governing sine calculation**, revealing the mathematical engine behind his trigonometry (Verse 12).
- A concluding verse extolling the **profound value of this knowledge** as a path to ultimate understanding (Verse 13).

These verses, often referred to collectively as the *Daśagītikā* ("Ten Gīti Verses"), were composed in the elegant *gīti* meter. This poetic structure made them ideal not just for calculation, but for memorization and oral transmission, ensuring that this essential cosmic knowledge could echo through generations.

They represent a masterful synthesis of observation, mathematics, and cosmology, establishing a revolutionary system destined to resonate across the centuries. This chapter will delve into these foundational verses one by one, exploring the ingenious numerical code, the seeds of trigonometry, and the integrated worldview they embody. We begin, appropriately, with Aryabhata's cosmic salutation.

3.1 The Cosmic Salutation – Aryabhata's Opening Verse (Gk.1)

Purpose: To invoke divine principles, state the author's name and the scope of the work, grounding the scientific endeavor within a philosophical context.

Before setting planets in motion or defining the flow of cosmic time, Aryabhata commences the *Aryabhatiya* with a single, resonant verse. More than mere formality, this opening stanza (Gītikāpāda 1) serves as both a profound invocation and a clear statement of purpose, encapsulating his vision and methodology.

Sanskrit:

प्रणिपत्यैकमनेकं कं सत्यां देवतां परं ब्रह्म ।

आर्यभटस्त्रीणि गदति गणितं कालक्रियां गोलम् ॥ १ ॥

IAST:

praṇipatyaikam anekaṃ kaṃ satyāṃ devatāṃ paraṃ brahma|

āryabhaṭas trīṇi gadati gaṇitaṃ kālakriyāṃ golam || 1 ||

This deliberate beginning, blending reverence with reason, sets the stage for the quantitative exploration of the heavens that follows.

Notes:

- **Praṇipatya:** "Having paid respectful obeisance"; signifies deep reverence.
- **Ka:** Identified with Brahma/Prajāpati, the creator.
- **Ekam/Anekam:** The One (ultimate reality) and the Many (phenomenal universe).
- **Paraṃ Brahma:** The Supreme Absolute.
- **Satyāṃ Devatām:** "The True Divinity," often an epithet for Paraṃ Brahma.
- **Āryabhaṭaḥ:** Self-reference, asserting authorship.
- **Trīṇi:** "Three things" – the scope of the work.
- **Gaṇitam:** Mathematics (arithmetic, algebra, geometry, trigonometry).
- **Kālakriyām:** Reckoning of Time (celestial mechanics, planetary positions over time).

- **Golam:** The Sphere (spherical astronomy, geometry of celestial sphere).

3.1.1 Bowing to the Infinite and the Manifold

The verse opens with a deep obeisance: *"praṇipatyaikam anekaṃ kaṃ satyāṃ devatāṃ paraṃ brahma"*— "Having paid respectful obeisance to Ka (identified with Brahma/Prajāpati), the One (*ekam*) and the Many (*anekam*), the True Divinity (*satyāṃ devatām*), the Supreme Brahman (*paraṃ brahma*)..."

This is not a casual nod but an intentional act of devotion. As the 7th-century commentator Bhaskara I highlights, the prefix *pra-* in *praṇipatya* signifies excellence and intensity—a profound prostration before commencing the work. Aryabhata invokes concepts central to Indic philosophical and cosmological thought:

- **Ka (Prajāpati/Brahma):** The creator deity, the architect of the cosmos in Vedic tradition.
- **Ekam (The One):** The ultimate, undifferentiated reality, Brahman, the source from which all emerges.
- **Anekam (The Many):** The diverse, phenomenal universe, the manifold expressions of the One.
- **Paraṃ Brahma:** The supreme, transcendent Absolute, encompassing both unity and diversity. *(Note: Satyāṃ Devatām, "the True Divinity," often considered an epithet for Paraṃ Brahma or a collective reference).*

By invoking these concepts, Aryabhata grounds his scientific undertaking within a profound cosmological framework. He acknowledges the ultimate unity underlying the apparent multiplicity of the cosmos—a universe whose intricate workings, though diverse (*anekam*), stem from a single, comprehensible (if ultimate) source (*ekam*), the very reality he seeks to model through number and geometry.

3.1.2 Three Keys to Unlock the Cosmos

Having established this reverential context, Aryabhata confidently declares his purpose: *"āryabhaṭas trīṇi gadati gaṇitaṃ kālakriyāṃ golaṃ"*— "...Aryabhata sets forth three things (*trīṇi*): mathematics (*gaṇitaṃ*), the reckoning of time (*kālakriyāṃ*), and the sphere (*golaṃ*)."

His self-reference (*āryabhaṭaḥ*) marks clear authorship. These "three" pillars correspond directly to the major sections and methodology of his work:

1. **Gaṇita (Mathematics):** Literally 'the science of calculation'. This encompasses arithmetic, algebra, geometry, and trigonometry – the essential language and tools required for quantitative astronomy. The unique numerical system introduced immediately following this verse is a prime example of the *Gaṇita* needed.
2. **Kālakriyā (Reckoning of Time):** The measurement and calculation (*kriyā*) related to time (*kāla*). This section deals with the units of time, from days to vast cosmic cycles (*yugas*, *kalpas*), and critically, the methods for calculating the mean positions of planets over time based on the parameters laid out in the *Daśagītikā*. It provides the dynamic framework.
3. **Gola (The Sphere):** Spherical astronomy. This section applies mathematics (especially trigonometry developed in *Gaṇita*) to the geometry of the celestial sphere (*khagola*), planetary orbits, eclipses, parallax, and the relationship between celestial phenomena and the observer on Earth. It provides the spatial and geometric context.

This triad represents Aryabhata's integrated approach: Mathematics provides the tools, Time-Reckoning provides the dynamic structure over which movements occur, and Spherical Astronomy provides the geometric model to map and predict those movements in space.

Key	Meaning	Role within Aryabhatiya
Gaṇita	Mathematics	Tools & language (arithmetic, algebra, trigonometry)
Kālakriyā	Reckoning of Time	Dynamic framework (cycles, mean positions)
Gola	The Sphere	Geometric/Spatial context (sphere, orbits, eclipses)

3.1.3 A Bridge from Reverence to Revolution

In Aryabhata's era (circa 500 CE), astronomy was deeply intertwined with ritual, calendrics, and cosmology. By beginning with an invocation that acknowledges the divine and the ultimate reality, Aryabhata connects his work to this existing worldview. Yet, this reverence immediately pivots to a declaration of a systematic, quantitative methodology (*trīṇi*).

This opening verse perfectly mirrors the spirit of the *Aryabhatiya* itself: a synthesis of the sacred and the scientific. It frames the rigorous mathematical investigation of the heavens not as a departure from tradition, but as a valid path—grounded in reverence—to understanding the intricate order of the cosmos. Having paid homage to the source of all phenomena, Aryabhata proceeds, in the verses that follow, to lay out the precise numerical parameters needed to unlock the workings of his mathematically conceived universe, starting with the very code used to express its immense scale.

3.2 The Alphabet of the Infinite – Aryabhata's Numerical Code (Gk.2)

Purpose: To introduce the unique alphabetical system used to encode large astronomical numbers concisely within metrical verse.

Following his cosmic salutation, Aryabhata, writing around 500 CE, immediately introduces a revolutionary computational tool in Verse 2 of the Gītikāpāda. This ingenious system, a testament to ancient Indian mathematical creativity, is the key to unlocking the vast astronomical numbers encoded in the subsequent verses:

Sanskrit:

वर्गाक्षराणि वर्गेऽवर्गेऽवर्गाक्षराणि कात् ङ्मौ यः ।

खद्विनवके स्वरा नव वर्गेऽवर्गे नवान्त्यवर्गे वा ॥ २ ॥

IAST:

vargākṣharāṇi varge'varge'vargākṣharāṇi kāt ṅmau yaḥ|

khadvinavake svarā nava varge'varge navāntyavarge vā || 2 ||

In this verse, Aryabhata unveils a unique method for encoding large numbers using the Sanskrit alphabet itself. Designed for extreme conciseness and likely facilitating oral transmission, this system encapsulates cosmic scales within elegant, memorable syllables.

Notes:

- **Vargākṣarāṇi:** "Classified consonants" (k-varga to p-varga, 1-25).
- **Avargākṣarāṇi:** "Unclassified consonants" (y to h, 30-100).
- **Kāt ṅmau yaḥ:** "From ka up to ma [are 1-25], and yaḥ [y onwards are 30+]."
- **Khadvinavake:** "Two nines of voids/zeros," implying 18 decimal places or powers of 100.
- **Svarā nava:** The nine vowels (a, i, u, ṛ, ḷ, e, ai, o, au), assigning place value.

3.2.1 Letters as Numbers: The Consonant Values

The system assigns numerical values to Sanskrit consonants, which are divided into two distinct groups based on traditional phonetic classification:

1. **Varga Letters (Classified Consonants):** These are the standard plosive and nasal consonants arranged in

the five 5-member phonetic groups (vargas) of the Sanskrit alphabet (ka-varga, ca-varga, ṭa-varga, ta-varga, and pa-varga). They represent the numbers 1 through 25 sequentially:

- ka-varga: k=1, kh=2, g=3, gh=4, ṅ=5
- ca-varga: c=6, ch=7, j=8, jh=9, ñ=10
- ṭa-varga: ṭ=11, ṭh=12, ḍ=13, ḍh=14, ṇ=15
- ta-varga: t=16, th=17, d=18, dh=19, n=20
- pa-varga: p=21, ph=22, b=23, bh=24, m=25

2. **Avarga Letters (Unclassified Consonants):** These are the remaining consonants (y, r, l, v, ś, ṣ, s, h), which follow the varga letters in the traditional alphabetic order. They represent multiples of ten, starting from 30:
    - y=30, r=40, l=50, v=60
    - ś=70, ṣ=80, s=90, h=100

The verse itself alludes to this structure with the phrase "kāt ṅmau yaḥ"—meaning approximately "from ka up to ma [are 1-25], and yaḥ [representing ya and subsequent letters, are 30 onwards].

	Guttural (Ka)	Palatal (Ca)	Retroflex (Ṭa)	Dental (Ta)	Labial (Pa)
Unvoiced/Unaspirated	k = 1	c = 6	ṭ = 11	t = 16	p = 21
Unvoiced/Aspirated	kh = 2	ch = 7	ṭh = 12	th = 17	ph = 22
Voiced/Unaspirated	g = 3	j = 8	ḍ = 13	d = 18	b = 23
Voiced/Aspirated	gh = 4	jh = 9	ḍh = 14	dh = 19	bh = 24
Nasal	ṅ = 5	ñ = 10	ṇ = 15	n = 20	m = 25

Table 3.1: Numerical Values of the Varga (Classified) Consonants in Aryabhata's Alphabetical Number System (Gk.2). These consonants (ka-varga through pa-varga) represent values from 1 to 25 sequentially, forming the base digits for encoding numbers in his cipher.

Letter (Avarga)	Value
y	30
r	40
l	50
v	60
ś	70
ṣ	80
s	90
h	100

Table 3.2: Numerical Values of the Avarga (Unclassified) Consonants in Aryabhata's Alphabetical Number System (Gk.2). These consonants (y to h) represent values from 30 to 100, providing larger base units for encoding numbers in his cipher, following the Varga letters. (See Table 3.1 for Varga letters).

3.2.2 Vowels and Place Value: The Mystery of the Powers

The true ingenuity of Aryabhata's numerical system emerges in its use of vowels to assign place value, transforming simple consonant values into components of vast numbers. Verse 2 states: "khadvinavake svarā nava..." – "the nine vowels (a, i, u, ṛ, ḷ, e, ai, o, au), [in their short and implicitly long forms,] correspond to the khadvinavake ('two nines of voids/zeros')." This phrase indicates a remarkable capacity to represent numbers across eighteen decimal places, establishing a sophisticated positional system commonly understood to be based on powers of 100. However, the precise mechanism for applying these vowel-based powers has prompted two main scholarly interpretations, creating a crucial ambiguity that requires careful navigation.

Interpretation A: The Additive Power-of-100 System

This is the most widely applied interpretation, particularly effective for decoding many astronomical constants in the Aryabhatiya, and is meticulously explained by the 7th-century commentator Bhaskara I. In this system, each of the nine principal vowels (and their long forms, which carry the same power) signifies a specific power of 100. This power acts as a multiplier for the numerical value of the consonant to which the vowel is attached. The powers are assigned as follows:

- a/ā = $\times 100^0 = \times 1$
- i/ī = $\times 100^1 = \times 100$
- u/ū = $\times 100^2 = \times 10{,}000$

- ṛ/ṝ = × 100^3 = × 1,000,000
- ḷ/ḹ = × 100^4 = × 100,000,000
- e = × 100^5 = × 10^{10} (ten billion)
- ai = × 100^6 = × 10^{12} (one trillion)
- o = × 100^7 = × 10^{14}
- au = × 100^8 = × 10^{16}

When employing this additive system, the total numerical value of a coded word is the sum of the values derived from its individual consonant-vowel syllables. Each syllable (e.g., 'khu', 'yṛ') contributes its consonant's base value multiplied by the power of 100 indicated by its vowel. This method consistently decodes many large astronomical constants, such as the Sun's revolutions per mahāyuga (encoded as khyughṛ, yielding 4,320,000, as will be demonstrated).

Vowel (Short/Long)	Power of 100 (Interpretation A)	Multiplier
a/ā	100^0	×1
i/ī	100^1	×100
u/ū	100^2	×10, 000
ṛ/ṝ	100^3	×1, 000, 000
ḷ/ḹ	100^4	×100, 000, 000
e	100^5	$\times 10^{10}$
ai	100^6	$\times 10^{12}$
o	100^7	$\times 10^{14}$
au	100^8	$\times 10^{16}$

Table 3.3: Vowel Values in Aryabhata's Alphabetical Number System (Gk.2), according to Interpretation A (Additive Power-of-100 System). The vowels (including short and long forms) determine the power of 100 by which the value of the preceding consonant is multiplied, enabling the encoding of large numbers within metrical verse. (See Tables 3.1 and 3.2 for consonant values)

Interpretation B: Odd/Even Place-Value System

While the additive system provides a foundational understanding of Aryabhata's encoding, Verse 2's seemingly simple phrase, "vargākṣarāṇi varge 'varge 'vargākṣarāṇi" ("Varga letters in varga [odd/square] places, avarga letters in avarga [even/non-square] places..."), hints at a more sophisticated structure: a place-value system akin to our modern understanding of decimal notation. This system, potentially read from right to left, likely complemented, or in specific instances superseded, the additive method.

Imagine a scenario where, alongside the additive system, Aryabhata utilized a positional notation, where the *location* of a letter within a coded word significantly influenced its value. This isn't a complete abandonment of the additive approach but rather a layering of additional meaning.

Vowels as Place Markers: Pairing Decimal Places

A key aspect of this interpretation involves the role of vowels. The nine vowels in the Sanskrit alphabet could function as markers, *not* representing individual numerical values, but rather *denoting pairs* of decimal places (powers of 10). For example:

- a/ā: Places 1-2 (10^0, 10^1 – units and tens)
- i/ī: Places 3-4 (10^2, 10^3 – hundreds and thousands)
- u/ū: Places 5-6 (10^4, 10^5 – ten-thousands and hundred-thousands)

Extending this pattern, the nine vowels could potentially cover up to eighteen decimal places. This assignment of vowels provides the framework for organizing the consonant values.

Consonants in Odd and Even Places: Assigning Values

Within this place-value framework, consonants take on the roles of numerical digits. Here's how the odd/even placement comes into play:

- **Varga Consonants (values 1-25):** Primarily fill odd-numbered decimal places (1st, 3rd, 5th, etc.).
- **Avarga Consonants (values 30-100):** Primarily fill even-numbered decimal places (2nd, 4th, 6th, etc.). In this context, the *tens digit* of the avarga consonant's value might be used (e.g., 3 for *ya*, 4 for *ra*).

Why This Complexity? The Case of Astronomical Constants
Why introduce such a complex system? The answer lies in the need to encode very large, precise numbers, especially critical astronomical constants. Decoding these constants with sufficient accuracy often proves impossible using only the simpler additive method.
A prime example is the number of the Moon's revolutions per *mahāyuga*, encoded as chayagiyiṅuśuchḷṛ and understood by scholars to represent 57,753,336. Achieving this value solely through additive methods is exceptionally difficult.

An Example - Decoding chayagiyiṅuśuchḷṛ:

Break into Syllables: cha + ya + gi + yi + ṅu + śu + chḷ + ṛ

Decode Each Syllable (Right to Left):

- **cha (Rightmost Syllable):**
  - Consonant ch = 6.

- Vowel a indicates decimal places 10^0 and 10^1.
- Since it's the rightmost syllable, assume the *odd* place (10^0) for now.
- Value = 6.

- **ya:**
  - Consonant y = 30. The "odd/even" rule suggests taking the tens digit, so 3.
  - Vowel a indicates decimal places 10^0 and 10^1.
  - The *even* place (10^1) is now used.
  - Value = 3 x 10 = 30.
- **gi:**
  - Consonant g = 3.
  - Vowel i indicates decimal places 10^2 and 10^3.
  - Since it is the next location to the previous location, use the odd place:
  - Value = 3 x 10^2 = 300.
- **yi:**
  - Consonant y = 30. Take the tens digit = 3
  - Vowel i indicates decimal places 10^2 and 10^3.
  - Even position
  - Value = 3 x 10^3 = 3000.
- **ṅu:**
  - Consonant ṅ = 5.
  - Vowel u indicates decimal places 10^4 and 10^5.
  - Since it is the next odd location to the previous location, use the odd place:
  - Value = 5 x 10^4 = 50,000
- **śu:**
  - Consonant ś = 70. Take the tens digit = 7
  - Vowel u indicates decimal places 10^4 and 10^5.

- Since it is the next even location to the previous location, use the even place:
- Value = 7×10^5 = 700,000

(Higher-Order Digits: A More Complex Application)

The remaining syllables, chḷ and ṛ, would then be used to generate the highest order digits '57', through a more complex application of these rules to encode 50,000,000 and 7,000,000.

Sum the Values (So Far): 700,000 + 50,000 + 3,000 + 300 + 30 + 6 = 753,336

The interpretation is thus: "57,753,336"

This example demonstrates how the Odd/Even Place System, even with its complexities, allows for encoding large numbers with what appear to be fairly compact syllable combinations. While a comprehensive and universally applicable rule set requires further investigation, the underlying principles offer a compelling alternative or supplement to the additive system for certain critical values.

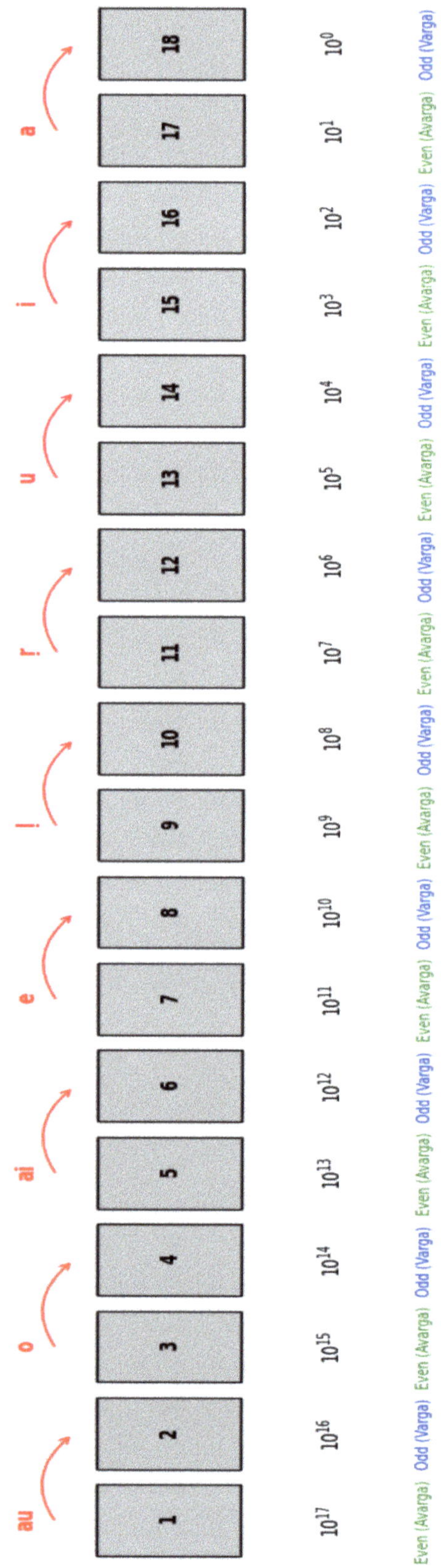

Figure 3.1: Aryabhata's Odd/Even Place System (Gk.2, Interpretation B). Vowels mark decimal place-pairs, with Varga/Avarga consonants in odd/even positions, enabling complex number decoding

Category	Letters (Examples)	Base Values	Vowel Role (Interpretation A - Additive)
Varga	k(1), kh(2)... ṅ(5)... m(25)	1-25	Multiplied by Vowel Power (a=1, i=100, u=10k...)
Avarga	y(30), r(40)... h(100)	30-100	Multiplied by Vowel Power (a=1, i=100, u=10k...)
Vowel	a/ā, i/ī, u/ū, ṛ/ṝ, etc.	N/A	Determines Power of 100 (100^0, 100^1, ... 100^8 for 18 places)

Table 3.4: Aryabhata's Alphabetical Number System (Gītikāpāda Verse 2 / Gk.2) – Interpretation A (Additive Power-of-100 System). This table summarizes the numerical values assigned to Varga (classified) and Avarga (unclassified) consonants, and illustrates how vowels assign powers of 100, which act as multipliers to form large numbers through an additive process.

3.2.3 Why the Ambiguity? The Case of the Moon's Revolutions

The necessity for considering Interpretation B becomes evident when attempting to decode certain key astronomical constants. For instance, trying to derive the Moon's revolutions per *mahāyuga*—encoded as *chayagiyiṅuśuchḷṛ*—using the simpler Additive System (Interpretation A) yields a sum far from the established value of 57,753,336. However, applying the more complex Positional Odd/Even Place System (Interpretation B), while intricate, allows scholars to reconstruct this correct value. This disparity strongly suggests that Aryabhata may have employed a more sophisticated place-value understanding, potentially using different applications of the vowel rules depending on the context, or perhaps relying on oral conventions that are now lost.

The exact mechanics intended by Aryabhata for all cases remain a subject of scholarly debate. Did he utilize both systems flexibly? Was there an oral key, transmitted alongside the text, that clarified the application for specific numbers? Or does the textual ambiguity itself reflect a system designed for computational flexibility or extreme compression? While Interpretation A, championed by Bhaskara I, is often sufficient for many constants, the necessity of invoking Interpretation B for critical values like the Moon's revolutions highlights the sophistication, and the inherent challenges, of fully reconstructing Aryabhata's ingenious method.

3.2.4 Decoding the Cosmos: An Example (Sun's Revolutions using Interpretation A)

To illustrate the power and elegance of the Additive Power-of-100 System (Interpretation A), let's decode Aryabhata's value for the Sun's revolutions in a *Mahāyuga* (4,320,000). This is given in Gītikāpāda Verse 3 by the coded syllables *khyughṛ*:

1. **Break into Syllables:** *khu* + *yu* + *ghṛ*

2. **Decode Each Syllable:**

   - ***khu***: Consonant *kh* = 2. Vowel *u* = $\times 100^2 = \times 10{,}000$. Value = $2 \times 10{,}000 = 20{,}000$.
   - ***yu***: Consonant *y* = 30. Vowel *u* = $\times 100^2 = \times 10{,}000$. Value = $30 \times 10{,}000 = 300{,}000$.
   - ***ghṛ***: Consonant *gh* = 4. Vowel *ṛ* = $\times 100^3 = \times 1{,}000{,}000$. Value = $4 \times 1{,}000{,}000 = 4{,}000{,}000$.

3. **Sum the Values:** 4,000,000 + 300,000 + 20,000 = **4,320,000**.

This example demonstrates how the system elegantly encodes a seven-digit number into just three compact syllables. Other immense numbers, such as Earth's sidereal rotations in a *mahāyuga* (1,582,237,500, coded as *ṅiśibuṇḷṛṣkhṛ* in Gk.3), are similarly compressed, showcasing the system's remarkable efficiency. The final phrase of Gītikāpāda Verse 2, "*...navāntyavarge vā*" ("...or at the end square/group"), remains somewhat enigmatic but might offer a clue for handling specific edge cases or permissible variations in applying the rules, perhaps relating to the consonant *m*(25) which completes the *varga* set.

3.2.5 Beyond Roman Numerals: Efficiency and Transmission

This unique alphabetical numerical code is foundational to the *Daśagītikā*. It transforms otherwise unwieldy astronomical data into a compact form, easily memorized, transmitted orally, and preserved across generations of manuscripts. It served as a linguistic ladder to comprehend cosmic scales, ingeniously blending mathematical innovation with the practical necessities of knowledge preservation in ancient India.

The sheer ingenuity required to handle the vast numbers needed for cosmic calculations led the 7th-century commentator Bhaskara I to suggest that this system arose not just from human intellect but "*brahmaṇaḥ prasādena*" ("by Brahmā's grace"). This reflects a perspective where profound insight might stem from deep contemplation or spiritual discipline (*tapas*) integrated with rational thought. The system forms the bedrock of the *Daśagītikā*, encoding its immense scales with an efficiency far surpassing contemporary systems like Roman numerals.

3.3 The Cosmic Pulse – Revolutions in a Yuga (Gk.3)

Purpose: To establish the fundamental average angular speeds of the Sun, Moon, Earth (rotation), and planets by stating their total revolutions over one *Mahāyuga*.

With his unique numerical code established (Gk.2), Aryabhata immediately puts it to work in Gītikāpāda Verse 3. This verse sets the cosmos in motion, defining its fundamental rhythms – the mean revolutions (*bhagaṇāḥ*) – of the Sun, Moon, Earth, and major planets over the vast expanse of a *Mahāyuga*. These revolution counts form the basis for calculating average planetary speeds and, consequently, their mean positions using methods like the *trairāśika* (Rule of Three), a process exemplified later.

Sanskrit:

युगरविभगणाः ख्युघृ शशि चयगियिङुशुछ्लृ कु ङिशिबुण्ऌ्ष्खृ प्राक् ।

शनि ढुङ्ढ्विघ्व गुरु ख्रिच्युभ कुज भद्लिझ्नुखृ भृगुबुधसौराः ॥ ३ ॥

IAST:

yugaravibhagaṇāḥ khyughṛ śaśi chayagiyiṅuśuchḷṛ ku ṅiśibuṇḷṛṣkhṛ prāk|

śani ḍhuṅvighva guru khrichyubha kuja bhadlijhnukhṛ bhṛgubudhasaurāḥ || 3 ||

Here, Aryabhata encodes the *bhagaṇāḥ* (revolutions) completed by key celestial bodies during one *Mahāyuga* (often simply called *yuga* in this context). This immense cycle, lasting 4,320,000 solar years (a duration itself defined by the Sun's coded revolution count, *khyughṛ*), serves as Aryabhata's fundamental 'master cycle' for tracking cosmic time and calculating average planetary speeds. Decoding these compact syllables reveals the precise, rhythmic dance underlying Aryabhata's mathematically conceived universe.

Notes:

- **Yugaravibhagaṇāḥ:** Revolutions (bhagaṇāḥ) of the Sun (ravi) in a Yuga.
- **Khyughṛ:** Sun's revolutions = 4,320,000 (implicitly defines Mahāyuga duration).
- **Śaśi:** Moon. **Chayagiyiṅuśuchḷṛ:** Moon's revolutions = 57,753,336.
- **Ku:** Earth. **Ṅiśibuṇḷṛṣkhṛ:** Earth's rotations = 1,582,237,500.
- **Prāk:** "Eastward" (referring to Earth's rotation).
- **Śani, Guru, Kuja:** Saturn, Jupiter, Mars (their revolutions also encoded).
- **Bhṛgubudhasaurāḥ:** Venus (Bhṛgu), Mercury (Budha) – their revolutions are in Gk.4, not Gk.3. "Saurāḥ"

might refer to planets in general or contextually to those listed.

3.3.1 Unveiling the Celestial Dance: Revolutions per Yuga

Gītikāpāda Verse 3 lists these crucial revolution counts. It is important to remember that Aryabhata, like his contemporaries, worked within a primarily geocentric framework for calculation, modeling the Sun, Moon, and planets as orbiting a central Earth. Thus, the Sun's 'revolutions' refer to its apparent annual path around the Earth. A revolutionary exception to this geocentric perspective on motion is his treatment of Earth's own rotation.

Celestial Body	Sanskrit Code (Approx.)	Revolutions in a Yuga	Notes
Sun (ravi)	khyughṛ	4,320,000	Defines Mahāyuga length in solar years (apparent orbits).
Moon (śaśi)	cayagiyiṅuśuchḷṛ	57,753,336	Sidereal orbits relative to stars. (Requires Interpretation B of Gk.2).
Earth (ku)	ṅiśibuṇḷṛṣkhṛ	1,582,237,500	Sidereal rotations prāk (eastward!), defining sidereal days. Key departure.
Saturn (śani)	ḍhuṅvighva	146,568	Slowest visible planet. (Standard value; variant ~146,564).
Jupiter (guru)	khricyubha	364,224	(Standard value).
Mars (kuja)	bhadlijhnukhṛ	2,296,832	(Standard value; variant ~2,296,824).
Venus/Mercury	bhṛgubudhasaurāḥ	(See Gk. 4)	Revolutions for Venus (bhṛgu) & Mercury (budha) given in Gk. 4.

Table 3.5: Revolution counts of key celestial bodies per Mahāyuga (4,320,000 solar years), as encoded by Aryabhata in Gītikāpāda Verse 3 (Gk.3). These values define the fundamental average speeds of these bodies within the Aryabhatan system and are crucial for calculating their mean positions and orbital periods. (Note: Values for Venus and Mercury are in Gk.4, Earth's rotation is eastward as specified in Gk.3, and Moon count requires a specific interpretation of Gk.2).

Earth's Groundbreaking Spin (prāk)

A truly remarkable element in this verse is "*ku ṅiśibuṇḷṛṣkhṛ prāk.*" Here, *Ku* refers to the Earth, *ṅiśibuṇḷṛṣkhṛ* encodes its staggering number of rotations in a *Yuga*, and the crucial term *prāk* specifies that these rotations are **eastward**. This explicitly attributes the daily westward motion of the stars, Sun, and Moon to the Earth's own eastward spin—a concept Aryabhata elaborates on in the Golapāda (Gola.9), distinguishing his model significantly from purely static-Earth cosmologies.

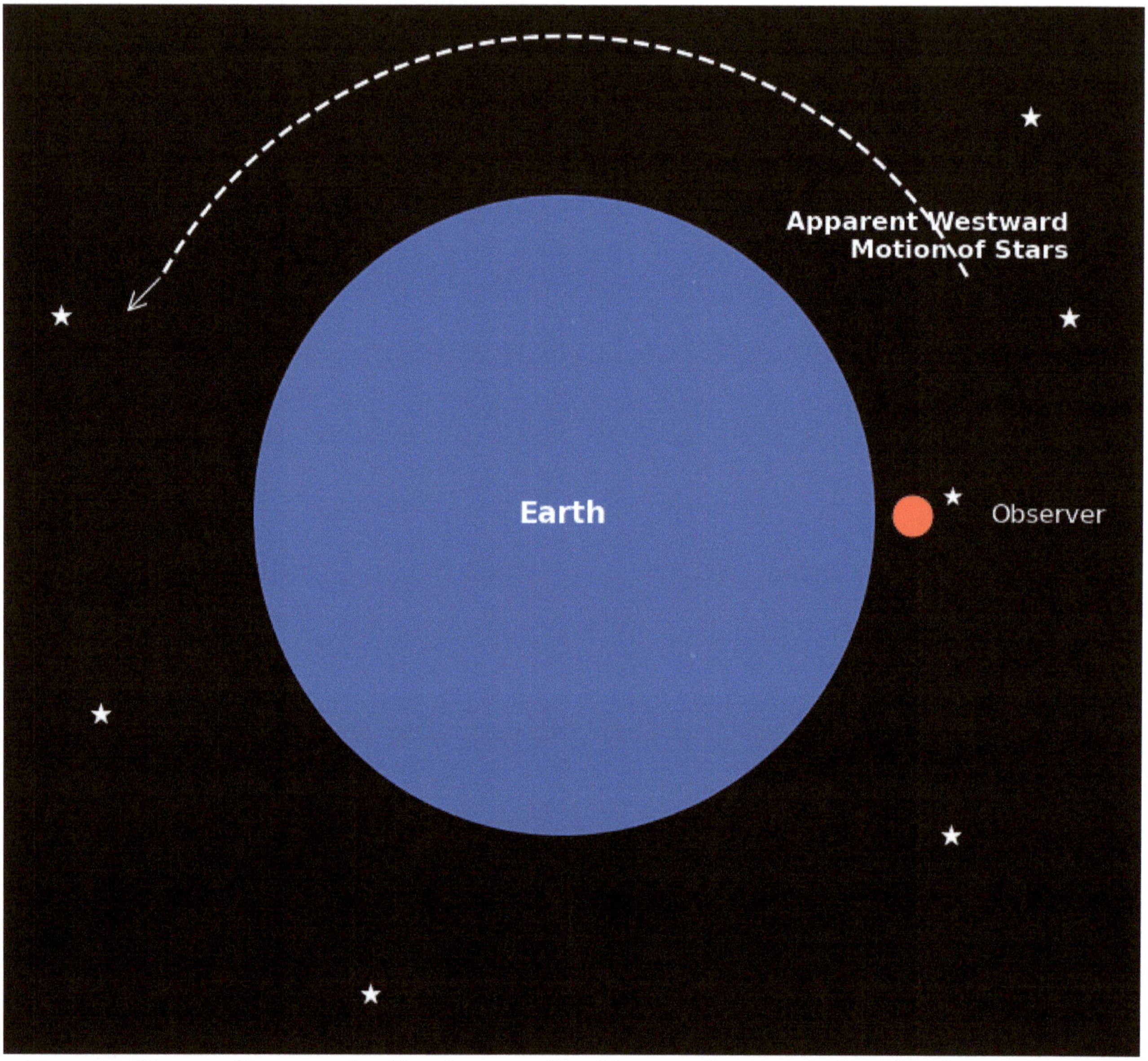

Figure 3.2: Diagram illustrating Aryabhata's concept of Earth's eastward rotation (indicated by the movement of the Observer on the Earth), as stated in Gītikāpāda Verse 3 (Gk.3) and elaborated in Golapāda Verse 9 (Gola.9). This rotation causes the apparent daily westward motion of the stationary stars (indicated by the yellow arrow and labels). This was a revolutionary departure from static-Earth models.

What does "Sidereal" mean?

The term "sidereal" originates from the Latin sidus (star). A sidereal measurement is made relative to the seemingly fixed background of distant stars. A **sidereal orbit or revolution** (such as those calculated for the Moon or planets in Aryabhata's system) refers to one complete 360° circle traversed by a celestial body against this stellar backdrop. A **sidereal day** (based on Earth's prāk or eastward rotations as defined in Gk.3) is the precise time it takes for the Earth to complete one full 360° spin relative to these distant stars. Aryabhata's parameters define this day with remarkable consistency, forming a fundamental basis for all his subsequent time-dependent calculations.

3.3.2 The Yuga and the Year: Cosmic Timescales

The Mahāyuga, spanning 4,320,000 years, provides the grand canvas for Aryabhata's astronomical system. His fundamental parameters for planetary and Earth revolutions (Gk.3) directly yield a precise length for the year:

- Number of Sidereal Days in a Yuga = 1,582,237,500 rotations (ku, Earth)
- Number of Solar Years in a Yuga = 4,320,000 revolutions (ravi, Sun)
- Length of Sidereal Year ≈ 1,582,237,500 sidereal days / 4,320,000 solar years ≈ **366.25868 sidereal days per solar year**.

This value represents the time taken for the Sun to complete one apparent revolution and return to the same position among the fixed stars (a sidereal year). It is remarkably close to the modern value (approximately 366.256 sidereal days relative to the fixed stars when measuring a solar year), showcasing the impressive accuracy of Aryabhata's foundational parameters.

3.3.3 Precision in Motion: Calculating Mean Sidereal Orbital Periods

An orbital period is the time required for a celestial body to complete one full orbit. Aryabhata's revolution counts per Mahāyuga allow for the calculation of highly accurate mean sidereal orbital periods for the planets and the Moon:

- **Moon:**(Sidereal Days/Yuga) / (Moon Sidereal Revolutions/Yuga)= 1,582,237,500 / 57,753,336 ≈ **27.396 sidereal days per revolution**(Modern mean sidereal period: ≈ 27.3217 days)
- **Saturn:**(Solar Years/Yuga) / (Saturn Sidereal Revolutions/Yuga)= 4,320,000 / 146,568 ≈ **29.474 solar years per revolution**(Modern mean sidereal period: ≈ 29.46 years)
- **Jupiter:**(Solar Years/Yuga) / (Jupiter Sidereal Revolutions/Yuga)= 4,320,000 / 364,224 ≈ **11.8608 solar years per revolution**(Modern mean sidereal period: ≈ 11.86 years)
- **Mars:**(Solar Years/Yuga) / (Mars Sidereal Revolutions/Yuga)= 4,320,000 / 2,296,832 ≈ **1.8808 solar years per revolution**(Modern mean sidereal period: ≈ 1.8809 years)

Body	Aryabhata's Mean Sidereal Period	Modern Mean Sidereal Period	Approx. Difference
Moon	~27.40 days	~27.32 days	~0.3%
Saturn	~29.474 years	~29.46 years	~0.05%
Jupiter	~11.861 years	~11.86 years	~0.008%
Mars	~1.8808 years	~1.8809 years	~0.005%

Table 3.6: Accuracy of Aryabhata's Mean Sidereal Orbital Periods Compared to Modern Values. This table compares the mean sidereal orbital periods for the Moon, Saturn, Jupiter, and Mars as derivable from Aryabhata's revolution counts (Gk.3) with modern scientifically accepted values, highlighting the impressive precision achieved in the Aryabhatiya.

Example: Mean Daily Motion

These revolution counts also yield mean daily motions. For example, the Moon's mean daily motion relative to the stars can be calculated as:

Moon Revolutions / Sidereal Days in a *Yuga*= 57,753,336 revolutions / 1,582,237,500 days≈ 0.036487 revolutions per day.

Converting to degrees:0.036487 revolutions/day × 360 degrees/revolution ≈ **13.135 degrees per day**.

(The modern mean sidereal motion is approximately 13.176 degrees per day). This calculated mean daily motion serves as the fundamental starting point for determining the Moon's position on any given day.

3.3.4 A Living Cosmos, A Practical System

The accuracy achieved using Aryabhata's *Yuga*-based revolution counts remains striking. These parameters were not merely abstract cosmological figures; they formed the core engine of a practical predictive system. As the commentator Bhaskara I emphasized, understanding celestial cycles was vital for various aspects of daily life—from agriculture and the timing of festivals to general timekeeping and astrology. By encoding these fundamental mean motions so compactly, Aryabhata provided the essential data for calculating planetary positions (*graha gaṇita*). Gītikāpāda Verse 3 defines the intricate cosmic pulse that Aryabhata then sought to anchor precisely in time (as detailed in Verse 5) and space (throughout subsequent verses), enabling the calculation of deviations from these mean motions to predict actual observed positions.

3.4 Anchoring the Cosmos – Orbits, Epoch, and Calculation (Gk.4)

Purpose: To provide crucial orbital reference points (apogees, nodes), the revolution counts for Mercury and Venus, and to define the precise starting moment (epoch) for all calculations within the system.

While Gītikāpāda Verse 3 established the fundamental average speeds (revolutions) of celestial bodies, predicting their actual observed positions requires more nuance. Planets appear to speed up and slow down in their paths, and the Moon visibly wobbles above and below the Sun's apparent course. Verse 4 of the Gītikāpāda provides crucial parameters to refine these motions and, critically, establishes the precise starting point in time and space for Aryabhata's entire computational system.

Sanskrit:

चन्द्रोच्च ज्रुखिध बुध सुगुशिथृण भृगु जषबिखुछृ शौरि गुरु कुजार्काणाम् ।

बुफिनच पातविलोमा बुधाह्नि अजार्कयोगात् प्रवृत्तचक्रः ॥ ४ ॥

IAST:

chandroccha jrukhidha budha suguśithṛṇa bhṛgu jaṣabikhuchṛ śauri guru kujārkāṇām|

buphinacha pātavilomā budhāhni ajārkayogāt pravṛttachakraḥ || 4 ||

This verse defines key orbital reference points, gives the missing revolution counts for Mercury and Venus, and sets the 'zero point' for Aryabhata's cosmic clock.

Notes:

- **Chandroccha:** Moon's Apogee (farthest point of its orbit). **Jrukhidha:** Its revolutions = 488,219 (or 488,203).
- **Budha, Bhṛgu:** Mercury, Venus (their revolutions encoded here).
- **Pāta:** Moon's Node (intersection of Moon's orbit with ecliptic). **Buphinacha:** Its revolutions = 232,226 (or 232,238).
- **Vilomā:** "Retrograde" (referring to the motion of the Moon's node).
- **Budhāhni:** "On a Wednesday" (stated starting day for epoch).
- **Ajārkayogāt:** "From the conjunction of the start of Aries (aja) and the Sun (arka)" (Vernal Equinox).
- **Pravṛttacakraḥ:** "The time cycle began."

3.4.1 Refining the Orbits: Apogees and Nodes

To model variations in planetary speed and enable accurate eclipse prediction within his geocentric framework, Aryabhata tracked specific, slowly moving reference points:

- **Chandroccha (Moon's Apogee):** The point in the Moon's orbit farthest from Earth. The Moon's motion relative to this apogee affects its apparent speed and is integral to the *manda* (slow anomaly) correction, which accounts for this variability.

- **Pāta (Moon's Node):** The points where the Moon's tilted orbit intersects the Ecliptic (the Sun's apparent annual path). These nodes are crucial for predicting eclipses, as eclipses can only occur when the Moon is near one of these intersection points at the time of New or Full Moon. Aryabhata notes that these nodes move *vilomā* (retrograde, or backward) along the ecliptic.

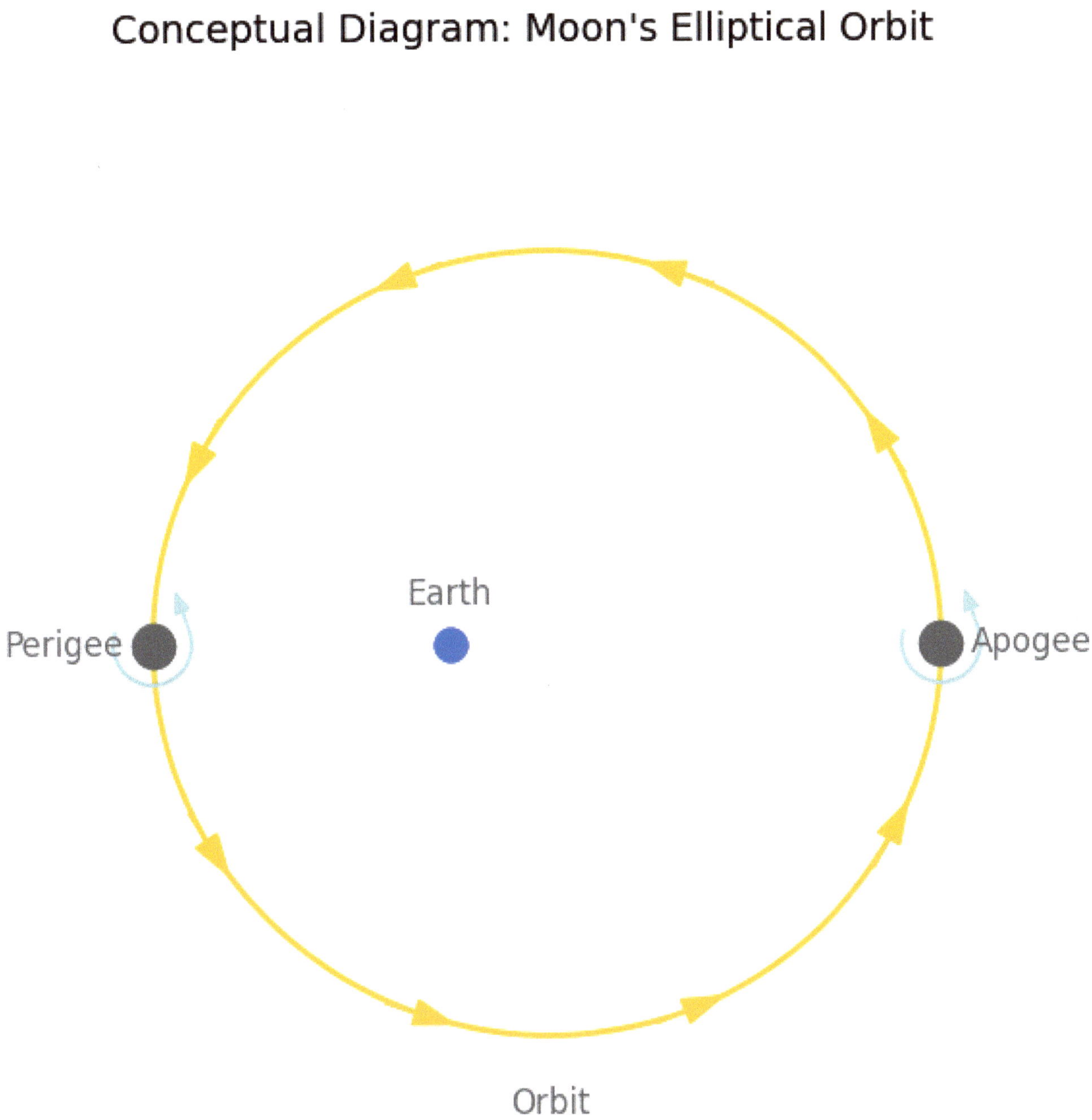

Figure 3.3 A simplified diagram illustrating the Moon's elliptical orbit around the Earth in a geocentric model, showing Perigee (closest) and Apogee (farthest) points. The varying distance causes the Moon's orbital speed to change (faster near Perigee, slower near Apogee), an irregularity accounted for by the Manda correction in Aryabhata's system.

Gītikāpāḍa Verse 4 encodes the number of revolutions these points, along with Mercury and Venus, complete in one *Mahāyuga* (4,320,000 solar years):

Parameter	Celestial Body	Cycles/Revs in Yuga	Sanskrit Code	Significance
Apogee (Uccha)	Moon	488,219	jrukhidha	Refines Moon's speed/position (manda correction). (Value 488,203 also cited).
Sidereal Revolutions	Mercury (budha)	17,937,060	suguśithṛṇa	Mean sidereal orbits (relative to stars). (Apparent motion needs Śīghroccha).
Sidereal Revolutions	Venus (bhṛgu)	7,022,376	jaṣabikhuchṛ	Mean sidereal orbits (relative to stars). (Variant 7,022,388 cited).
Apogees/Aphelia (Uccha)	Sun, Mars, Jup, Sat	(Effectively Zero)	śauri...arkāṇām	Implies apogees are fixed or slow, simplifying manda correction.
Node (pāta)	Moon (Ascending)	232,226	buphinacha	Crucial for eclipse prediction; moves vilomā (retrograde). (Variant 232,238 cited).

Table 3.7: Revolution counts per Mahāyuga (4,320,000 solar years) for the Moon's Apogee (Mandoccha), the Moon's Node (Pāta), Mercury, and Venus, as encoded by Aryabhata in Gītikāpāda Verse 4 (Gk.4). These parameters are crucial for calculating anomaly corrections (related to speed variations) and predicting eclipses (which occur near the Moon's nodes). (Note: Values may vary slightly depending on interpretation or textual variants).

Important Notes on Parameters:

1. **Moon Apogee:** Standard decoding yields 488,219. The value 488,203 (from variant readings or interpretations) is also frequently used in commentaries.

2. **Mercury/Venus Revolutions:** These are mean sidereal revolutions. Calculating their true positions requires applying further corrections, particularly the *Śīghroccha* (fast anomaly) correction.

3. **Node:** Standard decoding yields 232,226 (variant 232,238 also exists). This retrograde motion means the node completes one backward cycle in approximately 18.6 years.

3.4.2 The Lanka Dawn: A Cosmic Starting Point

To make calculations repeatable and universal, a precise starting point (**epoch**) is essential. Aryabhata defines his epoch in the latter part of the verse: *budhāhni ajārkayogāt pravṛttacakraḥ*:

- Generally understood as: "The time cycle (*chakraḥ*) began (*pravṛtta*) from the conjunction (*yogāt*) of the start of Aries (*aja*) and the Sun (*arka*) [i.e., Vernal Equinox], at Lanka, on a Wednesday (*budhāhni*)."
- **Lanka:** Aryabhata's prime meridian, a conceptual point on the equator (longitude 0° in his system).
- **Wednesday:** The specific starting weekday mentioned.
- **Start of Aries (*aja*)/Vernal Equinox:** The zero point of celestial longitude.
- **Epoch Date & Weekday Conflict:** Based on Verse 5's elapsed time, this corresponds to sunrise (or midnight prior) on **Friday, February 18, 3102 BCE** (Julian calendar), marking the beginning of the current Kali Yuga. The discrepancy between the verse stating "Wednesday" (*budhāhni*) and the calculation pointing to Friday is a well-known puzzle. Commentators suggest various resolutions (e.g., referring to a different time convention, or an error), but for practical computation, the defined moment (start of Kali Yuga, Friday 3102 BCE sunrise at Lanka) derived from the elapsed time in Verse 5 is the **operational epoch**.

This "Lanka Dawn" serves as the absolute zero point for all calculations of elapsed time and celestial longitude in Aryabhata's system.

3.4.3 From Cycles to Positions: The Rule of Three (Trairāśika)

How were these vast revolution counts used practically? The fundamental tool was the ***trairāśika*** (Rule of Three), a simple ratio calculation: *If A happens in time B, then X happens in time C, where X = (A × C) / B.* (See Gp.26 for its formal statement).

Example: Find Jupiter's approximate Mean Longitude (average position) after 1,000 civil days from the Kali Yuga epoch.

- Civil Day (*sāvana dina*): The practical day used in calendars (sunrise to sunrise).
- Knowns (using standard values from Gk.3):
  - Civil Days / Yuga = 1,577,917,500 (derived from Earth rotation parameters)
  - Jupiter Revolutions / Yuga = 364,224 (from Ch 3.3.1)
  - Time Elapsed (C) = 1,000 civil days
  - 1 Revolution = 360 degrees = 21,600 minutes of arc
- Proportion: (Jupiter Revs (X) / 1000 days (C)) = (364,224 Revs (A) / 1,577,917,500 days (B))
- Solve for Revolutions (X): X = (A × C) / B = (364,224 × 1,000) / 1,577,917,500 ≈ 0.230825 revolutions
- Convert to Arc: Mean Degrees Moved ≈ 0.230825 revs × 360 deg/rev ≈ **83.10 degrees** (or 4986 minutes of arc) from its starting position at 0° Aries.

This gives Jupiter's average position. Finding its true predicted position required further corrections (*manda* and *śīghra*) as detailed in the Kālakriyāpāda.

A Cosmos Taking Root

Verse 4 is pivotal. By quantifying the motion of crucial reference points like the Moon's apogee and node, providing the mean motions for Mercury and Venus, and establishing a precise starting epoch, Aryabhata gives his cosmic model a defined, calculable structure anchored in time and space.

3.5 The Grand Cosmic Clock – Yugas, Manus, and Kalpas (Gk.5)

Purpose: To define the nested hierarchy of vast cosmic time cycles (Yuga, Manvantara, Kalpa) and precisely locate the Kali Yuga epoch within this structure, providing the temporal scaffolding for all calculations.

Verse 4 was established *when* Aryabhata's cosmic clock started ticking – the Kali Yuga epoch. But where does this moment fall within the vast expanse of cosmic time conceived in Hindu cosmology? Verse 5 addresses this, mapping the nested cycles of Yugas, Manvantaras, and Kalpas, and precisely locating the epoch within this grand temporal structure.

Sanskrit:

काहो मनवो ढ मनुयुगाः श्ख गतास्ते च मनुयुगाः छ्ना च ।

कल्पादेर्युगपादा ग च गुरुदिवसाच्च भारतात् पूर्वम् ॥ ५ ॥

IAST:

kāho manavo ḍha manu yugāḥ śkha gatās te cha manu yugāḥ chnā cha|

kalpāder yugapādā ga cha gurudivasāccha bhāratāt pūrvam || 5 ||

This verse integrates traditional cosmological timescales into Aryabhata's precise numerical framework, defining the very heartbeat of Brahma's creation cycle and pinpointing "now" within it.

Notes:

- **Kāho:** "Day of Brahma" or Kalpa.
- **Manavo ḍha:** 14 (ḍha) Manus in a Kalpa.
- **Manu yugāḥ śkha:** 72 (śkha) Mahāyugas in a Manu.
- **(Kalpa = 14 × 72 = 1008 Mahāyugas).**
- **Gatās te cha manu yugāḥ chnā cha:** 27 (chnā) Mahāyugas of the current (7th) Manu had passed.
- **Kalpāder yugapādā ga cha:** 3 (ga) Yuga quarters (pāda) of the current Mahāyuga had passed from the start of the Kalpa.
- **Gurudivasāccha bhāratāt pūrvam:** "[Epoch reckoning] from Thursday, and before the Bharata [war]."

3.5.1 A Hierarchy of Cosmic Time

Aryabhata defines immense, nested time cycles using his alphabetical code, building upwards from the Mahāyuga (4,320,000 years, established implicitly by the Sun's revolutions in Verse 3):

Time Unit	Sanskrit Term	Duration (Years)	Relationship (Decoded from Verse 5)	Significance
Mahāyuga	Yuga	4,320,000	Base computational unit	A full cycle of 4 ages
Manvantara	Manu	311,040,000	72 Mahāyugas (śkha = 70+2 = 72)	Era ruled by a Manu
Kalpa	Kāhaḥ	4,354,560,000	14 Manvantaras (ḍha = 10+4 -> 14)*	A "Day of Brahma"

Table 3.8: Hierarchy of vast cosmic time cycles in Aryabhata's system, based on the parameters in Gītikāpāda Verse 5 (Gk.5). The Mahāyuga forms the base unit, nesting within Manvantaras and the immense Kalpa. This table uses the 1008 Mahāyuga duration for the Kalpa, derived from Gk.5, distinct from the 1000 Mahāyuga definition sometimes found in traditional texts (Kk.8).

Note on Kalpa Duration: Aryabhata's specific parameters (*śkha*=72 Yugas/Manu, *ḍha*=14 Manus/Kalpa) lead directly to a Kalpa of 14 × 72 = **1008 Mahāyugas**. This differs slightly from the 1000 Mahāyugas per Kalpa often found in Puranic texts. Aryabhata's choice of 1008 likely stems from a need for mathematical consistency within his system, ensuring that planetary cycles close appropriately over these vast periods. Bhaskara I consistently uses the 1008-yuga Kalpa for calculations derived from Aryabhata.

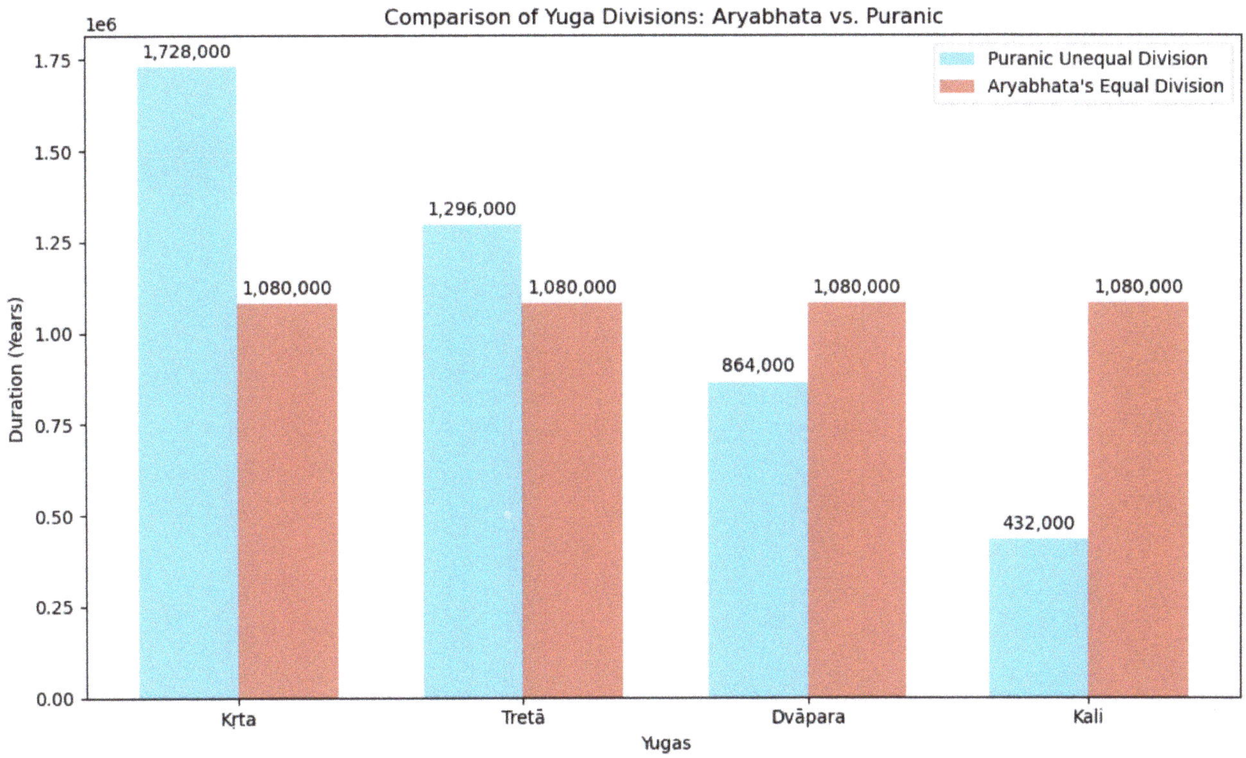

Figure 3.4: Comparison of Yuga divisions within a Mahāyuga (4,320,000 years). The traditional Puranic system uses unequal durations for Kṛta, Tretā, Dvāpara, and Kali Yugas (in a 4:3:2:1 ratio). Aryabhata's system (Gk.5) divides the Mahāyuga into four equal Yugapādas of 1,080,000 years each, a structure mathematically consistent with his astronomical parameters and defended by commentators like Bhaskara I.

3.5.2 Locating "Now" – The Start of the Kali Yuga

Having defined the structure, Aryabhata pinpoints the exact moment of his epoch (the start of the current Kali Yuga) relative to the beginning of the current Kalpa (*kalpādeḥ*):He states that up to this epoch moment:

1. Six Manvantaras had passed.

2. 27 Mahāyugas (*chnā* = 7+20 = 27) of the current (7th) Manvantara had passed.

3. 3 Yuga quarters (*ga* = 3) of the current (28th) Mahāyuga of this Manvantara had passed. These three elapsed quarters (*pāda*) represent the Kṛta Yuga, Treta Yuga, and Dvāpara Yuga of the current Mahāyuga cycle, leaving the Kali Yuga (the final quarter) about to begin.

Calculation of Elapsed Time (Start of Kalpa to Start of Kali Yuga):

- Time from previous 6 Manvantaras = 6 Manus × 72 Mahāyugas/Manu = 432 Mahāyugas.
- Time from elapsed Mahāyugas in current (7th) Manvantara = 27 Mahāyugas.
- Total elapsed full Mahāyugas = 432 + 27 = **459 Mahāyugas**.
- Time from elapsed Yuga Quarters within the current (28th) Mahāyuga:

- Aryabhata assumes four **equal quarters** (*yugapādas*) for the Mahāyuga, each lasting 4,320,000 / 4 = **1,080,000 years**. (This equal division is crucial for his system, defended by Bhaskara I against unequal Puranic yugas).
- Elapsed time from the 3 past quarters = 3 × 1,080,000 years = **3,240,000 years**.

- **Total Elapsed Years since Start of Kalpa:**
  - = (459 Mahāyugas × 4,320,000 years/Mahāyuga) + 3,240,000 years
  - = 1,982,880,000 + 3,240,000 = **1,986,120,000 years**.
- This precise number represents the vast duration, according to Aryabhata's parameters and equal yuga division, from the beginning of the current cosmic creation cycle (Kalpa) until the moment his astronomical clock starts ticking at the Kali Yuga epoch.

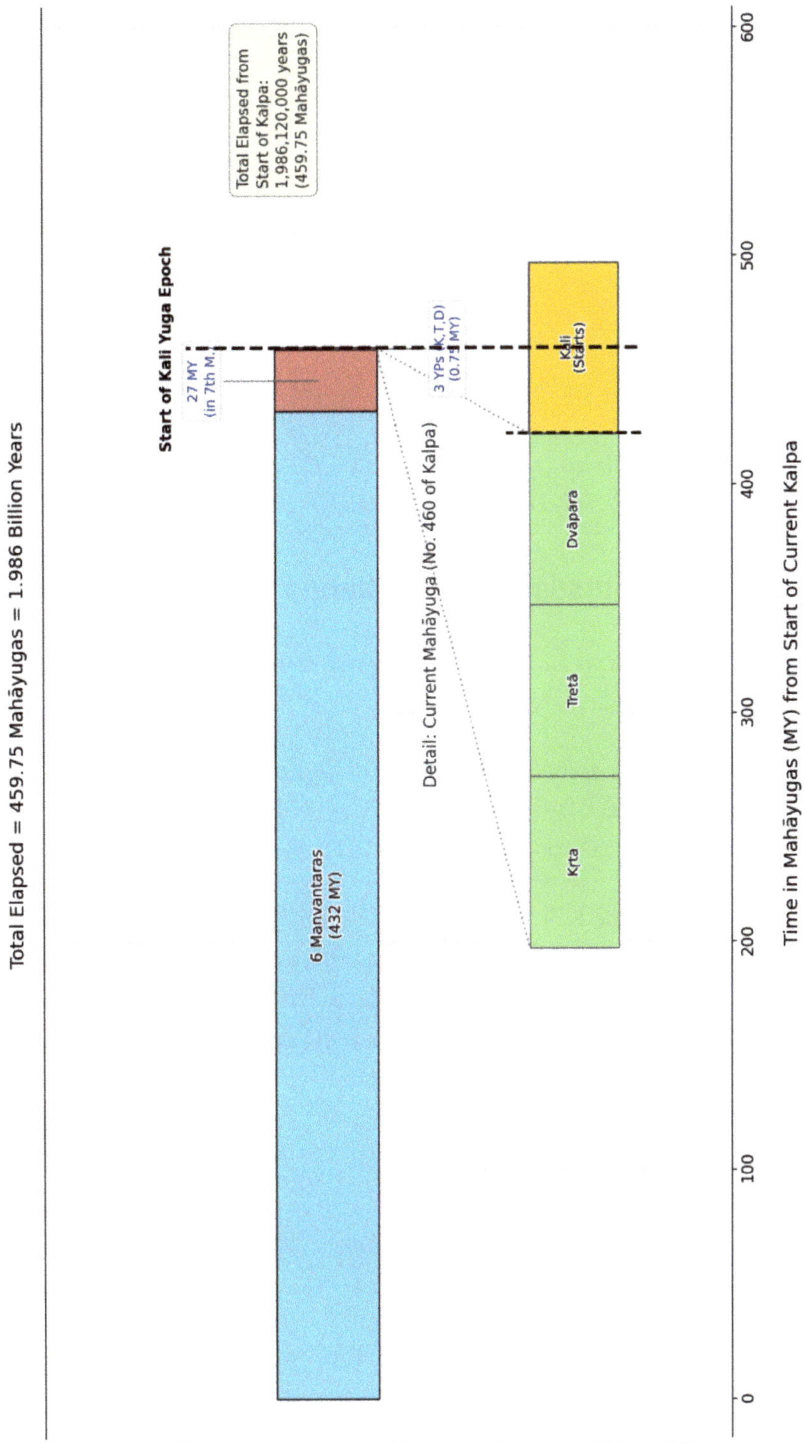

Figure 3.5: Aryabhata's Cosmic Timeline to Kali Yuga (Gk.5). Illustrates 459.75 Mahāyugas (1.986B years) elapsed from Kalpa start, segmented by Manvantaras and Yugapādas

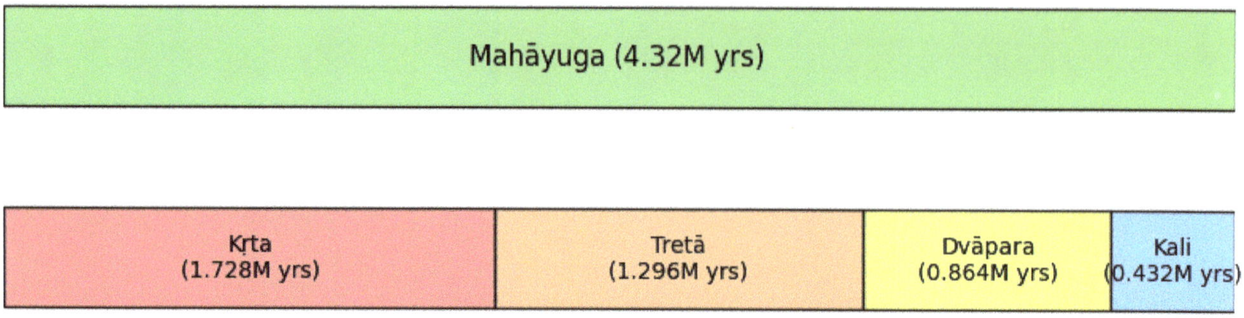

Figure 3.6: Divisions of the Mahāyuga (4.32 million years) and Related Time Concepts. The diagram illustrates Aryabhata's system (top), which divides the Mahāyuga into four equal Yugapādas (1.08 million years each) for astronomical calculation (Gītikāpāda Verse 5). This is contrasted with the traditional Puranic system (middle) using unequal durations (Kṛta, Tretā, Dvāpara, and Kali) in a 4:3:2:1 ratio.

3.5.3 Connecting to Culture: Thursday and the Bharata War

The verse concludes: *gurudivasācca bhāratāt pūrvam* – "[The epoch started] reckoning from Thursday (*gurudivasāt* = Jupiter's day), and occurred before the Bhārata [war]".

- This attempts to anchor the astronomically defined epoch (start of Kali Yuga, 3102 BCE, calculated as a Friday sunrise) to a specific preceding weekday (Thursday). (Recall the Friday/Wednesday/Thursday conflict discussed in 3.4). This remains a minor puzzle, potentially related to different day-reckoning conventions.

- Crucially, it places the start of the Kali Yuga **before** the great Mahabharata war. This represents a significant divergence from many Puranic traditions that tightly link the war's end (often associated with Krishna's departure) to the onset of the Kali Yuga. This difference underscores Aryabhata's commitment to establishing his astronomical timeline based on calculation and parameters, rather than strict adherence to potentially conflicting narrative chronologies.

3.5.4 Why It Matters: Anchoring All Calculations

Verse 5 provides the essential temporal scaffolding for the entire *Aryabhatiya*. It isn't just cosmological numerology; it's functionally critical:

- **Defines the "Zero Point":** By calculating the exact number of years elapsed from the absolute beginning of the Kalpa to the start of the Kali Yuga, Aryabhata ensures all planetary positions are consistent with this grander timescale.

- **Enables Long-Term Prediction:** Knowing the elapsed time within the Kalpa allows, in principle, calculations spanning vast durations.

- **Connects to Tradition while Innovating:** While relying on his own calculations for precision and consistency (e.g., equal yugas, 1008-yuga Kalpa), Aryabhata uses the traditional framework of Kalpas and Manvantaras, embedding his scientific system within the broader cultural understanding of cosmic time.

Verse 5 firmly places the starting point of Aryabhata's precise astronomical system within the immense, structured flow of cosmic time, providing the necessary foundation for calculating positions at any given moment relative to that epoch. Having defined the scales of time, Aryabhata turns next to the scales of space.

3.6 Measuring the Heavens' Vastness – The Size of Orbits (Gk.6)

Purpose: To establish the fundamental linear scale (in *yojanas*) of the celestial sphere, defining the size of the arena for celestial motion.

Having established the grand timescales culminating in the Kalpa (Verse 5), Aryabhata begins to map the spatial dimensions of his cosmos. Before defining the sizes of tangible bodies like the Earth and Moon, Verse 6 establishes the immense linear scale of the celestial sphere itself – the vast arena upon which the cosmic drama unfolds.

Sanskrit:

काहोर्मानं योजनैः ङ्यिला कक्ष्या ।

मन्दोच्चानां पातभगणाश्च तावन्तः ॥ ६ ॥

IAST:

kāhormānaṃ yojanaiḥ ṅyilā kakṣyā|

mandocchānāṃ pātabhagaṇāścha tāvantaḥ || 6 ||

This concise verse attempts to quantify the sheer size of the orbits or the celestial sphere in Aryabhata's chosen unit of distance, the *yojana*.

Notes:

- **Kāhormānaṃ:** Obscure term, possibly "measure of a day's rotation" or linking size to time/rotation.
- **Yojanaiḥ:** "In yojanas" (unit of distance).
- **Ṅyilā kakṣyā:** Circumference (kakṣyā) of the celestial sphere = ṅyilā.
- **Ṅyilā (interpreted value):** 3,300,750,000 Yojanas.
- **Mandocchānāṃ pātabhagaṇāścha tāvantaḥ:** Paths of apogees and nodes are also of that same (tāvantaḥ) size.

3.6.1 Decoding the Celestial Sphere's Circumference

The verse packs a universe of scale into few syllables, though interpretation is challenging and relies heavily on Bhaskara I and subsequent commentators:

1. ***kakṣyā***: Refers to the orbit or path of a celestial body. In this context, it signifies the **circumference of the celestial sphere**, or perhaps the mean orbit of the Sun, representing the overall linear scale of the system.

2. ***yojanaiḥ ṅyilā***: The size of this *kakṣyā* is given as *ṅyilā* Yojanas. Decoding *ṅyilā* via the standard rules (Interpretations A/B of Gk.2) proves difficult. However, commentators like Bhaskara I, and consistent application within the system (e.g., relating it to daily motion in Kālakriyāpāda 12), lead to the widely accepted interpretation that *ṅyilā* represents the enormous number **3,300,750,000 Yojanas.** This value is likely derived contextually or through methods not fully explicit in the verse.

3. ***kāhormānaṃ***: This term is obscure and debated. Interpretations include "in the measure of a day's rotation," "measure related to time/day," or linking the size to the speed of rotation. It seems intended to connect the vast linear size (*ṅyilā* Yojanas) to the fundamental daily cycle (*ahorātra*) of the heavens, possibly relating circumference to the constant daily linear motion described later in Kk.12.

4. The verse continues: ***mandocchānāṃ pātabhagaṇāḥ cha tāvantaḥ***, meaning 'And the paths traced by the **apogees** and **nodes** are also of that same size.' Here, 'apogees' (*mandocchāḥ*) refer to the calculated farthest points of planetary orbits from Earth, and 'nodes' (*pātāḥ*) are the theoretical points where planetary orbits cross the Sun's path (the ecliptic). These are not physical objects but abstract reference points that astronomers track because they move slowly over time. Aryabhata's assertion is that these abstract points, in their long-term movements, also describe great circles on the celestial sphere that share the same vast circumference (*ṅyilā*) he established earlier."

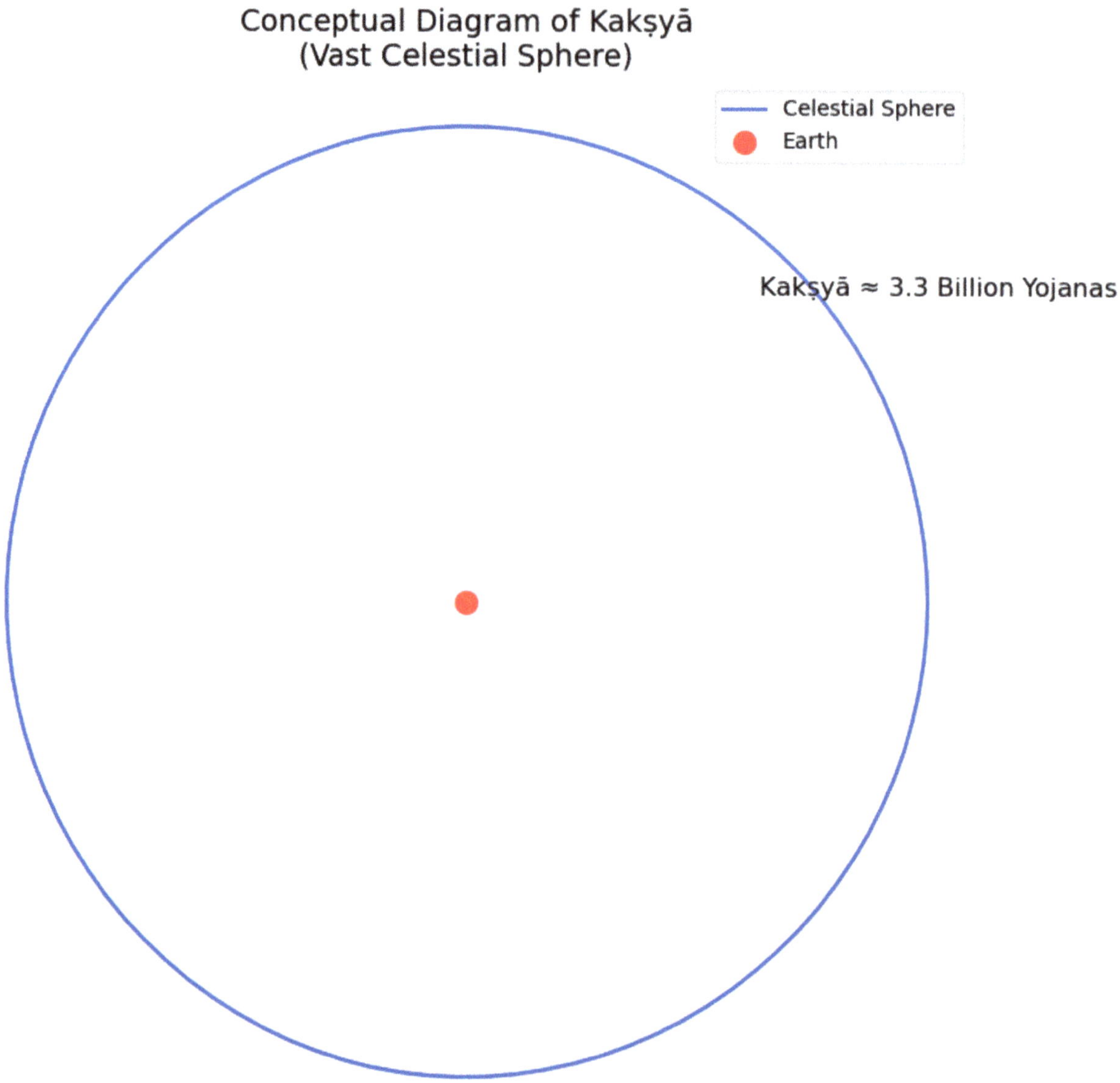

Figure 3.7: Conceptual diagram illustrating the immense scale of the celestial sphere (Kakṣyā) relative to the Earth at its center, as described in Gītikāpāda Verse 6 (Gk.6). Aryabhata quantified this vast circumference as approximately 3.3 billion Yojanas, establishing the fundamental linear scale of his geocentric cosmos. (Note: The exact modern equivalent of a Yojana is debated).

3.6.2 Significance: Establishing the Cosmic Arena in Yojanas

Verse 6 provides the fundamental linear scale (in Yojanas) for the celestial sphere within Aryabhata's geocentric model. While angular measures map positions, *ṅyilā* (interpreted as 3.3 billion *yojanas*) gives its vast physical circumference.

- **A Universe Measured:** It asserts that the entire cosmic stage, including the paths of abstract orbital points, is measurable in *yojanas*, bringing the heavens into the realm of quantifiable dimension.

- **Linking Size and Motion:** The term *kāhormānaṃ* hints at a deep connection between the linear size of the cosmos and its rotational dynamics, elaborated upon later (Kk.12).
- **Interpretational Challenges:** The precise numerical derivation and meaning of *ṅyilā* and *kāhormānaṃ* from the verse alone remain significant points of scholarly discussion, highlighting the reliance on commentary and systemic consistency.

By establishing this immense spatial framework, Verse 6 sets the stage for Verse 7, which will place the Earth, Sun, and Moon within this cosmic arena, defining their physical sizes relative to this backdrop and to each other.

3.7 Sizing the Earth, Sun, and Moon – The Measure of Worlds (Gk.7)

Purpose: To define the physical sizes (diameters) of the Earth, Sun, and Moon in *yojanas*, providing essential data for eclipse calculations and establishing the physical scale of the system.

Having established the immense scale of the celestial sphere (*kakṣyā*) in Yojanas (Verse 6), Aryabhata now anchors his model by defining the physical sizes of its most significant inhabitants: the Earth, Sun, and Moon. Verse 7 provides these crucial dimensions, setting the fundamental baseline and proportions for his measured universe.

Sanskrit:

नृषि योजनं ञिला भूव्यासोऽर्कशशाङ्कयोर्घञिङ ।

पणसैर्गुणिताः शेषा भृगुगुरुबुधशनिभौमानाम् ॥ ७ ॥

IAST:

nṛṣi yojanaṃ ñilā bhūvyāso'rkaśaśāṅkayor ghañiṅa|

paṇasairguṇitāḥ śeṣā bhṛgugurubudhaśanibhaumānām || 7 ||

Notes:

- **Nṛṣi yojanaṃ:** Defines the yojana (possibly based on human height, nṛ=man).
- **Ñilā bhūvyāsaḥ:** Earth's diameter (bhūvyāsaḥ) = ñilā = 1,050 yojanas.
- **Arkaśaśāṅkayoḥ ghañiṅa:** Diameters of Sun (arka) and Moon (śaśāṅka) encoded by ghañiṅa.
- **Sun Diameter (interpreted):** 4,410 yojanas.
- **Moon Diameter (interpreted):** 315 yojanas.
- **Paṇasaiḥ guṇitāḥ śeṣāḥ:** Diameters of remaining planets (Venus, Jupiter, Mercury, Saturn, Mars) obtained by multiplying by factors (paṇasaiḥ).
- **(Earth's diameter is foundational; Sun/Moon sizes crucial for eclipse calculations).**

3.7.1 Why Define These Sizes? Building the Physical Model

These dimensions are essential gears in Aryabhata's 'cosmic calculator':

- **Earth's Size (*bhūvyāsaḥ*):** Provides the fundamental baseline. It's the yardstick against which the cosmos is measured.

- **Sun and Moon Diameters (*arkaśaśāṅkayoḥ*):** Absolutely crucial for eclipse prediction. The geometry of eclipses depends directly on the relative physical sizes and distances of these three bodies.
- **Relative Planetary Sizes:** Provides a rule for the sizes of other planets, adding physical dimension to the model.

3.7.2 Setting the Scale: The Yojana and Earth's Diameter

Verse 7 first grounds the unit of cosmic distance, the **yojana**:

- ***nṛṣi yojanaṃ***: Defines the *yojana*. Often interpreted based on *nṛ* (man) suggesting 1 *yojana* ≈ 8,000 human heights (*puruṣa*) or similar measures. The exact modern equivalent is debated (estimates range widely, often cited as ~5-9 miles / 8-15 km), but internal consistency within the system is key.

Using this unit, Aryabhata gives his value for Earth's diameter:

- **Earth's Diameter (*bhūvyāsaḥ*) = *ñilā* = 1,050 yojanas.**
- Decoding: Using Interpretation A (additive) of the number system: *ñi* (10×100^1) + *lā* (50×100^0) = 1000 + 50 = 1050.
- Scale: Approx. 5,250 - 9,450 miles / 8,450 - 15,200 km. (1 Yojana = 5-7 Miles)
- Accuracy: This figure is **remarkably accurate** for the 5th century CE, likely derived from geometric reasoning based on observations (perhaps related to simultaneous gnomon measurements at different latitudes). Depending on the *yojana* length assumed, it's potentially within 1-15% of the modern mean diameter (~7918 miles / 12742 km). This accurate Earth diameter is a cornerstone of his system.

Celestial Body	Diameter (Yojanas - Gk.7)	Approx. Ratio to Earth Diameter
Earth	1050	1
Sun	4410	4.2
Moon	315	0.3
Venus	998	0.95
Jupiter	11519	10.97
Mercury	399	0.38
Saturn	9597	9.14
Mars	557	0.53

Table 3.9: Diameters of celestial bodies in Yojanas, as stated or implied by Aryabhata in Gītikāpāda Verse 7 (Gk.7). Earth's diameter (1,050 yojanas) provides the fundamental scale. The sizes of the Sun (4,410 yojanas) and Moon (315 yojanas) were crucial for eclipse calculations. (Note: The exact modern equivalent of a Yojana is debated; relative sizes are key. Values for other planets were often derived via factors from a base diameter).

3.7.3 Sizing the Sun and Moon

Next, the verse gives the diameters of the Sun and Moon using the code ***ghañiṅa***:

- **Sun Diameter (*arka*) = 4410 yojanas.**
  - Standard Interpretation: Consistently derived in commentaries. How it's decoded from *ghañiṅa* using the standard rules (Gk.2) is unclear, suggesting complex encoding, contextual derivation based on angular size and assumed distance, or possibly a textual variant. *(Angular Size (or Apparent Size) refers to how large an object appears to be in the sky, measured as an angle.)*
  - Significance: In Aryabhata's model, the Sun is ~4.2 times Earth's diameter (4410 / 1050). This primarily reflects its apparent angular size scaled up according to its assumed distance, rather than conveying the true vast physical size ratio known today.
- **Moon Diameter (*śaśāṅka*) = 315 yojanas.**
  - Standard Interpretation: Derived from the same code *ghañiṅa* or contextually by commentators. It's notable that 315 is exactly 1/14th of the Sun's diameter (4410 / 14 = 315).
  - Significance: Yields a Moon-to-Earth diameter ratio of 315 / 1050 = 0.3 or 30%. This is **strikingly close** to the modern value (~27.3%), suggesting careful observational grounding, perhaps from eclipse timings or angular size comparisons.

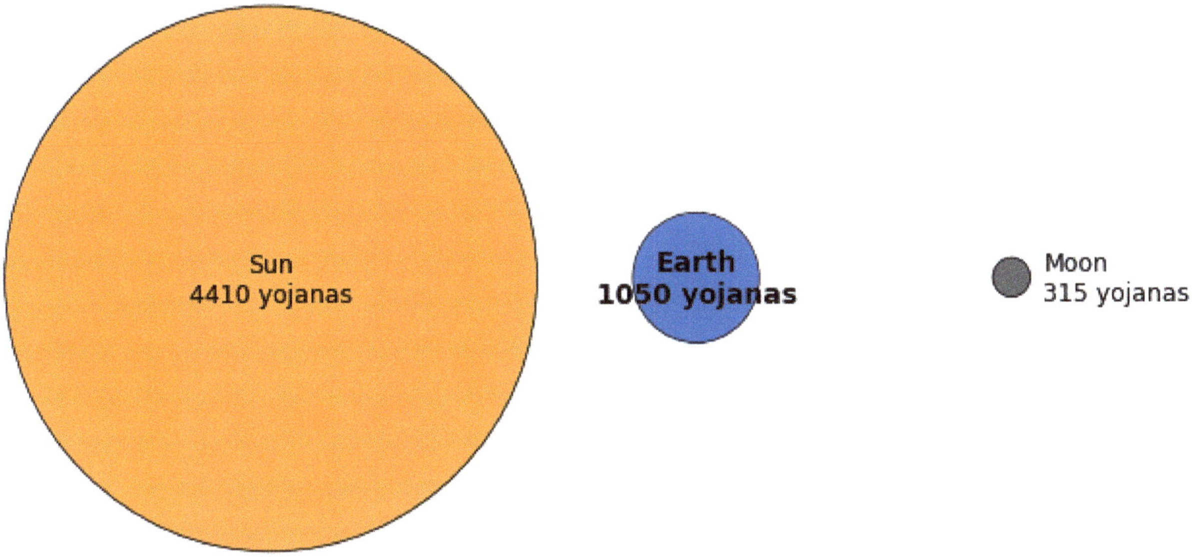

Figure 3.8: Comparative diagram illustrating the relative diameters of the Sun, Earth, and Moon based on the values given by Aryabhata in Gītikāpāda Verse 7 (Gk.7). Earth's diameter is given as 1,050 yojanas. Note the relative sizes, particularly the Moon's diameter being approximately 30% of Earth's, reflecting careful observation and calculation crucial for eclipse prediction.

3.7.4 Relative Sizes of Other Planets

The verse concludes with a rule for the diameters of the remaining planets:

- ***paṇasair guṇitāḥ śeṣāḥ bhṛgu-guru-budha-śani-bhaumānām***: "The remaining [diameters] of Venus, Jupiter, Mercury, Saturn, and Mars are obtained by multiplying [a base diameter] by factors (*paṇasaiḥ*)."
- Commentators like Bhaskara I interpret *paṇasaiḥ* as encoding specific fractions relative to a base diameter (likely Earth's). Standard interpretations yield approximate ratios: Venus ≈ 1/6th Earth's diameter, Jupiter ≈ 1/10th, Mars ≈ 1/15th, Saturn & Mercury ≈ 1/20th. These are rough relative sizes, likely based more on perceived brightness or tradition than precise measurement.

A Cosmos Measured in Size

Verse 7 populates Aryabhata's cosmos with objects of defined physical scale. By defining the *yojana*, providing a robust estimate for Earth's diameter, and giving the scaled diameters of the Sun and Moon crucial for eclipses, along with a method for other planets, Aryabhata establishes the fundamental physical dimensions necessary for his geometric model of the universe.

3.8 The Sun's Slanted Path – Mapping the Ecliptic (Gk.8)

Purpose: To describe the geometry of the ecliptic (Sun's annual path) relative to the celestial equator, explaining the cause of the seasons.

Having defined the fundamental linear scales of the cosmos and its key inhabitants (Verses 6 & 7), Aryabhata shifts focus to the crucial geometry of celestial motion. Verse 8 describes the fundamental pathway of the Sun across the sky – the **Ecliptic** – and its vital relationship to the Celestial Equator. This mapping explains the Sun's apparent annual north-south movement relative to the stars, the very basis for Earth's seasons.

Sanskrit:

मेषादेः कन्यान्तं सममुदगपमण्डलात् खगोलार्धम् ।

तौल्यादेर्मीनान्तं शेषार्धं दक्षिणेनैव ॥ ८ ॥

IAST:

meṣādeḥ kanyāntaṃ samam udagapamaṇḍalāt khagolārdham|

taulyādermīnāntaṃ śeṣārdhaṃ dakṣiṇenaiva || 8 ||

3.8.1 Tracing the Sun's Yearly Journey

This verse elegantly traces the Sun's apparent yearly path—the **ecliptic** (*apamaṇḍala*)—as it moves against the backdrop of the celestial sphere. Aryabhata divides this great celestial highway into two symmetrical, 180-degree halves, marking the Sun's journey north and south of the celestial equator.

- **The Northern Journey:** The first half of the year begins as the Sun enters the sign of Aries. From this point until it reaches the end of Virgo, the Sun travels along an arc that is **north** (*udag*) of the celestial equator. It reaches its peak northern distance from the equator at the summer solstice, marking the height of summer in the Northern Hemisphere.

- **The Southern Journey:** As the Sun enters Libra, it begins its journey along the second half of the ecliptic. From Libra through to the end of Pisces, its path deviates to the **south** (*dakṣiṇena*) of the celestial equator. It reaches its farthest southern point at the winter solstice.

The verse emphasizes that these two halves of the ecliptic are perfectly symmetrical (*samam*). The points where the Sun's path crosses the celestial equator are the **equinoxes**—the start of Aries (the Vernal Equinox) and the start of Libra (the Autumnal Equinox). This beautifully described geometric tilt is the fundamental cause of Earth's seasons.

3.8.2 Visualizing the Tilt: The Two Hoops

Imagine the celestial sphere. Draw two great circles:

- **Hoop 1:** The Celestial Equator (plane extending from Earth's equator).
- **Hoop 2:** The Ecliptic (plane of Earth's orbit around the Sun, or Sun's apparent path around Earth), tilted relative to the Equator.Verse 8 describes how these two hoops intersect at the equinox points (Aries 0°, Libra 0°). For six signs (Aries to Virgo), the Sun travels on the part of the Ecliptic north of the Equator hoop; for the other six (Libra to Pisces), it travels south. The angle between these hoops is the **Obliquity of the Ecliptic** (quantified as 24° in Verse 9).

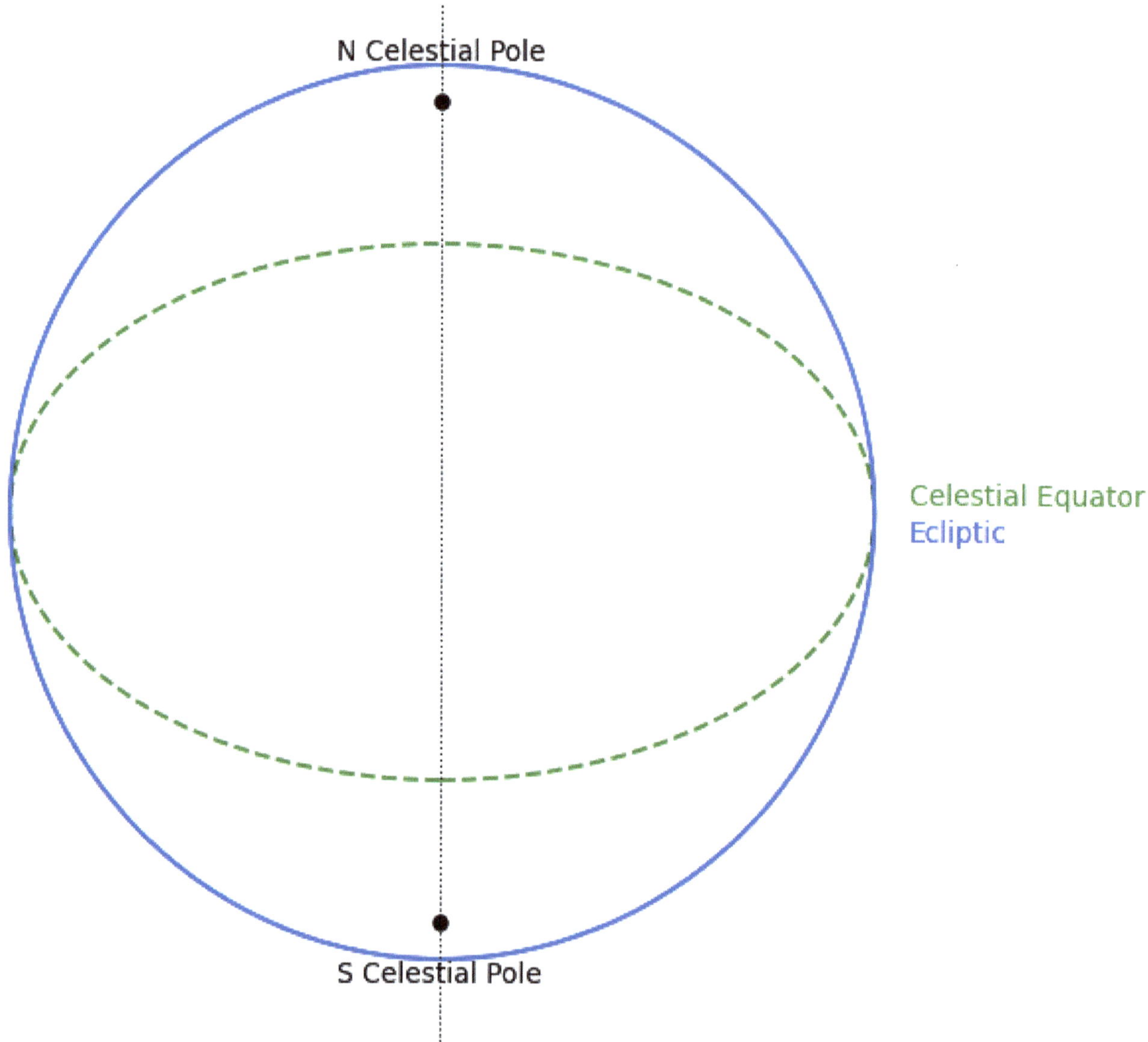

Figure 3.9: Diagram illustrating the fundamental relationship between the Celestial Equator (dashed green line) and the Ecliptic (solid blue line), representing the Sun's apparent annual path. The Ecliptic is tilted relative to the Equator, intersecting at the equinoxes and reaching maximum distance north and south (corresponding to the solstices). This tilt, quantified as the Obliquity of the Ecliptic (Gk.9), causes Earth's seasons (Gola.1). The diagram also shows the Celestial North and South Poles, around which the apparent daily rotation of the heavens occurs.

Significance: The Blueprint for the Seasons

This single verse, by defining the tilted path of the Sun, provides the essential geometric blueprint for understanding its motion and its profound consequences.

- **It Explains the Seasons:** The Sun's yearly journey north and south of the equator is the direct cause of the seasons. Its northward journey (*uttarāyaṇa*) brings summer to the Northern Hemisphere, while its southward journey (*dakṣiṇāyana*) brings winter.

- **It Defines the Celestial Coordinate System:** The relationship between the ecliptic (the Sun's path) and the celestial equator is the foundation for the entire celestial grid. The ecliptic becomes the reference plane for longitude, while the equator is the reference for declination.

- **It Is the Basis for Timekeeping:** Knowing the Sun's precise path and its changing declination throughout the year is essential for all practical timekeeping. It allows an astronomer to calculate the changing length of daylight, predict the exact times of sunrise and sunset, and use instruments like the gnomon to tell time with precision.

In summary, this verse lays out the qualitative blueprint of the Sun's slanted path. It is the very backbone of positional astronomy, setting the stage for the precise, quantitative measurements of tilt and celestial latitude that will follow.

3.9 Defining the Tilt and Wobble – Obliquity and Latitudes (Gk.9)

Purpose: To provide the numerical values for the Ecliptic's tilt (Obliquity) and the maximum deviations (latitudes) of the Moon and planets from the ecliptic path.

Verse 8 described the Sun's tilted path, the ecliptic. Now, in Verse 9, Aryabhata provides the precise numerical values for this fundamental tilt (the obliquity of the ecliptic) and for the maximum deviations (latitudes) of the Moon and planets from that ecliptic path. These angles are crucial for accurate positional astronomy and essential for predicting eclipses.

Sanskrit:

भापक्रमः २४ शशिविक्षेपांशार्धपञ्चम ४ १/२ ।

शनिगुरु कुजविक्षेपास्तद्विधार्धोऽथ भृगुबुधयोः ॥ ९

IAST:

bhāpakramaḥ 24 śaśivikṣepāṃśārdhapañcama 4 ½|

śanigurukujavikṣepās tad vidhārdho'tha bhṛgubudhayoḥ || 9 ||

Notes:

- **Bhāpakramaḥ 24:** Sun's (bhā) maximum declination (apakramaḥ) = Obliquity of the Ecliptic = 24 degrees.
- **Śaśivikṣepa aṃśa-ardha-pañcama 4 ½:** Moon's (śaśi) maximum latitude (vikṣepaḥ) is four and a half (ardha-pañchama) degrees (aṃśa) = 4.5°.
- **Śani guru kuja vikṣepāḥ tad vidhārdhaḥ:** Max latitudes of Saturn, Jupiter, Mars are half that (of Moon's) = 2.25°.
- **Atha bhṛgu budhayoḥ:** And likewise for Venus and Mercury (max latitude = 2.25°).

3.9.1 Quantifying Celestial Geometry: Degrees of Deviation

Having described the geometry of the heavens, Aryabhata now provides the precise numerical values for the key angles of his system. This verse concisely encodes the fundamental parameters of tilt and wobble that govern the motions of the Sun, Moon, and planets.

1. The Tilt of the Ecliptic (Obliquity): 24 Degrees

The verse first states that the maximum declination of the Sun is **24 degrees**. This is the **Obliquity of the Ecliptic**—the fundamental angle between the celestial equator and the ecliptic path. It quantifies the maximum distance, north or south, that the Sun ever reaches in its yearly journey.

Aryabhata's value of 24° was a standard and highly practical approximation for his era. While the true value was closer to 23.6°, using the clean integer 24° significantly simplified the trigonometric calculations needed to determine the seasons and the length of daylight.

2. The Wobble of the Moon (Max Latitude): 4.5 Degrees

Next, it states that the maximum latitude (*vikṣepa*) of the Moon is **4.5 degrees**. This is the measure of how far the Moon's tilted orbit allows it to wander north or south of the ecliptic highway.

This parameter is absolutely critical for one reason: **eclipse prediction**. Solar and lunar eclipses can only occur when the Moon's latitude is near zero, meaning it is crossing the ecliptic at the same time it aligns with the Sun and Earth. This single number is a key that unlocks the geometry of eclipses.

3. The Wobble of the Planets (Max Latitude): 2.25 Degrees

Finally, the verse provides the maximum latitudes for the planets.

- For **Saturn, Jupiter, and Mars**, the maximum latitude is given as half of the Moon's, or **2.25 degrees**.
- The verse then adds, "and likewise for **Venus and Mercury**," assigning them the same maximum latitude of **2.25 degrees**.

These values quantify the typical "wobble" of each planet above and below the ecliptic, providing the final angular parameters needed for a complete and accurate model of the solar system.

Parameter	Value (°)
Obliquity (Bhāpakrama)	24°
Moon Max Latitude (Śaśivikṣepa)	4.5°
Planet Max Latitude (Śani)	2.25°
Planet Max Latitude (Guru)	2.25°
Planet Max Latitude (Kuja)	2.25°
Planet Max Latitude (Bhṛgu)	2.25°
Planet Max Latitude (Budha)	2.25°

Table 3.10: Fundamental Angular Parameters: Obliquity of the Ecliptic and Maximum Celestial Latitudes, as given by Aryabhata in Gītikāpāda Verse 9 (Gk.9). These values are essential inputs for calculations involving the Sun's declination and the positions of the Moon and planets relative to the Ecliptic.

Why These Angles Are Crucial Calculation Inputs

These parameters are not merely descriptive; they are essential numerical inputs for the calculations performed later in the *Aryabhatiya*:

- **Obliquity (24°):**
    - Defines the limits of the Sun's north-south motion (declination).
    - Essential for calculating the Sun's declination on any given day of the year using trigonometric functions.
    - Needed for calculations involving the variation of daylight length, sunrise/sunset times (*chara* correction, Gola.27), and interpreting gnomon shadow lengths (Gola.31).
- **Maximum Latitudes (Moon 4.5°, Planets 2.25°):**
    - Account for the fact that the Moon and planets do not always move exactly on the ecliptic plane.
    - Moon's Latitude: Vital for determining whether a syzygy (new/full moon) will result in an eclipse. If the latitude is too large at that time, the Moon passes above or below the Sun (no solar eclipse) or the Earth's shadow (no lunar eclipse).
    - Planetary Latitudes: Necessary for calculating precise apparent positions (*sphuṭa graha*), as latitude affects the observed position against the background stars.

Verse 9 provides the limiting values for these crucial angular deviations. Calculating the actual declination or latitude of a body at a specific time requires applying these limits within a geometric model using trigonometry – the very trigonometry whose foundation Aryabhata lays in the next verse.

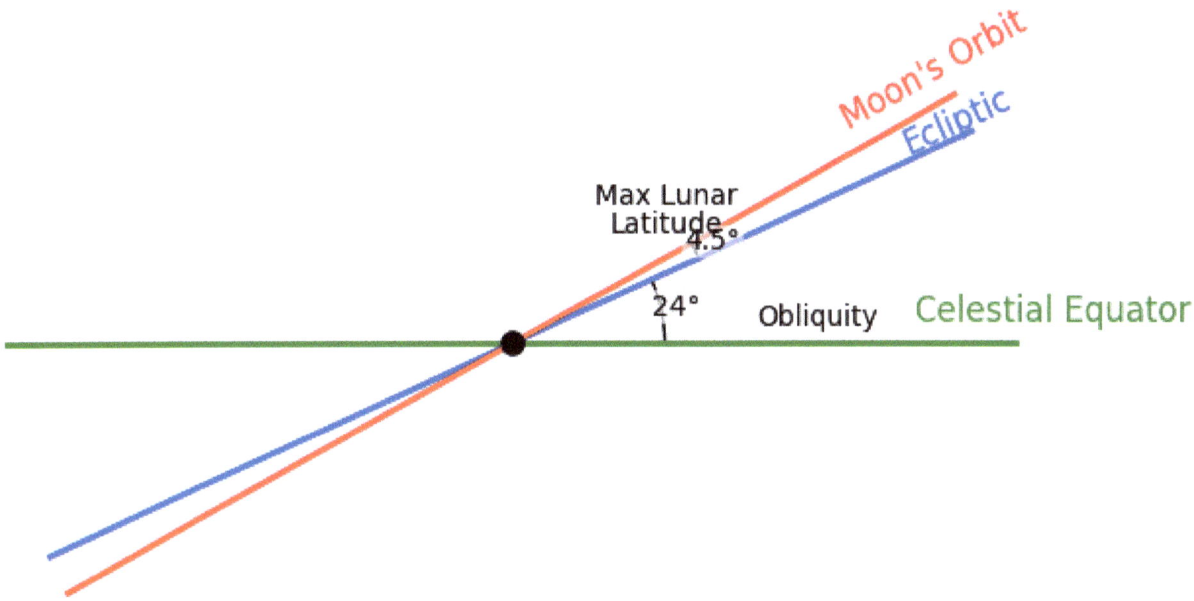

Figure 3.10: Geometric diagram illustrating the key angles quantified by Aryabhata in Gītikāpāda Verse 9 (Gk.9). The Obliquity (24°) is the angle between the Celestial Equator (green) and the Ecliptic (blue). The Moon's orbit (red) is tilted relative to the Ecliptic, with a maximum celestial latitude of approximately 4.5°. These fundamental angles govern seasons and are crucial inputs for astronomical calculations, particularly eclipse prediction.

3.10 The Seeds of Sine – Aryabhata's Table of Differences (Gk.10)

Purpose: To provide the fundamental numerical data, in the form of 24 R-Sine *differences*, required for all trigonometric calculations in the system.

Having set the geometric stage with the ecliptic's tilt and maximum latitudes, Aryabhata now introduces the core computational tool needed to navigate this celestial geometry: trigonometry, specifically through his unique formulation of the sine function. Verse 10 does not explicitly state trigonometric rules (that function is partly served by Verse 12 and applications in later chapters), but instead presents the fundamental numerical data – a list of 24 numbers derived from the sine function, ingeniously encoded in his alphabetical system. This table, or more accurately, the differences it contains, is a cornerstone of the *Aryabhatiya*'s computational power.

Sanskrit:

मखि भखि फखि धखि णखि ञखि ङखि हस्झ स्ककि किष्ग श्घकि किघ्व ।

घ्लकि किग्र हक्य धकि किच स्ग झश ङ्व क्ल प्त फ छ कला-अर्ध-ज्याः ॥ १० ॥

IAST:

makhi bhakhi phakhi dhakhi ṇakhi ñakhi ṅakhi hasjha skaki kiṣga śghaki kighva|

ghlaki kigra hakya dhaki kicha sga jhaśa ṅva kla pta pha cha kalā-ardha-jyāḥ || 10 ||

Notes:

- **Makhi, bhakhi, phakhi...cha:** 24 encoded alphabetical values representing numbers.
- **Kalā-ardha-jyāḥ:** "Elemental differences of R-Sines, measured in minutes of arc (kalā)."
- **24 R-Sine Differences (Δ):** 225, 224, 219, ..., 7.
- **Angular Interval:** Each difference corresponds to an arc of 3° 45' (225 minutes).
- **Implied Radius (R or Trijyā):** Sum of the 24 differences = 3438 minutes of arc (= R·sin(90°)).

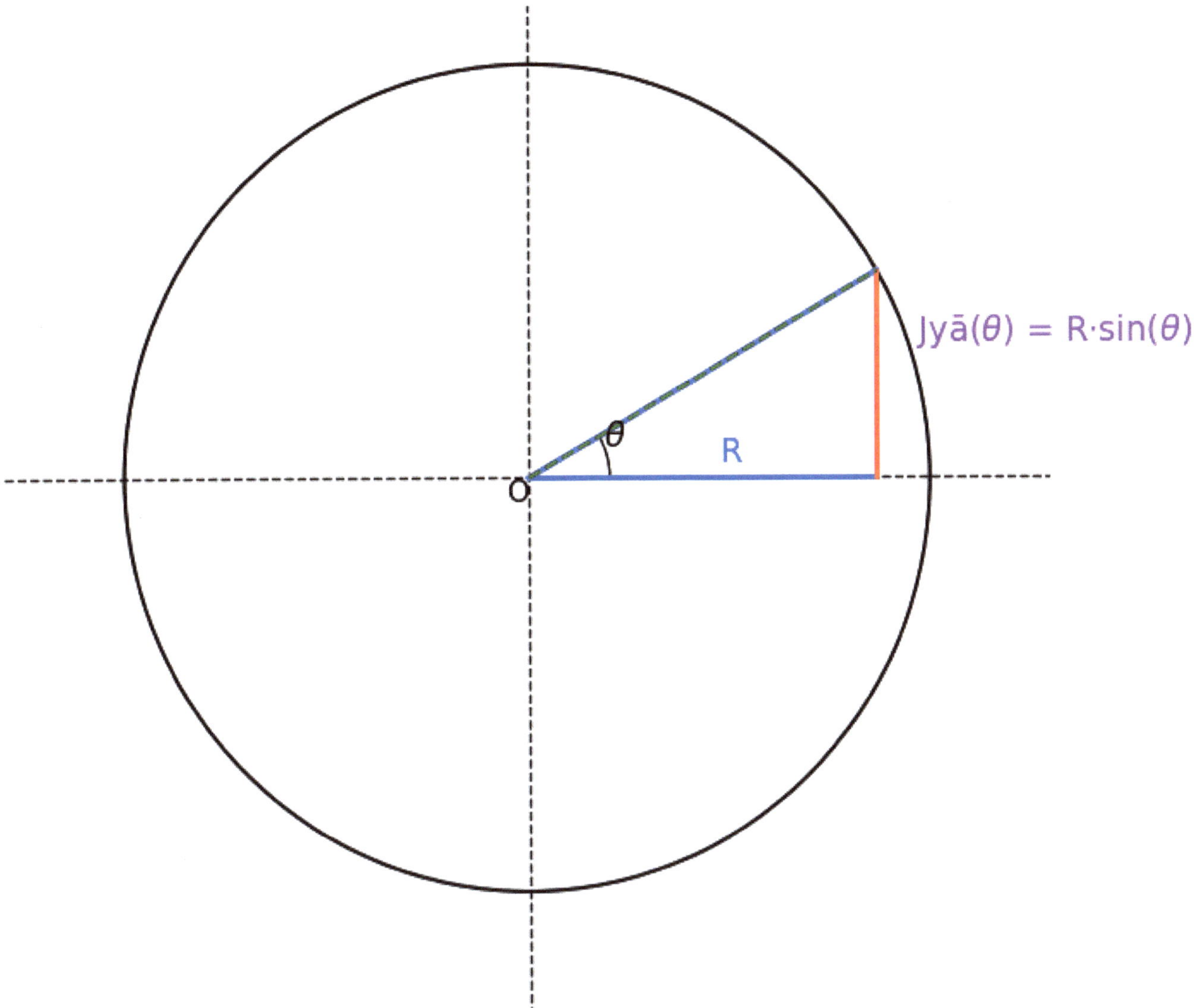

Figure 3.11: Geometric definition of the Indian sine function (Jyā or R·sinθ). In a circle of standard radius R (3438 minutes of arc), the Jyā of an angle θ is the length of the half-chord subtending that angle (shown as the vertical segment). This geometric concept forms the basis for Aryabhata's sine calculations and table (Gk.10 and Gk.12).

3.10.1 Decoding the Trigonometric Blueprint: The R-Sine Differences

Having set the geometric stage, Aryabhata now presents the numerical heart of his trigonometry. In this incredibly dense verse, he uses his unique alphabetical code to list 24 numbers. This list is the fundamental dataset for his entire sine function, a cornerstone of the *Aryabhatiya's* computational power.

Aryabhata describes these encoded values with the crucial term *kalā-ardha-jyāḥ*. To understand what he has given us, we must break this term down.

- ***Jyā*** (or *Jīvā*): Literally "bow-string," this is the Indian term for the sine. But unlike our modern sine, which is a dimensionless ratio, the *jyā* is a physical length. In a circle of a standard radius (R), the *jyā* of an angle is the length of the half-chord corresponding to that angle. It is equivalent to **R·sin(θ)**.
- ***Kalā***: This simply means a **minute of arc**, the unit of measurement for these R-Sine lengths.

- ***Ardha***: While it literally means "half," commentators like Bhāskara I explain that in this context, it refers to the "parts," "elements," or "segments" that are used to build the full sine table.

As we will see, these "elemental parts" are not the final R-Sine values themselves. They are something much more ingenious: the **successive differences** between the R-Sine values.

Therefore, the term *kalā-ardha-jyāḥ* can be understood as: **"the elemental differences of the R-Sines, measured in minutes of arc."** This verse does not give us the sine table directly; it gives us the foundational building blocks from which the entire table can be constructed.

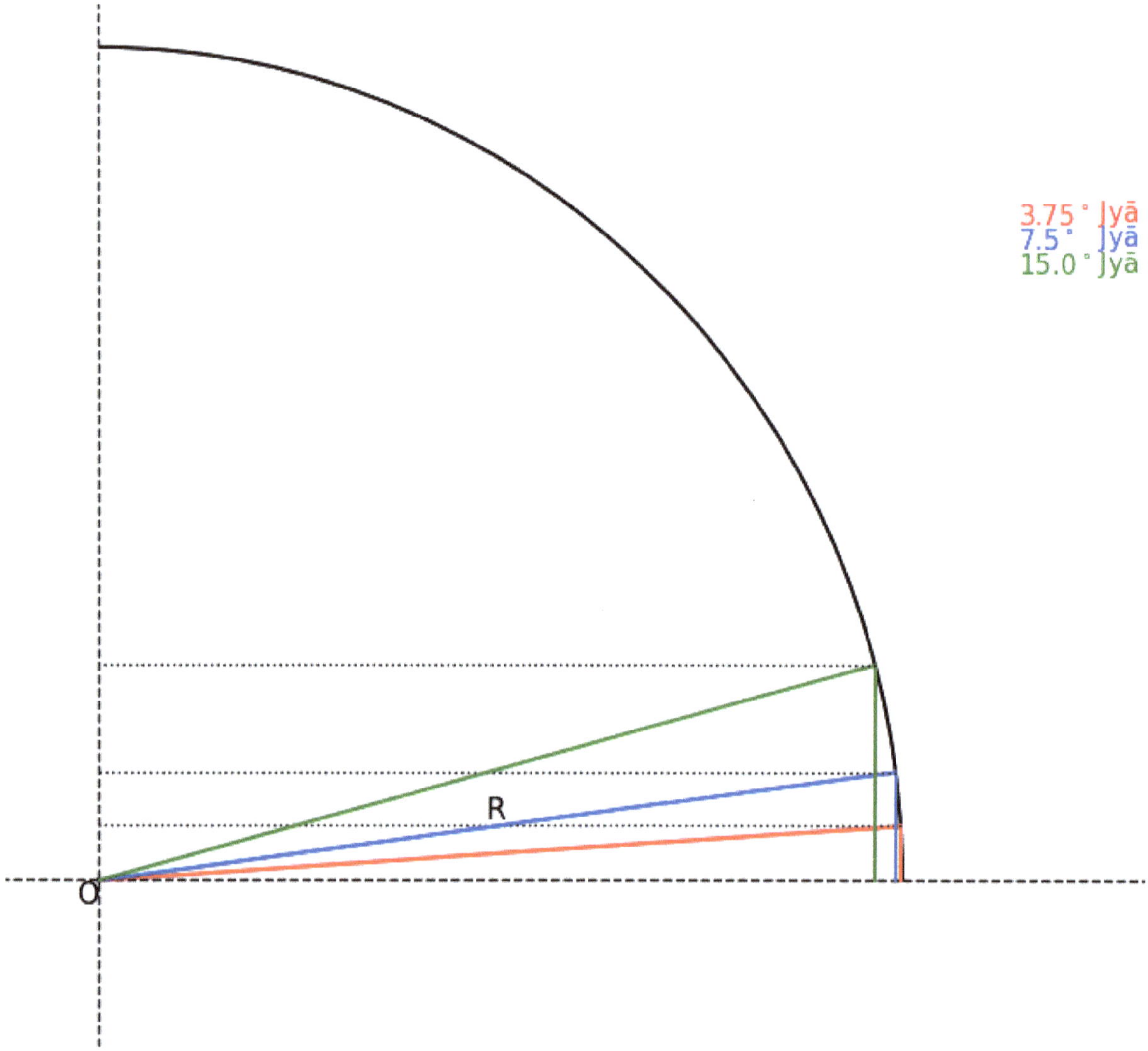

Figure 3.12: Geometric illustration of the Indian sine function (Jyā or R·sinθ). In Aryabhata's system, the sine of an angle is represented by the length of the half-chord in a circle of standard radius (R=3438′). This diagram shows how different angles correspond to different sine lengths. The values in Aryabhata's sine difference table (Gk.10) and the rule for their calculation (Gk.12) are based on this geometric concept

3.10.2 The 24 Encoded Differences

Each coded syllable in Gk.10 represents a number, which, when decoded using Aryabhata's alphabetical system, yields the following sequence of 24 values:

- *makhi* → 225
- *bhakhi* → 224
- *phakhi* → 219
- *dhakhi* → 215
- *ṇakhi* → 210
- *ñakhi* → 205
- *ṅakhi* → 199
- *hasjha* → 191
- *skaki* → 183
- *kiṣga* → 174
- *śghaki* → 164
- *kighva* → 154
- *ghlaki* → 143
- *kigra* → 131
- *hakya* → 119
- *dhaki* → 106
- *kica* → 93
- *sga* → 79
- *jhaśa* → 65
- *ṅva* → 51
- *kla* → 37
- *pta* → 22
- *pha* → 19 (Corrected value based on common reconstructions and commentary)

- *cha* → 7

3.10.3 Angular Intervals and the Crucial Insight: Differences, Not Cumulative Values

These 24 numbers correspond to 24 equal angular steps within a 90-degree quadrant. Each interval is therefore 90° / 24 = 3° 45′, which is equivalent to 225 minutes of arc.

The pivotal insight, crucial for understanding Aryabhata's method, is that these 24 numbers are **not** the cumulative R-Sine values ($Jyā_1$, $Jyā_2$, etc.) themselves. Instead, Aryabhata lists the **successive differences (Δ)** between these R-Sine values. Commentators refer to these as *jyā-piṇḍas* (sine-lumps/elements) or *khaṇḍa-jyās* (sine-segments/differences):

- Δ_1 = R·sin(1 × 3°45′) - R·sin(0°) = R·sin(225′) - 0 = 225
- Δ_2 = R·sin(2 × 3°45′) - R·sin(1 × 3°45′) = R·sin(450′) - R·sin(225′) = 224
- Δ_3 = R·sin(3 × 3°45′) - R·sin(2 × 3°45′) = R·sin(675′) - R·sin(450′) = 219...and so on, down to...
- Δ_{24} = R·sin(24 × 3°45′) - R·sin(23 × 3°45′) = R·sin(90°) - R·sin(86°15′) = 7

3.10.4 The Implied Standard Radius (R or Trijyā)

By summing all 24 of these listed differences, one obtains the R-Sine value for the final angle of the quadrant, which is 90 degrees:

Sum ($\Sigma\Delta_i$ for i=1 to 24) = 225 + 224 + 219 + ... + 19 + 7 = **3438**.

Since R·sin(90°) = R × 1 = R, this sum directly reveals Aryabhata's chosen standard radius (often called *trijyā*): **R = 3438 minutes of arc.**

This specific radius is mathematically significant and computationally convenient. It is a very close approximation of 21600 / (2π), where 21600 is the total number of minutes of arc in a full circle (360° × 60′°).

This choice greatly simplifies calculations involving the relationship between arc lengths and their corresponding R-Sines within his system.

n	Angle θ (n * 3°45')	R·sin Difference (Δn) [Gk.10 Valu	Cumulative R·sin(θ) (Jyā$_n$ = ΣΔ
1	3.75°	225	225
2	7.5°	224	449
3	11.25°	222	671
4	15.0°	219	890
5	18.75°	215	1105
6	22.5°	210	1315
7	26.25°	205	1520
8	30.0°	199	1719
9	33.75°	191	1910
10	37.5°	183	2093
11	41.25°	174	2267
12	45.0°	164	2431
13	48.75°	154	2585
14	52.5°	143	2728
15	56.25°	131	2859
16	60.0°	119	2978
17	63.75°	106	3084
18	67.5°	93	3177
19	71.25°	79	3256
20	75.0°	65	3321
21	78.75°	51	3372
22	82.5°	37	3409
23	86.25°	22	3431
24	90.0°	7	3438

Table 3.12: Aryabhata's Sine Difference Table, reconstructed from the 24 values encoded in Gītikāpāda Verse 10 (Gk.10). The first column lists the angular interval index (n, each representing 3°45'). The third column shows the R-Sine Difference (Δn) for that interval. The fourth column shows the cumulative R-Sine value (Jyā) obtained by summing the differences up to that interval (Jyā$_n$ = ΣΔ$_i$), with the final value reaching R=3438 minutes of arc.

3.10.5 Significance: A Revolution in Computation

The 24 differences encoded in Verse 10 represent more than just a list of numbers; they are a monumental achievement in the history of science. They are the blueprint for what is arguably the world's first formal table of R-Sines, and its creation was not a mere theoretical exercise. This table was the trigonometric engine that powered a new era of quantitative astronomy, unlocking a level of predictive power previously unattainable.

For the first time, Aryabhata's system gave astronomers the tools to:

- **Resolve Planetary Motion:** Accurately model the complex geometry of epicyclic motions to calculate the true positions of the planets (*sphuṭa graha*).
- **Predict Eclipses:** Determine the timing and circumstances of solar and lunar eclipses with unprecedented accuracy.
- **Solve Spherical Astronomy:** Master a wide range of problems involving the celestial sphere, such as calculating a star's declination or the ascensional difference needed for timekeeping.
- **Tell Time:** Apply trigonometry to terrestrial matters, like determining the time of day from the length of a gnomon's shadow.

The ingenuity on display in the verse is breathtaking. The decision to store successive differences rather than the full *jyā* values was a masterstroke of computational efficiency, simplifying both storage and calculation. This numerical data was then preserved through a sophisticated alphabetical cipher, a marvel of data compression for its time. Furthermore, the astute choice of radius R=3438—a value that ingeniously relates the circle's radius to its circumference in minutes of arc—demonstrates a level of mathematical insight and practical skill that was unparalleled.

The legacy of this work is immense. It became the bedrock of the entire Indian tradition of *jyā* calculations. Through scientific exchange, Aryabhata's tables heavily influenced Islamic astronomy, where the Sanskrit *jyā* was transliterated as *jayb*. When these Arabic texts were later translated into Latin, *jayb* (which also means "pocket" or "fold") was rendered as its Latin equivalent, *sinus*. This term entered the European mathematical lexicon, seeding the global development of modern trigonometry. Aryabhata's concise verse is, therefore, a crucial ancestor of the sine function used worldwide today.

In short, Verse 10 provides the finished engine for Aryabhata's astronomical system. Verse 12, as we shall see, will reveal the elegant mathematical principle that was used to forge it.

3.11 Linking Time, Spin, and Motion – The Prāna and the Minute (Gk.11)

Purpose: To establish the fundamental equivalence between the basic unit of time (*prāṇa*) and the basic unit of angular motion (*kalā*), unifying the system's temporal and spatial dimensions.

Having provided the geometric framework (Gk.8, Gk.9) and the trigonometric tools (Gk.10 data), Aryabhata returns in Verse 11 to a fundamental principle that elegantly unifies his system: the precise relationship between the passage of time and the apparent rotation of the heavens (caused, in his model, by Earth's spin). This verse establishes a crucial conversion factor that underpins all time-based predictions of celestial positions.

Sanskrit:

प्राणेनैति कलां भं खयुगांशे ग्रह जवो भवांशेऽर्कः ।

समलिप्ता भग्रहणा नवभिरल्पांशकैश्च समेत्य ॥ ११ ॥

IAST:

prāṇenaiti kalāṃ bhaṃ khayugāṃśe graha javo bhavāṃśe'rkaḥ|

samaliptā bhagrahaṇā navabhiralpāṃśakaiścha sametya || 11 ||

Notes:

- **Prāṇena eti kalāṃ bhaṃ:** "The sphere of stars (bhaṃ) moves (eti) one minute of arc (kalāṃ) in one breath (prāṇena)."
- **Prāṇa:** Unit of time ≈ 4 seconds (defined in Kk.1-2 as 1/6th of a vināḍikā; 1 day = 21,600 prāṇas).
- **Kalā:** Minute of arc (1/60th of a degree).
- **Fundamental Equivalence:** 1 prāṇa of time = 1 kalā of angular motion.
- **(Implies Sidereal Day = 21,600 prāṇas = 24 hours exactly).**
- **Khayugāṃśe graha javo...:** Obscure latter part, likely relating planetary speeds to fractions of Yuga/Zodiac.

3.11.1 The Cosmic Heartbeat: Linking Breath to Rotation

The most significant part of this verse, a conceptual cornerstone of Aryabhata's entire computational system, is its opening statement:

prāṇena eti kalāṃ bhaṃ

This translates to: **"The sphere of stars (*bhaṃ*) moves (*eti*) one minute of arc (*kalāṃ*) in one breath (*prāṇena*)."**

To understand the profound power of this assertion, we must break down its components. The *bhaṃ*, or "sphere of stars," represents the rotating celestial frame against which all motion is measured. Its movement is quantified in *kalās*, the standard minute of arc, which is 1/60th of a degree. A full 360° rotation of the celestial sphere therefore comprises 21,600 *kalās*.

The unit of time, the *prāṇa*, is where the ingenuity of the system truly lies. While the term literally means "breath," referring to the duration of a respiration cycle for a person at rest, Aryabhata standardizes it as a precise computational unit. Within the hierarchy of time established in his *Kālakriyāpāda* (1-2), we find that:

- **1 day (*divasa*) = 60 *nāḍīs***
- **1 *nāḍī* = 60 *vināḍikās***
- **1 *vināḍikā* = 6 *prāṇas***

This yields a total of 60 × 60 × 6 = 21,600 *prāṇas* in a single day. For the modern reader, this means one *prāṇa* corresponds to exactly four seconds (86,400 seconds / 21,600 = 4).

Herein lies the brilliance of Aryabhata's design. By defining a day as both 21,600 minutes of arc of celestial rotation and 21,600 *prāṇas* of time, he forges a direct and profoundly simple equivalence:

1 *kalā* (minute of arc) = 1 *prāṇa* (4 seconds of time)

This unique feature elegantly integrates the measurement of angular motion and the passage of time into a single, unified framework, providing a powerful and efficient foundation for all subsequent astronomical calculations.

3.11.2 Implications: Defining the Sidereal Day

The consequence of this elegant equivalence between time and arc is profound, for it establishes the very definition of the day within Aryabhata's system. The logic unfolds with mathematical certainty:

A full 360° rotation of the celestial sphere covers 21,600 *kalās*. Given the direct link of one *prāṇa* of time for every *kalā* of rotation, the time required for one full turn is therefore:

21,600 *kalās* × (1 *prāṇa* / 1 *kalā*) = 21,600 *prāṇas*

Since Aryabhata explicitly defines one standard day as consisting of 21,600 *prāṇas*, the conclusion is inescapable: one full rotation of the heavens—that is, one **sidereal day**—takes exactly one standard day.

When converted to modern units, this idealized day is also familiar:

21,600 *prāṇas* × (4 seconds / 1 *prāṇa*) = 86,400 seconds = 24 hours

Aryabhata's parameters therefore define the sidereal day—the period of Earth's rotation relative to the fixed stars—as exactly 24 hours. This created an absolutely consistent and self-contained framework for all time-dependent astronomical calculations. It is a testament to the system's design that this value differs only slightly from the modern measurement of the sidereal day, which is approximately 23 hours, 56 minutes, and 4 seconds.

Figure 3.13: The Prāṇa-Kalā Equivalence – Aryabhata's Keystone (Gk.11). This diagram illustrates the fundamental principle from Gītikāpāda Verse 11: one Prāṇa (a unit of time roughly equivalent to 4 modern seconds) corresponds directly to one Kalā (a minute of arc) of the apparent celestial sphere's rotation (or Earth's actual rotation in Aryabhata's model). This elegant 1:1 equivalence is a cornerstone of the Aryabhatiya, seamlessly integrating temporal measurement with angular displacement and forming a crucial conversion factor for all subsequent astronomical calculations.

Other Motion Relationships (Complex Interpretations)

In stark contrast to the lucid clarity of its opening, the second half of Verse 11 plunges into a passage of considerable obscurity. The line, beginning with *khayugāṃśe graha javo...*, has long been a subject of varied interpretation among both classical commentators and modern scholars.

While the phrase *graha javaḥ* almost certainly refers to the "mean velocity of the planets," the context in which it is placed is ambiguous. The term *khayugāṃśa* could signify a fraction of a great *Yuga*, while other readings suggest it relates to a *bhavāṃśa*, a 30-degree sign of the zodiac. Further complexity arises with the phrase *samaliptā bhagrahaṇā*, which might relate to mean planetary conjunctions or specific angular points within an eclipse cycle (*grahaṇa*). The final clause, *navabhiralpāṃśakaiścha sametya*, is particularly cryptic and has resisted a single, definitive explanation.

Given these ambiguities, the precise mathematical relationship Aryabhata intended to codify here remains a matter of scholarly debate. What is certain, however, is the foundational importance of the verse's unambiguous opening. The simple, powerful equivalence of the *prāṇa* and the *kalā* stands apart as a universally recognized and critical pillar of his entire system, even if the corollaries he derived from it are now partially lost to interpretation.

3.11.4 Significance: The Time-Angle Conversion Factor

Verse 11, particularly its first clause, functions as the master conversion gear within the intricate machinery of the *Aryabhatiya*. It is the crucial component that seamlessly links two fundamental domains: the measurement of elapsed time, reckoned in *prāṇas* and *nāḍikās*, and the measurement of angular position and motion upon the celestial sphere.

The implications of this direct correspondence are foundational, providing the system with its computational integrity:

- **It provides the essential formula for positional astronomy,** allowing an astronomer to calculate the angular distance a celestial body has traveled based on time elapsed since an epoch, or conversely, to determine the time of day from an observed celestial position.

- **It implicitly defines the sidereal day,** locking the fundamental period of rotation to exactly 21,600 *prāṇas* (24 hours). This establishes a bedrock of absolute consistency for all time-dependent calculations.

- **It quantifies the very rate of cosmic rotation,** specifying that the heavens appear to turn at a constant and predictable speed of one minute of arc for every four seconds of time.

By establishing this relationship, Aryabhata ensures that his mathematically modeled universe operates with an elegant, clockwork regularity. This single, powerful statement forges the critical link between the smallest measurable units of time and the grand, sweeping motion of the celestial sphere.

3.12 The Sine Calculation Rule – Unlocking the Table (Gk.12)

Purpose: To provide the mathematical algorithm or principle that governs the sine differences listed in Gk.10, allowing the entire table to be systematically calculated or verified.

In Verse 10, Aryabhata provided the foundational data for his trigonometry: 24 precise Sine Differences (*jyā-piṇḍas*). Now, in Verse 12, he reveals the mathematical algorithm or principle governing these differences. This rule allows for the systematic calculation—or at least verification—of the entire set of R-Sine values and demonstrates a deep understanding of the underlying mathematical structure of the sine function. It completes Aryabhata's computational toolkit for predictive astronomy presented in the Daśagītikā.

Sanskrit:

प्रथमाच्चापज्यार्धाद् यैरूनं खण्डितं द्वितीयर्धम् ।

तत् प्रथमज्यार्धांशैस्तैस्तैरूनानि शेषाणि ॥ १२ ॥

IAST:

prathamāc chāpajyārdhād yairūnaṃ khaṇḍitaṃ dvitīyardham|

tat prathamajyārdhāṃśais tais tairūnāni śeṣāṇi || 12 ||

Notes:

- **Prathamāc chāpajyārdhād:** "From the first R-Sine difference (Δ_1)."
- **Yairūnaṃ khaṇḍitaṃ dvitīyardham:** "By which the second R-Sine difference (Δ_2) is reduced/obtained."
- **Tat prathamajyārdhāṃśais tais tairūnāni śeṣāṇi:** "By those parts/fractions of the first R-Sine difference (Δ_1) are the remaining ones successively reduced."
- **Core Rule (Interpretation):** $\Delta_{n+1} \approx \Delta_n - (Jyā_n / \Delta_1)$, where $Jyā_n$ is the cumulative R-Sine up to step n.

3.12.1 Unveiling Aryabhata's Sine Table: From Differences to Values

Aryabhata's method for constructing his sine table is a two-part process. First, he provides a foundational set of 24 "Sine Differences" (Δ_n) in Verse 10. These are then used to build the actual R-Sine values.

Part 1: Calculating R-Sine Values ($Jyā_n$) by Summation

The most straightforward step, as mentioned in Chapter 3.10, is to generate the 24 cumulative R-Sine values (denoted $Jyā_n$, where $Jyā_n = R \cdot \sin(n \times 3°45')$) by successively adding the Sine Differences (Δ_n) from Verse 10:

- $Jy\bar{a}_1 = \Delta_1 = 225$
- $Jy\bar{a}_2 = Jy\bar{a}_1 + \Delta_2 = 225 + 224 = 449$
- $Jy\bar{a}_3 = Jy\bar{a}_2 + \Delta_3 = 449 + 219 = 668$
- ...and so forth, until $Jy\bar{a}_{24} = \Sigma\Delta_i$ (for i=1 to 24) = 3438, which corresponds to R·sin(90°).

This simple summation yields the R-Sine values essential for trigonometric calculations.

Part 2: Verse 12 – The Ingenious Rule for Generating the Sine Differences Themselves

Beyond simply using the pre-listed differences, Verse 12 reveals the sophisticated mathematical principle Aryabhata used to derive how these Sine Differences (Δ_n: 225, 224, 219, ...) decrease sequentially. It's a recursive method where the drop from one difference to the next is determined by the R-Sine values already calculated.

The Rule Simplified (Common Interpretation)

Let's define:

- Δ_n = The nth Sine Difference (e.g., Δ_1=225, Δ_2=224 from Verse 10).
- $Jy\bar{a}_n$ = The cumulative R-Sine up to the nth step (e.g., $Jy\bar{a}_1$=225, $Jy\bar{a}_2$=449).

Verse 12 essentially states that to get from one difference (Δ_n) to the next (Δ_{n+1}), you subtract a "correction factor." This correction factor increases with each step and is proportional to the cumulative R-Sine ($Jy\bar{a}_n$) calculated so far.

A widely accepted and effective interpretation, which closely reproduces Aryabhata's table, is:

The correction needed to find Δ_{n+1} from Δ_n is approximately $Jy\bar{a}_n$ divided by the very first difference ($\Delta_1 = 225$).

$\text{Correction}(n \to n+1) \approx Jy\bar{a}_n / \Delta_1 = Jy\bar{a}_n / 225$

Therefore, the (n+1)th difference is found recursively:

$\Delta_{n+1} \approx \Delta_n - (Jy\bar{a}_n / \Delta_1)$

Example: Testing Aryabhata's Rule

Let's see how this rule generates the first few differences, starting only with $\Delta_1 = 225$:

Step 1: From Δ_1 to Δ_2 (n=1)

- Given: $\Delta_1 = 225$
- Calculate $Jy\bar{a}_1$: $Jy\bar{a}_1 = \Delta_1 = 225$
- Calculate Correction(1→2): $Jy\bar{a}_1 / \Delta_1 = 225 / 225 = 1$
- Calculate Δ_2: $\Delta_2 \approx \Delta_1 - \text{Correction}(1\to2) = 225 - 1 = 224$
    - *(This perfectly matches the Δ_2 value from Verse 10.)*

Step 2: From Δ_2 to Δ_3 (n=2)

- We have: $\Delta_2 = 224$
- Calculate Jyā₂: $Jyā_2 = Jyā_1 + \Delta_2 = 225 + 224 = 449$
- Calculate Correction(2→3): $Jyā_2 / \Delta_1 = 449 / 225 \approx 1.9955...$
    - Rounding to the nearest integer (a practical step often assumed): Correction ≈ 2
- Calculate Δ_3: $\Delta_3 \approx \Delta_2 - \text{Correction}(2\to3) = 224 - 2 = 222$
    - *(This is close to the Verse 10 value of $\Delta_3 = 219$. Small discrepancies can arise from the approximation inherent in the rule, rounding choices, or varying interpretations of the Sanskrit phrasing like "prathamajyārdhāṃśais tais tair.")*

Step 3: From Δ_3 to Δ_4 (n=3)

- Using the more accurate table value for $\Delta_3 = 219$ (or one could proceed with the calculated $\Delta_3 = 222$):
- We have: $\Delta_3 = 219$
- Calculate Jyā₃: $Jyā_3 = Jyā_2 + \Delta_3 = 449 + 219 = 668$
- Calculate Correction(3→4): $Jyā_3 / \Delta_1 = 668 / 225 \approx 2.9688...$
    - Rounding: Correction ≈ 3
- Calculate Δ_4: $\Delta_4 \approx \Delta_3 - \text{Correction}(3\to4) = 219 - 3 = 216$
    - *(This is close to the Verse 10 value of $\Delta_4 = 215$.)*

3.12.2 The Deeper Mathematical Insight

Aryabhata's rule brilliantly captures a core property of the sine function. The amount subtracted at each step (the "Correction," which is Δ_n - Δ_{n+1}) is approximately proportional to the R-Sine value itself ($Jyā_n$). This relationship is a finite difference analog to the calculus principle that the second derivative of sin(x) is -sin(x). Thus, Aryabhata's method is a remarkable early approximation of this differential relationship, showcasing profound insight into the nature of the sine function well before the formal development of calculus.

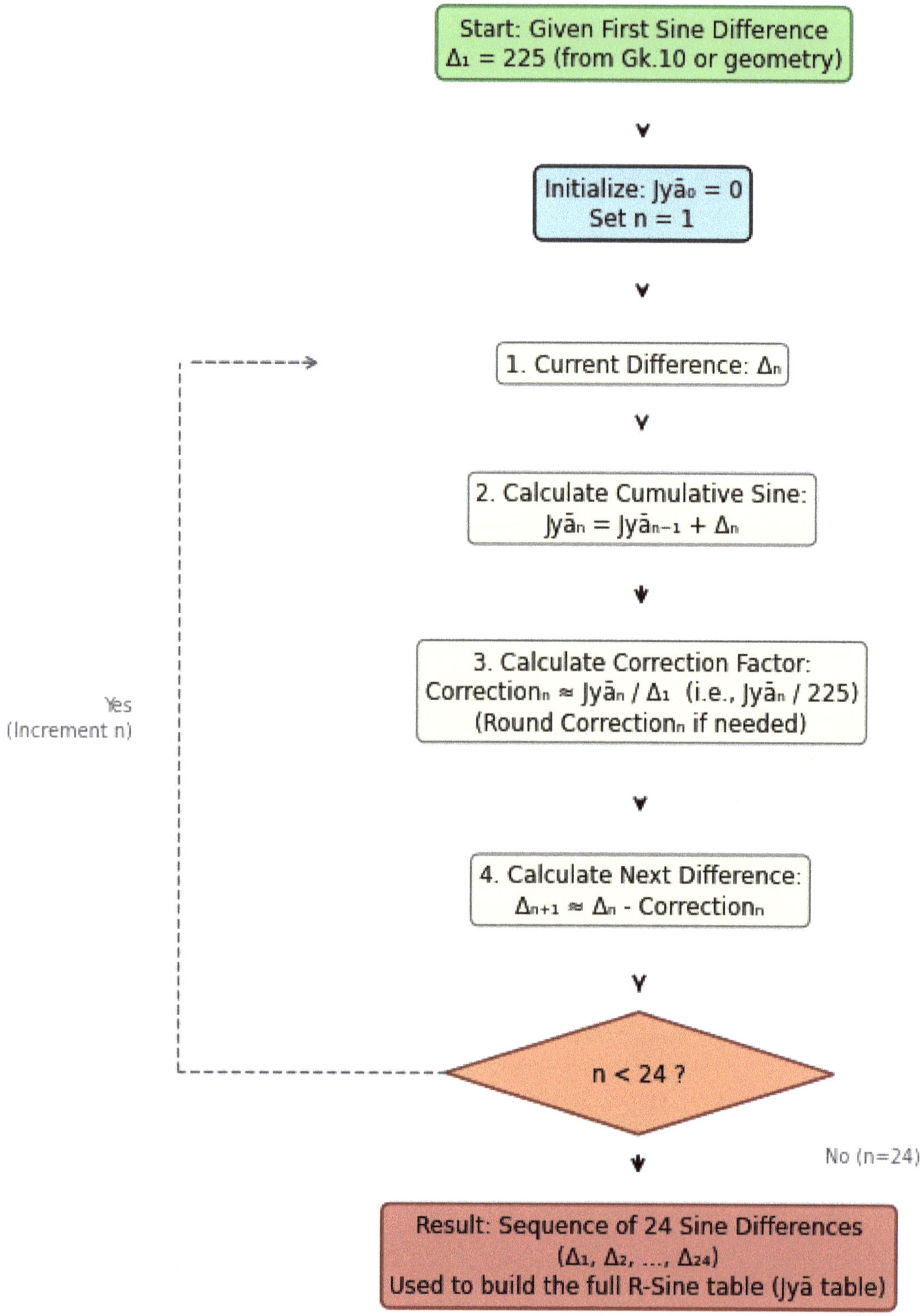

Figure 3.14: Flowchart illustrating the principle for generating Aryabhata's sine difference table, as described in Gītikāpāda Verse 12 (Gk.12). Starting with the first difference $\Delta(1)$, each subsequent difference $\Delta(n+1)$ is approximately calculated by subtracting a value proportional to the cumulative sine value up to the previous step (Jyā(n)) from the preceding difference $\Delta(n)$. This recursive rule allowed for the systematic construction of the entire table.

Significance: Beyond Computation

1. **Mathematical Insight:** Aryabhata didn't just list numbers; he discerned the functional relationship governing the sine differences. This anticipates concepts related to finite differences, numerical analysis, and the curvature of the sine function.

2. **Generative & Verification Tool:** The rule demonstrates how the entire sine table could, in principle, be generated recursively starting only from the first difference ($\Delta_1 = 225$) and the radius R (implied by the sum). It also serves as a method to verify the accuracy of the tabulated differences.

3. **A Complete Trigonometrical System:** Together, Verses 10 (data) and 12 (generative/verification rule) form a self-contained, remarkably sophisticated engine for sine calculations, which was critical for the predictive power of Aryabhata's astronomical models detailed in the Kālakriyāpāda and Golapāda.

3.13 The Fruit of Knowledge – The Phalaśruti (Gk.13)

Purpose: To articulate the profound value and purpose of mastering the preceding Gītikā verses, particularly the core "Daśagītikā" (ten gītis, usually Gk.3-12), framing this knowledge as a boat for navigating the complexities of cosmic understanding and as a path to ultimate comprehension.

With the foundational parameters and computational tools of his astronomical system established in the preceding twelve verses, Aryabhata shifts focus in the Gītikāpāda's concluding thirteenth verse. This final stanza takes the form of a *phalaśruti*, a traditional declaration of the "fruit" or benefit derived from engaging with the presented knowledge. In doing so, it elevates the preceding technical exposition beyond mere calculation, suggesting a deeper purpose to this knowledge beyond mere calculation.

Sanskrit:
दशगीतिकासूत्रमिदं भूग्रहचरितं भपञ्जरे ज्ञात्वा ।
ग्रहमगगणितं परमं ब्रह्म लभतेऽर्कलोकादि ॥ १३ ॥

Sanskrit:
दशगीतिकासूत्रमिदं भुवनमहिभानोर्ग्रहाणां च ।
प्रवृत्तिं ज्ञात्वास्मिन् ज्ञानानामुत्तमोत्तमं लभते ॥ १३ ॥

IAST (based on the first Sanskrit version:
daśagītikāsūtramidaṃ bhūgrahacharitam bhapañjare jñātvā|
grahabhagagaṇitaṃ paramaṃ brahma labhate'rkalokādi || 13 ||

IAST (for the alternative version):
daśagītikāsūtramidaṃ bhuvanamahibhānor grahāṇāṃ ca |
pravṛttiṃ jñātvāsmin jñānānām uttamottamaṃ labhate || 13 ||

(Note: There can be slight variations in the wording of this verse across different manuscripts and commentaries. The first version above is a commonly cited one. Second version is another popular rendering, perhaps closer to what Bhaskara I comments on.)

(For the purpose of this explanation, we'll lean towards the general meaning conveyed by most versions, which emphasizes understanding cosmic motions and attaining higher knowledge.)

This concluding verse distills the essence and reward of comprehending the Gītikāpāda.

Notes:

- **Daśagītikāsūtramidaṃ:** "This sūtra (aphoristic text) of ten gīti verses" (referring to Gk.3-12).
- **Bhūgrahacharitam bhapañjare jñātvā:** "Knowing the movement/character of the Earth and planets within the cage/sphere of stars."

- **Grahabhagagaṇitaṃ:** "The mathematics of planets and asterisms/stars."
- **Paramaṃ brahma labhate:** "Attains/obtains the Supreme Brahman." (Or similar phrases like "attains the highest knowledge" or "worlds of the Sun etc." depending on manuscript variant).
- **Phalaśruti:** Verse stating the merits/benefits of studying the preceding text.

3.13.1 The Essence of the "Ten Gīti Verses"

The verse begins by identifying the preceding core of the work as the *daśagītikāsūtram idaṃ*—"this *sūtra* of ten *gīti* verses." Each word here is significant:

- ***Daśagītikā***: As established previously, this is the name for the ten foundational verses (Gk.3–12) that set the numerical and astronomical parameters of the entire system.
- ***Sūtram***: This term signifies more than just a "text." It denotes a concise, aphoristic composition, a "thread" (*sūtra*) that weaves together profound knowledge, highlighting the dense and foundational nature of the verses.

The verse then states that by truly *jñātvā*—knowing, mastering, and deeply understanding—this foundational *Daśagītikā*, one comprehends the very workings of the cosmos. Aryabhata describes this knowledge with two powerful, parallel expressions: the *bhūgrahacharitam bhapañjare*, or "the movement of the Earth and the planets within the cage of stars," and the *bhuvanamahibhānor grahāṇāṃ ca pravṛttiṃ*, "the true nature and motion of the cosmos, the Earth, the Sun, and the planets." The first phrase offers a striking visual metaphor of a structured, mechanical universe, while the second speaks to a deeper understanding of its fundamental principles.

3.13.2 Navigating the Ocean of Cosmic Knowledge

The core benefit is often framed metaphorically. One common interpretation (from slightly different phrasing of Gk.13 not fully captured above but often discussed by commentators) is that this knowledge acts as a "nauḥ" (boat) to cross the "jñānasamudram sadāsajjñānam" (the ocean of knowledge, distinguishing true from false understanding, or knowledge of the existent from the non-existent).

- **Discerning Truth:** The precise astronomical parameters and methods presented in the Gītikāpāda allow one to discern the true, predictable order of the cosmos from erroneous or confused notions. It provides a rational framework for understanding celestial phenomena.
- **Mastery of Subject:** It promises mastery over "grahabhagagaṇitaṃ" (the mathematics of planets and asterisms/stars), meaning a comprehensive understanding of astronomical calculation.

3.13.3 Attaining the Ultimate Understanding

The ultimate fruit is profound. The verse declares that one who truly grasps this knowledge "paramaṃ brahma labhate" (attains/obtains the Supreme Brahman) or "svayaṃbhuvamadhigacchati" (attains the Self-Existent, an epithet for Brahman or Prajāpati). Some versions mention attaining worlds like that of the Sun ("arkalokādi").

- **Connecting Science and Philosophy:** This powerfully links the rigorous scientific pursuit outlined in the Gītikāpāda to the highest philosophical and spiritual goals of Indian traditions. Understanding the intricate, lawful order of the physical cosmos is presented as a path to comprehending the ultimate reality that underpins it.

- **Echoing the Invocation:** This brings the Gītikāpāda full circle, back to the invocation of Brahman in Verse 1. The scientific endeavor, grounded in reverence, leads back to a deeper appreciation of that ultimate source.

- **Highest Knowledge:** Other readings suggest one "jñānānām uttamottamaṃ labhate" (attains the highest of all knowledges), placing this astronomical-mathematical knowledge at a pinnacle.

Significance: The Seal of Value

Verse 13 serves several crucial functions:

- **Motivation and Validation:** It provides a strong motivation for the diligent study of the preceding technical verses by highlighting the profound rewards.

- **Epistemological Claim:** It asserts the power of this rational, mathematical approach to astronomy to lead to true understanding and dispel ignorance about the cosmos.

- **Philosophical Integration:** It seamlessly integrates the scientific knowledge of the Aryabhatiya within the broader Indic philosophical quest for ultimate truth (Brahman).

- **Concluding the Section:** It acts as a formal colophon, signaling the completion of this foundational Gītikāpāda section and affirming its significance.

- **Reinforcing Authority:** By claiming such high rewards, it subtly reinforces the authority and importance of Aryabhata's system.

In essence, Gītikāpāda Verse 13 elevates Aryabhata's astronomical teachings from a mere set of rules and numbers to a profound system of knowledge capable of guiding the seeker through the complexities of the universe to a state of enlightened understanding. It underscores the holistic vision where rigorous science and deep philosophical inquiry are not separate but complementary paths to truth.

3.14 Conclusion: The Foundation is Laid

The thirteen verses of the *Gītikāpāda* stand as a testament to Aryabhata's genius for synthesis, concise expression, and profound scientific vision. In this opening section, he does not merely set a stage; he erects the very pillars upon which his entire astronomical system rests and frames its pursuit as a path to ultimate understanding.

The work opens with a reverent invocation (Gk.1) before immediately introducing a novel alphabetical code (Gk.2)—a crucial tool for encoding the immense numbers of cosmic calculation and ensuring their efficient transmission. With this instrument in hand, Aryabhata proceeds to lay out the fundamental parameters of his universe. He defines its rhythms through the revolution counts of the planets, Sun, and Moon within the vast cycle of the *Mahāyuga* (Gk.3), a model in which he boldly includes the Earth's own eastward rotation. He charts the subtle motions of key orbital points like the lunar apogee and nodes (Gk.4) and anchors his cosmos in a grand hierarchy of time, precisely locating the commencement of the *Kali Yuga* (Gk.5). The physical scale of this universe is then defined in *yojanas*, from the celestial sphere down to the Earth, Sun, and Moon (Gk.6-7), and its geometry is mapped through the inclination of the ecliptic and celestial latitudes (Gk.8-9).

Crucially, the *Gītikāpāda* also forges the powerful trigonometry needed to make this cosmic model predictive. It presents the core data for his sine table as a list of 24 differences (Gk.10), establishes the vital link between time and angular motion through the *prāṇa-kalā* equivalence (Gk.11), and reveals the elegant mathematical rule governing the sine differences themselves (Gk.12).

Finally, Aryabhata seals this foundational section with a *phalaśruti* (Gk.13), declaring that mastery of these "ten *gīti* verses"—this knowledge of cosmic motion and its mathematical underpinnings—is a pursuit that dispels ignorance and can lead the dedicated seeker to the highest form of knowledge, even to an understanding of the Supreme Brahman.

Like a master architect, Aryabhata laid a vast, precise, and purposeful foundation in these opening verses. Thus, the *Gītikāpāda* reveals itself to be no mere prologue but the compact, encoded nucleus of a scientific revolution. He crafted a framework where planetary rhythms could be mapped, measured, and refined into true positions using the mathematical tools forged in the subsequent *Gaṇitapāda* and applied across time and space in the *Kālakriyāpāda* and *Golapāda*.

Its blend of computation, cosmology, and philosophical aspiration, amplified and defended by commentators like Bhaskara I, resonated through centuries of Indian astronomy and beyond, leaving an indelible mark on the global history of science. These thirteen verses remain a powerful example of how observation, mathematics, and a reverence for cosmic order could intertwine in the intellectual ferment of Gupta-era India, creating a system whose echoes we still perceive today.

***

Chapter Four

The Mathematical Engine – Ganitapāda

The Science of Calculation: Numbers, Roots, Ratios, and Aryabhata's π

4.1 Introduction: The Mathematical Heart of the Aryabhatiya

Having laid the cosmological and astronomical foundations in the *Daśagītikā* (Chapter 3), Aryabhata turns in the Gaṇitapāda to forging the essential mathematical tools required to operate his system. This second section of the *Aryabhatiya*, comprising 33 concise verses (*ślokas*), forms the core of his mathematical legacy and provides a profound glimpse into the sophisticated computational techniques available in 5th-century India. Composed around 499 CE in Kusumapura, this chapter is far more than a mere compilation of formulas; it reveals a mind adept at abstraction, algorithmic thinking, and geometric reasoning, driven by both practical necessity and the demands of astronomical precision.

Aryabhata masterfully interweaves arithmetic, geometry, algebra, and the seeds of trigonometry within these verses. Many problems addressed are practical, touching on commerce, measurement (*kṣetra-vyavahāra*), and construction, perhaps implicitly linked to traditions like the precise layouts of Vedic altars found in the older *Śulba Sūtras*. Others provide the essential mathematical engine for the complex astronomical calculations presented later in the Kālakriyāpāda and Golapāda. This chapter delves into his groundbreaking contributions, exploring:

- **Foundational Innovations:** The revolutionary decimal place-value system (Verse 2), the bedrock of efficient calculation.
- **Geometric Mastery:** Principles for calculating areas (triangles, squares, circles, trapezoids), volumes (cubes, spheres – with a notable inaccuracy), and understanding properties of shapes like intersecting circles and inscribed figures (Verses 3, 6-9, 11, 12, 13, 18).
- **Algorithmic Prowess:** Detailed, systematic algorithms for extracting square and cube roots (Verses 4, 5), prefiguring modern computational methods.
- **Algebraic Techniques:** Methods for handling arithmetic progressions (sums, terms, finding 'n'), sums of squares and cubes, proportions (Rule of Three, inverse, compound), mixtures, interest, and basic algebraic identities (Verses 19-31).
- **Trigonometric Seeds & Precision:** The remarkably accurate calculation of π (Verse 10) and geometric rules involving the half-chord (*jyā*, foundational to sine) and the 'arrow' (*śara*, versine) pivotal for astronomical

modelling (Verses 11, 12, 13).

- **Practical Applications:** The use of the gnomon (*śaṅku*) and its shadow (*chāyā*), along with similar triangles, for measurement (Verses 14-16), and a clear statement of the Pythagorean theorem (Verse 17).
- **The Pinnacle Algorithm:** The ingenious *kuṭṭākāra* (pulverizer) method for solving linear indeterminate equations, crucial for astronomical cycle calculations (Verses 32-33).

This chapter unveils the brilliance of Aryabhata's mathematical innovations as presented in the Gaṇitapāda, helping to understand their profound and enduring legacy, which paved the way for calculations underpinning fields from calendrics to modern technology.

4.2 Invocation and the Decimal Revolution (Gp.1-2)

The Gaṇitapāda section commences, as is traditional in classical Indian scientific texts, with an invocatory verse (*maṅgalācharaṇa*). This opening verse (Gp.1) pays homage to cosmic principles and celestial bodies, immediately grounding the mathematical knowledge that follows within its primary intended domain: astronomy.

4.2.1 Ganitapāda: Invocation

Sanskrit:

ब्रह्मकुशशिबुधभृगुरविकुजगुरुकोणभगणान् नमस्कृत्य ।

आर्यभटस्त्विह निगदति कुसुमपुरेऽभ्यर्चितं ज्ञानम् ॥ १ ॥

IAST:

brahmakuśaśibudhabhṛguravikujagurukoṇabhagaṇān namaskṛtya |

āryabhaṭastviha nigadati kusumapure'bhyarcitaṁ jñānam || 1 ||

Translation: "Having paid reverence to Brahman (Creator), Earth (*ku*), Moon (*śaśi*), Mercury (*budha*), Venus (*bhṛgu*), Sun (*ravi*), Mars (*kuja*), Jupiter (*guru*), Saturn (*koṇa*, identified with Saturn), and the asterisms (*bhagaṇa*, referring to the sphere of stars or their revolutions), Aryabhata sets forth here in Kusumapura the knowledge respected [there]."

Notes:

- **Brahman, Ku, Śaśi, Budha, Bhṛgu, Ravi, Kuja, Guru, Koṇa, Bhagaṇān:** Invocation to the Creator, Earth, Moon, Mercury, Venus, Sun, Mars, Jupiter, Saturn (Koṇa), and Asterisms/Star-sphere.
- **Namaskṛtya:** "Having paid reverence."

- **Āryabhaṭaḥ:** Self-reference, authorship.
- **Iha nigadati:** "Here sets forth/declares."
- **Kusumapura:** Location (Pataliputra/Patna).
- **Abhyarcitaṁ jñānam:** "Respected knowledge" (referring to mathematics).

Explanation and Significance: This verse serves three key purposes:

1. **Invocation:** It honors Brahman and the major celestial bodies relevant to Indian astronomy, linking the mathematics to its cosmological application.
2. **Authorship and Location:** It clearly identifies the author as Aryabhata and the setting as Kusumapura (likely Pataliputra/Patna).
3. **Statement of Purpose:** Aryabhata declares his intention to present (*nigadati*) this respected (*abhyarcitaṁ*) knowledge (*jñānam*) of mathematics.

Having set the stage and paid due reverence, Aryabhata immediately turns to the foundational tool upon which all subsequent calculations depend: the revolutionary concept of the decimal place-value system, elegantly laid out in Verse 2.

4.2.2 The Powers of Ten: Articulating Decimal Place Value

Sanskrit:

एकं च दश च शतं च सहस्रं त्वयुतनियुते तथा प्रयुतम् ।

कोट्यर्बुदं च वृन्दं स्थानात् स्थानं दशगुणं स्यात् ॥ २ ॥

IAST:

ekaṁ cha daśa cha śataṁ cha sahasraṁ tvayutaniyute tathā prayutam|

koṭyarbudaṁ cha vṛndaṁ sthānāt sthānaṁ daśaguṇaṁ syāt || 2 ||

Translation: "One (*ekaṁ*), ten (*daśa*), hundred (*śataṁ*), thousand (*sahasraṁ*), ten thousand (*ayuta*), hundred thousand (*niyuta*), million (*prayuta*), ten million (*koṭi*), hundred million (*arbuda*), and billion (*vṛnda*)—each place (*sthāna*) is ten times the preceding place (*sthānāt sthānaṁ daśaguṇaṁ syāt*)."

Notes:

- **Ekaṁ, Daśa, Śataṁ, Sahasraṁ, Ayuta, Niyuta, Prayuta, Koṭi, Arbuda, Vṛnda:** Sanskrit names for one, ten, hundred, thousand, ten thousand, hundred thousand, million, ten million, hundred million, billion.

- **Sthānāt sthānaṁ daśaguṇaṁ syāt:** "Each place is ten times the preceding place."

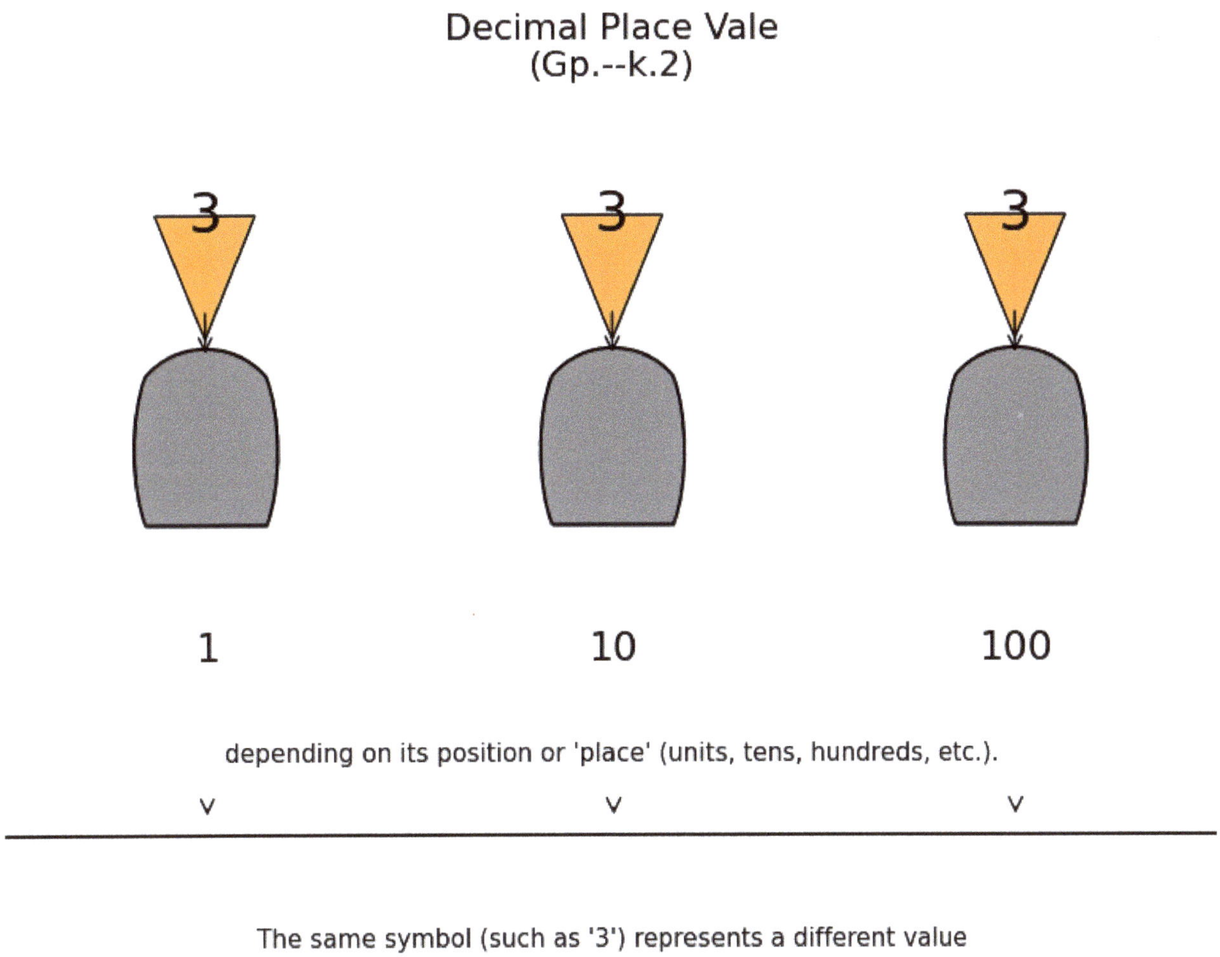

Figure 4.1: Illustration of the decimal place-value principle ("sthānāt sthānaṁ daśaguṇaṁ syāt") as articulated by Aryabhata in Gaṇitapāda Verse 2. The same symbol (such as '3') represents a different value depending on its position or 'place' (units, tens, hundreds, etc.).

4.2.3 The Decimal Breakthrough

This single verse (Gp.2) marked a seismic shift in mathematical thought. While place value concepts might have existed earlier in India or elsewhere in less systematic forms (e.g., Babylonian sexagesimal), Aryabhata's clear articulation within a computational framework was pivotal. His system uniquely combined simplicity (base 10) with the power of positional value. The genius lay in the explicit statement of the principle of *sthāna* (place): the same digit '1' could represent one, ten, or one hundred, depending solely on its position—a concept embodying abstraction and remarkable efficiency.

More than a mere notational convenience, the decimal system liberated mathematics, transforming numbers from simple counters into powerful tools for complex calculations. Crucially, it implicitly required a concept or placeholder for zero (*śūnya*) to give positional notation its full power, even though zero is not explicitly named or symbolized as a distinct numeral in this verse. A place without a value (like the tens place in '305') needed a way to be marked empty, a role zero fulfills. This innovation was essential for deciphering cosmic cycles—lunar months, planetary revolutions (yugas spanning millions of years as seen in Gk.3-5), and the intricate dance of eclipses.

Bhaskara I, Aryabhata's invaluable 7th-century commentator, lauded the decimal system as *lekhāgamanyāyena*—a shared scholarly convention. He marveled at its abstraction, noting: "Numbers lack intrinsic weight; their power blooms from the system we craft." This highlights the philosophical depth: a digit's meaning emerges solely from its designated place. This abstraction enabled Aryabhata to effortlessly handle vast numbers, like a *mahāyuga* (4,320,000 years from Gk.3), essential for astronomical precision.

Aryabhata's clear formulation (or systematic presentation) of the decimal place-value system traveled westward, primarily via Islamic scholars. By the 8th century CE, scholars like Al-Khwarizmi introduced "Hindu numerals" (which by then included a fully developed zero symbol) to Baghdad. Fibonacci's *Liber Abaci* (1202 CE) subsequently brought the system to Europe, sparking an arithmetic revolution. Today, it underpins nearly all modern numeracy.

Timeline of Decimal System Transmission (Key Milestones):

- **c. 500 CE:** Aryabhata clearly articulates the decimal place-value principle in the *Aryabhatiya* (Gp.2).
- **c. 628 CE:** Brahmagupta uses zero explicitly in calculations in his *Brāhmasphuṭasiddhānta*.
- **approx. 700-800 CE:** The system travels to Southeast Asia and westward via trade routes.
- **c. 825 CE:** Al-Khwarizmi describes the "Hindu art of reckoning" (*ḥisāb al-hindī*) in Baghdad.
- **10th Century CE:** Gerbert of Aurillac (later Pope Sylvester II) encounters the numerals, likely in Spain.
- **1202 CE:** Fibonacci published *Liber Abaci* in Italy, strongly advocating for and popularizing the system in Europe.
- **15th-16th Centuries CE:** Widespread adoption across Europe accelerates the Scientific Revolution.

Feature	Indian Decimal (Aryabhata)	Roman	Babylonian (Sexagesimal)	Chinese Rod	Mayan (Vigesimal)
Base	10	Hybrid	60	10	20
Place Value	Yes (Explicit)	No (Additive)	Yes	Yes	Yes (Vertical)
Zero Represent.	Implicit (Placeholder)	No	Yes (Placeholder symbol)	Yes (Space/Symbol)	Yes (Shell symbol)
Digits Needed	10 (0-9 evolved later)	7 (I,V,X,...)	~59 symbols + Place	9 rods + Place	3 symbols + Place
Efficiency	Very High	Low	High (Fractions good)	High	High (Vertical calc)
Example (305)	३०५ (Devanagari example)	CCCV	5,5 (5*60 + 5)	算三百O五 (Example)	15.5 (15*20 + 5)

Table 4.1: Comparative Features of Ancient Numeral Systems, Highlighting Aryabhata's Indian Decimal System

Notes on the Time Periods:

- **Indian Decimal (Aryabhata):** c. 500 CE is specific to Aryabhata's clear articulation. The numerals themselves (Brahmi numerals) were evolving earlier.
- **Roman:** While originating earlier, the system we commonly recognize was prominent from around the 1st century BCE through the medieval period in Europe.
- **Babylonian (Sexagesimal):** Used for a very long period, with its peak in the Old Babylonian period (c. 2000-1600 BCE) and continuing in astronomical use much later.
- **Chinese Rod Numerals:** Used from the Warring States period (c. 475-221 BCE) well into the Ming Dynasty (1368–1644 CE).
- **Mayan (Vigesimal):** Primarily associated with the Classic Maya period, though its origins are earlier.

Sanskrit Name (Gp.2)	Power of 10	Numerical Value
Eka	10^0	1
Daśa	10^1	10
Śata	10^2	100
Sahasra	10^3	1,000
Ayuta	10^4	10,000
Niyuta	10^5	100,000
Prayuta	10^6	1,000,000
Koṭi	10^7	10,000,000
Arbuda	10^8	100,000,000
Vṛnda	10^9	1,000,000,000

Table 4.2: Sanskrit names for the powers of ten, as listed by Aryabhata in Gaṇitapāda Verse 2 (Gp.2). These terms define the decimal place-value system, where the value of a digit depends on its position (sthāna), increasing tenfold for each place to the left.

4.3 Shapes, Numbers, and the Art of Extraction (Gp.3-9)

Moving from the abstract power of numbers and place value, Aryabhata turns to geometry (*kṣetra-vyavahāra*), demonstrating how numbers describe and measure the tangible world. Geometry was vital in ancient India, underpinning land measurement, complex temple architecture, and the precise layouts of Vedic fire altars—a tradition documented centuries earlier in the *Śulba Sūtras*, which used knotted ropes (*śulba*) for construction. Aryabhata provided systematic computational methods for these practices, covering basic shapes, areas, volumes, and essential root extraction algorithms.

4.3.1 Square and Cube Definitions

Sanskrit:

वर्गः समचतुरश्रः फलं च सदृशद्वयसंवर्गः ।

घनः स द्वादशाश्रिः स्यात् सदृशत्रयसंवर्गो ॥ ३

IAST:

vargaḥ samachaturaśraḥ phalaṁ cha sadṛśadvayasaṁvargaḥ |

ghanaḥ sa dvādaśāśriḥ syāt sadṛśatrayasaṁvargo || 3 ||

Translation/Explanation: "A square (*varga*) is a four-sided figure with equal sides (*samachaturaśraḥ*), and its area (*phala*) is the product of two equal sides (*sadṛśadvayasaṁvargaḥ*). The product of three equal quantities is a cube (*ghana*), and it [a cube as a solid] has twelve edges (*dvādaśāśriḥ*)." This verse defines the basic concepts of squaring and cubing a number geometrically and arithmetically, linking the numerical operation to the spatial form.

Notes:

- **Vargaḥ samachaturaśraḥ:** "A square is a four-sided figure with equal sides."
- **Phalaṁ sadṛśadvayasaṁvargaḥ:** "Its area is the product of two equal sides" (Area = side × side).
- **Ghanaḥ dvādaśāśriḥ:** "A cube [solid] has twelve edges."
- **Sadṛśatrayasaṁvargo:** "The product of three equal quantities is a cube [volume]" (Volume = side × side × side).

Examples:

- **Square Field:** A square field 10 units on a side has an area of 10 × 10 = 100 square units.

- **Wooden Cube:** A wooden block 3 units on each side is a cube with volume 3 × 3 × 3 = 27 cubic units.

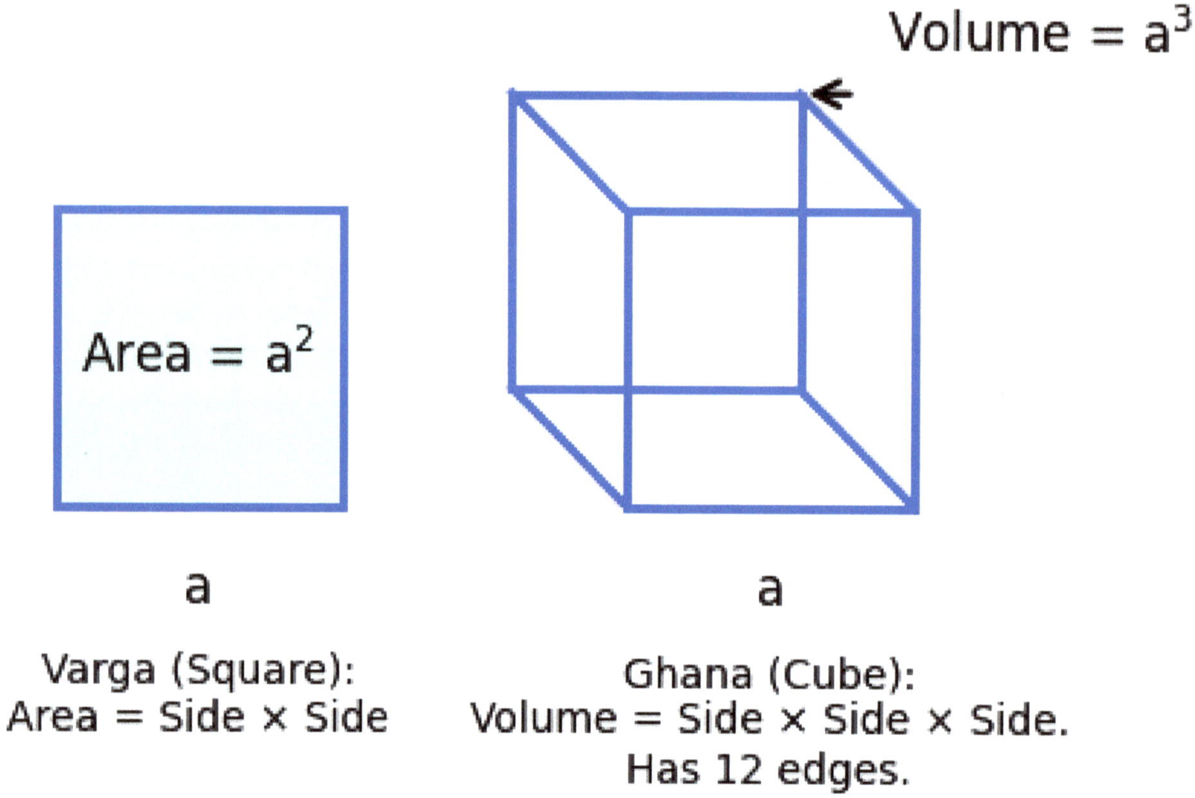

Figure 4.2: Geometric definition of Varga (Square) and Ghana (Cube), as described by Aryabhata in Gaṇitapāda Verse 3 (Gp.3). The diagram illustrates that a Varga is a square figure whose area is the product of its sides, and a Ghana is a cube (a solid with 12 edges) whose volume is the product of three equal sides. These definitions link arithmetic operations to geometric forms.

4.3.2 Square Root Algorithm

Sanskrit:

भागं हरेदवर्गात् नित्यं द्विगुणेन वर्गमूलेन ।

वर्गाद्वर्गे शुद्धे लब्धं स्थानान्तरे मूलम् ॥ ४ ॥

IAST:

bhāgaṁ hared avargāt nityaṁ dviguṇena vargamūlena |

vargādvarge śuddhe labdhaṁ sthānāntare mūlam || 4 ||

Translation: "Always divide the non-square [part of the number] by twice the square root [already obtained so far]. After subtracting the square [of the quotient, in context: the product formed using the quotient] from the [current] square-place [part of the number], the quotient obtained (*labdhaṁ*) is the root [digit] in the next place."

Notes:

- **Bhāgaṁ hared avargāt:** "Divide the non-square [part]."
- **Nityaṁ dviguṇena vargamūlena:** "Always by twice the square root [obtained so far]."
- **Vargādvarge śuddhe:** "After subtracting the square [of the quotient, contextually (trial divisor)×(quotient)]."
- **Labdhaṁ sthānāntare mūlam:** "The quotient obtained is the root [digit] in the next place."

4.3.2.1 Decoding the Algorithm: A Step-by-Step Guide (Example: √676)

Aryabhata presents a remarkably systematic and efficient method for extracting the square root of a number, an algorithm deeply reliant on the decimal place-value system. His concise verse outlines an iterative process very similar to the long-division method for square roots taught today.

Step-by-Step Guide

Step 1: Group the Digits

To find the square root of 676, start by grouping the number's digits in pairs from the right. For 676, this gives us:

- 6 | 76 The leftmost group may have one or two digits.

Step 2: Find the First Digit of the Square Root

Look at the leftmost group (6). Find the largest whole number whose square is less than or equal to this group. For 6:

- $2^2 = 4$ (fits, since $4 \leq 6$)
- $3^2 = 9$ (too large) So, the first digit of the square root is 2.

Step 3: Subtract and Bring Down the Next Pair

Subtract the square (4) from the first group (6):

- 6 - 4 = 2 Bring down the next pair of digits (76), forming a new number: 276.
  Current square root: 2

Step 4: Form the Trial Divisor

Double the current square root (2):

- 2 × 2 = 4 This 4 is the starting point for the trial divisor.

Step 5: Find the Next Digit

Find the largest digit (let's call it x) to add to the trial divisor (4) to form a new number (4x), such that 4x multiplied by x is less than or equal to the current number (276). Try $x = 6$:

- Trial divisor = 46
- 46 × 6 = 276 (fits, since $276 \leq 276$) Trying $x = 7$:
- Trial divisor = 47
- 47 × 7 = 329 (too large) So, $x = 6$ is correct.

Step 6: Subtract the Product

Subtract the product (46 × 6 = 276) from the current number:

- 276 - 276 = 0 Add the digit $x = 6$ to the square root.
  Current square root: 26

Step 7: Repeat or Stop

If there's a remainder and more digit pairs to bring down, repeat steps 4–6 using the new square root (26) to form the next trial divisor (2 × 26 = 52). Since the remainder is 0 and there are no more digits, the process stops.
The square root of 676 is 26.

4.3.2.2 Interpretation and Significance

Aryabhata's method is a brilliant example of ancient Indian mathematics. It uses the place-value system to break down the problem into manageable steps, much like modern long-division for square roots. His approach, refined by later mathematicians like Bhaskara I, shows how to handle each stage by doubling the current root and carefully subtracting products. This technique, developed centuries ago, highlights the advanced computational skills of ancient India and remains a foundation for how we calculate square roots today.

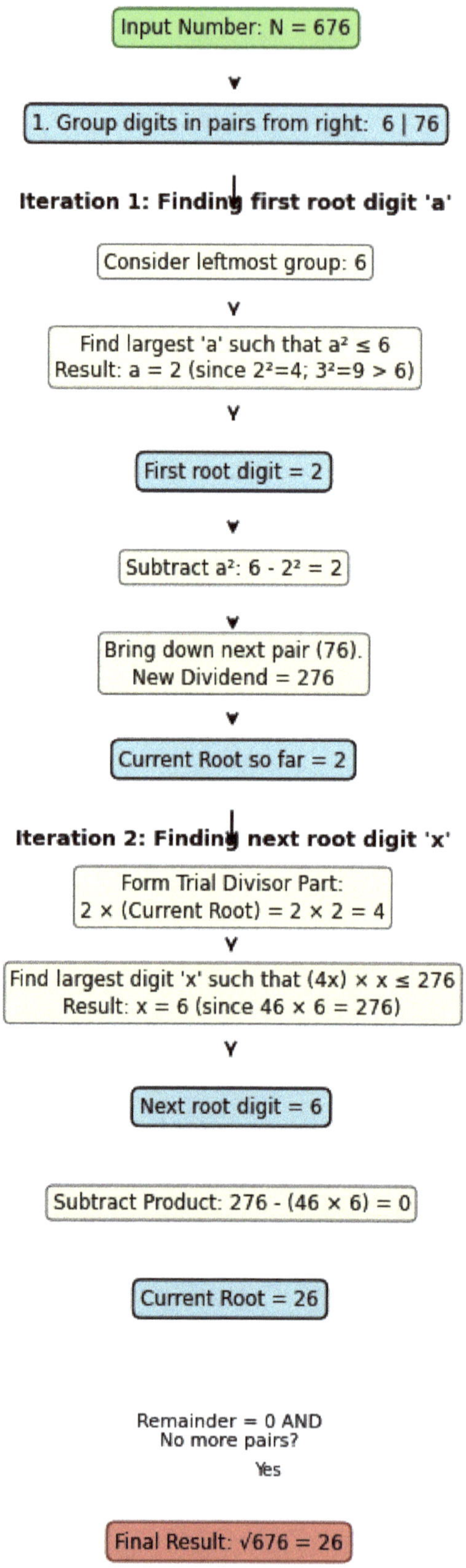

Figure 4.3: Aryabhata's Square Root Algorithm (Gaṇitapāda, Gp.4): Flowchart of the iterative, decimal-based process, with √676 example.

4.3.3 Cube Root Algorithm

Sanskrit:

अघनाद् भजेद् द्वितीयात् त्रिगुणेन घनस्य मूलवर्गेण ।

वर्गस्त्रिपूर्वगुणितः शोध्यः प्रथमाद् घनश्च घनात् ॥ ५ ॥

IAST:

aghanād bhajed dvitīyāt triguṇena ghanasya mūlavargeṇa |

vargas tripūrvaguṇitaḥ śodhyaḥ prathamād ghanaś cha ghanāt || 5 ||

Translation: "Divide the second non-cube [place digits, from right] by three times the square of the cube root [part already obtained]. Subtract the square of the quotient, multiplied by three times the previous [part of the root, considering its place value], from the [next] non-cube [place digits], and [subtract] the cube [of the quotient] from the [last] cube [place digits]."

Notes:

- **Aghanād bhajed dvitīyāt:** "Divide the second non-cube [place digits]."
- **Triguṇena ghanasya mūlavargeṇa:** "By three times the square of the cube root [part obtained]."
- **Vargastripūrvaguṇitaḥ śodhyaḥ prathamād:** "Subtract the square of the quotient, multiplied by three times the previous [root part], from the [next] non-cube [place]."
- **Ghanaścha ghanāt:** "And [subtract] the cube [of the quotient] from the [last] cube [place]."

4.3.3.1 Decoding the Algorithm: The Mechanics of Cube Extraction

At the heart of Aryabhata's method for finding a cube root is a simple but brilliant insight: a number's cube can be built, and therefore dismantled, piece by piece. The algorithm essentially "peels away" the components of a cube by cleverly using the algebraic expansion of a binomial:

$$(a + b)^3 = a^3 + 3a^2b + 3ab^2 + b^3$$

Imagine the number we want to find, like 15. We can think of it as being composed of a tens part ($a = 10$) and a units part ($b = 5$). Aryabhata's procedure first identifies the largest part of the root ('a'), subtracts its cube (a^3), and then systematically works through the remaining terms ($3a^2b$, $3ab^2$, and b^3) to find the next part of the root ('b').

Let's walk through the process with an example: finding the cube root of **3375**.

Step 1: Group the Digits
Starting from the right, we group the number's digits in threes. This grouping corresponds to the powers of 1000 (10^3) in the decimal system. **3375** becomes **3 | 375**

Step 2: Find the First Digit of the Root
Look at the leftmost group (in this case, 3). What is the largest whole number whose cube is less than or equal to 3? $1^3 = 1$
$2^3 = 8$ (This is too large)

The answer is **1**. This is the first digit of our cube root.

Step 3: Subtract and Bring Down
Subtract the cube of our new digit (1^3) from the first group (3) and bring down the next group of digits to form our new working number. $3 - 1^3 = 2$
Bring down the 375 to get: **2375**

So far, our root is **1**. We now need to find the next digit using the remainder, 2375.

Step 4: Estimate the Next Digit
This is where the algorithm reveals its genius. Our remaining number, 2375, is composed of the last three terms of our formula: $3a^2b + 3ab^2 + b^3$. The largest of these is the $3a^2b$ term. We can use it to estimate our next digit, 'b'.

In our formula, 'a' is the root part we've already found (1), but we must account for its place value—it represents the tens place, so a = 10. The term we use for estimation is $3a^2$, which is $3 \times 10^2 = 300$.

To get a trial digit for 'b', we divide our working number by this value: $2375 \div 300 \approx 7$

This gives us a rough estimate. Through a quick trial (or experience), we find the correct digit is **5**. Let's proceed with b = 5.

Step 5: Systematically Subtract the Three Terms
Now, we subtract the three remaining parts of the algebraic expansion from our working number, 2375, using a = 10 and b = 5.

1. **Subtract $3a^2b$:**
   $3 \times (10)^2 \times 5 = 3 \times 100 \times 5 = \mathbf{1500}$; $2375 - 1500 = 875$

2. **Subtract $3ab^2$:**
   $3 \times 10 \times (5)^2 = 30 \times 25 = \mathbf{750}$; $875 - 750 = 125$

3. **Subtract b^3:**
   $5^3 = \mathbf{125}$; $125 - 125 = \mathbf{0}$

Step 6: Finalize the Root
The remainder is 0, which confirms that 5 was the correct second digit. By combining our first digit (1) and our second digit (5), we get the final answer.

The cube root of 3375 is 15.

If the number had been larger (e.g., having three groups of digits), we would simply repeat steps 4 and 5, treating our current root (15) as the new 'a' to find the next digit.

4.3.3.2 A Mark of Mathematical Genius

Aryabhata's method is far more than a clever trick; it is a profound demonstration of mathematical insight. Its flawless execution depends on a complete mastery of the decimal place-value system and an implicit understanding of algebraic principles, centuries before algebra was formalized in the way we know it today. The algorithm's iterative nature—breaking a large, complex problem into a series of predictable, manageable steps—is the very essence of modern algorithmic thinking.

This was not merely an academic exercise. Such algorithms were powerful tools for solving real-world problems. Whether calculating the dimensions of a temple granary from its volume or performing the demanding calculations of astronomy, root extraction was essential. For Aryabhata, whose primary work was focused on the heavens, high-precision methods like this were indispensable for modeling the movements of the planets. This algorithm stands as a testament to a sophisticated mathematical culture, prefiguring by nearly a millennium the methods that would eventually become standard across the globe.

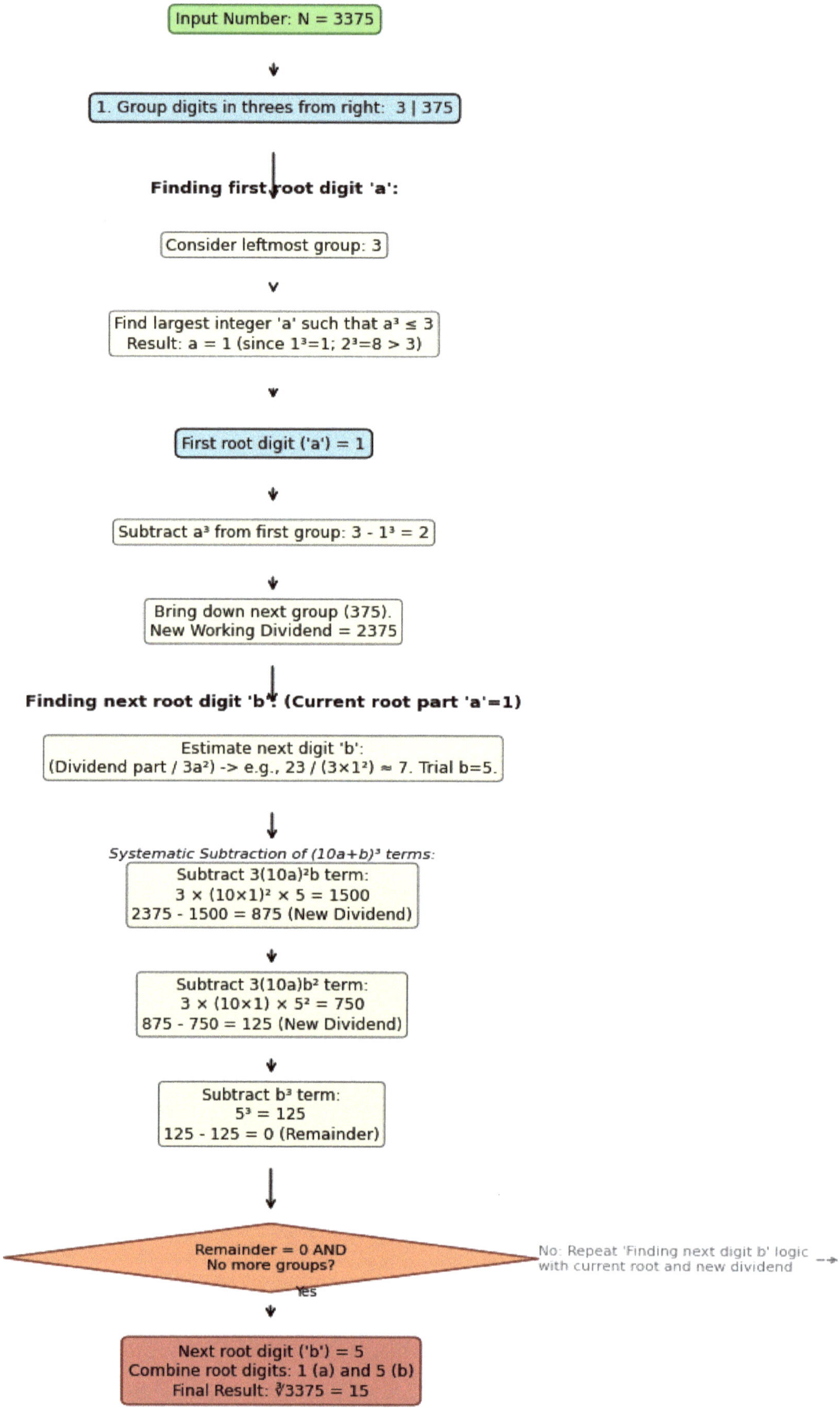

Figure 4.4: Aryabhata's Cube Root Algorithm (Gp.5): Iterative decimal method, illustrated with $\sqrt[3]{3375}$.

4.3.4 Triangle Area, Solid Volume Calculation

Sanskrit:

त्रिभुजस्य फलशरीरं समदलकोटीभु जार्धसंवर्गः ।

ऊर्ध्वभुजा तत्समवर्गार्धं स घनः षडश्रिरिति ॥ ६ ॥

IAST:

tribhujasya phalaśarīraṁ samadalakoṭībhu jārdhasaṁvargaḥ |

ūrdhvabhujā tatsamavargārdhaṁ sa ghanaḥ ṣaḍaśririti || 6 ||

Translation and Interpretation:

This verse addresses the calculation of area for a triangle and then transitions to the volume of a specific, though ambiguously described, solid.

- **Triangle Area:** The first part clearly states: "The area (*phalaśarīraṁ*) of a triangle is half the product of the perpendicular height (*samadalakoṭī*) and the base (*bhuja*)."This correctly gives the standard formula: **Area = (1/2) × base × height**.

- **Solid Volume (Obscure Reference):**The second part of the verse, "The volume (*ghana*) of a six-edged solid (*ṣaḍaśri*)..." followed by the computational instruction *ūrdhvabhujā tatsamavargārdhaṁ*, is significantly more obscure. The term *ṣaḍaśri* literally means "six-edged," which could refer to various solids (e.g., a triangular prism, a pyramid with a triangular or quadrilateral base if edges are counted in a specific way).Commentators like Bhaskara I interpret this phrase contextually, attempting to reconcile it with known solids and plausible calculation methods. Common scholarly interpretations suggest Aryabhata might be referring to:
    - **A right triangular prism:** In this case, the volume would be Base Area (of the triangle) × Height (of the prism).
    - **A right pyramid (possibly triangular base):** Here, the volume would be (1/3) × Base Area × Height. The exact solid intended by Aryabhata and the precise meaning of *ūrdhvabhujā tatsamavargārdhaṁ* remain uncertain from the verse alone, highlighting the crucial role of the commentarial tradition and comparative analysis in interpreting such condensed ancient texts.

Notes:

- **Tribhujasya phalaśarīraṁ:** "The area of a triangle."
- **Samadalakoṭībhu jārdhasaṁvargaḥ:** "Half the product of the perpendicular height (koṭī) and the base

(bhuja)." (Area = ½ × base × height).

- **Ūrdhvabhujā tatsamavargārdhaṁ sa ghanaḥ ṣaḍaśririti:** Obscure second part concerning volume of a "six-edged solid (ṣaḍaśri)." Interpretations vary (e.g., triangular prism, pyramid).

Examples:

- **Triangular Garden:** If the base is 10m and the perpendicular height is 6m, the Area = (1/2) × 10m × 6m = 30 m^2.
- **Hypothetical Solid Volume:** If we assume a triangular prism with the base area calculated above (30 m^2) and a prism height of 5m:
    - Volume (Prism interpretation) = 30 m^2 × 5m = 150 m^3.
    - If interpreted as a pyramid with the same base and height: Volume (Pyramid interpretation) = (1/3) × 30 m^2 × 5m = 50 m^3.

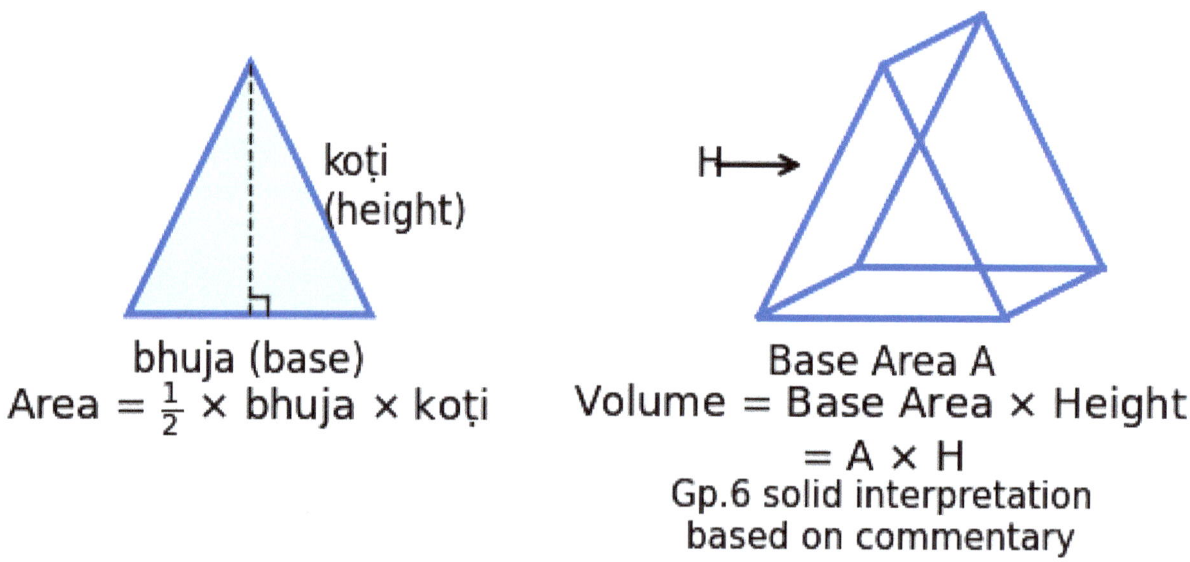

Figure 4.5: Geometric definitions for calculating the area of a triangle (left) and the volume of a solid (right), as described in Gaṇitapāda Verse 6 (Gp.6). Triangle area is given by half the product of the base (bhūja) and height (koṭi). The diagram on the right illustrates a common interpretation from commentary regarding the volume of a solid (such as a prism or pyramid) based on its base area and height.

4.3.5 Circle Area, Sphere Volume Calculation

Sanskrit:

समपरिणाहस्यार्धं विष्कम्भार्धहतमेव वृत्तफलम् ।

तन्निजमूलेन हतं घनगोलफलं निरवशेषम् ॥ ७ ॥

IAST:

samapariṇāhasyārdhaṁ viṣkambhārdhahatameva vṛttaphalam |

tannijamūlena hataṁ ghanagolaphalaṁ niravaśeṣam || 7 ||

This verse offers rules for two fundamental geometric figures: the circle and the sphere, providing insights into Aryabhata's mensuration techniques.

Notes:

- **Samapariṇāhasyārdhaṁ viṣkambhārdhahatameva vṛttaphalam:** "Half the circumference multiplied by half the diameter (radius) is the area of the circle." (Area = $(C/2) \times (D/2) = \pi R^2$).
- **Tannijamūlena hataṁ ghanagolaphalaṁ niravaśeṣam:** "That [area] multiplied by its own square root gives the exact volume of the sphere." (V = Area × √Area = $\pi R^2 \times \sqrt{\pi} R = \pi^{3/2} R^3$).
- **Niravaśeṣam:** "Exact," "complete."

4.3.5.1 Circle Area – A Foundation of Accuracy

The first half of the verse accurately details the calculation of a circle's area:

"Half the circumference (samapariṇāha[sya arthaṁ]) multiplied by half the diameter (viṣkambhārdhahatameva) is indeed the area of the circle (vṛttaphalam)."

Decoding the Rule: Aryabhata's instruction translates directly to a correct formula for the area of a circle:

- "Half the circumference" (C/2)
- "Multiplied by half the diameter" (D/2, which is the radius R)So, Area = (C/2) × (D/2). Since $C = 2\pi R$ and $D = 2R$, this becomes:

$$\textbf{Area} = (2\pi R / 2) \times (2R / 2) = \pi R \times R = \pi R^2$$

This correct formulation was essential for any calculations involving circular areas, especially when combined with his highly accurate value of π (Gp.10).

Part 2: Sphere Volume – A Puzzling Formulation

The verse then transitions to the volume of a sphere, presenting a rule that has been a significant point of discussion among scholars:

> **"That [area] multiplied by its own square root gives the exact (niravaśeṣam) volume of the sphere (ghanagolaphalam)."**

Decoding the Rule:

- **"That [area]"**: This clearly refers to the area of the circle (πR^2) established in the first part of the verse.
- **"multiplied by its own square root"**: This means (Area of Circle) × √(Area of Circle).
- **Mathematical Translation**:
    - Volume (V) = $(\pi R^2) \times \sqrt{(\pi R^2)}$
        - $V = \pi R^2 \times (\sqrt{\pi} \times R)$
        - $V = \pi^3/^2 * R^3$

The "Niravaśeṣam" Enigma and the Discrepancy:

Aryabhata's formula for the sphere's volume ($V \approx 5.568\ R^3$, if using his $\pi \approx 3.1416$) is notably different from the correct modern formula, $V = (4/3)\pi R^3$ (which is approximately $4.189\ R^3$).

The discrepancy becomes particularly intriguing due to Aryabhata's assertion that his result is *niravaśeṣam* – "exact," "complete," or "leaving no remainder." This claim of exactness for an incorrect formula has puzzled commentators and historians for centuries.

- **Historical Perspective – Bhaskara I's Awareness:** The astute 7th-century commentator Bhaskara I, in his *Aryabhatiya-bhashya*, was aware of this issue. He acknowledged that Aryabhata's formula yielded a different value for the sphere's volume than those derived from other traditional methods, some of which (likely based on principles of dissection or summation of infinitesimals that approached the correct $(4/3)\pi$ factor) were more accurate. Bhaskara I doesn't directly refute Aryabhata but carefully notes the existence of alternative, more commonly accepted results.
- **Scholarly Speculation on the Cause:** The reason for Aryabhata's unique and incorrect formulation remains a subject of scholarly debate. Possibilities include:
    - A novel, but flawed, theoretical geometric derivation or analogy.
    - A misinterpretation or incorrect generalization of an older, perhaps approximate, rule.

- A genuine mathematical error in reasoning, particularly challenging in three-dimensional geometry without the tools of calculus.*(Textual corruption is generally considered less likely for such a central formula).*

- **Impact within Aryabhata's System:** While a significant mathematical error, the direct impact of this specific sphere volume formula on the primary astronomical calculations within the *Aryabhatiya* (such as predicting planetary longitudes or eclipse timings) was likely minimal. For these, the diameters, angular sizes, and orbital parameters of celestial bodies were far more critical than their precise volumetric measures.

This particular rule in the Gaṇitapāda serves as a valuable historical data point, reminding us that the path of scientific discovery is not always straightforward. Aryabhata's assertion of "exactness" for this formula makes it a fascinating case study, prompting consideration of the diverse methods of inquiry and the inherent challenges faced by early mathematicians. It also underscores the vital role of the commentarial tradition in contextualizing, and sometimes critically engaging with, foundational texts.

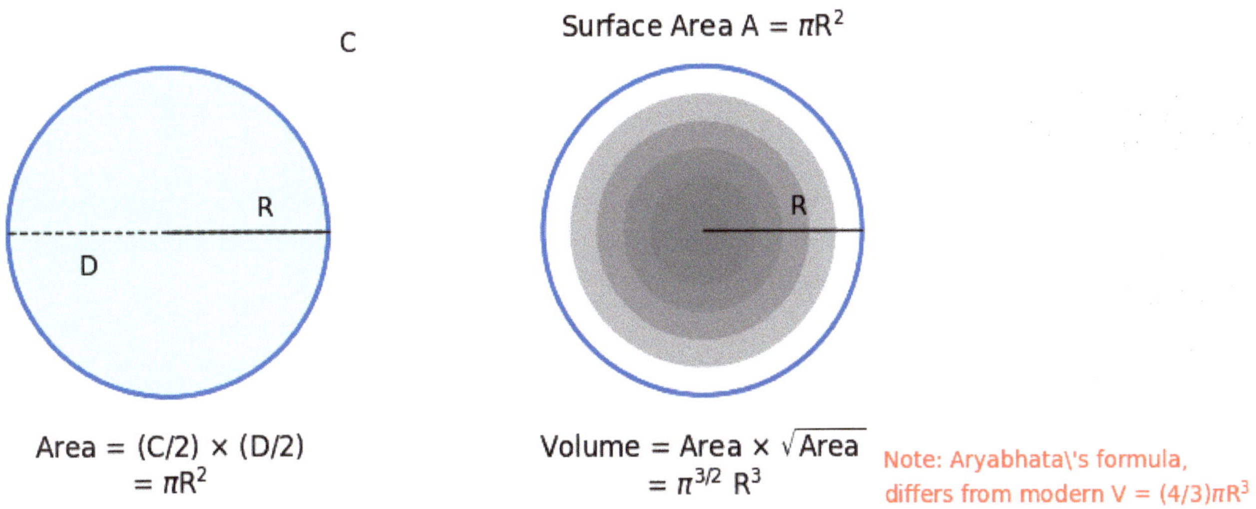

Figure 4.6: Geometric formulas for calculating the area of a circle (left) and the volume of a sphere (right), as described in Gaṇitapāda Verse 7 (Gp.7). Aryabhata correctly gave the circle area as half the circumference multiplied by half the diameter. His unique formula for sphere volume, however, differs from the modern formula, a notable point discussed in relation to this verse.

4.3.6 Trapezoid Area Calculation

Sanskrit:

आयामगुणे पार्श्वे तद्योगहृते च लम्बके खाते ।

आयामगुणे मध्ये क्षेत्रफलमेतद् विसदृशेऽपि ॥ ८ ॥

IAST:

āyāmaguṇe pārśve tadyogahṛte cha lambake khāte |

āyāmaguṇe madhye kṣetraphalametad visadṛśe'pi || 8 ||

Translation and Interpretation (Trapezoid Area):

While the Sanskrit phrasing of this verse is somewhat obscure, potentially alluding to practical contexts such as excavation (*khāte* = ditch/pit), the principles described for finding the area (*kṣetraphalam*) of a quadrilateral with unequal parallel sides (*visadṛśa*) are consistent with the standard formula for the area of a trapezoid:

Area = (1/2) × (Sum of parallel sides) × Height.

The terms within the verse likely map to the components of this formula, understood within a specific geometric or practical context familiar to Aryabhata's audience:

- ***āyāmaguṇe pārśve*** (potentially "length multiplied by [each] side" or related to the parallel sides).
- ***tadyogahṛte*** (potentially "divided by their sum," which, when combined with other terms, could lead to averaging the parallel sides).
- ***lambake*** (clearly refers to the perpendicular height).
- ***āyāmaguṇe madhye*** (perhaps "length/height multiplied by the mean [of the parallel sides]").

Despite the terse wording, the verse is generally interpreted as providing a rule equivalent to the modern trapezoid area formula.

Example:

- **Temple Ramp (Trapezoidal Face):**
    - Parallel sides: 5m and 3m

- Perpendicular Height: 10m
- Area = (1/2) × (5m + 3m) × 10m = (1/2) × 8m × 10m = **40 m²**.

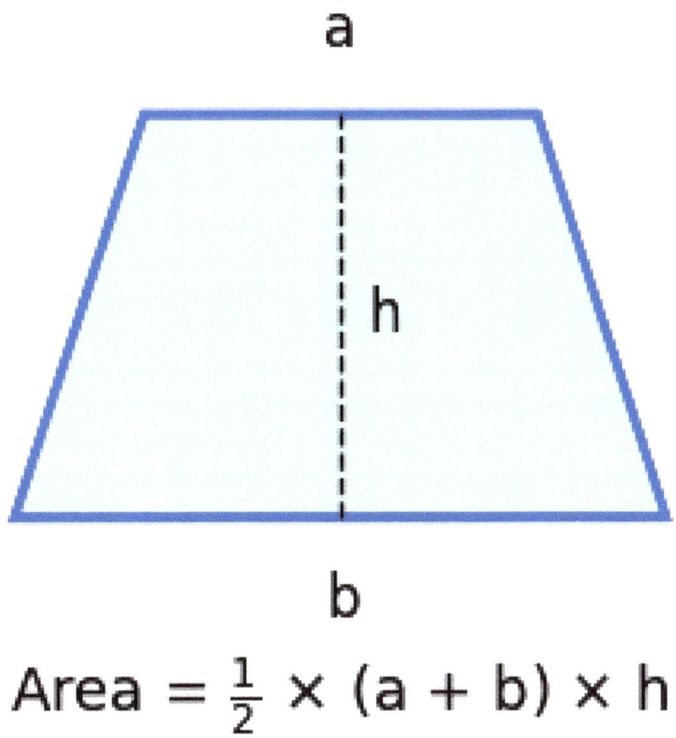

Figure 4.7: Geometric definition for calculating the area of a trapezoid (a quadrilateral with parallel sides 'a' and 'b'), as described in Gaṇitapāda Verse 8 (Gp.8). The area is given by half the sum of the parallel sides multiplied by the perpendicular height (h).

4.3.7 General Area Concept and Chord Introduction

Sanskrit:

सर्वेषां क्षेत्राणां प्रसाध्य पार्श्वे फलं तदभ्यासः ।

वृत्ते धनुषः कार्या प्रधिना शरसंयुतेनार्धम् ॥ ९ ॥

IAST:

sarveṣāṁ kṣetrāṇāṁ prasādhya pārśve phalaṁ tadabhyāsaḥ|

vṛtte dhanuṣaḥ kāryā pradhinā śarasaṁyutenārdham || 9 ||

Notes:

- **Sarveṣāṁ kṣetrāṇāṁ prasādhya pārśve phalaṁ tadabhyāsaḥ:** "For all plane figures, having determined sides, the area is from their systematic calculation." (General principle for area).
- **Vṛtte dhanuṣaḥ kāryā pradhinā śarasaṁyutenārdham:** "In a circle, concerning the segment ('bow', dhanuṣaḥ), calculations involve the chord (pradhinā/jyā) and 'arrow' (śara)."
- **Jyā:** Half-chord (R·sinθ).
- **Śara:** Arrow/versine (R(1-cosθ)).

Translation & Interpretation:

This verse acts as a crucial pivot, forming a conceptual bridge between the general principles of area calculation and the specific geometry of the circle, which is the gateway to Aryabhata's trigonometry.

The first line offers a broad maxim for "all plane figures" (*sarveṣāṁ kṣetrāṇām*): having determined their defining dimensions, "the area (*phalam*) is derived from their systematic calculation" (*tadabhyāsaḥ*). This is a formal declaration of the principle underlying the preceding verses: the area of any shape can be found by applying the appropriate computational rules to its sides and heights. The term *tadabhyāsaḥ* implies not just a calculation, but the methodical application of an established procedure.

Having stated this general rule, Aryabhata then focuses his lens specifically upon the circle (*vṛtte*). He introduces the analysis of a circular segment, which he visualizes as a "bow" (*dhanuṣaḥ*), and highlights two geometric components critical for its measurement:

- **Pradhi/Jyā (Chord/Half-Chord):** While *pradhi* can mean circumference, in the context of a bow (*dhanuṣaḥ*), it clearly refers to the "bow-string" or chord. As is foundational to his work, Aryabhata's system primarily uses the *jyā* (half-chord), which is equivalent to the modern **R·sinθ**.

- **Śara (Arrow):** This is the height of the circular segment—the perpendicular distance from the midpoint of the chord to the arc. It is the geometric basis for the versine function, equivalent to $\mathbf{R(1\text{-}cos\theta)}$.

The verse's concluding phrase, *śarasaṁyutenārdham*, hints at the intricate formulas that interrelate these components, often involving operations with "half the quantity combined with the *śara*."

Significance: A Gateway to Trigonometry

Verse 9 thus serves as a critical gateway. It affirms the power of systematic calculation for all areas and then strategically introduces the *jyā* and the *śara*. These are not merely descriptive parts of a circle; they are the fundamental geometric units of Aryabhata's unique formulation of trigonometry. It is precisely the relationship between the *jyā*, the *śara*, and the radius that allows for the numerical quantification of angles and arcs—the essential skill for the astronomical calculations that are this treatise's ultimate goal.

These geometric principles were not confined to abstract mathematics. They were integral to the work of surveyors in land measurement, architects in construction (echoing the precision of the earlier *Śulba Sūtras*), and priests in determining auspicious timings. Having laid this foundational groundwork for measuring the shapes of the earth, Aryabhata now turns to the tools required to measure the heavens, introducing one of his most celebrated and enduring contributions.

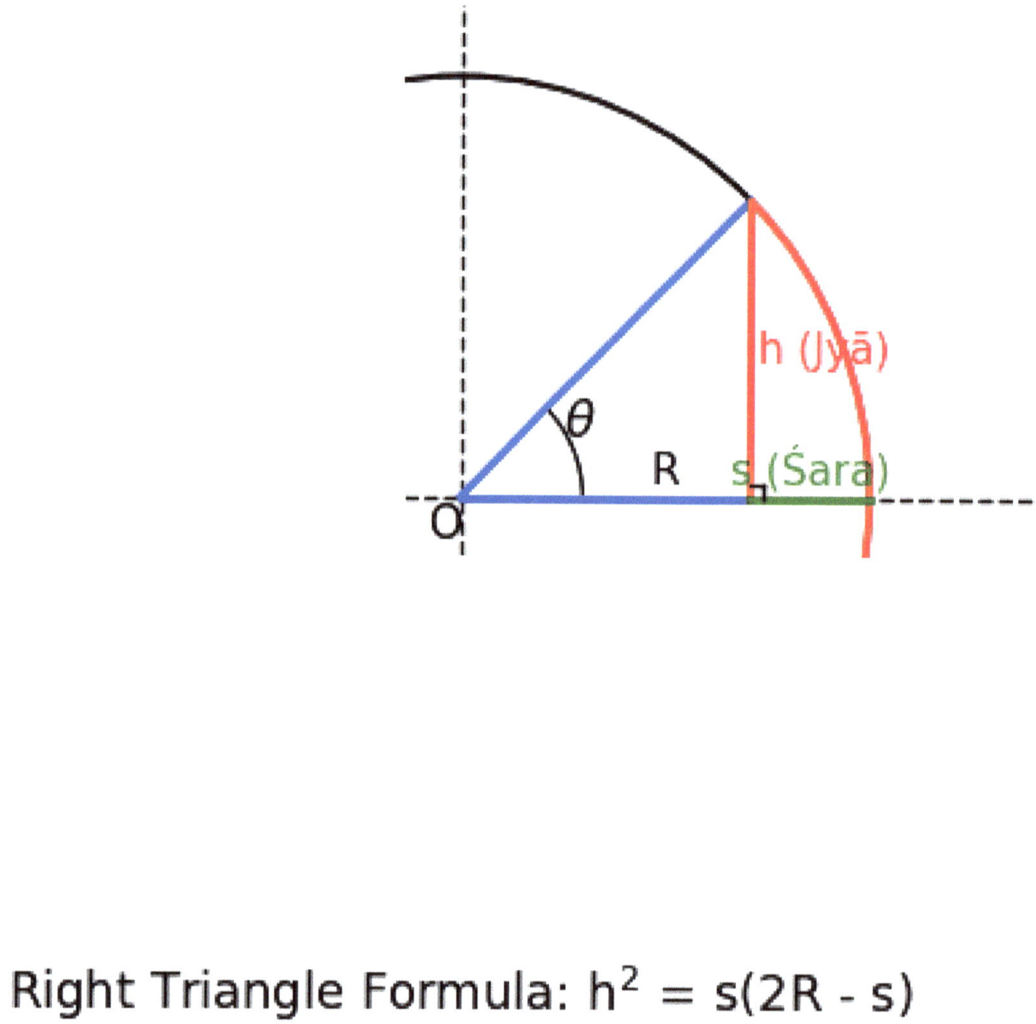

Figure 4.8: Geometric definition of the Indian sine (Jyā) and Versine (Śara or 'Arrow'). Jyā (h) is the length of the half-chord of an arc (red) in a circle of radius R. Śara (s) is the height of the segment. The diagram illustrates the relationship between h and s, along with the diameter (2R), captured by the formula $h^2 = s(2R - s)$ (stated by Aryabhata in Gp.12 and Gp.17), fundamental to his trigonometry.

4.4 Pi, Circles, and Trigonometric Foundations (Gp.10-13)

This sequence of verses marks a pivotal moment in the Gaṇitapāda, providing Aryabhata's famous approximation for π, essential formulas for circle calculations, and geometric theorems involving chords and arcs that lay the groundwork for trigonometric applications.

4.4.1 Aryabhata's Pi (π) (Gp.10)

Sanskrit:

चतुरधिकं शतमष्टगुणं द्वाषष्टिस्तथा सहस्राणाम् ।

अयुतद्वयविष्कम्भस्य आसन्नो वृत्तपरिणाहः ॥ १० ॥

IAST:

chaturadhikaṃ śatamaṣṭaguṇaṃ dvāṣaṣṭistathā sahasrāṇām|

ayutadvayaviṣkambhasya āsanno vṛttapariṇāhaḥ || 10 ||

Translation: "Add four to one hundred, multiply by eight, and then add sixty-two thousand: [the result] is the approximate (*āsannaḥ*) circumference (*vṛttapariṇāhaḥ*) of a circle whose diameter (*viṣkambha*) is twenty thousand (*ayutadvaya*)."

Notes:

- **Chaturadhikaṁ śatamaṣṭaguṇaṁ dvāṣaṣṭistathā sahasrāṇām:** "(100+4) × 8 + 62000 = 62832." (This is Circumference, C).
- **Ayutadvayaviṣkambhasya:** "Of a diameter (viṣkambha) of twenty thousand (20,000)." (This is Diameter, D).
- **Āsanno vṛttapariṇāhaḥ:** "[This is] the approximate circumference."
- **π = C/D = 62832 / 20000 = 3.1416.**
- **Āsannaḥ:** "Approximate," "near" – correctly notes π's irrational nature.

Calculation and Value:

1. Add four to one hundred: 100 + 4 = 104
2. Multiply by eight: 104 × 8 = 832

3. Add sixty-two thousand: 832 + 62,000 = 62,832

4. This (62,832) is the circumference (C) for a diameter (D) of 20,000.

5. Since $\pi = C / D$, Aryabhata's value for π is: $\boldsymbol{\pi} = \mathbf{62{,}832 / 20{,}000 = 3.1416}$

Significance:

- **Remarkable Accuracy:** This value (3.1416) is exceptionally accurate for its time, correct to four decimal places. It significantly improved upon earlier Indian approximations (such as $\sqrt{10} \approx 3.1622$, sometimes used for practical mensuration, or the simpler value of 3) and contemporary or earlier Greek values (like Archimedes' approximation of $22/7 \approx 3.1428$). Aryabhata's value remained unsurpassed in Europe for over a millennium.
- **Computational Use:** This precise value was essential for Aryabhata's astronomical calculations, particularly those involving circular orbits, angular conversions, and the construction of his accurate sine (*jyā*) table.
- **(Approximate):** Importantly, Aryabhata explicitly calls this value *āsannaḥ* ("approximate," "near"). This correctly acknowledges that π is an irrational number whose exact value cannot be captured by a finite fraction, demonstrating a profound insight into the nature of this fundamental constant.

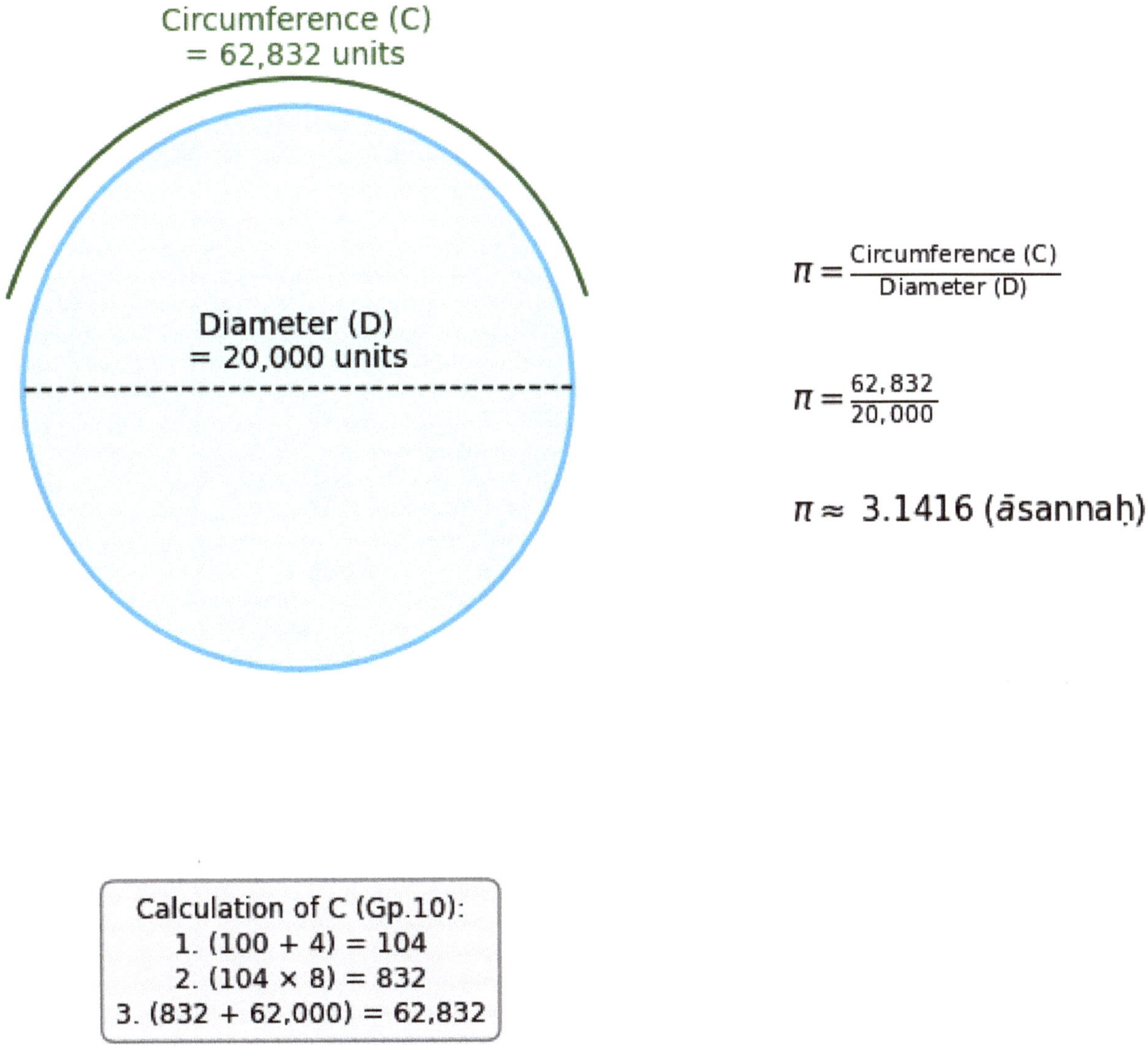

Figure 4.9: Aryabhata's Derivation of π from Proportions (Gaṇitapāda Verse 10). The diagram illustrates a circle with a diameter (D) of 20,000 units, as specified by Aryabhata. The verse then outlines a calculation (inset box) for its circumference (C), yielding 62,832 units. Aryabhata's remarkably accurate approximation of π is then derived from the fundamental ratio C/D, resulting in $\pi \approx 3.1416$. He importantly qualified this value as 'āsannaḥ' (approximate), indicating an understanding of π's nature.

4.4.2 A Key Sine Value and Circle Area from Half-Chords

Sanskrit:

परिधेः षड्भागांशो ज्या विष्कम्भार्धतुल्या स्यात् ।

समवृत्तस्यार्धज्यास् तास् तु समानीय परिधेयाः ॥ ११ ॥

IAST:

paridheḥ ṣaḍbhāgāṃśo jyā viṣkambhārdhatulyā syāt|

samavṛttasyārdhajyās tās tu samānīya paridheyāḥ || 11 ||

Notes:

- **Paridheḥ ṣaḍbhāgāṃśo jyā viṣkambhārdhatulyā syāt:** "The jyā [half-chord] of one-sixth the circumference (arc of 60°) is equal to half the diameter (radius, R)." (R·sin(30°) = R/2).
- **Samavṛttasyārdhajyās tās tu samānīya paridheyāḥ:** "The half-chords (ardhajyās) of a circle – from these, by systematic composition, [the area/properties] are to be known." (Suggests area derivation from half-chords).

Interpretation:

This verse presents two distinct but related geometric statements concerning the circle. The first provides a precise, foundational trigonometric value, while the second hints at a deeper conceptual basis for understanding the circle's area through its constituent half-chords.

A Foundational Sine Value: R·sin(30°) = R/2

The first line of the verse is an unambiguous statement of geometric fact:

> **"The *jyā* [half-chord] corresponding to one-sixth part of the circumference is equal to half the diameter."**

To a geometer, the meaning is immediate. "One-sixth part of the circumference" describes an arc of 360°/6 = 60°. The *jyā* of this 60° arc corresponds to the modern R·sin(30°). "Half the diameter" is, of course, the radius (R). Thus, Aryabhata states a powerful identity: **R·sin(30°) = R/2**, which simplifies to the fundamental trigonometric truth that **sin(30°) = 1/2**.

This was no mere geometric curiosity. For a system built on an iterative method of differences, this easily derivable and certain value was an unshakeable anchor point. It provided a crucial checkpoint for verifying the accuracy of the entire sine table (Gk.10, Gk.12), ensuring the integrity of the whole computational structure.

Circle Area from Half-Chords: A Conceptual Hint

In contrast to the clarity of the first line, the second is more condensed and has prompted significant scholarly interpretation. It reads: "From the half-chords (*ardhajyās*) of a uniform circle, by systematic composition, [its properties] are to be known."

Commentators like Bhaskara I interpret this as a conceptual reinforcement of the circle area formula given earlier (Gp.7). While that verse provides the direct computational rule, this line suggests a deeper understanding of *why* the formula works. It implies that by understanding the properties of the *jyās* (sines) distributed around the circle, one can, through a process of summation or "systematic composition," arrive at the circle's area.

This line of reasoning is a conceptual forerunner to integral calculus. While Aryabhata does not detail an explicit infinitesimal method, the idea of using the collection of half-chords (sines) to determine the area of the whole is profoundly insightful. It anticipates the logic of integration, where an area is found by summing an infinite number of infinitesimally small triangles, whose dimensions are themselves defined by sines and cosines.

Overall Significance

Verse 11 is therefore pivotal. It furnishes a cornerstone trigonometric value, R·sin(30°), which was instrumental in the construction and verification of Aryabhata's sine table. Furthermore, it alludes to the fundamental role of the sine function in comprehensively measuring the properties of a circle, including its area, through a method that prefigures modern calculus. This verse powerfully underscores the profound interconnectedness of geometry, trigonometry, and mensuration within Aryabhata's singular mathematical framework.

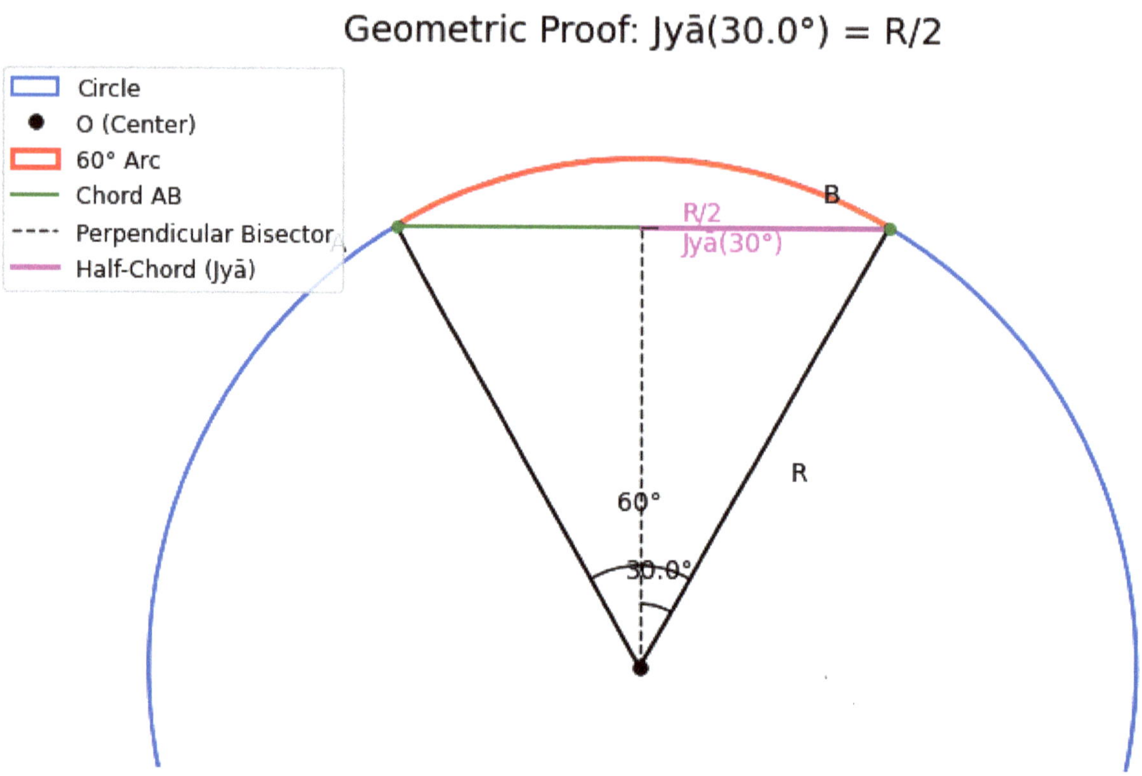

Figure 4.10: Geometric diagram illustrating the relationship stated by Aryabhata in Gaṇitapāda Verse 11 (Gp.11): the sine (Jyā) corresponding to a 60° arc is equal to R/2. A triangle formed by the two radii and the chord of a 60° arc is equilateral (sides of length R). The half-chord (Jyā of 30°) is half the length of the chord, thus R/2. This fundamental identity was likely used in constructing or verifying Aryabhata's sine table.

4.4.3 The Chord-Arrow Theorem: Geometry of Circular Segments

Sanskrit:

परिधेः अर्धं ज्यायस् तद् वृत्ते शरसमासगुणं कार्मुकम् ।

ज्याधनुषोः एवं स्यात् सर्वत्र क्षेत्रनिदर्शनम् ॥ १२ ॥

IAST:

paridher ardhaṃ jyāyas tad vṛtte śarasamāsaguṇaṃ kārmukam|

jyādhanuṣoḥ evaṃ syāt sarvatra kṣetranidarśanam || 12 ||

Notes:

- **Paridher ardhaṃ jyāyas tad vṛtte śarasamāsaguṇaṃ kārmukam:** (Concise phrasing for the theorem).
- **Relationship:** $h^2 = s(D - s)$ or $h^2 = s(2R - s)$, where h=half-chord (jyā), s=arrow (śara), D=diameter, R=radius.
- **Jyādhanuṣoḥ evaṃ syāt sarvatra kṣetranidarśanam:** "Thus should be the calculation for chords and arcs in all figures." (Universal applicability).

Interpretation and The Theorem

This compact verse unlocks the fundamental geometry of the circular segment. Let us decode its terms. The *jyā* (*h*) is the **half-chord**, while the *śara* (*s*) is the **"arrow"**—the height of the segment measured from the midpoint of the full chord to the arc. The verse describes a powerful relationship between these elements, a theorem that is the heart of circular measurement.

It states that the square of the half-chord (h^2) is equal to the product of the arrow (s) and the remaining portion of the diameter (D – s). In modern mathematical terms:

$$\mathbf{h^2 = s \cdot (2R - s)}$$

This simple but profound formula provides a direct way to relate the linear measure of a chord to the dimensions of its corresponding arc.

A Powerful Computational Tool

More than a static geometric fact, this theorem works in concert with the Pythagorean theorem to become a dynamic computational tool. By applying the Pythagorean theorem to the right triangle formed by the radius (R), the half-chord (*h*), and the line segment (R-s), one can establish a direct link between the sine and cosine analogues.

To a modern eye, this is deeply significant. The *jyā* (*h*) is equivalent to **R·sin(θ)**. Its Pythagorean counterpart, the line segment $\sqrt{(R^2 - h^2)}$, is none other than **R·cos(θ)**. This relationship was crucial, as it allowed Aryabhata to derive the value of the *śara* (related to the versed sine, or *utkrama-jyā*) directly from the *jyā* (sine). This was a foundational step in constructing a complete and accurate trigonometric system.

A Universal Principle

Aryabhata concludes with a statement of broad significance, emphasizing that this relationship is not a mere curiosity but a universal principle (*sarvatra*—everywhere) that applies to any and all problems involving circular geometry. It is the "demonstration concerning chords and arcs in all figures."

Astronomical Significance

This geometric machinery was not an abstract exercise; it was indispensable for Aryabhata's astronomy. Its applications were direct and practical:

- **Calculating Eclipses:** To determine the magnitude of a solar or lunar eclipse, an astronomer must calculate the obscured portion of the celestial body. This "bitten" area is a circular segment, and its height—the *śara*—is the key variable that this theorem unlocks.

- **Planetary Models:** The theorem was essential for converting between angular positions and linear distances within the geometric models of planetary orbits, such as epicycles.

- **Spherical Astronomy:** It provided a method for solving problems on the celestial sphere, where great circles behave according to these same geometric rules.

Verse 12 is therefore a critical link in Aryabhata's system, providing the precise geometric tools needed to turn observational data into a predictive, mathematical science of the cosmos.

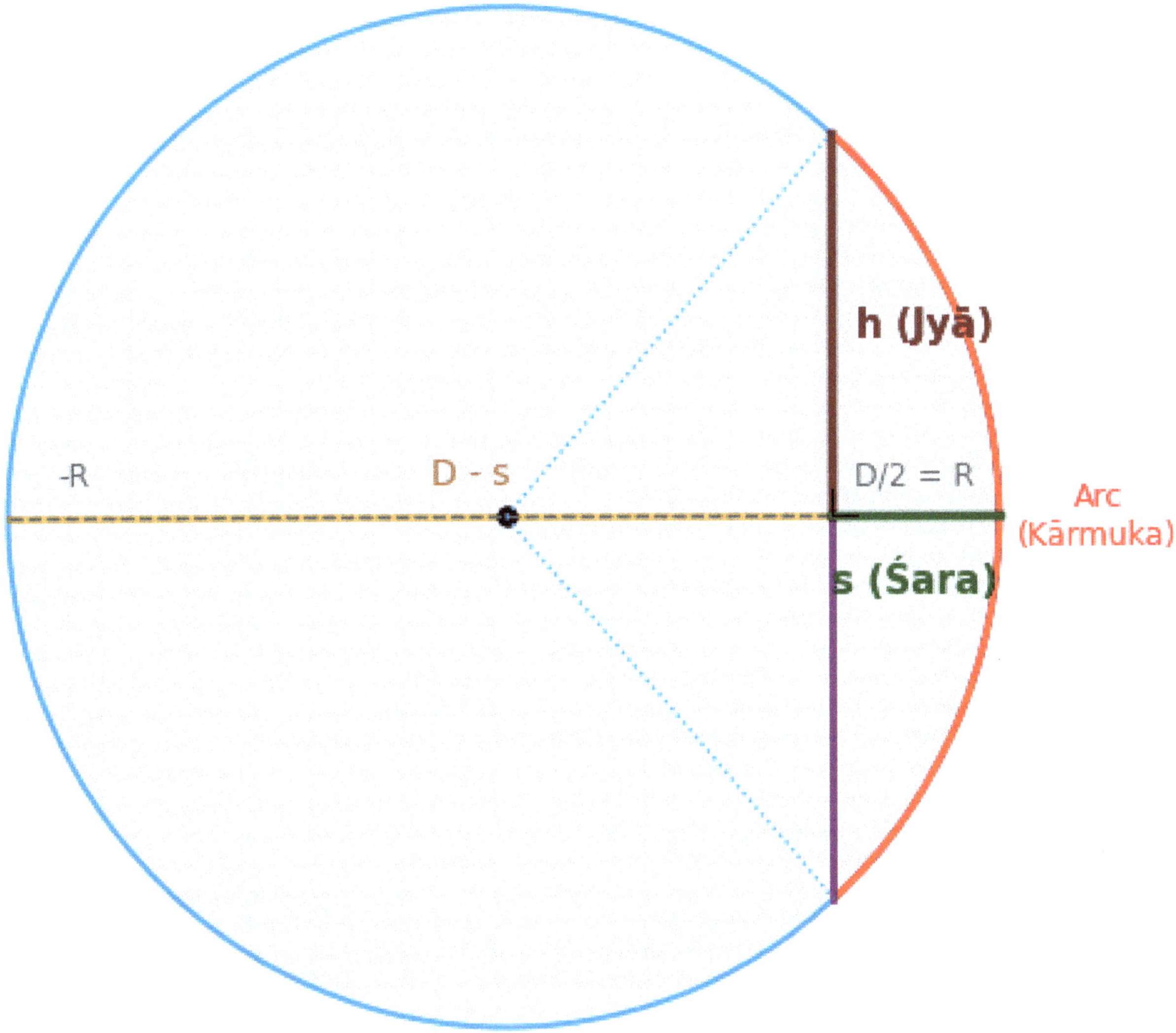

$$h^2 = s \cdot (D - s)$$
$$\text{or } h^2 = s \cdot (2R - s)$$

Figure 4.11: The Half-Chord (Jyā) and Arrow (Śara) Theorem from Gp.12. This diagram illustrates the fundamental geometric relationship stated by Aryabhata in Gaṇitapāda Verse 12. For a circular segment defined by an arc (kārmuka), its half-chord ('h', Jyā) and arrow ('s', Śara) are related to the circle's diameter (D, or 2R) by the formula: $h^2=s(D-s)$ This theorem is pivotal for Aryabhata's system of trigonometry and its applications in astronomy, particularly for calculations involving circular segments such as in eclipse predictions.

4.4.4 Intersecting Circles – Finding Segment Heights

Sanskrit:

ग्रासोनयोगदलकृतिं ग्रासगुणां द्विगुणाद् हृत्वा ।

सम्पातशरौ लब्धौ द्वयोः परस्परयोगे तु ॥ १३ ॥

IAST:

grāsonayogadalakṛtiṃ grāsaguṇāṃ dviguṇād hṛtvā |

sampātaśarau labdhau dvayoḥ parasparayoge tu || 13 ||

Notes:

- **Grāsonayogadalakṛtiṃ grāsaguṇāṃ dviguṇād hṛtvā:** Technical terms related to overlap (grāsa), sum, half, square, multiplication, division.
- **Sampātaśarau labdhau:** "The arrows at intersection (śara$_1$, śara$_2$) are obtained."
- **Dvayoḥ parasparayoge tu:** "For two [circles] in mutual conjunction/intersection."

Interpretation:

Predicting the precise extent of an eclipse presents a critical geometric challenge: how to calculate the overlap when two celestial disks intersect. This verse addresses the core of that problem, providing a method to find the heights (*śara*, "arrows") of the circular segments formed by the intersection. Whether it is the Moon crossing the Sun's disk or passing through Earth's shadow, the depth of these segments is the key to an eclipse's magnitude.

Decoding Aryabhata's Method

To solve this, Aryabhata provides a highly condensed procedural rule for finding these crucial *sampātaśarau*, or "arrows at the intersection." The Sanskrit phrasing is highly technical, using terms like *grāsonayogadalakṛtiṃ* ("the square of half the sum [of the radii] diminished by the overlap"). It resists a simple, literal translation and is best understood through geometric reconstruction, as guided by traditional commentaries.

The underlying geometry to find these segment heights (*śara$_1$* and *śara$_2$*) involves a multi-step process:

1. **Find the Common Chord:** The intersection of two circles is defined by a common chord. The length of half this chord (let's call it m) is the essential intermediate value. This length can be calculated from the radii of the two circles (R_1 and R_2) and the distance between their centers (d) using established geometric principles.
2. **Calculate the *Śaras*:** Once this common half-chord (m) is known, the *śara* for each circle is found by applying

the very principle established in Verse 12: $\mathbf{s = R - \sqrt{(R^2 - m^2)}}$.

Aryabhata's verbal instructions—with terms like *grāsa* (overlap) and commands for multiplication (*guṇāṃ*) and division (*hṛtvā*)—outline the precise computational pathway to arrive at these values without using modern algebraic notation.

Astronomical Application: The Key to Eclipse Magnitude

This calculation is the indispensable key to determining the magnitude (*grāsa*) of an eclipse. The magnitude, which represents the fraction of the eclipsed body's diameter that is covered, depends directly on these segment heights. The total obscured depth is related to the sum of the two *śaras*, and comparing this depth to the eclipsed body's full diameter gives the final magnitude (a process detailed further in the *Golapāda*, Gola.44).

Verse 13, therefore, provides a critical geometric algorithm that forms a cornerstone of Aryabhata's sophisticated eclipse-prediction methodology. It is a powerful demonstration of his ability to translate complex spatial problems into concise, actionable computational procedures.

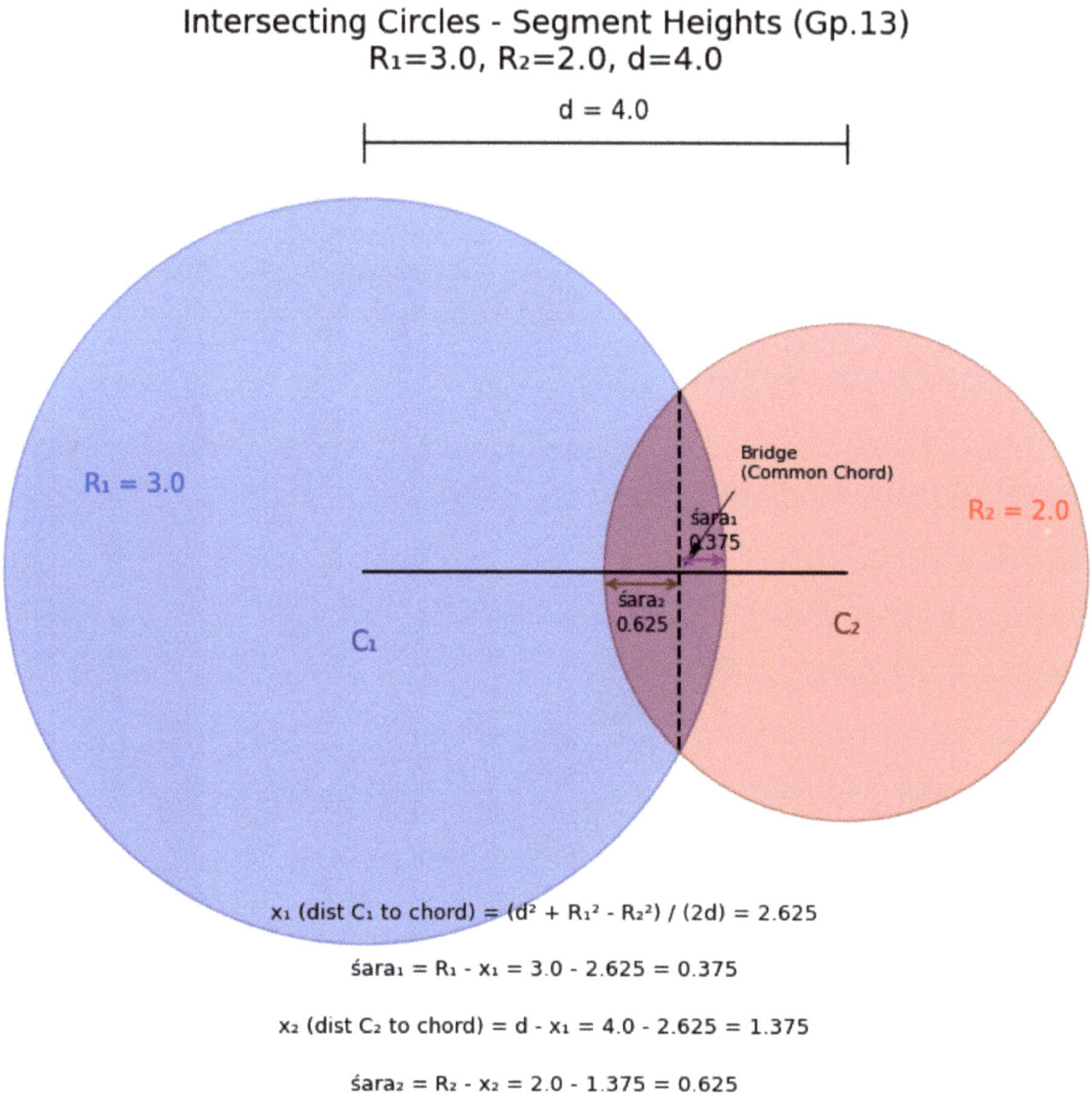

Figure 4.12: Visualizing the śara (segment height) calculation for intersecting circles, fundamental to Aryabhata's eclipse predictions (Gaṇitapāda, Verse 13). The diagram shows the common chord ("Bridge"), the distances from each center to this chord (x_1 and x_2), and the resulting śaras (0.375 and 0.625) for circles with R_1=3.0, R_2=2.0, and d=4.0.

4.5 Measurement by Shadow: The Gnomon (Śanku) (Gp.14-17)

Aryabhata applies geometric and trigonometric principles to the practical task of measurement using the ***śaṅku*** (gnomon) and its shadow (***chāyā***), techniques vital for timekeeping and surveying.

4.5.1 Finding Distance or Height using Gnomon

Sanskrit:

शङ्कोः प्रमाणवर्गं छायावर्गेण संयुतं कृत्वा ।

यत् तस्य वर्गमूलं विष्कम्भार्धं स्ववृत्तस्य ॥ १४ ॥

IAST:

śaṅkoḥ pramāṇavargaṃ chāyāvargena saṃyutaṃ kṛtvā|

yat tasya vargamūlaṃ viṣkambhārdhaṃ svavṛttasya || 14 ||

Notes:

- **Śaṅkoḥ pramāṇavargaṃ chāyāvargena saṃyutaṃ kṛtvā:** "Square of gnomon height (G^2) + square of shadow length (S^2)."
- **Yat tasya vargamūlaṃ viṣkambhārdhaṃ svavṛttasya:** "The square root of that sum is the radius of its circle (hypotenuse/karṇa, K)."
- **Calculation:** $K = \sqrt{(G^2 + S^2)}$. (Implicitly uses Pythagorean theorem).

Translation: "Having added the square of the gnomon's height (*śaṅkoḥ pramāṇavargaṃ*) to the square of the shadow length (*chāyāvargena*), the square root of that sum (*tasya vargamūlaṃ*) is the radius (*viṣkambhārdhaṃ*) of its circle [the hypotenuse/slant distance]."

Interpretation:

This verse addresses the fundamental geometry of the shadow cast by a gnomon. It describes the right-angled triangle formed by the vertical gnomon (height *G*), its horizontal shadow (*S*), and the sunbeam connecting the top of the gnomon to the tip of the shadow. The verse provides the rule for calculating the length of this sunbeam—the hypotenuse, which is termed the **karṇa**.

The rule is a direct application of the Pythagorean theorem:

$\text{Karṇa} = \sqrt{(G^2 + S^2)}$

While seemingly simple, calculating the *karṇa* is a crucial intermediate step. With all three sides of the gnomon-shadow triangle known, it becomes possible to determine angles—such as the Sun's altitude—through the trigonometric relationships established earlier. Furthermore, this value is essential for solving problems involving similar triangles, a technique Aryabhata will employ in the very next verses (Gp.15-16). This verse, therefore, forges a key link in the chain of practical geometric reasoning.

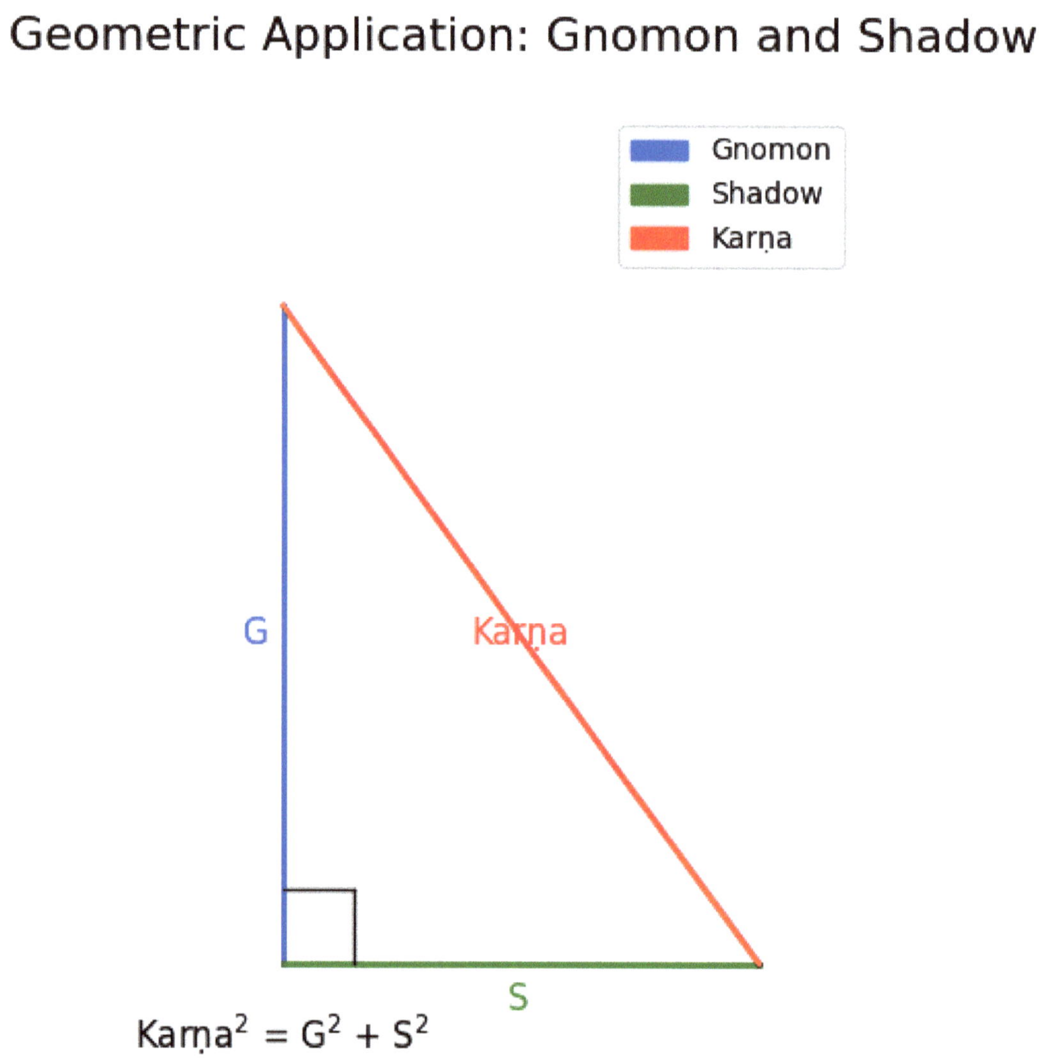

Figure 4.13: Geometric diagram illustrating the right triangle formed by a Gnomon (Śaṅku, height G) and its Shadow (Chāyā, length S). The hypotenuse is called the Karṇa. The diagram shows the relationship between these sides as stated by the Pythagorean theorem (Gp.17), and this geometry is used to calculate the Karṇa from G and S (Gp.14), a step fundamental to various gnomon-based astronomical measurements and time calculations.

4.5.2 Shadow Calculation for Lamp/Object

Sanskrit:

शङ्कुगुणं शङ्कुभुजाविवरं शङ्कुभुजयोर् विशेषहृतम् ।

यल् लब्धं सा छाया ज्ञेया शङ्कोः स्वमूलाद् हि ॥ १५ ॥

IAST:

śaṅkuguṇaṃ śaṅkubhujāvivaraṃ śaṅkubhujayor viśeṣahṛtam|

yal labdhaṃ sā chāyā jñeyā śaṅkoḥ svamūlād hi || 15 ||

Notes:

- **Śaṅkuguṇaṃ śaṅkubhujāvivaraṃ:** "Gnomon height (G) multiplied by the horizontal distance (D) [between gnomon and lamp base]."
- **Śaṅkubhujayor viśeṣahṛtam:** "Divided by the difference in heights (L-G) [of lamp (L) and gnomon (G)]."
- **Yal labdhaṃ sā chāyā jñeyā:** "The result is the shadow length (S)."
- **Formula:** S = (G × D) / (L - G). (Based on similar triangles).

Interpretation:

Aryabhata now applies the geometry of the gnomon to solve a new and practical problem: If a lamp of a known height is placed at a certain distance from a gnomon, what will be the length of the shadow it casts?

The solution lies in the elegant and powerful principle of similar triangles. Let *L* be the height of the lamp, *G* be the height of the gnomon, and *D* be the horizontal distance between their bases. The length of the gnomon's shadow (*S*) is found by proportion:

$$\mathbf{S = (G \times D) / (L - G)}$$

The verse's instructions map onto this formula with surgical precision. It directs the user to multiply the gnomon's height (*G*) by the distance between the two bases (*śaṅkubhujāvivaraṃ*, or *D*), forming the numerator. This product is then divided by the "difference between the gnomon and the [lamp's] height" (*śaṅkubhujayor viśeṣahṛtam*, or *L-G*). The result (*labdhaṃ*) is precisely the length of the shadow extending from the gnomon's own base (*svamūlāt*).

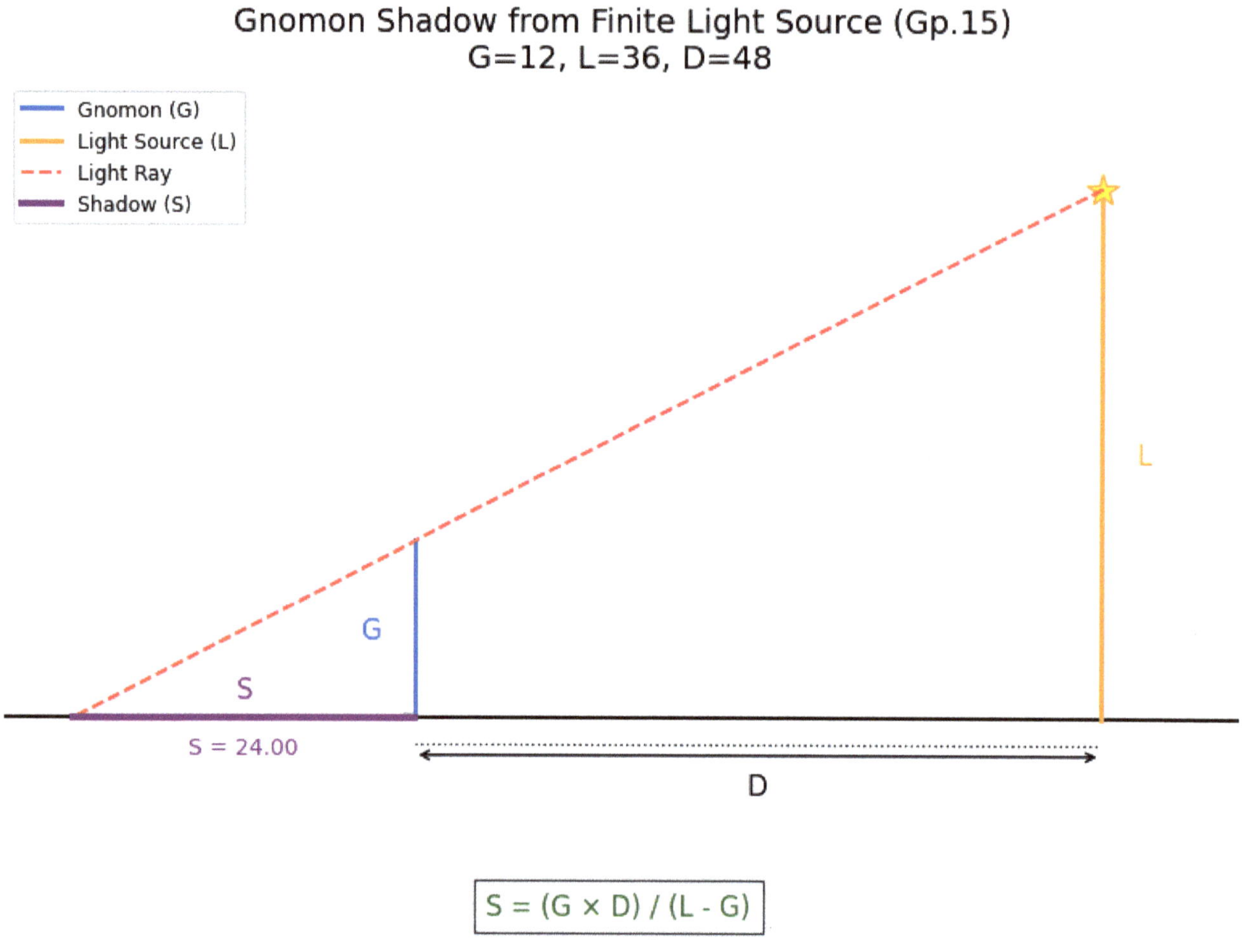

Figure 4.14: Diagram illustrating the geometry of a Gnomon (Śaṅku) and its shadow (Chāyā). The height of the gnomon (G), the shadow length (S), and the Sun's ray form a right triangle, used in various astronomical calculations.

4.5.3 Finding Distance from Two Gnomon Shadows

Sanskrit:

छायागुणितं छायाग्रविवरमूनेन भाजिता कोटी ।

शङ्कुगुणा सा कोटी स्वभुजा भक्ता फलं भवति ॥ १६ ॥

IAST:

chāyāguṇitaṃ chāyāgravivaramūnena bhājitā koṭī|

śaṅkuguṇā sā koṭī svabhujā bhaktā phalaṃ bhavati || 16 ||

Notes:

- **Chāyāguṇitaṃ chāyāgravivaramūnena bhājitā koṭī:** "Shadow (S_2) × distance between shadow-tips (ΔD), divided by difference of shadows ($|S_1-S_2|$) = Koṭī (distance from S_2 tip to object base)."
- **Śaṅkuguṇā sā koṭī svabhujā bhaktā phalaṃ bhavati:** "That Koṭī × Gnomon height (G), divided by its shadow (S_2), becomes the height (H/phala) [of the object]."
- **Formulas:** Koṭī = $(S_2 \times \Delta D) / |S_1-S_2|$; $H = (G \times (\text{Koṭī} + S_2)) / S_2$.

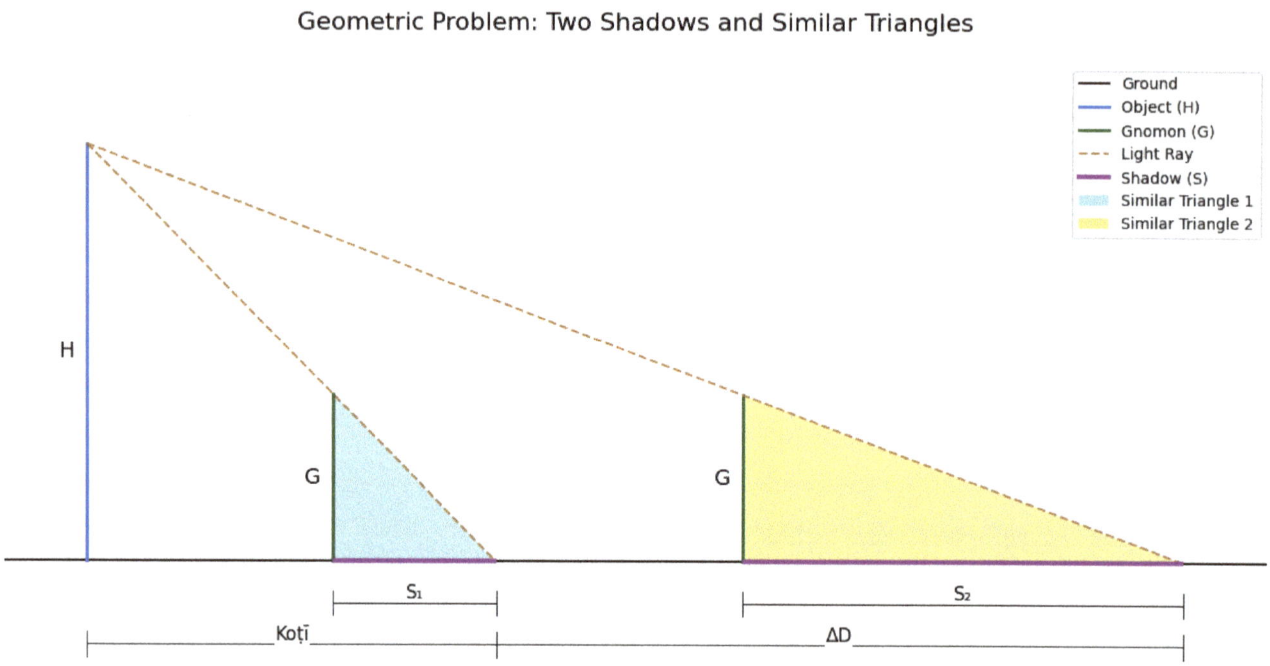

Figure 4.15: Geometric diagram illustrating the problem of determining the height of an object (or light source, H) and its distance (Koṭī) using two Gnomons (Śaṅku) of known height (G) and their corresponding shadow lengths (S_1, S_2). The setup relies on the principle of similar triangles (highlighted as Triangle 1 and Triangle 2). This method is described by Aryabhata in Gaṇitapāda Verse 16 (Gp.16).

Interpretation:

Aryabhata now tackles a classic and practical challenge: how to determine the height of a tall object and its distance from an observer, using only a gnomon of known height and the ground upon which it stands. The verse outlines an ingenious solution that relies on the elegant power of similar triangles, requiring two separate shadow measurements.

The Method

The procedure involves taking two shadow measurements from a single gnomon (G), typically by moving the gnomon along a line toward the object. Let us define the key terms:

- ***G***: The height of the gnomon (*śaṅku*).
- S_1 and S_2: The lengths of the first and second shadows.
- ΔD: The distance between the tips of the two shadows (*chāyāgravivaram*).
- ***Koṭī***: The horizontal distance from the tip of the second shadow (S_2) to the base of the tall object.
- ***H*** (*Phala*): The height of the inaccessible object.

1. Finding the Distance (*Koṭī*)

The verse first describes how to find the horizontal distance from the second observation point to the object's base. The instruction is to multiply the second shadow's length (S_2) by the distance between the shadow-tips (ΔD), and then divide this product by the difference between the shadow lengths. This translates to the formula:

$$\text{Koṭī} = (S_2 \times \Delta D) / |S_1 - S_2|$$

The verse's instruction to divide by the *ūnena* ("the lesser" or "the difference") implies using the magnitude of the difference, ensuring the formula works regardless of which shadow is longer.

2. Finding the Height (*Phala*)

Once the *koṭī* is determined, the height of the object (*H*, or *phala*) is found through another direct application of similar triangles. The verse explains that this *koṭī* (plus its corresponding shadow length) relates to the object's height in the same proportion as the gnomon's shadow relates to the gnomon's height. The formula is:

$$H = (G \times (\text{Koṭī} + S_2)) / S_2$$

This calculation effectively scales up the known dimensions of the small gnomon-shadow triangle to find the unknown height of the large object-distance triangle.

Significance

This method is a powerful example of practical geometry in action, demonstrating how to measure inaccessible heights and distances with the simplest of tools. A classic problem in the history of applied mathematics, this technique was likely used in surveying, architectural planning, and military reconnaissance. It showcases a sophisticated application of geometric principles, transforming simple observations into a precise understanding of the world's dimensions.

4.5.4 Pythagorean Theorem Stated

Sanskrit:

यश् चैव भुजावर्गः कोटिवर्गश् च कर्णवर्गः सः ।

वृत्ते शरसंवर्गोऽर्धज्यावर्गः स खलु धनुषोः ॥ १७॥

IAST:

yaś chaiva bhujāvargaḥ koṭivargaś cha karṇavargaḥ saḥ|

vṛtte śarasaṁvargo'rdhajyāvargaḥ sa khalu dhanuṣoḥ || 17 ||

Notes:

- **Yaś chaiva bhujāvargaḥ koṭivargaś cha karṇavargaḥ saḥ:** "Square of base (bhuja^2) + square of perpendicular (koṭi^2) = square of hypotenuse (karṇa^2)." ($a^2 + b^2 = c^2$).
- **Vṛtte śarasaṁvargo'rdhajyāvargaḥ sa khalu dhanuṣoḥ:** "In a circle, product involving arrow (s(D-s)) = square of half-chord (h^2)." (Reiteration of Gp.12).

Interpretation:

This concise verse serves to formally state two cornerstone theorems of geometry. In codifying these principles, Aryabhata is not merely listing facts; he is cementing the logical foundations of his entire mathematical edifice.

The Pythagorean Theorem

First, the verse gives a clear and direct statement of the theorem governing right-angled triangles: "Indeed, the square of the base (*bhujāvargaḥ*) plus the square of the perpendicular (*koṭivargaḥ*) is the square of the hypotenuse (*karṇavargaḥ*)."

For a right-angled triangle with sides *a* (base, *bhuja*) and *b* (perpendicular, *koṭi*), and hypotenuse *c* (*karṇa*), this translates to the universal principle:

$$\mathbf{a^2 + b^2 = c^2}$$

This theorem, whose knowledge in India dates back to the *Śulba Sūtras*, is the engine behind countless calculations in the *Aryabhatiya*, from finding the *karṇa* of a gnomon's shadow (Gp.14) to the fundamental constructions of trigonometry.

The Chord-Arrow Relationship Revisited

Next, Aryabhata reconfirms the fundamental relationship for circular segments: "In a circle, the product involving the arrow is indeed the square of the half-chord of the [corresponding] arc."

The term *śarasaṁvargaḥ*, or "product involving the arrow," concisely denotes the product of the arrow (s) and the remaining portion of the diameter (D-s). The statement thus reiterates the pivotal theorem from Gp.12:

$$\mathbf{h^2 = s(D - s)}$$

This relationship is the lynchpin of Aryabhata's trigonometry. It allows for the interconversion between linear measures within a circle and the angular quantities they represent, a process essential for astronomical models and eclipse calculations.

Significance

Verse 17 serves as a powerful affirmation of the indispensable truths upon which Aryabhata's system is built. The Pythagorean theorem governs the world of right triangles, which are essential for all gnomon-based and trigonometric problems. The chord-arrow relationship unlocks the geometry of the circle, which is the key to measuring the heavens. Together, these two pillars support a framework that bridges the geometry of the Earth with the mechanics of the cosmos.

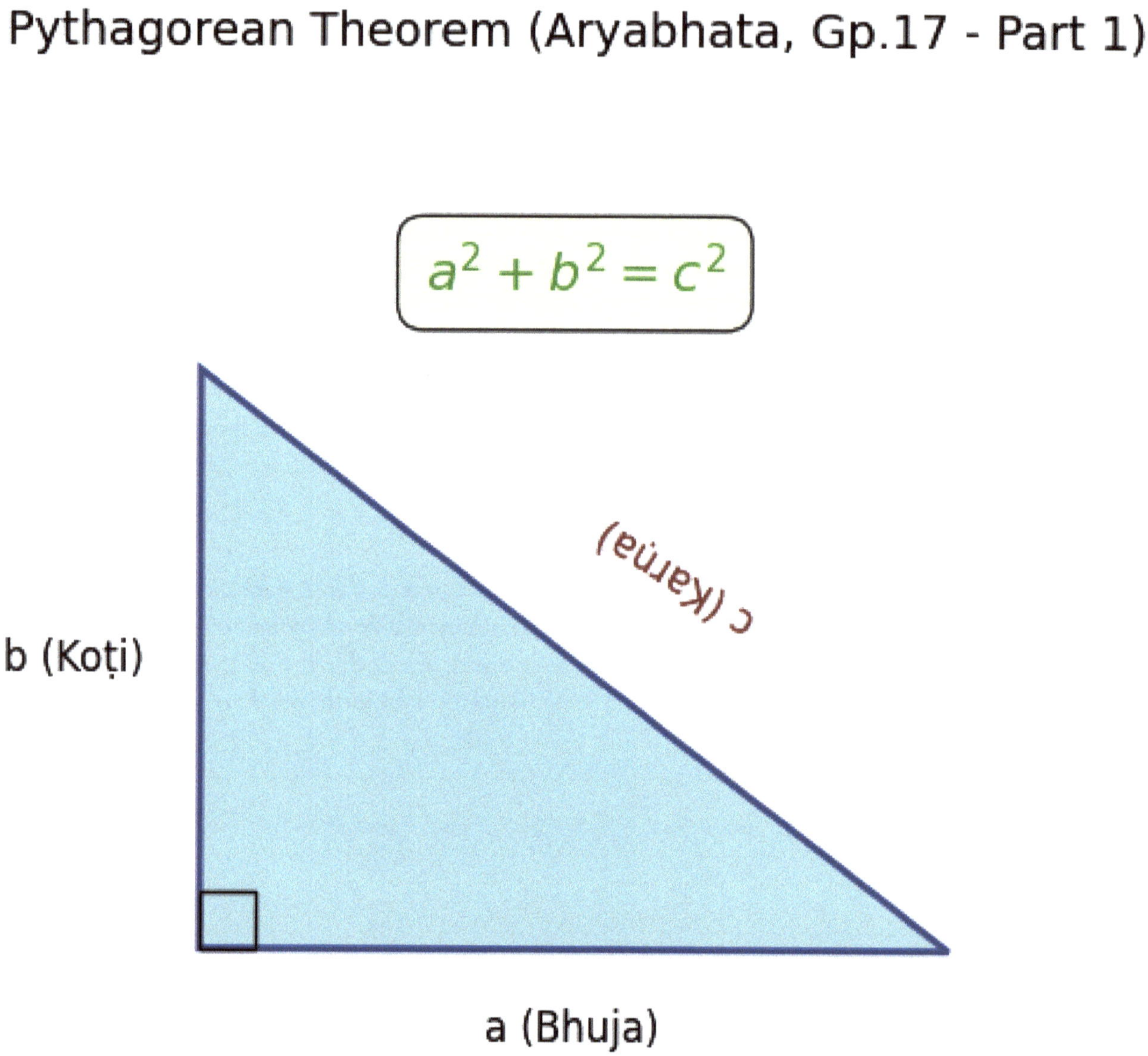

Figure 4.16: The Pythagorean Theorem as Stated in Gp.17 Aryabhata's Gaṇitapāda Verse 17 first affirms the fundamental geometric relationship for a right-angled triangle: the sum of the squares of the base (Bhuja, 'a') and the perpendicular (Koṭi, 'b') is equal to the square of the hypotenuse (Karṇa, 'c'), i.e.,$a^2 + b^2 = c^2$ This theorem is foundational to many calculations in the Aryabhatiya.

4.5.5 Area of Intersecting Circles Segments

Sanskrit:

ग्रासोनयोर् व्यासगुणं शरार्धं मिश्रं तयोर् मिश्रधनूषु चिह्नम् ।

एवम् परस्परगुणौ विष्कम्भार्धभक्तफलम् इष्टम् ॥ १८ ॥

IAST:

grāsonayor vyāsaguṇaṃ śarārdhaṃ miśraṃ tayor miśradhanūṣu chihnam|

evam parasparaguṇau viṣkambhārḍhabhaktaphalam iṣṭam || 18 ||

Interpretation (Conceptual):

Following his method for calculating the depth of an eclipse (Gp.13), Aryabhata now tackles an even more complex geometric problem: calculating the *area* of the lens-shaped region, or "lune," formed when two circles intersect. While knowing the depth (*śara*) is essential for an eclipse's magnitude, knowing the area provides a far more nuanced understanding of the event's visual appearance and the total amount of light obscured.

Decoding a Dense Verse

Verse 18 is exceptionally condensed and has posed a famous challenge to commentators and modern scholars alike. Its terse, technical phrasing—a cascade of terms like *grāsonayor vyāsaguṇaṃ śarārdhaṃ miśraṃ*—resists a simple, step-by-step translation into a single modern formula. This obscurity is not a sign of confusion, but rather an indication of the problem's complexity.

Reconstructing the Likely Method

It is highly probable that the verse outlines a sophisticated algorithm for finding the area of the overlapping lune, a procedure that would build upon the results from the preceding verses. The methodology likely involved:

1. **Calculating the Segment Heights:** First, obtaining the *śara* values for each circle using the method from Gp.13.
2. **Finding the Area of Each Segment:** Next, calculating the area of the two circular segments that form the lune. This was a non-trivial problem in itself, likely involving the segment heights (*śaras*) and their corresponding half-chords.
3. **Summing the Areas:** The total area of the lune would be the sum of these two segment areas.

Aryabhata's concluding phrase—"thus, by mutual products and divisions involving half-diameters, the desired result is obtained"—strongly suggests a systematic, multi-step computational procedure rather than a single, elegant formula.

Astronomical and Mathematical Significance

Verse 18, despite its interpretive challenges, points toward Aryabhata's deep engagement with advanced geometric problems crucial for a comprehensive analysis of eclipses. It stands as a testament to the depth of reasoning pursued in his school, pushing beyond simple metrics to quantify complex, two-dimensional shapes. While the precise operational details require careful reconstruction, the verse's very existence demonstrates a commitment to solving the most difficult geometric problems demanded by a sophisticated and predictive astronomical science.

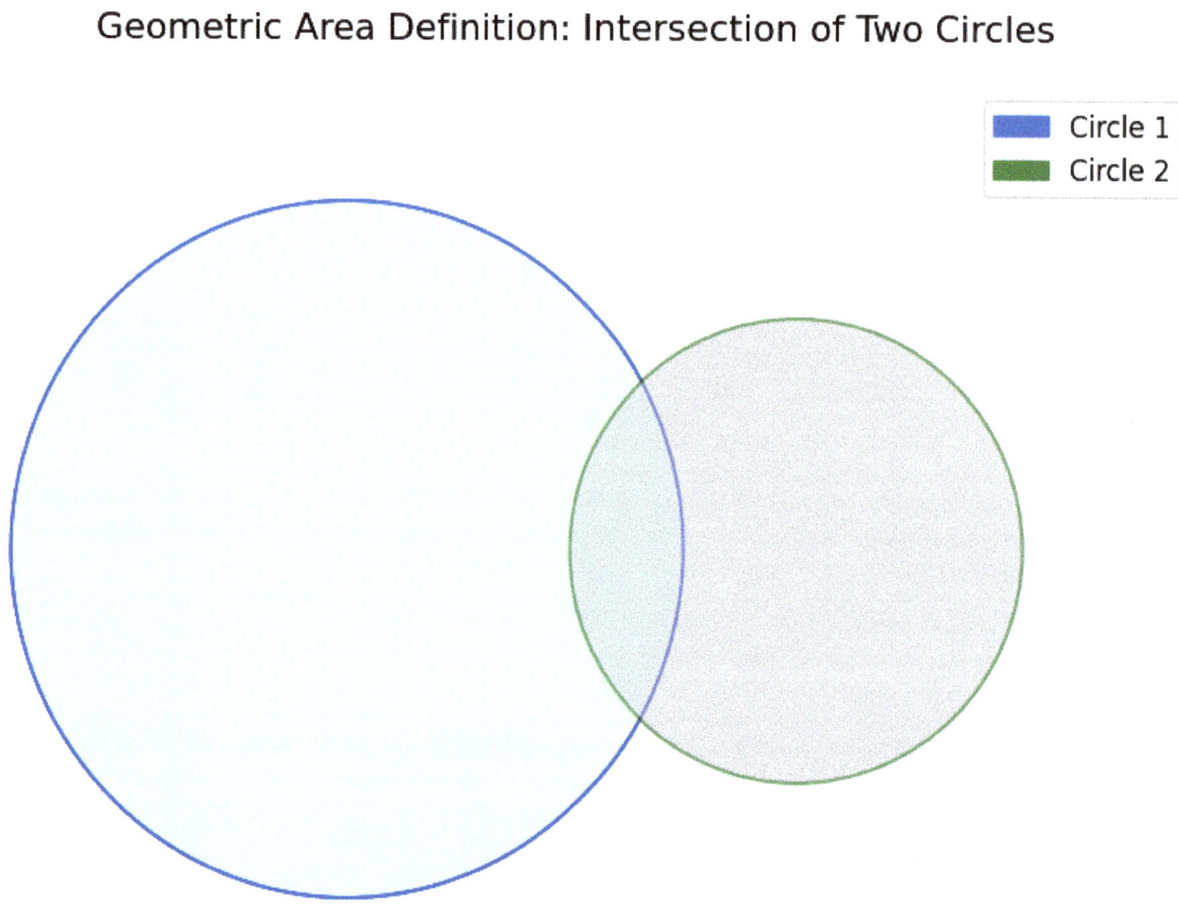

Figure 4.17: Aryabhata's Gp.18: Area of Intersecting Circles. Illustrates calculating this area using chord/segment principles (Gp.9, 12-13), vital for determining eclipse obscuration.

4.6 Sequences, Proportions, and Foundational Algebra (Gp.19-31)

Having laid down principles for basic arithmetic, geometry, and root extraction, Aryabhata shifts focus in Gaṇitapāda Verses 19 through 31 to a suite of indispensable algebraic and computational tools. This section explores arithmetic progressions (*śreḍhī-vyavahāra*), methods for summing series (including sums of squares and cubes), fundamental algebraic identities, the cornerstone Rule of Three (*trairāśika*) for proportions and its variations, rules for mixtures, and calculations related to interest. These techniques were not abstract mathematical exercises; they were vital for organizing astronomical data (such as the regularly changing sine differences or planetary positions over time), calculating cumulative effects, solving practical commercial problems that were part of the era's applied mathematics, and providing intermediate steps within more complex astronomical algorithms.

4.6.1 Mastering Arithmetic Progressions (AP)

Sanskrit:

इष्टं व्येकं दलितं सपूर्वम् उत्तरगुणं समुखमध्यमम् ।

इष्टगुणितम् इष्टधनं त्वथवाऽऽद्यन्तौ पदार्धहतम् ॥ १९ ॥

IAST:

iṣṭaṁ vyekaṁ dalitaṁ sapūrvaṁ uttaraguṇaṁ samukhamadhyamam|

iṣṭaguṇitaṁ iṣṭadhanaṁ tvathavā ”dyantau padārdhahatam || 19 ||

Translation & Explanation: "The desired number [of terms] (iṣṭaṁ) minus one (vyekaṁ), halved (dalitaṁ), multiplied by the common difference (uttaraguṇaṁ), added to the first term (samukham) yields the middle term value (madhyamam). Multiplied by the desired number [of terms] (iṣṭaguṇitaṁ) is the sum (iṣṭadhana); or, the first and last terms [summed] (ādyantau) multiplied by half the number of terms (padārdhahatam)."

The Sum of a Series: Mastering Arithmetic Progression

How can you find the sum of a long series of numbers, like 1 + 3 + 5 + ... + 99, without the tedious work of adding each one? In this verse, Aryabhata presents two elegant and efficient methods for summing an arithmetic series—a sequence of numbers where the difference between each term is constant.

His instructions read like a concise recipe: "The desired number of terms minus one, halved, multiplied by the common difference, and added to the a first term, gives the middle value. This multiplied by the number of terms is the sum. Or, the sum of the first and last terms, multiplied by half the number of terms, gives the total."

Let's unpack these two powerful techniques.

Method 1: The Middle Way

Aryabhata's first method is based on a simple, intuitive idea: the sum of a balanced series is just the average value multiplied by the number of items. Here, the "average value" is the value of the term exactly in the middle of the series.

To find this "middle value" (*madhyamam*), Aryabhata gives a precise formula:

Middle Value = First Term + [(Number of Terms – 1) / 2] × Common Difference

Once you have this middle value, the total sum is easy:

Sum = Middle Value × Number of Terms

This method is logical and always works, providing a reliable way to find the sum by first finding the series's central point.

Method 2: The Elegant Shortcut

Aryabhata's second method is a brilliant shortcut that is still the standard formula taught in mathematics classes today. The insight here is that you can find the average value of the entire series by simply averaging the very first and very last terms.

Average Value = (First Term + Last Term) / 2

Once you have this average, the total sum is found just as before:

Sum = Average Value × Number of Terms

This can be written in its most famous form:

Sum = (Number of Terms / 2) × (First Term + Last Term)

This approach is incredibly efficient, as it only requires knowing the beginning, the end, and the total count of terms in the series.

Why This Mattered for Astronomy

This was no mere textbook exercise for Aryabhata. The ability to quickly sum a series was a fundamental tool for an astronomer. Any quantity that changes by a fixed amount over regular time intervals—a common occurrence in the heavens—forms an arithmetic progression.

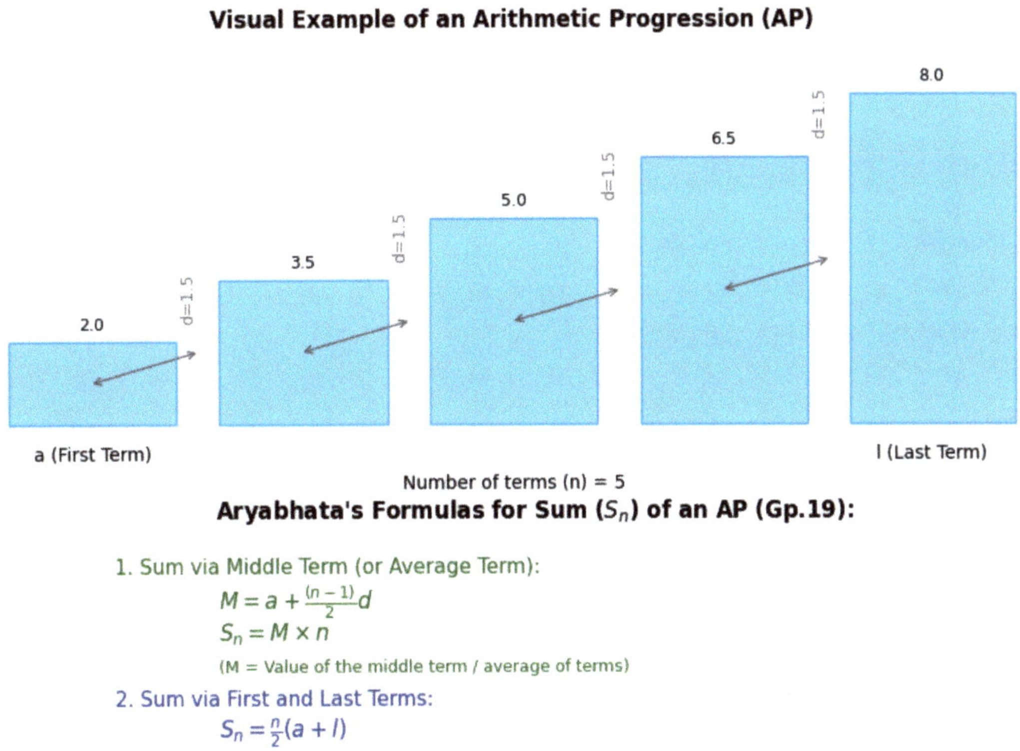

Figure 4.18: Aryabhata's Arithmetic Progression Sum (Gp.19). Visualizes an AP (first term a, common difference d, last term l, n terms) with Aryabhata's two methods from Gaṇitapāda Verse 19 for its sum (S_n): (1) middle term × n, and (2) (n/2) × (a + l)

4.6.2 Sums of Series (Triangular Numbers)

Sanskrit:

सैकपदगच्छगुणिताद् गच्छाद् द्विगुणात् पदत्रिवर्गाप्तम् ।

षड्भक्तं चितिघनसा कृतिचितिघनः स्यात् पदघनानाम् ॥ २० ॥

IAST:

saikapadagacchaguṇitād gacchād dviguṇāt padatrivargāptam|

ṣaḍbhaktaṁ chitighanasā kṛtichitighanaḥ syāt padaghanānām || 20 ||

Beyond the Basics: Summing a Series of Series

Having mastered the simple arithmetic series, Aryabhata now delves into a more complex and abstract challenge: what if you want to sum the sums themselves? This is like moving from a simple line of bricks to calculating the total number of bricks in a layered, triangular wall.

The Sanskrit verse addressing this is extremely condensed, a common feature of ancient mathematical texts. The clues, however, are clear enough for us to reconstruct his method.

Decoding the "Sum-Pile"

The verse contains two critical clues: the term ***chitighana***, which translates to "sum-pile," and the instruction to divide by six, ***ṣaḍbhaktaṁ***.

A "sum-pile" is a beautiful visual for what we now call the sum of **triangular numbers**. A triangular number is the sum of all whole numbers up to a certain point. For example:

- The 1st triangular number is 1.
- The 2nd is 1 + 2 = 3.
- The 3rd is 1 + 2 + 3 = 6.
- The 4th is 1 + 2 + 3 + 4 = 10.
  ...and so on.

You can picture these as triangular stacks of stones or cannonballs. A *chitighana* is the total number of stones in a stack of these triangular layers—a pyramid.

Aryabhata's formula for this "sum-pile" is elegantly confirmed by his instruction to "divide by six." This points directly to the classic formula for summing the first 'n' triangular numbers:

Sum of a "Pile" = [n × (n + 1) × (n + 2)] / 6

For example, to find the sum of the first four triangular numbers (1 + 3 + 6 + 10), we can use Aryabhata's formula with n=4:

Sum = [4 × (4 + 1) × (4 + 2)] / 6 = (4 × 5 × 6) / 6 = 120 / 6 = 20

This matches our manual sum of 1 + 3 + 6 + 10 = 20.

A Glimpse of Deeper Mathematics

The verse also contains a cryptic phrase that seems to hint at the relationship between sums of squares and sums of cubes. This appears to be a deliberate teaser for an even more advanced topic that Aryabhata addresses in a later verse, demonstrating the logical structure and depth of his work.

Significance: From Practical to Pure Mathematics

These formulas for higher-order series represent a significant leap in mathematical abstraction. While a simple series might be used to calculate total travel over several days, summing a series of series is a step into pure number theory. It shows an interest in the fundamental properties of numbers themselves, not just their application to immediate, real-world problems.

This ability to analyze complex, cumulative effects was essential for advanced mathematical thought and laid the groundwork for understanding quantities that change in non-linear or accelerating ways—a cornerstone of higher mathematics and physics.

Understanding 'Chitighana' and 'Ṣaḍbhaktaṁ' in Gp.20

Visualizing Triangular Numbers ($T_k = k(k+1)/2$)

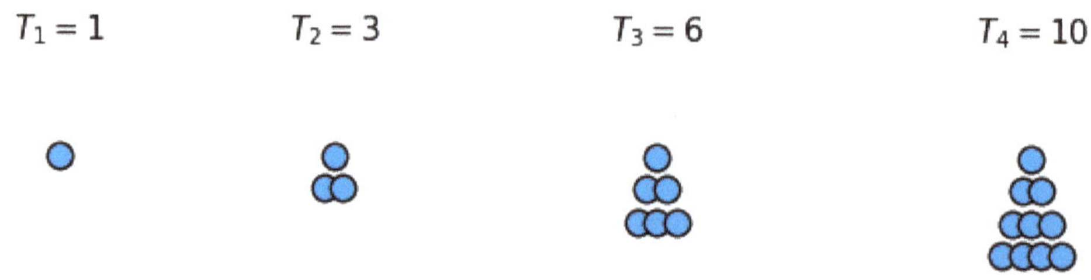

Example: $T_1 = 1, T_2 = 3, T_3 = 6, T_4 = 10$

Aryabhata's Formula for Chitighana (Sum of Triangular Numbers - Gp.20 Interpretation)

The term 'chitighana' (sum-pile) and the mention of 'ṣaḍbhaktaṁ' (divided by six) in Gp.20 strongly suggest a formula for the sum of the first 'n' triangular numbers (S_{T_n}):

$$S_{T_n} = \sum_{k=1}^{n} T_k = \sum_{k=1}^{n} \frac{k(k+1)}{2} = \frac{\mathbf{n(n+1)(n+2)}}{\mathbf{6}}$$

Example for n=3:

$S_{T_3} = T_1 + T_2 + T_3 = 1 + 3 + 6 = 10$

Using formula: $S_{T_3} = \frac{3(3+1)(3+2)}{6} = \frac{3 \times 4 \times 5}{6} = \frac{60}{6} = 10$

Figure 4.19: Aryabhata's Gp.20: Sum of Triangular Numbers (Chitighana). Interprets Gp.20 (via 'chitighana', 'ṣaḍbhaktaṁ') for summing triangular numbers (S_Tn = n(n+1)(n+2)/6), with visual examples of T_k, showcasing his higher-order series work.

4.6.3 Finding the Number of Terms in an Arithmetic Progression

Sanskrit:

गच्छोऽष्टोत्तरगुणिताद् द्विगुणाद्युत्तरविशेषवर्गयुतात् ।

मूलं द्विगुणाद्यूनं स्वोत्तरभजितं सरूपार्धम् ॥ २१ ॥

IAST:

gaccho 'ṣṭottaraguṇitād dviguṇādyuttaraviśeṣavargayutāt|

mūlaṁ dviguṇādyūnaṁ svottarabhajitaṁ sarūpārdham || 21 ||

Working Backwards: Finding the Number of Terms

We have seen how to find the sum of a series if we know the number of terms. But Aryabhata now tackles a far more difficult question: what if you know the final sum, but you don't know how many terms it took to get there?

For example, if you know a planetary body moved a total distance of 1,000 units, and you know its starting speed and daily acceleration, how many days did its journey take? This "reverse problem" is a significant leap in mathematical complexity. It requires solving what we now call a **quadratic equation**.

A Verbal Recipe for the Quadratic Formula

Aryabhata's verse is a stunning demonstration of algebraic mastery. He provides a step-by-step recipe to find the number of terms (n) when the sum (S), the first term (a), and the common difference (d) are known. His verbal instructions are a precise match for the modern quadratic formula.

The standard formula for the sum of a series is:

$$\mathbf{S = (n/2) \times [2a + (n-1)d]}$$

When you rearrange this to solve for n, you get a quadratic equation:

$$\mathbf{dn^2 + (2a - d)n - 2S = 0}$$

To solve this equation for n is no simple task. Yet, Aryabhata's verbal instructions lay out the exact path, which is equivalent to the well-known solution:

$$\mathbf{n = [\sqrt{(2a - d)^2 + 8dS} - (2a - d)] / 2d}$$

His verse essentially says:

1. Multiply the sum (**S**) by eight times the common difference (**8d**).
2. Add this to the square of the quantity "**twice the first term minus the difference**" (i.e., $(2a-d)^2$).
3. Take the **square root** of that entire result.
4. From this, subtract the quantity "**twice the first term minus the difference**" (2a-d).
5. Finally, **divide** the result by **twice the common difference** (2d).

A Leap in Algebraic Thinking

This is a profound achievement. Aryabhata is not just manipulating a formula; he is providing a universal algorithm to solve a class of quadratic equations. He accomplished this without the convenience of modern symbolic algebra (like *x*, *y*, or *n*), instead encoding the entire complex procedure in a compact and perfectly ordered set of verbal instructions.

This ability was crucial for moving beyond simple calculation to genuine problem-solving. It allowed an astronomer or mathematician to determine the unknown *duration* or *number of events* in a process, a fundamental requirement for any predictive science.

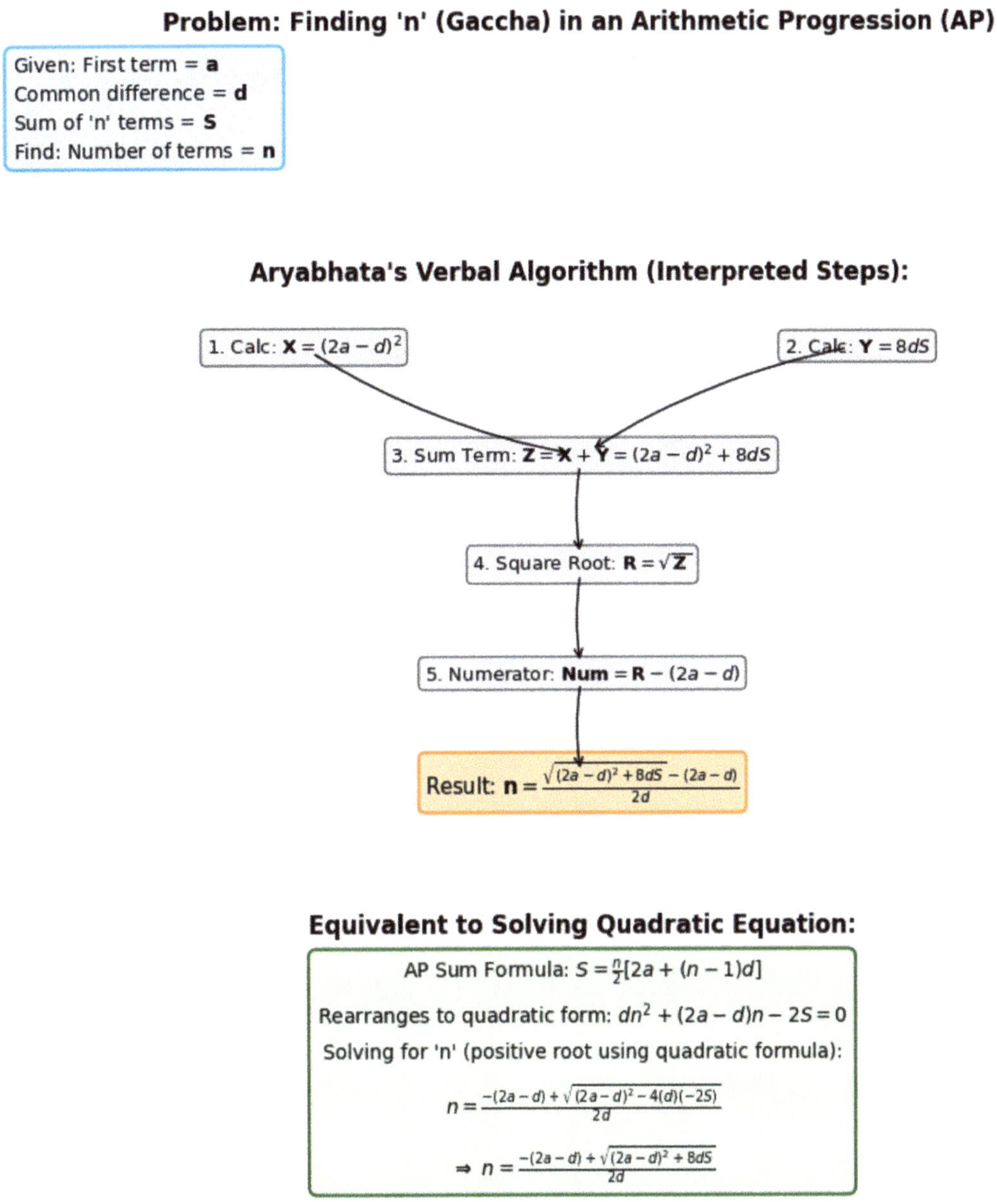

Figure 4.20: Aryabhata's Method for 'n' in an AP (Gp.21). Illustrates finding an AP's number of terms ('n' or 'gaccha') given its sum (S), first term (a), and common difference (d). It visually deconstructs Aryabhata's verbal steps (Gp.21), showing their mathematical equivalence to applying the quadratic formula to the AP sum equation for 'n'.

4.6.4 Sums of Squares and Cubes

Sanskrit:

एकोत्तराद्युपचितेर्गलितं पदवर्गसंगुणं पदयुक् ।

षड्भक्तं तत्कृतिचितिघनः स चितिवर्गश्चैकपदात् ॥ २२ ॥

IAST:

ekottarādyupacitergalitaṁ padavargasaṁguṇaṁ padayuk|

ṣaḍbhaktaṁ tatkṛtichitighanaḥ sa citivargaścaikapadāt || 22 ||

The Harmony of Squares and Cubes

Having laid the foundations of arithmetic series, Aryabhata now reaches the summit, revealing two famous and beautiful formulas for summing more complex progressions: the sum of squares and the sum of cubes. The Sanskrit verses are characteristically dense, but their clues point to a deep understanding of the hidden architecture of numbers.

1. The Sum of Squares: A "Square-Pile"

First, Aryabhata provides a method for summing the squares of the first 'n' whole numbers ($1^2 + 2^2 + 3^2 + ...$). He uses the term ***kṛti-chitighana***, which can be thought of as a "sum-pile of squares."

The crucial clue in his verse is the phrase ***ṣaḍbhaktaṁ***—"divided by six." This instruction is the key that unlocks the formula, which we recognize today as the standard for the sum of squares:

Sum of Squares = [n × (n+1) × (2n+1)] / 6

This formula provides a powerful shortcut to finding the sum of what would otherwise be a series of tedious calculations.

2. The Sum of Cubes: A Hidden Masterpiece

Next, Aryabhata presents a true gem of number theory. He describes the result for summing the cubes of the first 'n' numbers ($1^3 + 2^3 + 3^3 + ...$) with the elegant phrase ***citivarga***—"the square of the sum-pile."

This means something astonishing: **The sum of the cubes of a series of numbers is the same as the square of their simple sum.**

Let's test this incredible claim with the first four numbers (1, 2, 3, 4):

- **The sum of their cubes:**

$1^3 + 2^3 + 3^3 + 4^3 = 1 + 8 + 27 + 64 = \mathbf{100}$

- **The simple sum of the numbers:**
  $1 + 2 + 3 + 4 = \mathbf{10}$

- **The square of that sum:**
  $10^2 = \mathbf{100}$

The results match perfectly. This beautiful identity, known today as Nicomachus's Theorem, is expressed in modern notation as:

$$\textbf{Sum of Cubes} = [\, n \times (n+1) / 2 \,]^2$$

Significance: From Tools to Treasures

Why were these formulas so important? Unlike the rule for finding the area of a triangle, the direct, practical applications of summing squares and cubes are less immediately obvious. This is a different kind of mathematics.

These formulas represent a leap from practical calculation into the realm of **pure number theory**. The earlier methods were tools, like a carpenter's saw and hammer, for solving problems in the physical world. These formulas are treasures, like a perfectly cut jewel, admired for their intrinsic beauty and symmetry.

This fascination with the fundamental properties of numbers themselves demonstrates a high level of mathematical abstraction. It shows an intellectual curiosity that goes beyond "what works" to ask "what is true and beautiful." These concepts became foundational in the history of mathematics and were essential for the later development of more complex series, such as those used in the advanced calculus of the Kerala School centuries later.

Aryabhata's Formulas for Sums of Squares and Cubes (Gp.22)

A: Sum of the First 'n' Squares (S_{n^2} or Kṛti-citighana)

Series: $1^2 + 2^2 + 3^2 + \ldots + n^2$

Aryabhata's Formula (Interpreted):

$$S_{n^2} = \frac{n(n+1)(2n+1)}{6}$$

(Corresponds to Gp.22 phrase '...divided by six (ṣaḍbhaktaṁ)...')

Example for n=3:

$1^2 + 2^2 + 3^2 = 1 + 4 + 9 = 14$

Formula: $S_{3^2} = \frac{3(3+1)(2 \cdot 3 + 1)}{6} = \frac{3 \cdot 4 \cdot 7}{6} = 14$

B: Sum of the First 'n' Cubes (S_{n^3})

Series: $1^3 + 2^3 + 3^3 + \ldots + n^3$

Aryabhata's Relation (Interpreted from '...citivargaścaikapadāt...'):

Sum of first 'n' integers: $S_n = \frac{n(n+1)}{2}$

$$S_{n^3} = \left(\frac{n(n+1)}{2}\right)^2 = (S_n)^2$$

Example for n=3:

$1^3 + 2^3 + 3^3 = 1 + 8 + 27 = 36$

Sum of integers $S_3 = \frac{3(3+1)}{2} = 6$

Formula: $S_{3^3} = (S_3)^2 = 6^2 = 36$

Figure 4.21: Aryabhata's Formulas for Sums of Squares and Cubes (Gp.22)

4.6.5 Algebraic Problems – Solving for Two Unknowns

Sanskrit:

सङ्कलितं द्विगुणं वर्गेणान्तरयुतेन संभजेत् ।

द्विगुणं योगविहीनं वर्गान्तरभक्तमन्तरं ज्ञेयम् ॥ २३ ॥

IAST:

saṅkalitaṁ dviguṇaṁ vargeṇāntarayutena saṁbhajet|

dviguṇaṁ yogavihīnaṁ vargāntarabhaktam antaraṁ jñeyam || 23 ||

Solving for the Unknown: An Algebraic Puzzle

Imagine you are given a mathematical puzzle. You have two secret numbers, but instead of being told their values, you are only given clues about them, such as:

- Their sum and their product.
- Their sum and their difference.
- The sum of their squares and their difference.

How do you find the original numbers?

In this verse, Aryabhata provides a powerful and systematic method for solving such problems. His approach reveals a deep understanding of algebra and relies on a single, brilliant insight.

The Master Key: Finding the Sum and Difference

Aryabhata's central strategy is a "master key" to unlocking any of these puzzles: if you can determine just two simple values—the **sum** of the numbers (let's call it $\mathbf{S = x + y}$) and their **difference** ($\mathbf{D = x - y}$)—the individual numbers can be found instantly.

Once these two keys are found, the solution is straightforward:

- The larger number is $\mathbf{(S + D) / 2}$
- The smaller number is $\mathbf{(S - D) / 2}$

The logic is simple and elegant. Adding the sum (x+y) and the difference (x-y) together gives (x+y) + (x-y) = 2x. Halving this result reveals x. Similarly, subtracting the difference from the sum gives (x+y) – (x-y) = 2y, which is then halved to find y.

An Algebraic Toolkit

The true genius of Aryabhata's method lies in how he finds the sum (S) and difference (D) when they aren't given directly. He uses fundamental algebraic identities as a flexible toolkit to transform the given clues into the "master keys" he needs.

His method relies on the interplay between identities like:

- $(x + y)^2 = x^2 + 2xy + y^2$
- $(x - y)^2 = x^2 - 2xy + y^2$

These identities act like conversion tools. For example, if he knew the sum of the numbers (x+y) and the sum of their squares (x^2+y^2), he could use the first identity to quickly find their product (xy). Once he had the product, he could use the second identity to find the difference (x-y), thus obtaining both "master keys" needed to solve the puzzle.

Significance: The Foundation of Solving Systems

This is far more than a clever trick for solving riddles. It demonstrates a complete command of algebraic manipulation and forms the bedrock of solving what we now call **systems of equations**. This ability to work with multiple interacting variables and solve for several unknowns at once is essential for any advanced science, from modeling the complex motions of the planets to countless applications in modern physics and engineering.

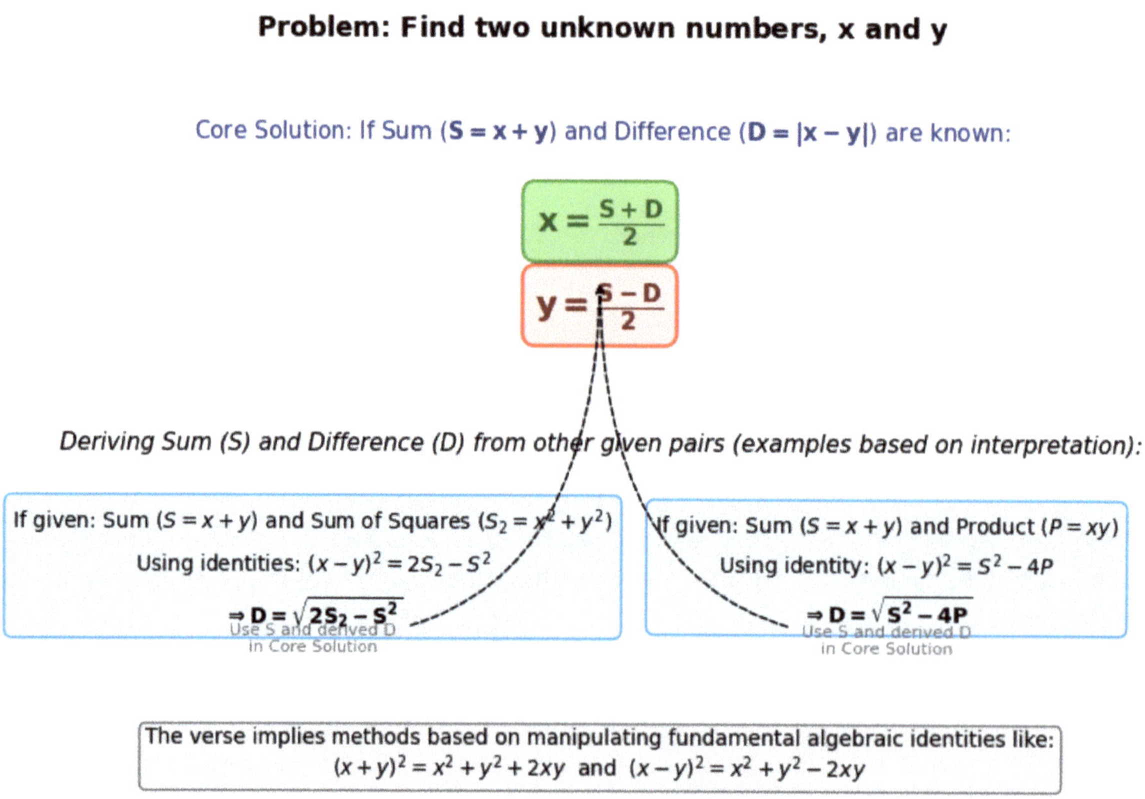

Figure 4.22: Aryabhata's Gp.23: Solving for two unknowns (x, y) using their sum (S) and difference (D), with methods to derive D from other given pairs (e.g., S & product, or S & sum of squares).

4.6.6 Finding Factors (Implicit Quadratic Solution)

Sanskrit:

मूलं द्विगुणं प्रक्षेप्य वर्गं वर्गेण संयुतम् ।

गुणकारान्तरभक्तं द्वेधा लब्धं गुणकारौ ॥ २४ ॥

IAST:

mūlaṁ dviguṇaṁ prakṣipya vargaṁ vargeṇa saṁyutam|

guṇakārāntarabhaktaṁ dvedhā labdhaṁ guṇakārau || 24 ||

A Recipe for Solving Riddles: Finding Numbers from their Product and Difference

Here is another classic algebraic puzzle: you are told the difference between two secret numbers and also what they multiply to. How can you find them?

For example, imagine two numbers whose difference is 5 and whose product is 36. You could guess and check, but Aryabhata provides a direct and flawless recipe that works every time.

His verbal instructions lay out the following steps:

1. **Square the Difference (D):** First, take the difference between the two numbers and square it.
   - In our example: $5^2 = 25$
2. **Add Four Times the Product (P):** Next, add four times the product of the two numbers to your result.
   - In our example: $25 + (4 \times 36) = 25 + 144 = 169$
3. **Take the Square Root:** Now, find the square root of this new sum. This number is your "key."
   - In our example: $\sqrt{169} = 13$
4. **Add and Subtract the Difference:** To find the two original numbers, simply add and subtract the original difference (D) from your key value.
   - Larger number = 13 + 5 = 18
   - Smaller number = 13 – 5 = 8

5. **Halve the Results:** Finally, take half of each result. These are your two secret numbers.

   - Larger number = 18 / 2 = **9**
   - Smaller number = 8 / 2 = **4**

We can check our answer: the product (9 × 4) is 36, and the difference (9 – 4) is 5. The recipe works perfectly.

The Deeper Mathematics: Solving a Quadratic Equation

This elegant recipe is more than just a clever trick; it is a brilliant method for solving a quadratic equation without ever writing it down. Finding two numbers from their difference (D) and product (P) is mathematically equivalent to finding the roots of the equation:

$$\mathbf{z^2 - Dz - P = 0}$$

Aryabhata's verbal recipe is a step-by-step implementation of the quadratic formula, tailored to this specific problem. In modern terms, his solution is:

- Larger Number = $[\, D + \sqrt{(D^2 + 4P)}\,] / 2$
- Smaller Number = $[\, -D + \sqrt{(D^2 + 4P)}\,] / 2$
- This kind of problem-solving was essential for practical geometry, such as finding the length and width of a rectangular field when its area (the product) and the difference between its sides were known. It stands as another powerful testament to the sophisticated level of algebraic skill present in Aryabhata's time.

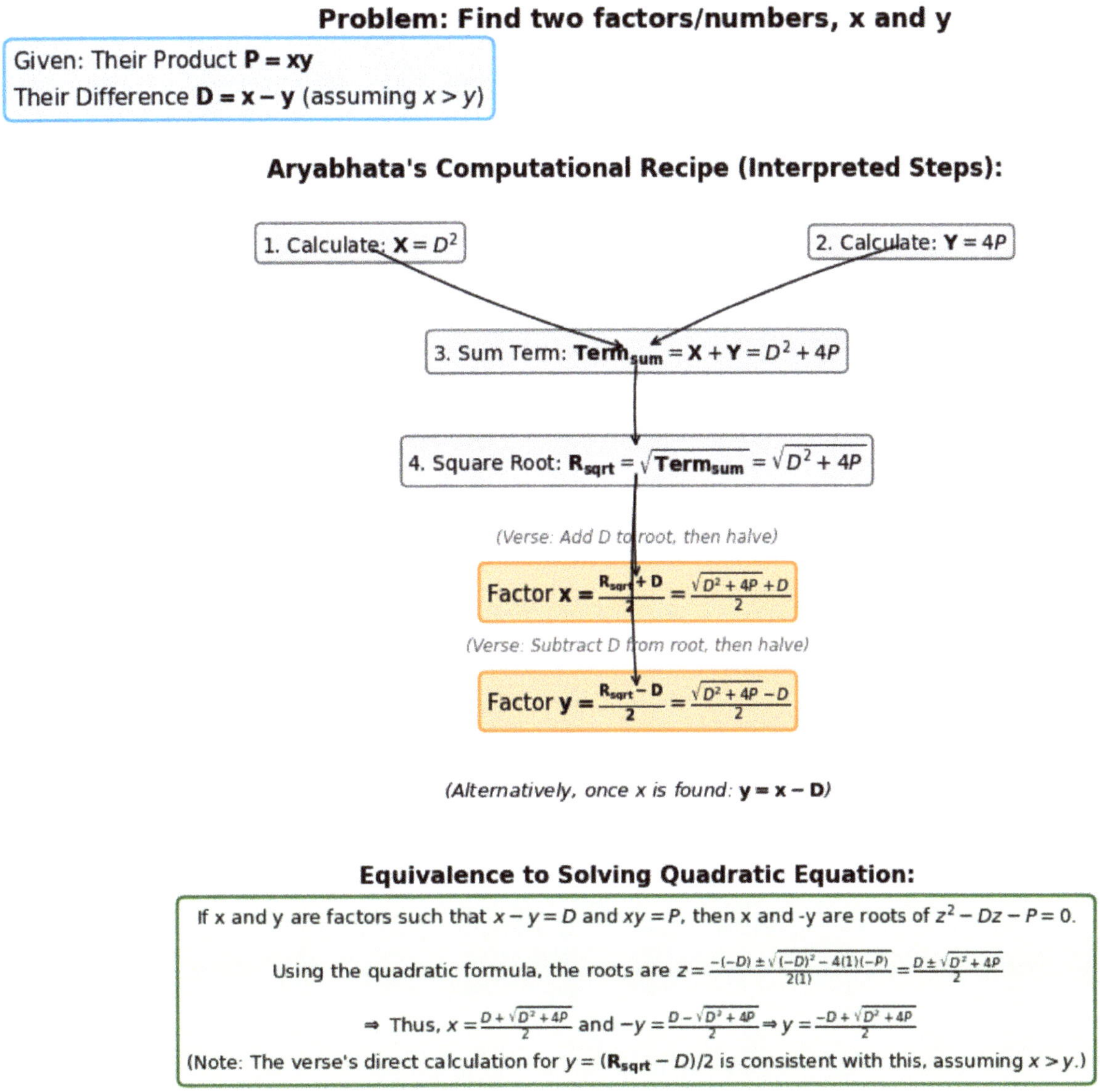

Figure 4.23: Aryabhata's Gp.24: Finding two factors (x, y) from their product (P) and difference (D). Illustrates the computational steps involving an intermediate square root, and its equivalence to solving a quadratic equation.

4.6.7 Rule for Interest Calculations

Sanskrit:

मूलफलं सफलं कालमूलगुणम् अर्धमूलकृतियुतम् ।

मूलं मूलार्धोनं कालहृतं स्यात् फलं शेषम् ॥ २५

IAST:

mūlaphalaṁ saphalaṁ kālamūlaguṇam ardhamūlakṛtiyutam|

mūlaṁ mūlārdhonaṁ kālahṛtaṁ syāt phalaṁ śeṣam || 25 ||

Translation & Explanation (Interpretive): "The principal-interest (mūlaphalaṁ) combined amount (saphalaṁ)... multiplied by time and principal (kālamūlaguṇam)... added to the square of half the principal (ardhamūlakṛtiyutam)... take the root (mūlaṁ)... subtract half principal (mūlārdhonaṁ)... divided by time (kālahṛtaṁ) is the interest (phalaṁ)."

The Mathematics of Money: Solving for Interest

While Aryabhata's focus was often on the heavens, his mathematics was equally at home on Earth, particularly in the practical world of commerce. This verse tackles a classic financial problem, demonstrating how to handle calculations involving interest on a loan or investment.

The verse begins by referencing the basic components of such a problem: the principal (*mūla*, the initial amount), the time (*kāla*), and the interest (*phala*). But its true brilliance lies in the solution it offers for a much harder, "reverse" problem.

A Financial Puzzle

Imagine this scenario: you know the starting amount of a loan (the principal), the final amount that was paid back, and the duration of the loan. The one thing you *don't* know is the interest rate that was applied. How would you calculate it?

This is a non-trivial problem because the unknown value (the interest rate) is intertwined with other variables in the calculation. It requires more than simple arithmetic; it requires algebra.

An Algebraic Recipe for Finding the Interest Rate

Aryabhata's verse lays out a precise algebraic recipe to solve this exact puzzle. His verbal instructions, though highly condensed, describe a sequence of operations that will flawlessly reveal the unknown interest. The steps, which involve multiplying by the principal and time, adding the square of half the principal, and taking a square root, are a clear sign of a sophisticated algebraic maneuver.

What Aryabhata has encoded in his compact verse is a method for solving a **quadratic equation**.

The Deeper Mathematics

Like a master chef who hides complex techniques in a simple recipe, Aryabhata has provided an algorithm that solves the underlying quadratic equation without ever needing to write it down in modern symbols. Problems involving interest rates, where the rate is applied over time to a principal, often lead to this type of equation. The steps he prescribes—especially the tell-tale "add a square and take the root"—are the unmistakable signature of the quadratic formula at work.

Significance: Practical Mathematics in Action

This verse is a prime example of *gaṇita*—the Indian tradition of practical mathematics. It demonstrates that the same powerful algebraic reasoning used to model the orbits of the planets could be applied with equal precision to solve the financial problems of merchants, lenders, and everyday people.

It shows that for Aryabhata, mathematics was not an abstract game. It was a universal tool, capable of bringing clarity and predictive power to nearly any field of human endeavor, from the celestial to the commercial.

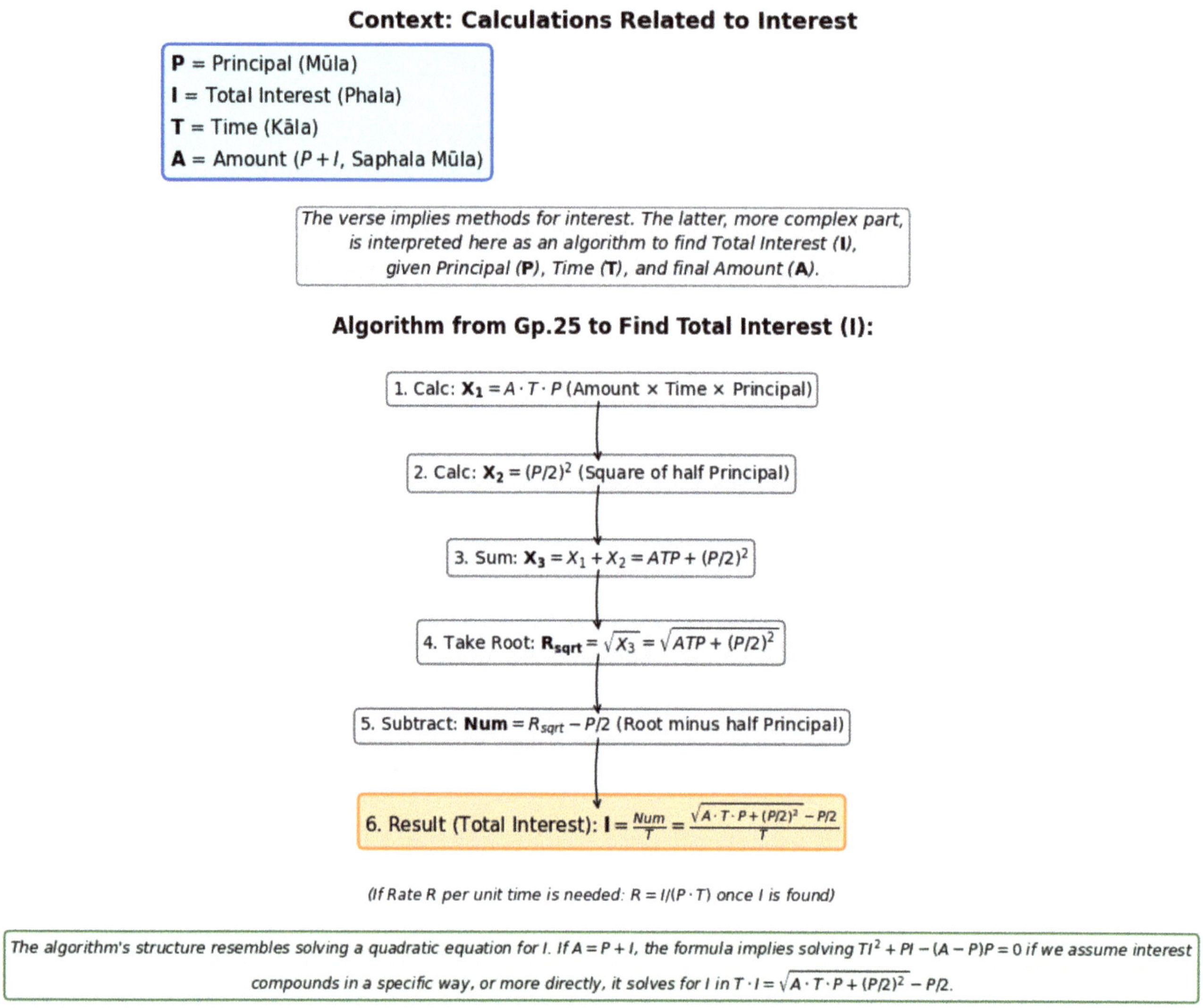

Figure 4.24: Algorithm for Interest (I). Illustrates Gp.25 (latter part) as steps (multiplications, additions, sqrt, subtraction, division) to determine I when Principal (P), Time (T), and Amount (A) are known, suggesting a quadratic relationship.

4.6.8 The Rule of Three (Trairāśika) – A Cornerstone of Proportional Reasoning

Sanskrit:

त्रैराशिकफलराशिं तमथेच्छाराशिना हतं कृत्वा ।

लब्धं प्रमाणभजितं तस्मादिच्छाफलमिदं स्यात् ॥ २६ ॥

IAST:

trairāśikaphalarāśiṁ tamathecchārāśinā hataṁ kṛtvā|

labdhaṁ pramāṇabhajitaṁ tasmādicchāphalamidaṁ syāt || 26 ||

The Rule of Three: The Engine of Proportionality

If three artisans can craft five statues in a week, how many statues can nine artisans craft in the same time? This simple question holds the key to one of the most powerful and frequently used tools in all of practical mathematics: the **Rule of Three**, or *Trairāśika* in Sanskrit.

In this verse, Aryabhata lays out a simple, foolproof recipe for solving any problem involving direct proportion. It was the mathematical workhorse of the ancient world, used for everything from calculating prices in the marketplace to modeling the movements of the planets.

The Three Knowns and the One Unknown

Aryabhata's method involves three known quantities, which he names poetically, to find one unknown result.

1. **The Standard (*Pramāṇa*):** The baseline cause or input. (e.g., 3 artisans)
2. **The Fruit (*Phala*):** The result produced by the standard. (e.g., 5 statues)
3. **The Requisition (*Icchā*):** The "desired" new cause or input. (e.g., 9 artisans)

The goal is to find the *icchāphalam*—the "desired fruit" corresponding to the new input.

A Simple, Timeless Recipe

Aryabhata's verse gives a clear set of instructions: "Multiply the 'Fruit' by the 'Requisition.' Divide the result by the 'Standard'; this yields the desired 'Fruit'."

In modern terms, if A gives B, what does C give?

Answer (X) = (C × B) / A

Let's solve our artisan puzzle using Aryabhata's recipe:

1. Multiply the "Fruit" (5 statues) by the "Requisition" (9 artisans):
    - $5 \times 9 = 45$
2. Divide this result by the "Standard" (3 artisans):
    - $45 / 3 = 15$

The answer is **15 statues**. The method is simple, elegant, and universally applicable.

An Indispensable Tool for the Astronomer

For Aryabhata, the Rule of Three was far more than a tool for commerce. It was the fundamental engine for scaling the cosmos. Astronomical tables often gave the total revolutions of a planet over a vast time period—a *yuga* of millions of years. To find the planet's position for a specific number of days (*ahargaṇa*), he needed a reliable way to scale down these immense numbers.

The Rule of Three was the answer. It was essential for:

- **Calculating Planetary Positions:** Scaling down long-term planetary motions to find their positions on any given day.
- **Unit Conversion:** Converting between different units of time, distance, or angular measurement.
- **Geometry:** Solving problems involving similar triangles and other figures where proportions are key.

This simple rule formed the bridge between raw, large-scale data and precise, daily predictions. It was, in essence, the mathematical heartbeat of Aryabhata's astronomical system.

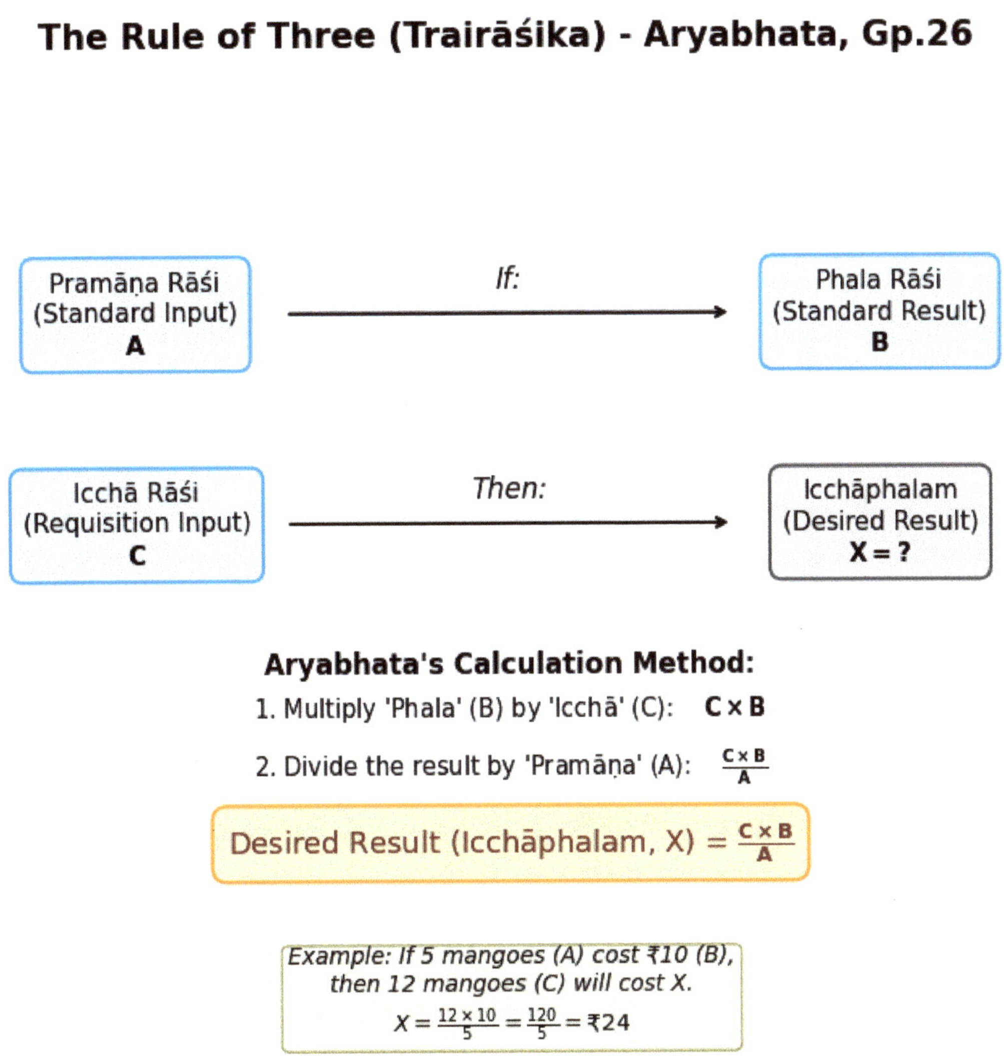

*Figure 4.25: Aryabhata's Rule of Three (Trairāśika, Gp.26). Illustrates direct proportion: finding Desired Result (X) from Standard Input (A), Standard Result (B), and Requisition Input (C) via X = (C*B)/A. Includes example.*

4.6.9 Inverse Rule of Three and Introduction to Compound Proportion

Sanskrit:

राशित्रयाद्विपर्ययकृतेः फलमिच्छागुणं प्रमाणहृतम् ।

पञ्चनवसप्तराशौ पर्ययतो वा फलं भवति ॥ २७ ॥

IAST:

rāśitrayādviparyayakṛteḥ phalamicchāguṇaṁ pramāṇahṛtam|

pañcanavasaptarāśau paryayato vā phalaṁ bhavati || 27 ||

The Rule of Three: The Advanced Toolkit

The simple Rule of Three is powerful, but what happens when the logic is reversed, or when a problem involves more than three quantities? In a follow-up verse, Aryabhata expands his toolkit, providing methods for these more complex scenarios.

1. The Inverse Rule: When More Means Less

Sometimes, an increase in one quantity causes a *decrease* in another. Consider this classic problem: If 2 workers can build a wall in 6 days, how long will it take 4 workers?

Using the simple Rule of Three would give the wrong answer (12 days). Common sense tells us that more workers should mean *less* time. This is a problem of **inverse proportion**.

Aryabhata provides a simple modification for these cases, which he calls *viparyayakṛteḥ*, or "by inverse operation." The recipe is slightly different: Instead of multiplying the 'Fruit' by the 'Requisition,' you multiply the 'Standard' by the 'Fruit.'

Let's solve our worker puzzle with this Inverse Rule:

1. **Standard (A):** 2 workers
2. **Fruit (B):** 6 days
3. **Requisition (C):** 4 workers

The Inverse Recipe:

1. Multiply the 'Standard' (2 workers) by the 'Fruit' (6 days):
   - $2 \times 6 = 12$
2. Divide this result by the 'Requisition' (4 workers):
   - $12 / 4 = 3$

The answer is **3 days**. The inverse rule handled the logic perfectly. This simple but critical modification made the Rule of Three a truly universal tool for proportionality.

2. The Compound Rule: Juggling Multiple Variables

What if a problem involves five, seven, or even more quantities?

If 3 artisans (Cause 1) working for 8 days (Cause 2) can craft 5 statues (Effect), how many statues can 9 artisans working for 12 days produce?

This is a problem of **compound proportion**. To solve it, Aryabhata describes a brilliant method of "transference" or "alternation" (*paryayato*). While the verse is extremely concise, commentators explain it as a systematic way to organize and solve the problem.

The method involves two key steps:

1. **Organize the Data:** All the "causes" (like artisans, days, hours) and the "effect" (statues) from the first scenario are placed on one side. The causes from the second scenario and its unknown effect are placed on the other.
2. **The "Transference" Maneuver:** To solve for the unknown effect, the "causes" from its side are swapped with the "effect" from the other side. Then, all the numbers on the "cause" side are multiplied together and divided by the product of the numbers on the "effect" side.

This may sound complex, but it is a highly efficient algorithm—a repeatable recipe that reduces a complicated problem with many variables into a single, straightforward calculation.

Significance: The Power of Algorithmic Thinking

With these additions, Aryabhata provides a complete and robust system for solving nearly any problem of proportion, simple or complex. It demonstrates a sophisticated level of algorithmic thinking—the ability to create a clear, step-by-step process that can be applied to a wide class of problems. For science and commerce, where real-world situations often involve multiple changing factors, this was an indispensable mathematical technology.

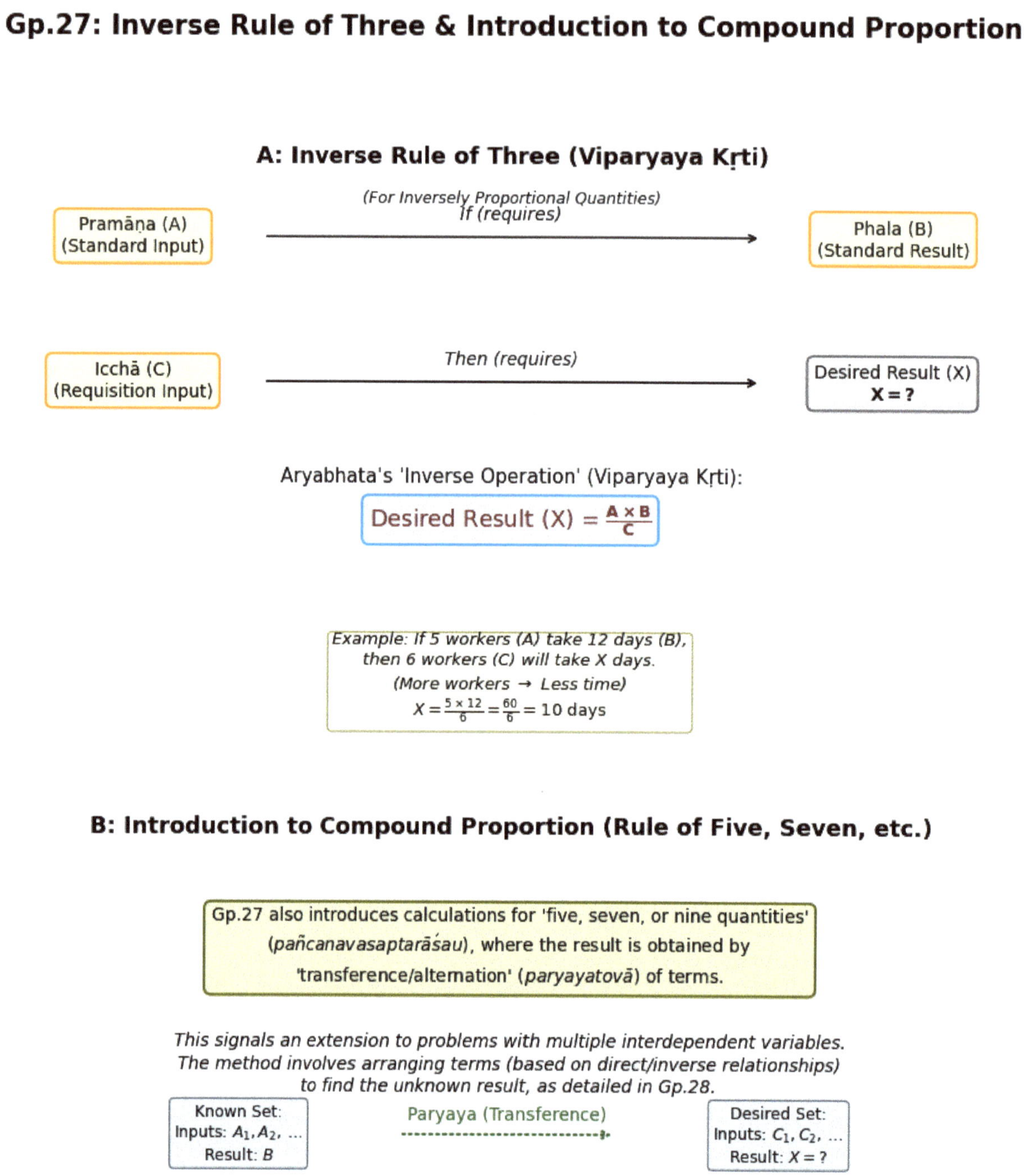

*Figure 4.26: Gp.27: Inverse Rule of Three & Compound Proportion Intro. (A) Illustrates Inverse Rule (Viparyaya Kṛti: X=(A*B)/C, e.g., workers/days). (B) Introduces compound proportion for 'five, seven, or nine quantities' via 'transference/alternation' (paryāyatovā), leading to Gp.28.*

4.6.10 Method for Compound Proportion

Sanskrit:

फलं प्रक्षेप्य भागहारं गुणकाराशिभिर्हतम् ।

भागहारा विभज्यन्ते प्रमाणफलराशिना ॥ २८ ॥

IAST:

phalaṃ prakṣepya bhāgahāraṃ guṇakārāśibhirhatam|

bhāgahārā vibhajyante pramāṇaphalarāśinā || 28 ||

The Ultimate Algorithm: Juggling Multiple Variables

Real-world problems rarely involve just one changing factor. An astronomer might need to account for a planet's speed, its distance, *and* the passage of time. A merchant might need to calculate costs based on the number of workers, the days they work, *and* the hours per day.

To solve these multi-layered problems, Aryabhata provides a master algorithm. It's a systematic procedure that elegantly handles any number of variables by sorting them into two groups of numbers and then performing a single, final calculation.

Let's walk through this powerful method with an example:

Problem: If **3 artisans**, working **8 hours** a day for **10 days**, can produce **90 statues**, how many statues can **5 artisans**, working **6 hours** a day for **12 days**, produce?

The Method: A Step-by-Step Guide

Step 1: Create Two Piles of Numbers

We start by creating two piles of numbers. We can call them the **Numerator Pile** (the numbers we will multiply on top) and the **Denominator Pile** (the numbers we will multiply on the bottom).

The very first number in our **Numerator Pile** is always the known result from the first scenario.

- **Numerator Pile starts with: [90]**

Step 2: The "Transference" Logic

Next, we process each *type* of cause (artisans, hours, days) one by one to decide where its numbers go. This is the heart of the algorithm, what Aryabhata calls *paryaya*, or "transference."

For each cause, we ask a simple question: **Does more of this cause lead to *more* of the result, or *less*?**

- If **more cause -> more result**, the relationship is **direct**.
- If **more cause -> less result**, the relationship is **inverse**.

The Transference Rule:

- For a **direct** relationship, the *new* value goes in the Numerator Pile and the *old* value goes in the Denominator Pile.
- For an **inverse** relationship, the *old* value goes in the Numerator Pile and the *new* value goes in the Denominator Pile.

Let's apply this to our problem:

1. **Artisans:** More artisans produce *more* statues. This is **direct**.
    - So, the new value (5) goes in the Numerator Pile.
    - The old value (3) goes in the Denominator Pile.
2. **Hours:** More hours of work produce *more* statues. This is **direct**.
    - So, the new value (6) goes in the Numerator Pile.
    - The old value (8) goes in the Denominator Pile.
3. **Days:** More days of work produce *more* statues. This is **direct**.
    - So, the new value (12) goes in the Numerator Pile.
    - The old value (10) goes in the Denominator Pile.

Step 3: The Final Calculation

After processing all the causes, our two piles of numbers are complete:

- **Numerator Pile:** [90, 5, 6, 12]
- **Denominator Pile:** [3, 8, 10]

The beauty of this method is that it reduces a complex problem to a single calculation. To find the answer, we simply: **Multiply all the numbers in the Numerator Pile and divide by the product of all the numbers in the Denominator Pile.**

- Answer = (90 × 5 × 6 × 12) / (3 × 8 × 10)
- Answer = 32,400 / 240
- **Answer = 135 statues**

This powerful algorithm turns a potentially confusing puzzle into an orderly and predictable process. It is a testament to the sophisticated procedural thinking of Aryabhata and the mathematical tradition he represented, providing a universal engine for solving problems of compound proportion.

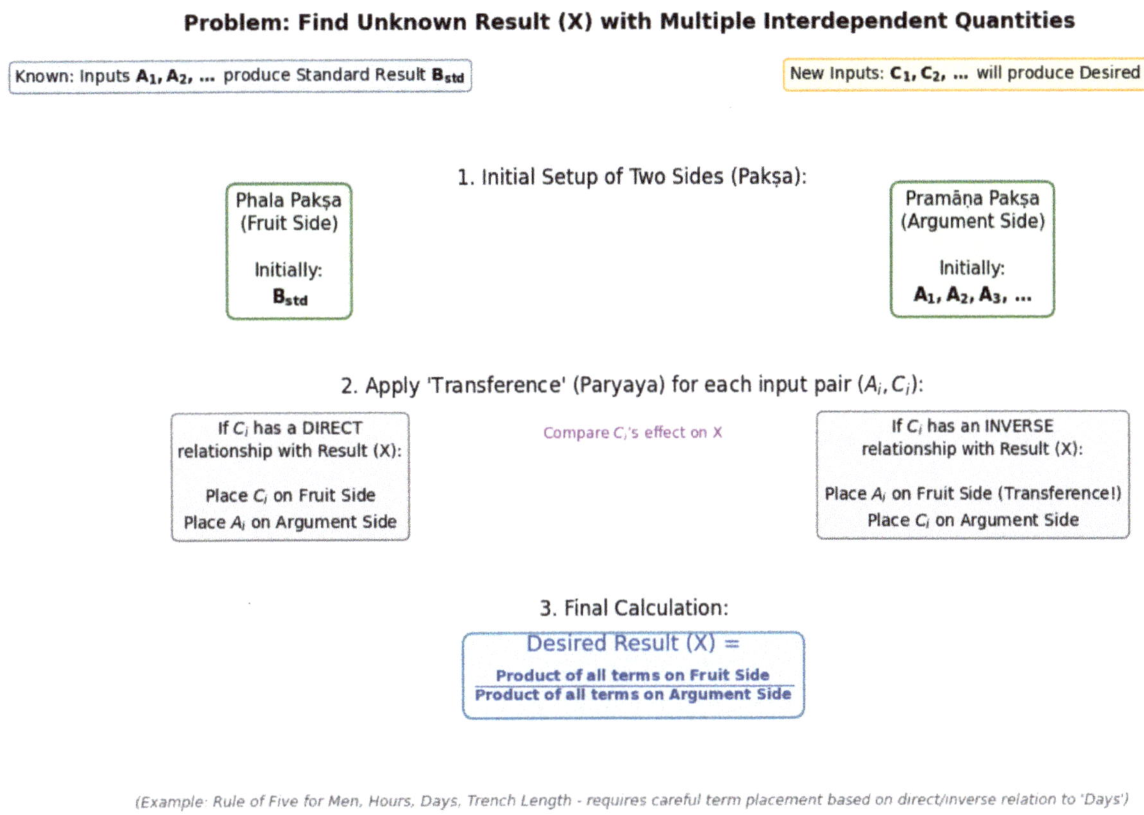

Figure 4.27: Gp.28: Aryabhata's Compound Proportion (Paryaya). Illustrates the 'transference' method: setting up Fruit/Argument Sides, allocating terms based on direct/inverse relationships, and calculating Desired Result (X) as (Product Fruit Side) / (Product Argument Side).

4.6.12 Rule of Mixtures (Alligation)

Sanskrit:

मूल्यं फलेन संनादति योगेन फलयोगहृतं तन्मूल्यम् ।

द्रव्यविशेषान् प्रक्षेपविधिं संक्षेपतः ॥ २९ ॥

IAST:

mūlyaṃ phalena saṃnādati yogena phalayoghṛtaṃ tanmūlyam|

dravyaviśeṣān prakṣepavidhiṃ saṃkṣepataḥ || 29 ||

Translation

The value (*mūlya*) multiplied by the quantity (*phalena*), summed and divided by the sum of quantities (*phalayoghṛtaṃ*), yields the average value of the mixture (*tanmūlyam*). Thus, briefly, the rule for mixing different substances (*dravyaviśeṣān prakṣepavidhiṃ*).

Interpretation

In this verse, Aryabhata turns his attention from the abstract realm of geometry to the pragmatic challenges of commerce, codifying the "rule of mixtures." This powerful principle provides a method for determining the average value of a blended commodity—a cornerstone calculation for trade and resource management.

The rule itself is an elegant application of what is now known as the weighted average. The average value is found by summing the products of each component's value and its quantity, and then dividing this total value by the total quantity:

$$\textbf{Average Value} = \Sigma(\text{Value}_i \times \text{Quantity}_i) / \Sigma(\text{Quantity}_i)$$

The practical utility of this rule is immediately apparent. Consider a merchant mixing two qualities of grain: 10 units of a superior grain valued at 5 coins per unit, and 20 units of a lesser grain valued at 3 coins per unit. The total value of the mixture is the sum of the parts: (10 × 5) + (20 × 3), which equals 110 coins. The total quantity is 30 units. Applying Aryabhata's rule, the average price of the mixture is therefore 110 divided by 30, or approximately 3.67 coins per unit.

This method is more than a simple commercial shortcut; it is a clear articulation of a fundamental algebraic concept. Verse Gp.29 showcases Aryabhata's remarkable ability to distill complex relationships into concise, generalizable rules. By demonstrating the application of rigorous mathematics to real-world problems, this verse bridges the gap between theory and practice, underscoring the *Aryabhatiya*'s role not just as a theoretical treatise, but as a practical handbook whose influence on Indian mathematical traditions was both profound and enduring.

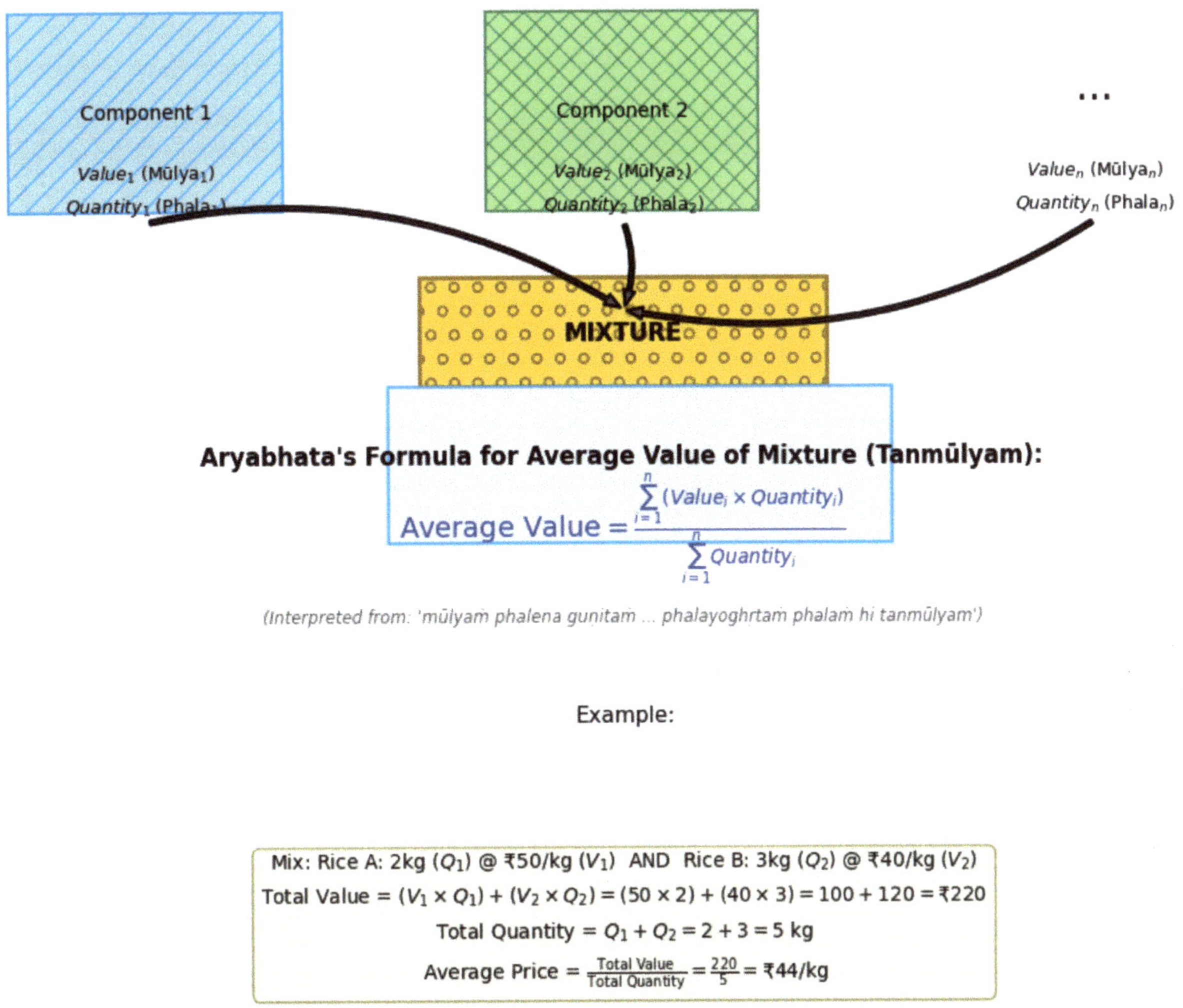

Figure 4.28: Aryabhata's Gp.29: Rule of Mixtures (Miśraka-Vyavahāra). Illustrates calculating a mixture's average value (weighted average) by summing (component value_i × quantity_i) and dividing by total quantity. Includes rice example.

4.6.13 Solving for Unknowns in Proportions or Sequences

Sanskrit:

क्षेपा विशोध्य गच्छान्तरभक्तास्ते भवन्ति ये क्षेपाः ।

यद्येको राशिः स्यादेकस्याज्ञानतो राशिः ॥ ३० ॥

IAST:

kṣepā viśodhya gacchāntarabhaktāste bhavanti ye kṣepāḥ |

yadyeko rāśiḥ syādekasyājñānato rāśiḥ || 30 ||

The Mathematical Detective: Finding the Missing Number

Imagine you are given a sequence of numbers that grows by a steady amount, but you only know a few of the values.

Imagine you are told that the 5th term in a series is 17, and the 12th term is 38. How could you figure out the common difference that separates each step in the series?

This is a classic "mathematical detective" problem, and in this verse, Aryabhata provides a master strategy for solving it. He explains how to find an unknown quantity (*kṣepa*) by using the known values as clues.

The Method: Using Differences as Clues

Aryabhata's verse describes a beautifully simple and logical process:

1. **Find the Difference in the Values:** First, calculate the total difference between the two known values.
2. **Find the Difference in the Positions:** Next, calculate how many "steps" or positions separate the two known terms.
3. **Divide to Find the Unknown:** Finally, divide the difference in values (from Step 1) by the difference in positions (from Step 2). The result is the common difference—the secret value that defines the entire series.

Let's solve our example using this method:

Solving our example:

1. Difference in Values: 38 – 17 = **21**
2. Difference in Positions: 12th term – 5th term = **7 steps**
3. The Common Difference: 21 / 7 = **3**

The common difference for the series is **3**. With this key, we could now reconstruct any term in the entire sequence.

Significance: The Power of Interpolation

This verse describes a general and powerful algebraic technique for solving for an unknown in a linear system. While it sounds abstract, its practical importance, especially for an astronomer, cannot be overstated. This is the core principle behind *interpolation*—the art of finding an unknown value that lies *between* two known values.

For Aryabhata, this was an indispensable tool. His astronomical tables might list a planet's position for Day 5 and Day 10, but to predict an eclipse, he might need to know its precise position on Day 7. The method described in this verse allowed him to "read between the lines" of his own data with mathematical precision.

This is a perfect example of Aryabhata moving beyond simple, fixed formulas. He is providing a flexible, logical procedure for reasoning about unknowns, a hallmark of sophisticated algebraic thought that is fundamental to all modern science.

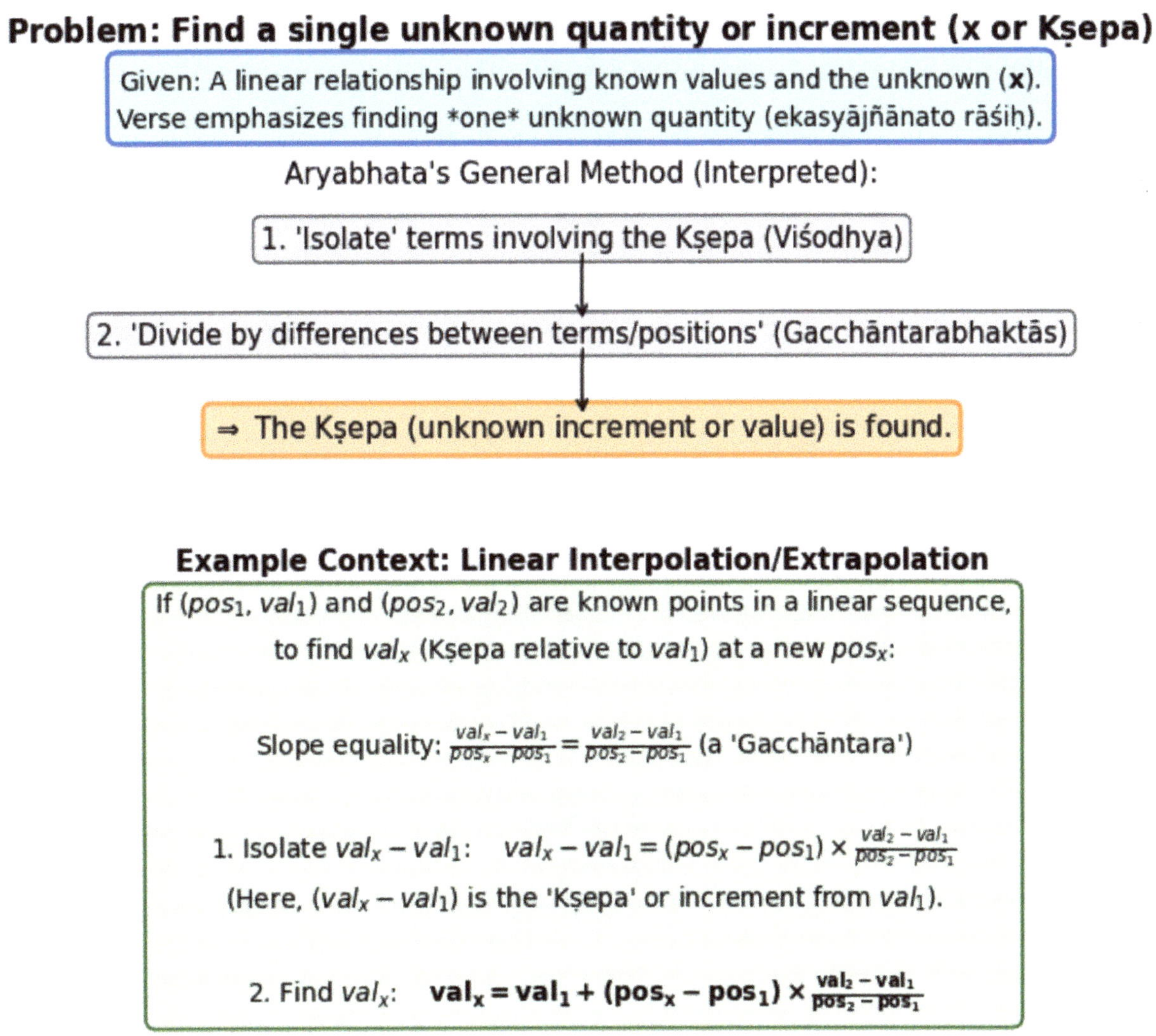

Figure 4.29: Gp.30: Aryabhata's 'Kṣepa' Principle for Unknowns. Illustrates isolating terms and dividing by differences to find an unknown (x), exemplified by linear interpolation/extrapolation for linear relationships.

4.6.14 Solving for Variables in Interest Equations

Sanskrit:

मूलगुणं कालेन विभक्तं तत्फलयुतेन मूलेन ।

फलमूनं कालेन स्याद् यद्यूनं तत्फलं ज्ञेयम् ॥ ३१ ॥

IAST:

mūlaguṇaṁ kālena vibhaktaṁ tatphalayutena mūlena|

phalamūnaṁ kālena syād yadyūnaṃ tatphalaṁ jñeyam || 31 ||

Unraveling the Financial Knot: Solving for the Unknown

Imagine you are reviewing an old loan agreement. You can see the original amount borrowed (the principal), the total interest paid, and the interest rate, but the crucial detail of *how long* the loan lasted is missing. How could you find the missing time?

This is a common financial puzzle, and in this verse, Aryabhata provides the key to solving it. He demonstrates how to unravel the simple interest formula to find any single missing piece, whether it's the principal, the rate, or the time.

The Method: Algebra in Action

The standard formula for simple interest is a balanced mathematical sentence:

Interest = Principal × Rate × Time

If one of these components is unknown, the equation is like a knot. Aryabhata's method is a systematic way to untie that knot. His verbal instructions describe the process of rearranging the mathematical sentence until the unknown value stands alone on one side.

Let's use his logic to solve our puzzle of finding the missing time:

Goal: Find the Time.

1. **Start with the known fact:** The total Interest paid.
2. **Undo the multiplications:** To isolate the Time, we need to reverse the effect of multiplying by the Principal and the Rate. We do this by dividing.

3. **The solution:** Divide the total Interest by the Principal, and then divide that result by the Rate.

The rearranged formula becomes:

Time = Interest / (Principal × Rate)

Aryabhata's verse describes these same kinds of operations—multiplication, division, and subtraction—as a step-by-step recipe for isolating any target variable. He provides a universal algebraic strategy, not just a single-purpose formula.

Significance: A System for Problem-Solving

This verse reinforces Aryabhata's focus on creating systematic methods for real-world problems. It shows that his algebraic reasoning was not confined to geometry or astronomy but was a powerful, all-purpose tool.

By providing a clear method for untangling financial calculations, he demonstrates a profound understanding of algebra as a practical engine of logic. This ability to manipulate and reorder mathematical relationships is the very foundation of modern scientific and financial analysis.

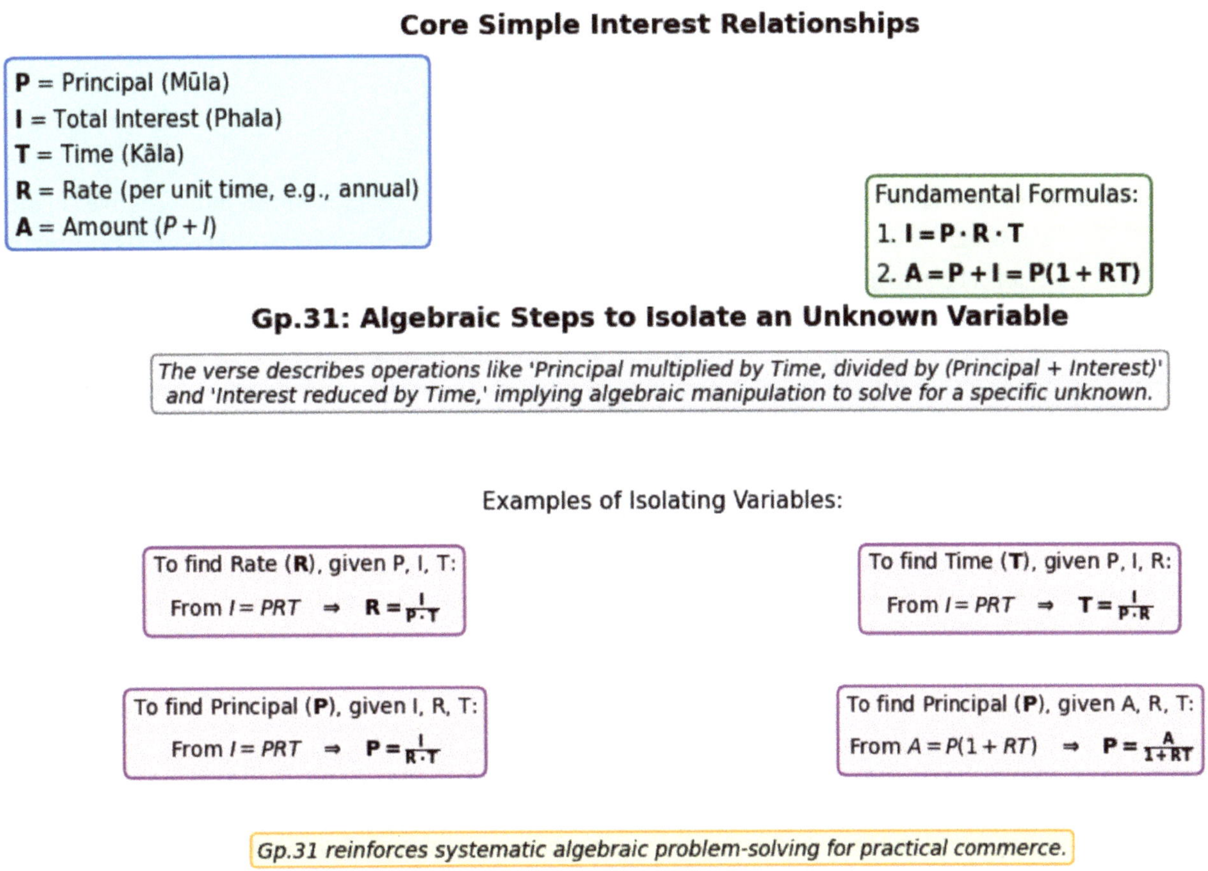

Figure 4.30: Gp.31: Algebraic Solutions for Simple Interest. Interprets Gp.31 by defining terms (P,R,T,I,A) and fundamental formulas (I=PRT, A=P+I), then illustrating formula rearrangements to solve for unknowns like R, T, or P.

4.7 The Pulverizer: Aryabhata's Pinnacle Algorithm (Kuttākāra) (Gp.32-33)

The Pulverizer: Aryabhata's Master Algorithm

Of all Aryabhata's contributions, one stands above the rest, a testament to his profound mathematical genius: the **Kuṭṭākāra**. Often translated as "The Pulverizer," this powerful and elegant algorithm is presented in the final two verses of the Mathematics chapter. It stands as his most celebrated and sophisticated tool.

4.7.1 What is the Pulverizer?

At its heart, the Pulverizer is a method for solving a special kind of puzzle known as a linear indeterminate equation. These are equations that can have multiple, or even infinite, possible whole-number solutions. The challenge is not just to find *a* solution, but to find the *right* one that fits the constraints of a problem.

The most important form of this puzzle, especially for an astronomer, can be stated like this:

Find a whole number **N** such that when you multiply it by a known number **A**, and then divide the result by another known number **B**, you are left with a specific remainder **C**.

This "Great Remainder Puzzle" is the core problem that the Pulverizer was designed to crack.

4.7.1.1 An Essential Tool for a Clockwork Universe

For an astronomer in Aryabhata's time, the Pulverizer was not an abstract curiosity; it was an indispensable engine for making sense of the heavens. Indian astronomy operated on immense timescales called *yugas*, vast epochs lasting millions of years. The astronomical constants—like the number of revolutions a planet makes—were defined over these entire epochs.

This created a fundamental problem: how do you connect a cosmic blueprint spanning millions of years to the practical question of where a planet is *today*?

The Pulverizer was the bridge. It allowed an astronomer to synchronize different celestial cycles and convert the vast, long-term data into precise, daily calculations. The "Great Remainder Puzzle" appeared constantly in this work:

- **N** was the unknown number of days that had passed since the start of the age (*ahargaṇa*).
- **A** and **B** were the planetary constants (e.g., a planet makes **A** revolutions in **B** days).
- **C** was the observed position of the planet in its current cycle (the remainder).

By solving this puzzle, Aryabhata could determine the exact number of days that corresponded to a planet's observed position. He could work backwards from an observation to find the time, or work forwards to predict the exact moment when two planets would appear at the same point in the sky (a conjunction).

Without a reliable method to solve these remainder problems, the entire predictive system would grind to a halt. The Pulverizer was the mathematical key that unlocked Aryabhata's clockwork universe, making it a predictable and calculable system.

Verses 32-33: Unveiling the Algorithm

The Sanskrit verses outlining the *Kuṭṭākāra* algorithm are remarkably concise and require careful interpretation, often relying on the elucidations provided by later commentators like Bhaskara I. Here are the original verses with their IAST transliteration:

Sanskrit :

अधिकग्रभागहारं छिन्याद् ऊनग्रभागहारेण ।

शेषपरस्परभक्तं मतिगुणम् अग्रान्तरे क्षिप्तम् ॥ ३२ ॥

IAST:

adhikāgrabhāgahāraṃ chindyād ūnagrabhāgahāreṇa|

śeṣaparasparabhaktaṃ matiguṇam agrāntare kṣiptam ||32||

Sanskrit:

अध उपरि गुणितम् अन्त्ययुग् ऊनच्छेदभाजिते शेषम् ।

अधिकच्छेदगुणं द्विच्छेदाग्रम् अधिकग्रयुतम् ॥ ३३ ॥

IAST:

adha upari guṇitam antyayug ūnacchedabhājite śeṣam|

adhikacchedaguṇaṃ dvicchedāgram adhikāgrayutam ||33||

4.7.2 Decoding the Pulverizer: A Journey into the Algorithm

While the original Sanskrit verses are incredibly dense, scholars have decoded the Pulverizer's general procedure. It can be understood as a brilliant two-phase process that is strikingly similar to a modern method known as the Extended Euclidean Algorithm.

4.7.2.1 Phase 1: The Forward Reduction

This first phase is a process of "pulverizing" or reducing two numbers down to their essential core. It begins with the two key numbers from the puzzle (let's call them **A** and **B**). The larger number is divided by the smaller, and the remainder is noted. Then, the process is repeated: the previous divisor is divided by the new remainder.

This chain of divisions continues until the remainder becomes zero. The last non-zero remainder in this chain is the **Greatest Common Divisor (GCD)** of the original two numbers. Throughout this process, the quotients from each division are carefully recorded in a sequence, forming a "chain" (*valli*) that will be crucial for the next phase.

An Important Condition: For a solution to the puzzle to even exist, this GCD must be able to divide the third number in our puzzle, the remainder **C**.

4.7.2.2 Phase 2: The Backward Construction

Once the numbers have been "pulverized" down to their GCD, the algorithm reverses course to build the solution back up. This is a recursive process, meaning each step depends on the one before it.

Starting with a simple chosen number (often 0 or 1), the algorithm uses the chain of quotients from the first phase—*but in reverse order*—to build a new series of numbers. The Sanskrit phrases in the verse, like "the lower multiplied by the one above it and added to the one below that," hint at this step-by-step upward construction.

The final numbers generated by this backward process are the raw materials for the solution. Through one final set of operations described in the verses, these numbers are transformed into the answer to the original puzzle—the precise integer **N** that was sought.

4.7.3 The Pulverizer in Action: A Modern Example

While following Aryabhata's exact verbal recipe requires specialized study, we can demonstrate the power of the underlying method with a modern approach to solve the same kind of puzzle.

The Problem: Find the smallest whole number **N** such that when you multiply it by 137 and divide by 60, you are left with a remainder of 10.

In mathematical terms, this is written as: **137N = 10 (mod 60)**

Step 1: Simplify the Problem

First, we can simplify the large number 137. When 137 is divided by 60, the remainder is 17 (since $137 = 2 \times 60 + 17$). For problems involving remainders, we only care about the remainder itself. So, our problem becomes much simpler:

- **17N = 10 (mod 60)**

Step 2: Find the "Secret Key" (The Modular Inverse)

To isolate N, we need to find a "secret key" number that effectively undoes the multiplication by 17. This key is a number that, when multiplied by 17, gives a remainder of 1 when divided by 60. This is called the *modular multiplicative inverse.*

Using a process equivalent to Aryabhata's own—the "Forward Reduction" and "Backward Construction" steps—we can find this key. For the numbers 17 and 60, the key turns out to be **53**.

The magic of this key is that **17 × 53** gives a remainder of **1** when divided by 60. This allows us to cancel out the 17 in our problem.

Step 3: Use the Key to Solve for N

Now, we multiply both sides of our simplified problem by the secret key (53): **53 × (17N)= 53 × 10 (mod 60)**

Because 53 × 17 leaves a remainder of 1, the entire left side of the equation simplifies to just **1N**, or **N**. The right side becomes 530.

- **N = 530 (mod 60)**

Step 4: Find the Final Answer

The puzzle is now almost solved. We just need to find the remainder of 530 when it's divided by 60.

- **530 = 8 × 60 + 50**

The remainder is 50.

So, the smallest positive whole number that solves our puzzle is **N = 50**.

This example demonstrates the kind of precise, non-obvious integer solutions the Pulverizer was designed to find—solutions that were absolutely essential for the accuracy of Aryabhata's astronomical calculations.

4.7.4 Significance: A Bridge from Ancient Stars to Modern Security

The Kuṭṭākāra algorithm is far more than a historical curiosity. Its legacy is a stunning testament to the enduring power of mathematical discovery.

1. The Keystone of Ancient Astronomy
For Aryabhata, the Pulverizer was the engine that made his entire astronomical system work. It allowed him to connect vast cosmic cycles to practical, daily timekeeping, enabling him to create accurate calendars and predict the intricate dance of the planets. Without it, his science would have been impossible.

2. A Monumental Mathematical Breakthrough
The Pulverizer represents a massive leap in algebraic thought. The ability to create a systematic, repeatable algorithm to find whole-number solutions for these complex puzzles was exceptionally advanced for its time and demonstrated a mastery of number theory distinct from developments elsewhere in the world.

3. The Unseen Guardian of the Digital Age
Perhaps most astonishingly, the core logic of the Pulverizer—the Extended Euclidean Algorithm—is not just relevant today; it is a cornerstone of our modern world. It is a critical component of the **RSA encryption algorithm**, which secures countless emails, financial transactions, and digital communications across the globe every day.

The process of finding a "secret key" (the modular inverse) is exactly what RSA uses to create public and private keys for encryption. This creates a direct, unbroken line from a mathematical algorithm developed in 5th-century India to the very foundation of modern digital security.

The Pulverizer, therefore, is not just a relic of the past. It is a powerful example of how ancient intellectual achievements can lay the groundwork for concepts that continue to shape and secure our modern lives.

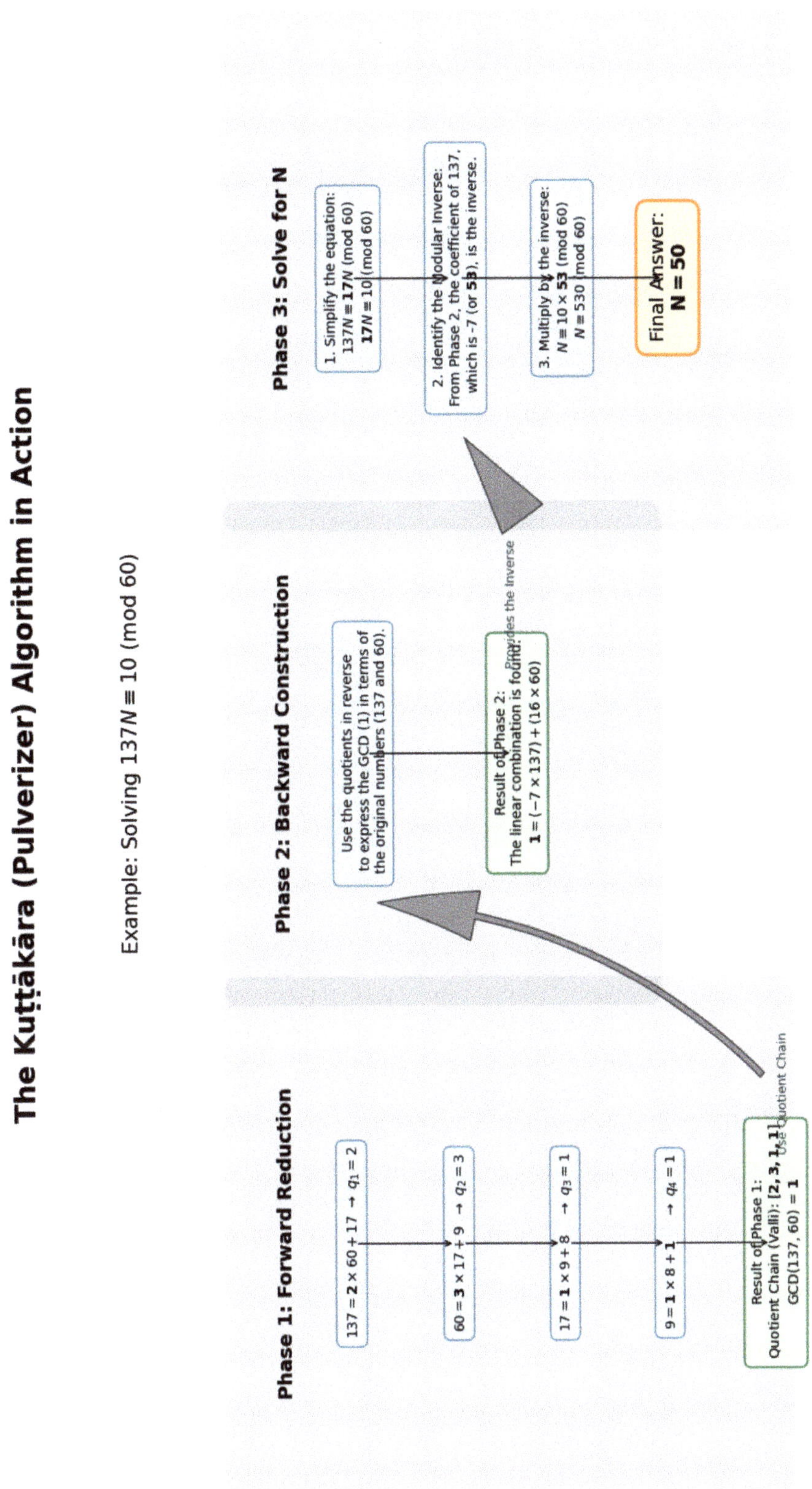

Figure 4.31: Aryabhata's Kuṭṭākāra (Gp.32-33). Details the two-stage (Euclidean reduction; backward recursion) method for solving indeterminate equations like 137N = 10 (mod 60), vital for astronomical calculations.

4.8 Conclusion: The Power of Ganita

Decoding the *Gaṇitapāda* reveals Aryabhata in his element: a mathematician of extraordinary scope and skill, meticulously crafting the computational engine for his astronomical vision. The journey begins with his foundational command of the decimal place-value system (Gp.2)—a concept so fundamental it is easy to overlook its revolutionary power. From there, he lays out elegant algorithms for extracting square and cube roots (Gp.4-5), essential prerequisites for the quantitative precision his astronomy demanded.

We witness his geometric acumen in the precise area formulas and his approach to volumes (Gp.6-9), where even his noted misstep with the sphere's volume (Gp.7) offers a fascinating glimpse into the authentic challenges of early mathematical exploration. This section is crowned by his famous, startlingly accurate value for π (Gp.10) and the key theorems on chords and arcs (Gp.11-13) that form the heart of his trigonometric system. His genius, however, lay not just in abstract principles but in their practical application, as seen in his elegant use of the gnomon for measurement (Gp.14-17) and his direct confrontation with complex problems like intersecting circles (Gp.18).

Aryabhata's versatility extends further still. He demonstrates a strong command over arithmetic series (Gp.19-22) and the indispensable rules of proportion (Gp.26-28), crucial for scaling cosmic data from the page to the heavens. He even provides methods for the pragmatic calculations of commerce, such as interest (Gp.25) and mixtures (Gp.29). But the true jewel of this chapter is undoubtedly the *Kuṭṭākāra* algorithm (Gp.32-33). This "Pulverizer" is a sophisticated algebraic tool, ingeniously designed to solve the linear indeterminate equations that were the very key he needed for his most demanding astronomical calculations, such as determining the *ahargaṇa* and untangling immense time cycles.

What becomes clear is that the *Gaṇita* is not a collection of disparate topics, but a cohesive toolkit forged for a singular purpose: to serve his astronomical quest. These diverse tools—from basic arithmetic to the seeds of trigonometry and advanced algebra—were the essential components he required to construct the predictive models of planetary motion in the *Kālakriyāpāda* and to geometrically map the celestial sphere in the *Golapāda*. The mathematical engine built within these thirty-three verses is a monument to Aryabhata's genius, a unified vision where every theorem and algorithm serves its part in the grand challenge of mapping the heavens.

***

Chapter Five

Charting Cosmic Rhythms – Kālakriyāpāda

The Mathematical Dance of Time: Cycles, Calendars, and Planetary Prediction

5.1 Introduction: Measuring the Immeasurable in India's Golden Age

5.1.1 The Intellectual Crucible: India in 499 CE

As the 5th century drew to a close, time in India was both a cosmic rhythm and a mathematical puzzle. The subcontinent, particularly the regions under the influence of the Gupta Empire, pulsed with a vibrant intellectual energy. This "Golden Age" was a crucible of discovery, where political stability and economic prosperity fueled profound advances in philosophy, science, and mathematics.

Great cities like Pataliputra and the renowned astronomical hub of Ujjain were centers of spirited scholarly debate. Monastic universities like Nalanda attracted the finest minds from across Asia, who came to engage with Vedic studies, Buddhist philosophy, logic, and medicine. But above all, this era was defined by its mastery of numbers.

This was the world that perfected the decimal place-value system, including the revolutionary concept of zero (*śūnya*). It was a world that saw the flourishing of sophisticated algebra and the development of trigonometry into a powerful, predictive tool. These were not abstract disciplines; they were the essential toolkit for *Jyotiṣa*, the integrated science of astronomy and astrology that sought to map the heavens and understand its influence on human affairs.

It was in this fertile intellectual milieu, around the year 499 CE, that Aryabhata composed his masterpiece, the *Aryabhatiya*. Equipped with powerful mathematics and grounded in a keen observational tradition, he created a work that was at once a masterful synthesis of centuries of knowledge, a critical correction of past errors, and a bold leap into the future.

In this third chapter, the *Kālakriyāpāda*, or "The Measure of Time," Aryabhata lays out the grand temporal framework of his system. He defines the vast cosmic cycles and the smaller units of earthly time, building the essential clockwork needed to measure the movements of the cosmos. This is the calendar and the chronicle of his universe.

5.1.2 Philosophical Grounding and Astronomical Focus: Defining Time

How does humanity grasp the relentless, invisible flow of time (*kāla*)? Indian philosophers debated its nature: Was it an independent substance (*dravya*) as Nyāya-Vaiśeṣika suggested, central to cosmic evolution (*prakṛti*) as per Sāṃkhya, or merely a conceptual label for momentary flux (*kṣaṇikavāda*) as in some Buddhist schools? While Aryabhata's work doesn't explicitly engage these metaphysical debates, his pragmatic, mathematical approach suggests an affinity with schools emphasizing empirical observation and rational structure.

The Kālakriyāpāda, the third *pāda* of the *Aryabhatiya*, thus takes a distinctly practical, astronomical path to understanding time. Bhaskara I (fl. c. 629 CE), our indispensable guide through his commentary, opens his discussion of this section with a reverent homage to Śiva as Mahākāla (Great Time), acknowledging the profound nature of *kāla*. Yet, he swiftly pivots to astronomy's specific needs, asserting that for the purpose of knowing time within astronomical science (*jyotiṣaśāstre tu kālajñānārthaṃ*), the calculation of planetary motions (*grahagaṇitam ārabhayate*) must begin.

For Aryabhata, time manifests undeniably through celestial motion (*gati*). The intricate dance of the Sun and Moon defines observable units like days, months, and years; the steady revolutions of the planets against the backdrop of stars mark out longer durations. Action (*kriyā*), in this astronomical context, *is* motion; as Bhaskara I might observe, "This celestial motion is what we shall measure to understand time." The Kālakriyāpāda, therefore, meticulously constructs the essential temporal machinery for Aryabhata's system. It defines fundamental time units (from the fleeting *prāṇa* to vast cosmic ages), quantifies the grand cycles established in the *Daśagītikā*, reconciles diverse timekeeping systems (lunar, solar, civil, sidereal), establishes calendrical rules, and crucially, details the computational methods—including the sophisticated epicyclic model—to track the celestial clockwork with precision. This chapter lays the indispensable temporal foundation upon which the spatial geometry of the *Golapāda* and its calculations of true celestial positions are built.

5.2 Basic Time Units and the Division of Space (Kk.1-2)

Precision in astronomy begins with clear, unambiguous definitions of the units used to measure time and space (angles). In the opening verses of the Kālakriyāpāda, Aryabhata establishes the fundamental hierarchies for both, creating a unified metrology essential for all subsequent calculations.

5.2.1 Time's Hierarchy

Sanskrit:

वर्षं द्वादशमासाः त्रिंशद्दिवसः भवेत् स मासस्तु ।

षष्टिर्नाड्यो दिवसः षष्टिश्च विनाडिका नाडी ॥ १ ॥

IAST:

varṣaṃ dvādaśamāsāḥ triṃśaddivasaḥ bhavet sa māsastu|

ṣaṣṭirnāḍyo divasaḥ ṣaṣṭiścha vināḍikā nāḍī || 1 ||

To build a predictive science of the heavens, one must first define its most basic dimension: time. Aryabhata begins this task with characteristic precision, establishing in this verse the fundamental units of his astronomical clock.

Translation: "A year is twelve months; that month indeed is of thirty days. A day is sixty *nāḍīs*; and a *nāḍī* is sixty *vināḍikās*."

The verse lays out a clear, cascading hierarchy. A year (*varṣaṃ*) is defined as twelve months (*dvādaśamāsāḥ*), and each month (*māsaḥ*) is standardized for computational purposes to thirty days (*triṃśaddivasaḥ*). This civil or "mean" day (*divasaḥ*) is then subdivided into sixty *nāḍīs*, and each *nāḍī*, in turn, is composed of sixty *vināḍikās*.

As the great commentator Bhaskara I emphasizes, the purpose of these explicit definitions was to create a standardized scientific framework, distinct from the variable vernacular or regional time units used in daily life. This level of precision was non-negotiable for the system's primary applications: the accurate construction of the calendar (*pañcāṅga*), the correct timing of religious rituals (*yajñakāla*), and the complex calculations required for horoscopy (*Horāśāstra*).

This verse, therefore, is not merely a list of conversions; it is the establishment of a universal, scientific clock, the essential temporal framework for the clockwork universe Aryabhata sought to describe.

Term	Definition	Approx. Modern Equivalent	Notes
Varṣa	12 Months	1 Year	Primarily the (Solar) Year, based on the Sun's motion (Kk.5).
Māsa	30 Days (nominal)	~1 Month	Computational Month used in the system; differs from synodic/solar months.
Divasa	60 Nāḍīs	1 Day	The Civil (Sāvana) Day, typically reckoned sunrise to sunrise (Kk.5, Kk.11).
Nāḍī	1/60th Day	24 minutes	Also Ghaṭikā; fundamental unit measured by water clock or gnomon shadow.
Vināḍikā	1/60th Nāḍī	24 seconds	Also Vighaṭikā, Pala; primary small unit for detailed calculations.

5.2.2 Sidereal Pulse and Spatial Order – The Prāna-Kalā Keystone

This verse moves from the broader time units of Verse 1 to the smallest practical measures, linking them to observable human actions like breath and speech. More profoundly, it establishes a parallel hierarchy for dividing angular space, thereby setting up the crucial equivalence between time and angle that underpins Aryabhata's entire computational astronomy.

Sanskrit:

गुर्वक्षराणि षष्टिः विनाडिकार्क्षी षडेव वा प्राणाः ।

एवं कालविभागः क्षेत्रविभागस्तथा भगणात् ॥ २ ॥

IAST:

gurvakṣarāṇi ṣaṣṭiḥ vināḍikārkṣī ṣaḍeva vā prāṇāḥ|

evaṃ kālavibhāgaḥ kṣetravibhāgastathā bhagaṇāt || 2 ||

Translation: "Sixty long syllables (*gurvakṣarāṇi*) are a sidereal (*ārkṣī*) *vināḍikā*, or [alternatively, a *vināḍikā* is] six breaths (*prāṇāḥ*). Thus time (*kāla*) is divided; so too is the field (*kṣetra* - space/circle) divided, starting from the revolution (*bhagaṇāt*)."

The Human Scale of Time

Having defined the larger units of time, Aryabhata now zooms in on the smallest, grounding his abstract astronomical clock in tangible, human scales. He offers two ways to feel the length of a *vināḍikā* (which we now know is 24 seconds). A *vināḍikā* is the time it takes to either:

1. Pronounce sixty long syllables.

2. Take six normal human breaths (*prāṇāḥ*).

This second definition is particularly insightful. By equating six breaths to one *vināḍikā*, he establishes a standard measure for a single breath, or *prāṇa*:

1 breath (*prāṇa*) = 4 seconds

This simple, human-centric definition is the secret key to his entire computational system.

The Master Stroke: Uniting Time and Space

Herein lies the genius of Aryabhata's system. He uses the *prāṇa* to create a direct and elegant link between the measurement of time and the measurement of space on the celestial sphere.

As he states elsewhere, the heavens appear to rotate one minute of arc (*kalā*) in the time it takes to draw one breath (*prāṇa*).

1 *prāṇa* of Time = 1 *kalā* of Rotation

This is the computational master stroke of the *Aryabhatiya*. It means that time and angular motion are no longer two separate domains; they are perfectly interchangeable. If you know how many breaths have passed, you know exactly how many minutes of arc the sky has turned. This simple equivalence forms the bedrock of his predictive astronomy, allowing for seamless conversions between the speed of a planet and the time it takes to travel.

A Unified System: The Parallel Hierarchies

The verse concludes by stating a profound symmetry: "Thus time is divided; so too is the field (space/circle) divided." Aryabhata reveals that his system for dividing time is a perfect mirror of his system for dividing the circle.

The Division of Time (per day):

- 1 Day = 60 *nāḍīs*
- 1 *Nāḍī* = 60 *vināḍikās*
- 1 *Vināḍikā* = 6 *prāṇas* (breaths)

This gives a total of **21,600 breaths** in one day.

The Division of Space (the circle):

- 1 Circle = 360 degrees
- 1 Degree = 60 *kalā* (minutes of arc)

This gives a total of **21,600 minutes of arc** in one circle.

This is no coincidence. The numbers are identical by design. Aryabhata has masterfully structured his units so that the total number of breaths in a day is exactly equal to the total number of minutes of arc in a circle.

This unified framework, interconnected by the brilliant *prāṇa-kalā* equivalence, is the engine of his mathematical universe. It provides a remarkably coherent and computationally efficient system where time directly translates to angular motion, forming the basis for calculating exactly where the celestial bodies were, are, and will be.

Notes:

- **Gurvakṣarāṇi ṣaṣṭiḥ vināḍikārkṣī:** "Sixty long syllables are a sidereal vināḍikā."

- **Ṣaḍeva vā prāṇāḥ:** "Or [a vināḍikā is] six breaths (prāṇāḥ)." (1 prāṇa ≈ 4 seconds).
- **Evaṃ kālavibhāgaḥ:** "Thus time is divided."
- **Kṣetravibhāgastathā bhagaṇāt:** "So too is the field (space/circle) divided, starting from the revolution (bhagaṇa)."

Sanskrit Term	Definition	Approx. Modern Equivalent
Varṣa	12 Māsa	1 Year
Māsa (Nominal)	30 Divasa	~30 days
Divasa	60 Nāḍī	1 Day
Nāḍī	60 Vināḍikā	24 minutes
Vināḍikā	6 Prāṇa	24 seconds
Prāṇa	Base unit (1 breath)	4 seconds

Table 5.1: Fundamental Time Units in Aryabhata's System, as defined in Kālakriyāpāda Verses 1 and 2 (Kk.1-2). These units, from the Varṣa (Year) down to the Prāṇa (Breath), form a precise, hierarchical framework for astronomical calculations and timekeeping, based on sexagesimal divisions and observed phenomena.

Sanskrit Term	Definition	Value
Bhagaṇa	12 Rāśi	360°
Rāśi	30 Aṃśa	30°
Aṃśa/Bhāga	60 Kalā	1°
Kalā/Liptā	60 Vikalā	1'
Vikalā/Viliptā	Base unit	1"

Table 5.2: Fundamental Angular Units in Aryabhata's System. As described in Kālakriyāpāda Verse 2 (Kk.2), angular space is divided hierarchically from a full revolution (Bhagaṇa) down to small units like the Vikalā (arcsecond), using a sexagesimal system parallel to the time units (Figure 5.1). Context: Comparison with Babylonian/Greek Systems.

5.3 Planetary Cycles and Conjunctions (Kk.3-4)

Applying the defined units of time and space, and drawing upon the fundamental planetary revolution counts established in the Daśagītikā (specifically Gk.3 and Gk.4), Aryabhata now quantifies key cosmic rhythms and relative cycles in Verses 3 and 4 of the Kālakriyāpāda.

5.3.1 Verse 3: Conjunctions of Planets and the Vyatīpāta Cycle (Kk.3)

Sanskrit:

भगणा द्वयोर् द्वयोर् ये विशेषशेषा युगे द्वियोगास्ते ।

रविशशिनक्षत्रगणाः सम्मिश्राश्च व्यतीपाताः ॥ ३ ॥

IAST:

bhagaṇā dvayor dvayor ye viśeṣaśeṣā yuge dviyogāste|

raviśaśinakṣatragaṇāḥ sammiśrāścha vyatīpātāḥ || 3 ||

Translation:(a) "The revolutions (bhagaṇāḥ) of any two [planets], their difference (viśeṣaśeṣāḥ) in a yuga, are their conjunctions (dviyogāḥ) in that yuga."(b) "The sum of the Sun's and Moon's [sidereal] revolutions (raviśaśinakṣatragaṇāḥ = Sun revs + Moon revs), combined (sammiśrāḥ), are the vyatīpātāḥ [in a yuga]."

Cosmic Rhythms: Conjunctions and Cycles

Having defined the units of time, Aryabhata now uses his vast cosmic timeline, the *Mahāyuga*, to calculate the rhythms of how the planets interact with each other. This verse provides the rules for two distinct types of celestial meetings.

1. The Meeting of Two Planets: Sidereal Conjunctions

First, Aryabhata addresses a simple question: how many times do any two planets, moving at different speeds, "lap" each other and align in the sky over the course of a *Mahāyuga*?

This is a **sidereal conjunction**, an alignment of two planets as seen against the backdrop of the distant stars. The method for calculating this is elegantly simple:

The number of conjunctions between two planets is simply the difference between their total revolution counts in a *Mahāyuga*.

For example, let's consider Mars and Jupiter:

- Mars completes **2,296,832** revolutions in a *Mahāyuga*.
- Jupiter completes **364,224** revolutions in the same period.

The number of times Mars laps Jupiter is:

2,296,832 – 364,224 = 1,932,608 conjunctions per *Mahāyuga*.

This straightforward calculation provides a precise measure of the relative rhythm of any two celestial bodies.

2. The *Vyatīpāta* Cycle: A Special Alignment

Next, the verse defines a more complex and traditionally significant cycle known as the ***Vyatīpāta***. While this term has various astrological meanings, often referring to an inauspicious alignment of the Sun and Moon, Aryabhata defines it here in a purely computational way.

He states that the number of *Vyatīpāta* cycles in a *Mahāyuga* is the **sum** of the total revolutions of the Sun and the Moon.

- Sun's revolutions: **4,320,000**
- Moon's revolutions: **57,753,336**

The total number of *Vyatīpāta* cycles is:

4,320,000 + 57,753,336 = 62,073,336 per *Mahāyuga*.

By defining this cycle purely through the sum of their mean motions, Aryabhata transforms a complex, traditionally-defined phenomenon into a predictable, mathematical period. While the exact astronomical configuration this number was meant to track is a subject of scholarly discussion, it demonstrates his method of integrating traditional concepts into a rigorous, computational framework.

Notes:

- **Bhagaṇā dvayor dvayor ye viśeṣaśeṣā yuge dviyogāste:** "Revolutions of any two [planets], their difference in a yuga, are their conjunctions in that yuga." (Number of conjunctions = Rev_1 - Rev_2).
- **Raviśaśinakṣatragaṇāḥ sammiśrāścha vyatīpātāḥ:** "Sum of Sun's and Moon's [sidereal] revolutions combined are the vyatīpātāḥ [in a yuga]."

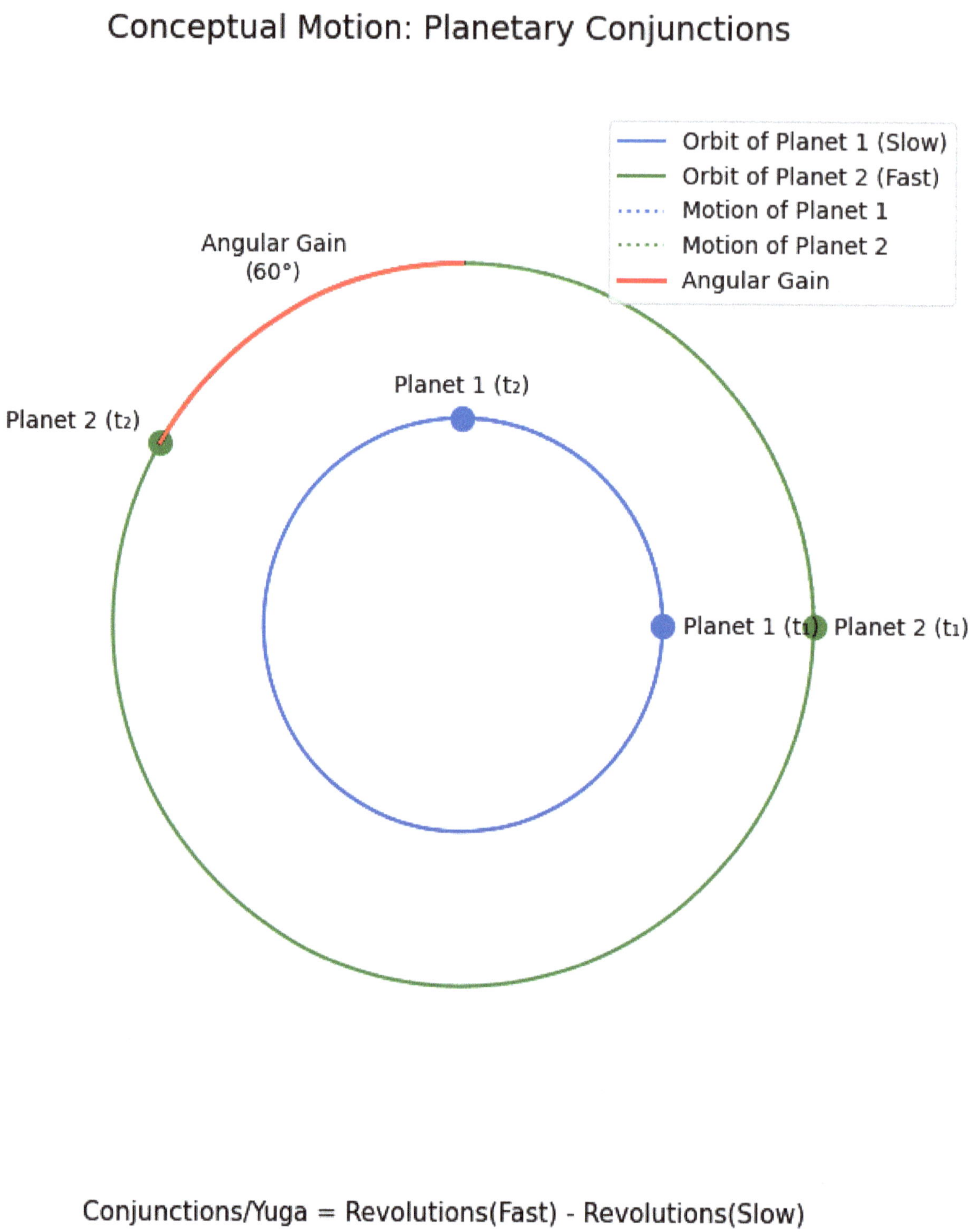

Figure 5.1: Conceptual diagram illustrating how the difference in orbital speeds between two planets (Planet 2, faster, and Planet 1, slower) leads to conjunctions. Each time the faster planet completes one full revolution more than the slower one, they align. The total number of conjunctions between two planets in a Mahāyuga is therefore the difference between their total revolution counts in that period, as stated in Kālakriyāpāda Verse 3 (Kk.3).

5.3.2 Apogee/Anomaly Cycles and Jupiter Years

Sanskrit:

स्वोच्चभगणाः स्वभगणैर्विशेषिताः स्वोच्चनीचपरिवर्ताः ।

गुरुभगणा राशिगुणा अश्वयुजाद्या गुरोरब्दाः ॥ ४ ॥

IAST:

svocchabhagaṇāḥ svabhagaṇairviśeṣitāḥ svocchanīchaparivartāḥ|

gurubhagaṇā rāśiguṇā aśvayujādyā gurorabdāḥ || 4 ||

Translation: "Revolutions of its apogee (svocchabhagaṇāḥ), subtracted from the planet's own [sidereal] revolutions (svabhagaṇaiḥ viśeṣitāḥ = difference taken), are the number of its apogee-perigee cycles (svocchanīchaparivartāḥ)." "Jupiter's [sidereal] revolutions (gurubhagaṇāḥ) multiplied by the [twelve zodiacal] signs (rāśiguṇāḥ), are Jupiter's years (gurorabdāḥ), beginning from Aśvayuj (the start of Aries)."

The Imperfect Circle: Correcting for Planetary Speed

If you watch a planet over many months, you will notice that it doesn't move at a perfectly steady speed. It seems to speed up and slow down in a regular, predictable rhythm. This deviation from perfectly uniform motion is called the **anomaly**. In this verse, Aryabhata provides the method for calculating the number of these anomaly cycles, a crucial step in correcting a planet's average position to find its true, observable position.

The Race Car Analogy

Imagine a race car on an oval track.

- It completes one full lap each time it crosses the finish line. This is like its **sidereal revolution**.
- However, its speed changes. It moves fastest on the straightaway closest to the stands and slowest on the far side of the track. The point of slowest speed is its **apogee**, or *Mandoccha*.

The number of times the car goes from "fastest" back to "fastest" is its **anomaly cycle**. This is not quite the same as a full lap. Aryabhata's model accounts for this by imagining that the apogee itself moves very slowly around the track.

Calculating the Anomaly Cycles

To find the number of anomaly cycles for any planet in a *Mahāyuga*, Aryabhata gives a simple rule:

Anomaly Cycles = Planet's Sidereal Revolutions – Apogee's Revolutions

This difference gives the precise number of times the planet completes a full cycle relative to its own point of slowest motion.

Two Types of Anomaly

Aryabhata's system uses this principle to model two distinct types of observed anomalies:

1. **The Slow Anomaly (*Manda*):** This correction accounts for the planet's changing speed in its own orbit, as described in the race car analogy. It corrects for the fact that orbits are not perfect circles with the Earth at the center.

2. **The Fast Anomaly (*Śīghra*):** This is a more complex correction that accounts for the planet's motion as seen from a moving Earth. It brillianty models the synodic cycle—the rhythm of a planet's appearance relative to the Sun—including the strange phenomenon of **retrograde motion**, where a planet appears to temporarily move backward in the sky.

By calculating the number of these *manda* and *śīghra* cycles, Aryabhata created the fundamental parameters for his system of corrections. This allowed him to move from a simple model of average speeds to a sophisticated system that could accurately predict the true, non-uniform motions of the planets as they are actually observed in the sky.

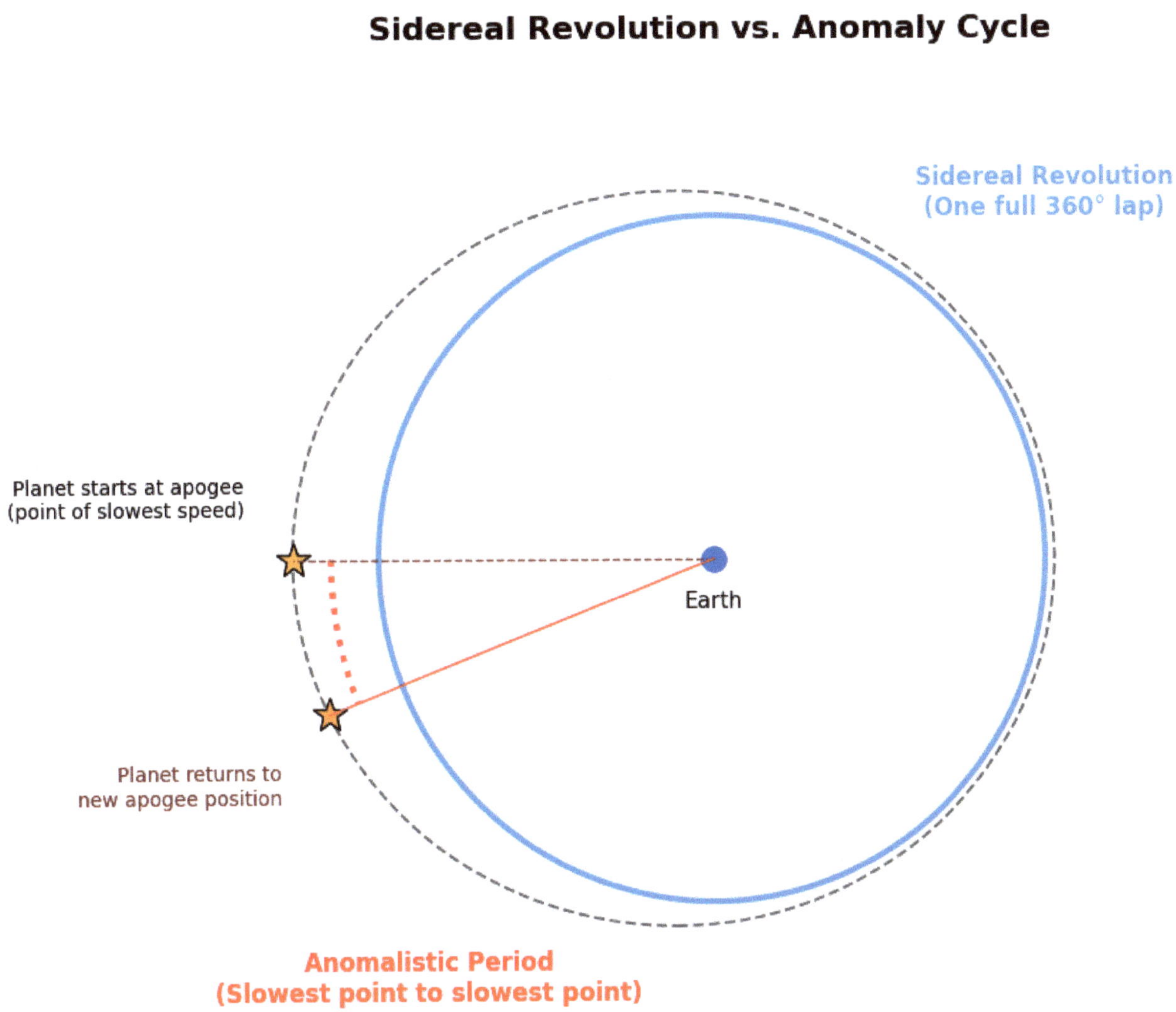

Figure 5.2: A diagram showing the difference between a planet's sidereal revolution (a full 360° orbit) and its anomalistic period. As the apogee—the point of slowest speed—moves forward, the planet must travel slightly more than 360° to return to it.

The Great Year of Jupiter

Beyond the cycles of the Sun and Moon, ancient Indian calendrics often used a larger rhythm based on the slow, majestic journey of the planet Jupiter, or *Guru*. In this verse, Aryabhata provides the precise astronomical definition for this traditional unit of time, the "Jupiter Year."

What is a Jupiter Year?

A **Jupiter Year** (*gurorabda*) is not the time it takes Jupiter to make one full orbit. Instead, it is defined as:

> **The average time it takes for Jupiter to travel through one single sign of the zodiac (30 degrees).**

Since there are 12 signs in the zodiac, it takes 12 of these "Jupiter Years" for the planet to complete one full revolution against the stars.

The Calculation

Using the revolution counts from the *Gītikāpāda*, Aryabhata can calculate the total number of these "Jupiter Years" in a *Mahāyuga*:

- Jupiter's total revolutions: **364,224**
- Signs per revolution: **12**
- Total Jupiter Years = 364,224 × 12 = **4,370,688 per *Mahāyuga***.

Significance: A 12-Year Cycle
Because Jupiter's actual orbital period is about 11.86 Earth-years, this definition creates a grand cycle of approximately **12 solar years**. This 12-year Jovian cycle held deep cultural and calendrical significance in India. It was often used in naming individual years within a larger 60-year cycle, a system with parallels to the Chinese zodiac.
By providing a precise astronomical calculation for this cycle, Aryabhata integrated a traditional and culturally important time unit into his rigorous, mathematical framework.
These verses demonstrate how Aryabhata used his foundational parameters to derive the key cyclical periods essential for both tracking planetary motions and constructing calendars. However, accurate timekeeping over long periods requires carefully distinguishing between different types of days, months, and years. The next section clarifies these distinct measures of time.

Notes:

- **Svocchabhagaṇāḥ svabhagaṇairviśeṣitāḥ svocchanīchaparivartāḥ:** "Revolutions of its apogee (Uccha), subtracted from planet's own revolutions, are its apogee-perigee cycles (anomaly/kendra cycles)." (No. of anomaly cycles = Planet Revs - Uccha Revs).
- **Gurubhagaṇā rāśiguṇā aśvayujādyā gurorabdāḥ:** "Jupiter's revolutions multiplied by the [twelve] signs, are Jupiter's years, beginning from Aśvayuj (Aries)." (Number of Jupiter Years = Jupiter Revs × 12).

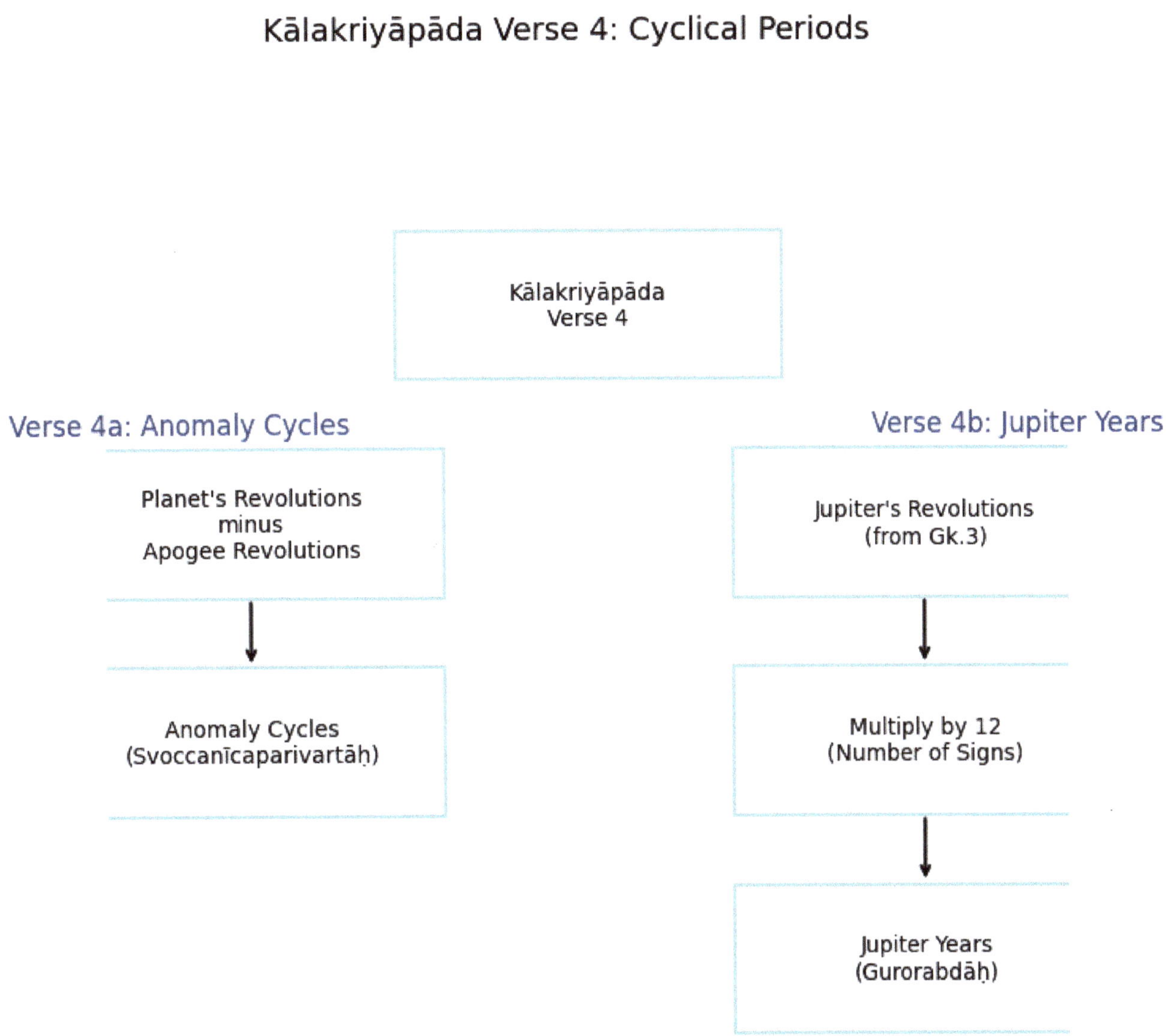

Figure 5.3: Dual Calculations in Aryabhata's Kk.4. (a) Anomaly cycles derived from (Planet Revs - Uccha Revs)/Yuga. (b) 'Jupiter Years' defined as (Jupiter Revs × 12 signs)/Yuga.

5.4 Four Types of Time Measurement (Kk.5)

Accurate astronomical calculation and calendrics depend on clearly distinguishing between different ways of measuring time, each based on a different celestial reference frame. Verse 5 explicitly defines the four crucial, interacting time systems used in Indian astronomy and Aryabhata's framework.

Time System	Sanskrit Name	Basis (from Kk.5)	Key Unit
Solar	Saura	Sun rev	Solar Year
Lunar	Cāndra	Sun-Moon conj	Synodic Month
Civil	Sāvana	Sunrise-Sunrise/Ravibhūyoga	Civil Day
Sidereal	Nākṣatra	Earth rot/Bhūdivasa	Sidereal Day

Table 5.3: Aryabhata's Four Fundamental Time Systems (Kk.5). Defines Solar, Lunar, Civil, and Sidereal time, each based on distinct celestial motions and crucial for accurate calendrics and astronomy.

5.4.1 The Fourfold Framework

Sanskrit:

रविभगणा रव्यब्दा रविशशियोगा भवन्ति शशिमासाः ।

रविभूयोगा दिवसा भूदिवसाश्चापि नाक्षत्राः ॥ ५ ॥

IAST:

ravibhagaṇāḥ ravyabdāḥ raviśaśiyogāḥ bhavanti śaśimāsāḥ|

ravibhūyogāḥ divasā bhūdivasāścāpi nākṣatrāḥ || 5 ||

Having established the hierarchical units of time, Aryabhata now anchors them to their astronomical origins. In this verse, he succinctly defines the physical basis for each of the primary temporal cycles:

Translation: "Sun's revolutions are solar years; Sun-Moon conjunctions are lunar months. Sun-Earth conjunctions are civil days; Earth's rotations are sidereal [days]."

This compact statement confirms the observational foundation of his system:

- A **solar year** (*ravyabda*) is defined by one complete revolution of the Sun against the backdrop of the stars (*ravibhagaṇa*)—what we would call a sidereal year.

- A **lunar month** (*śaśimāsa*) is the period between two successive conjunctions of the Sun and Moon (*raviśaśiyoga*)—the synodic month.

- A **civil day** (*divasa*) is defined by successive "Sun-Earth conjunctions" (*ravibhūyoga*). As commentators like Bhaskara I clarify, this is interpreted practically as the interval between two consecutive sunrises for an observer on Earth, known as the *sāvana* day.

- A **sidereal day** (*nākṣatra ḍivasa*) is defined by one full rotation of the Earth relative to the fixed stars (*bhūdivasa*), a direct reference to the Earth's rotation count given in the *Gītikāpāda*.

To ensure system-wide consistency, Aryabhata's framework, as elaborated by Bhaskara I, establishes a common starting point for both the civil (*sāvana*) and sidereal (*nākṣatra*) days: sunrise at the prime meridian of Lanka. This verse thus demonstrates how abstract units of time are tethered to the observable, repeating cycles of the cosmos.

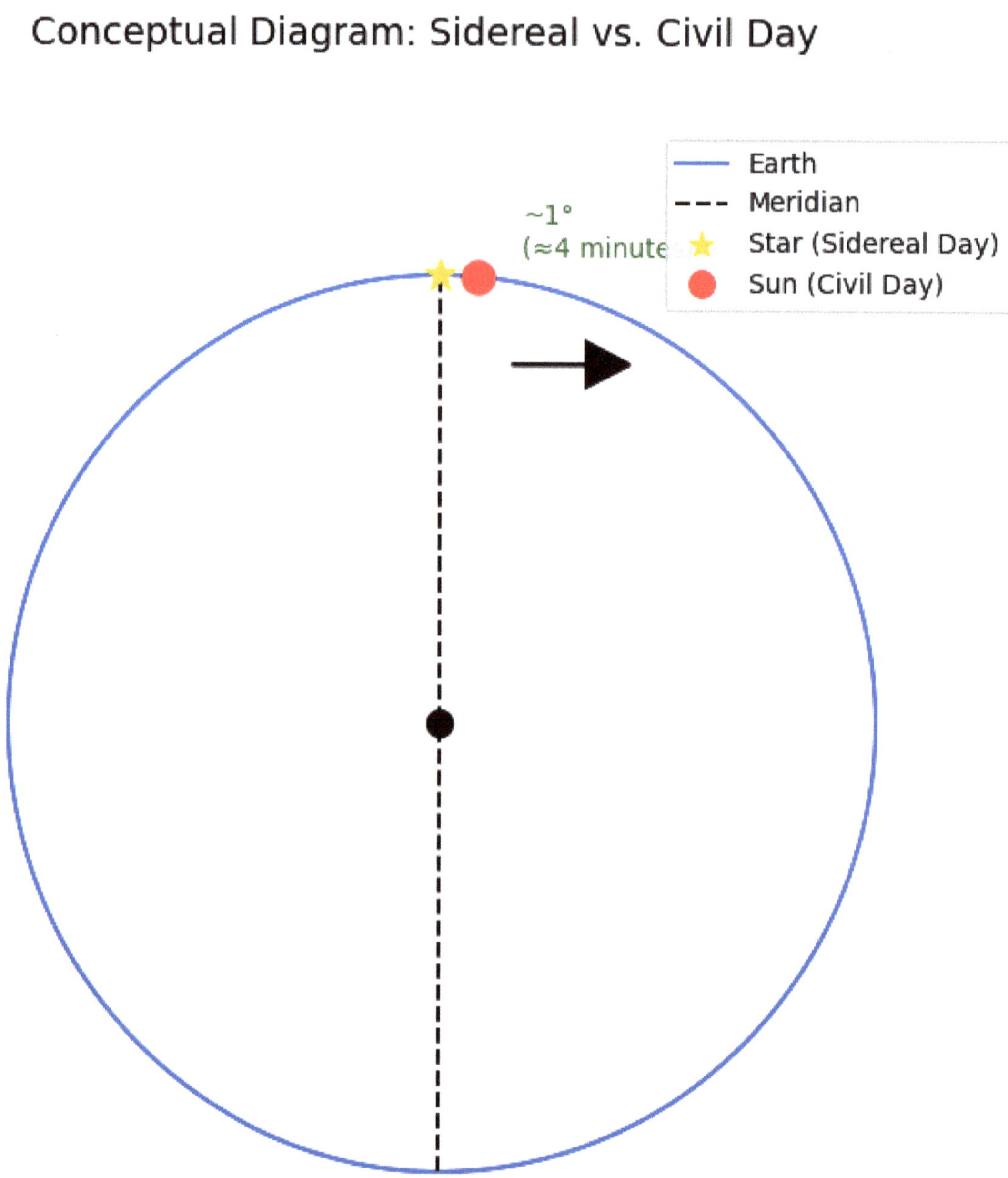

Figure 5.4: Sidereal vs. Civil (Sāvana) Day (Aryabhata, Kk.5). A Sidereal Day is Earth's rotation relative to stars. A Civil Day (sunrise to sunrise) is ~4 minutes longer as Earth 'catches up' to the Sun's orbital motion. This distinction is key for astronomical timekeeping.

The Interplay of Cosmic Clocks

These four distinct temporal systems—Sāvana, Nākṣatra, Cāndra, and Saura—do not march in perfect lockstep. The genius of Aryabhata's framework lies in precisely quantifying their interplay, revealing the slight but cumulative differences that are the key to all calendrical science.

- The **Sāvana (civil) day** is slightly longer than the **Nākṣatra (sidereal) day**. This is because in the time it takes the Earth to complete one full rotation against the stars, the Sun has also advanced slightly eastward along the ecliptic. The Earth must therefore rotate a little further to bring the Sun back to its meridian, making the solar-based day longer than the star-based day.

- A similar dynamic governs the lunar cycle. The **Cāndra (synodic) month** of approximately 29.53 days is longer than the Moon's true sidereal period of 27.32 days. After completing one orbit relative to the stars, the Moon must travel further to "catch up" with the moving Sun and re-establish the conjunction that defines the new moon.

The fundamental consequence of these misalignments is the central problem of all luni-solar calendrics: the solar year does not contain a whole number of lunar months or civil days. To prevent the calendar from drifting out of sync with the Moon's phases or the Sun's seasons, a system of periodic adjustments is essential.

This necessitates the complex machinery of intercalation (the addition of an extra month, *adhikamāsa*) and omission (*kṣayatithi*) to maintain long-term alignment. The critical insight is that Aryabhata's system does more than just describe this problem; it provides the precise numerical parameters required to solve it. The revolution counts established in the *Gītikāpāda* are the raw data from which the exact rates of these necessary adjustments can be calculated, forming the mathematical backbone of the Indian calendar.

Notes:

- **Ravibhagaṇāḥ ravyabdāḥ:** "Sun's revolutions are solar years (Saura)."
- **Raviśaśiyogāḥ bhavanti śaśimāsāḥ:** "Sun-Moon conjunctions are lunar months (Cāndra/Synodic)."
- **Ravibhūyogāḥ divasā:** "Sun-Earth conjunctions (successive sunrises) are civil days (Sāvana)."
- **Bhūdivasāścāpi nākṣatrāḥ:** "Earth's rotations (relative to stars) are sidereal days (Nākṣatra)."

5.5 Harmonizing Luni-Solar Time: Intercalation and Omission (Kk.6)

Indian calendrical systems are typically luni-solar, designed to keep months aligned with the Moon's phases while ensuring the year remains synchronized with the solar seasons. This necessitates periodic adjustments because twelve lunar months (approximately 354 days) are shorter than one solar year (approximately 365.25 days), and the average length of a lunar day (tithi) differs slightly from that of a civil (sāvana) day. These adjustments involve inserting extra "leap" months (adhimāsa) and occasionally omitting lunar days (avamarātra or kṣayatithi). Kālakriyāpāda Verse 6 (Kk.6) provides the computational rule for determining these adjustments based on the elapsed time since the epoch, with the long-term rates for such adjustments being derived directly from Aryabhata's fundamental parameters established in the Daśagītikā.

5.5.1 The Core Calculation of Adjustments (Kk.6)

Sanskrit:
अधिमासकशेषं यद् दिवसाभ्यस्तं युगाधिमासोनम् ।
अधिमासकावमानि च तद् विज्ञेयं यथोद्दिष्टम् ॥ ६ ॥

IAST:
adhimāsakaśeṣaṃ yad divasābhyastaṃ yugādhimāsonam |
adhimāsakāvamāni cha tad vijñeyaṃ yathoddiṣṭam || 6 ||

(Note: Manuscript variations exist. This is a plausible reconstruction. More commonly cited interpretations focus on using divisors and multipliers derived from yuga parameters to find the accumulated adhimāsas and avamas from the ahargaṇa. The verse essentially outlines finding the 'remainder' or cumulative count of these adjustments.)

With the various time systems defined, Aryabhata provides the computational engine needed to synchronize them. The central challenge of a luni-solar calendar is knowing precisely when to insert an intercalary month (*adhikamāsa*) or omit a lunar day (*avamarātra*). Verse Kk.6 provides the solution, a method powered by the single most important input for calendrical calculations: the **ahargaṇa**, or the total count of civil days elapsed since the epoch.

The verse outlines a rule of elegant proportionality:

Conceptual Translation: "The remainder for intercalary months is found by multiplying the elapsed days by the total intercalary months in a *yuga* and dividing by the total days in a *yuga*. The same method applies to omitted lunar days. These are to be known as instructed."

This is a powerful algorithm. By taking the *ahargaṇa* for any given date, an astronomer can multiply it by the total number of *adhimāsas* in a *Mahāyuga* and then divide by the total number of civil days in that same epoch. The integer part of the result reveals how many intercalary months have passed, while the remainder (*śeṣa*) indicates how close the calendar is to needing the next one. A parallel calculation is performed for the omitted lunar days (*avamarātra*).

5.5.2 Deriving the Long-Term Adjustment Rates

The power of this rule lies in the precision of its inputs. The multipliers and divisors are not arbitrary; they are derived directly from the fundamental cosmic parameters laid out in the *Gītikāpāda* for one *Mahāyuga* (4,320,000 years).

1. Intercalary Months (*Adhimāsa*) per Yuga:

- First, the number of synodic (lunar) months is found by subtracting the Sun's revolutions from the Moon's: 57,753,336 – 4,320,000 = **53,433,336 lunar months**.
- The number of solar months is simply the number of years × 12: 4,320,000 × 12 = **51,840,000 solar months**.
- The difference between these two yields the total number of intercalary months per *yuga*: 53,433,336 – 51,840,000 = **1,593,336 *adhimāsas***.
  - This means, on average, an extra month is added roughly every 2.7 solar years.

2. Omitted Lunar Days (*Avamarātra*) per Yuga:

- The total number of lunar days (*tithis*) is the number of lunar months × 30: 53,433,336 × 30 = **1,603,000,080 *tithis***.
- The total number of civil (*sāvana*) days in a *yuga* is a standard value in Aryabhata's system: **1,577,917,500 civil days**.
- The difference between the count of *tithis* and civil days gives the number of omitted days: 1,603,000,080 – 1,577,917,500 = **25,082,580 *avamarātras***.
  - This staggering number reflects the constant, subtle adjustment required to align the lunar day count with the solar day count.

5.5.3 From Cosmic Cycles to Practical Calendars

This meticulous computational framework was the engine for constructing and maintaining accurate luni-solar calendars (*pañcāṅgas*). Armed with the *ahargaṇa* for a given date, an astronomer could use the proportional rule of Kk.6 to determine exactly how many *adhimāsas* and *avamarātras* should have occurred since the beginning of the age. This ensured that religious festivals tied to specific lunar phases remained broadly aligned with the correct seasons determined by the solar year.

In this way, Aryabhata's system masterfully connects the immediate needs of daily timekeeping to the grand, predictable cycles of his cosmology, creating a calendar whose accuracy is anchored in the vast and unwavering rhythms of the universe.

Notes (for Kk.6):

- **Adhimāsakaśeṣaṁ / Avamāna[śeṣaṁ]:** "Remainder/count related to intercalary months" / "Remainder/count related to omitted lunar days."

- **Divasābhyastaṃ:** "Multiplied by the [elapsed] days" (i.e., by the ahargaṇa).
- **Yugādhimāsa[onam]:** "Yuga intercalary months [as a factor, with 'ūnam' implying division by total yuga days]."
- **Ahargaṇa (Implicit Input):** The calculation of these adjustments relies fundamentally on the ahargaṇa.
- **Yuga Parameters (Implicit Basis):** The rates/divisors used are derived from the revolution counts over a Mahāyuga (Gk.3-5, Kk.5).

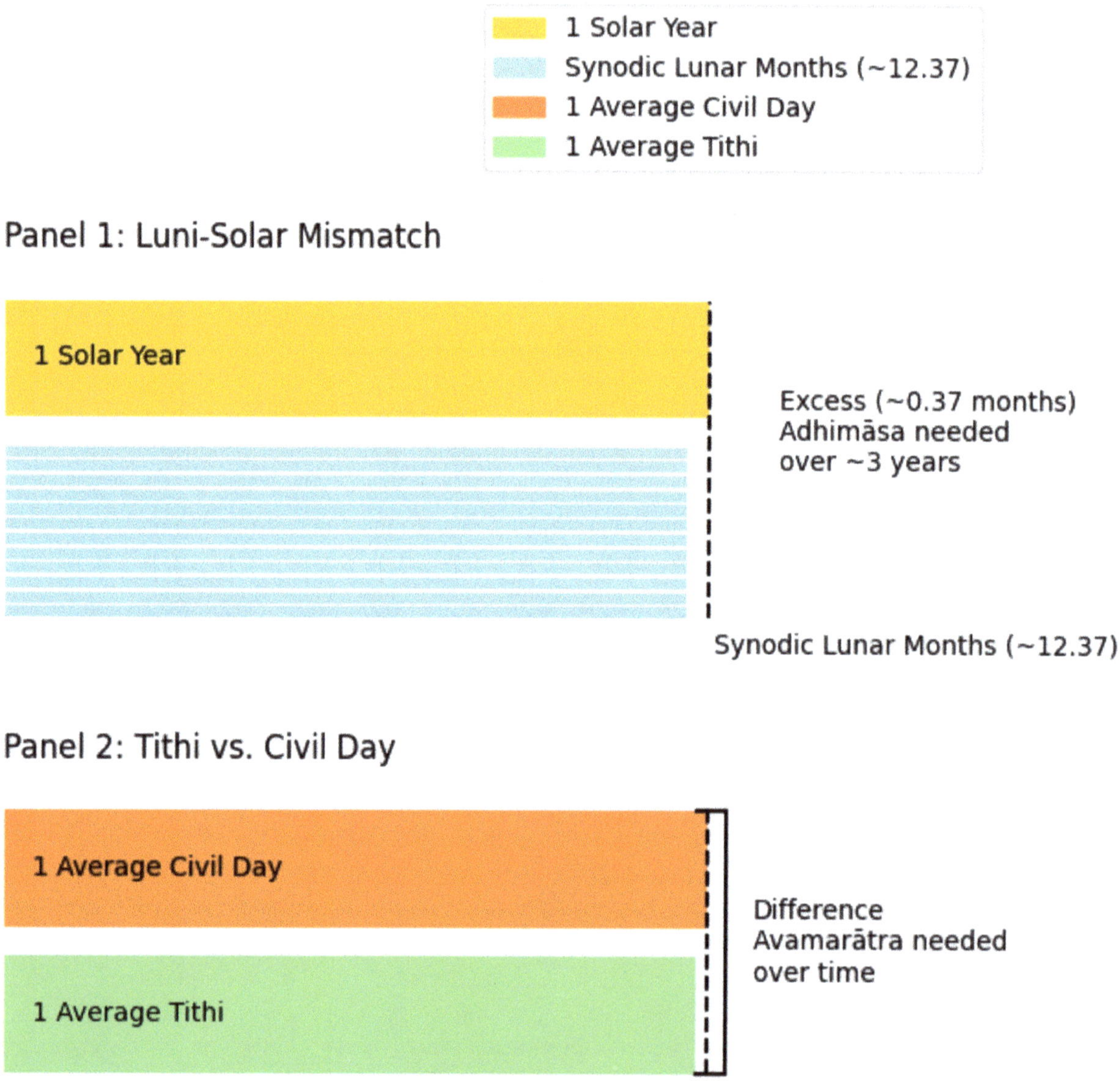

Figure 5.5: Luni-solar calendar mismatches requiring adjustments. Panel 1: Solar Year > 12 Synodic Lunar Months (necessitating intercalary Adhimāsa). Panel 2: Tithi < Civil Day (necessitating omitted Avamarātra/Kṣayatithi). Aryabhata's Mahāyuga parameters precisely quantified these differences.

5.6 Yugas and Brahmā's Day – Cosmic Timescales (Kk.7-8 Context & Gk.5)

Having established the need for calendrical adjustments based on differential rates over a Mahāyuga, Aryabhata's system places these calculations within an even grander cosmological framework. The following verses define the vast timescales of the Mahāyuga, its subdivisions, and the overarching Kalpa cycle, connecting practical computational astronomy to a profound vision of cosmic time. This framework is crucial for ensuring the long-term consistency of his astronomical model.

5.6.1 Proportional Calculation within Yuga Divisions

This verse provides the principle for proportionally calculating various celestial cycles within any desired subdivision of the Mahāyuga. It is this very principle that commentators like Bhaskara I leverage to argue for Aryabhata's system of *equal* yuga quarters, which is essential for the mathematical coherence of his planetary model.

Sanskrit:

युगग्रहभगणान् मासान् वर्षाण्य् अयनम् ऋतून् अहोरात्रान् ।

इष्टान् यथेष्टभागात् कृत्वा रूपाद् यथोक्तफलम् ॥ ७ ॥

IAST:

yugagrahabhagaṇān māsān varṣāṇy ayanam ṛtūn ahorātrān|

iṣṭān yatheṣṭabhāgāt kṛtvā rūpād yathoktaphalam || 7 ||

Translation: "The revolutions of planets in a yuga (*yugagrahabhagaṇān*), months (*māsān*), years (*varṣāṇi*), solstitial courses (*ayanam*), seasons (*ṛtūn*), [and] day-night cycles (*ahorātrān*); having determined these for desired [periods] from a desired fraction [of a yuga] (*yatheṣṭabhāgāt*) (or starting from unity [a single yuga as the base], *rūpād*), the result is as stated [i.e., can be calculated proportionally, *yathoktaphalam*]."

The Uniformity of Time: The Case for Equal *Yuga* Quarters

In traditional Puranic lore, the great cosmic ages were not equal. They followed a descending ratio of 4:3:2:1, where the first age (*Kṛta Yuga*) was a long golden era, and the final age (*Kali Yuga*, our own) was the shortest and most corrupt.

In this pivotal verse, Aryabhata makes a simple but profound statement that directly challenges this ancient tradition. He declares that all celestial cycles can be calculated for any fraction of a *Mahāyuga* through simple proportion.

The Principle of Proportionality

The core message is one of uniform, predictable motion. If you know that a planet completes a certain number of revolutions in a full *Mahāyuga* of 4,320,000 years, you can find how many revolutions it completes in one-quarter of that time by simply dividing by four.

This principle seems obvious, but its implication is revolutionary: **for this mathematical proportionality to work, the "quarters" of the *Mahāyuga* must themselves be of equal length.**

This means that Aryabhata's system requires a *Mahāyuga* divided into four equal *yugapādas* (quarters), each lasting exactly 1,080,000 years. This directly contradicts the traditional 4:3:2:1 ratio.

Bhāskara I's "Smoking Gun" Argument

This verse became the "smoking gun" that the 7th-century commentator Bhāskara I used to defend Aryabhata's radical new system. Bhāskara I argued with mathematical certainty that the Puranic system of unequal ages was completely incompatible with Aryabhata's meticulously defined planetary revolutions.

He showed that only if the four ages are equal can the planets maintain their precise, clockwork-like rhythm and return to their starting alignment at the end of each *Mahāyuga*.

In this brilliant interpretive move, Bhāskara I makes it clear that Aryabhata was making a deliberate choice. He was privileging the internal mathematical consistency of his predictive, scientific model over the established, mythological chronology.

This verse, therefore, is not just a simple statement about proportionality. It is a quiet declaration of scientific principle. It establishes that in Aryabhata's cosmos, time is not a moral drama of declining ages, but a uniform, consistent, and, most importantly, **calculable** dimension.

Figure 5.6: Kk.7: Proportional calculation within Yuga divisions and its implication of equal Yugapādas for Aryabhata's Mahāyuga structure.

5.6.2 The Kalpa – A Day of Brahmā

Sanskrit:

युगसहस्रं कल्पो ब्राह्मो दिवसो भवेत् स विज्ञेयः ।

रात्रिस् तस्य तावत्य् एव प्रभवो विनाशश् च ॥ ८ ॥

IAST:

yugasahasraṃ kalpo brāhmo divaso bhavet sa vijñeyaḥ|

rātris tasya tāvaty eva prabhavo vināśaś cha || 8 ||

With this verse, Aryabhata expands the temporal horizon to its furthest limit, defining the vast cosmological cycle known as the **Kalpa**.

Translation: "A thousand *yugas* is a *Kalpa*, known as a Day of Brahmā. His Night is just as long; during the Day is creation and during the Night is dissolution."

Here, the *Kalpa*—the span of active cosmic creation (*prabhavaḥ*)—is defined as lasting 1,000 *Mahāyugas*, or 4.32 billion solar years. It is followed by a "Night of Brahmā," an equally long period of cosmic rest and dissolution (*vināśaḥ*). The full cycle of this cosmic breath spans a staggering 8.64 billion years, a framework of "deep time" that situates all of human history within an almost unimaginable expanse.

The Tension Between Cosmology and Calculation: 1000 vs. 1008

A point of deep scholarly interest arises here. While this verse (Kk.8) explicitly defines the *Kalpa* as 1000 *Mahāyugas*, aligning with many traditional Puranic sources, Aryabhata's own core parameters in the *Daśagītikā* (Gk.5) imply a different value. The structure of 14 *Manvantaras*, each containing 72 *Mahāyugas*, mathematically necessitates a *Kalpa* of 14 × 72 = **1008 *Mahāyugas***.

Commentators like Bhaskara I navigate this apparent discrepancy with practical wisdom. They often cite the 1000-*yuga* definition when discussing general cosmology but rely implicitly on the mathematically derived 1008-*yuga* structure when performing precise astronomical calculations. This highlights a crucial principle: when rigorous computation is required, the internal mathematical consistency of Aryabhata's astronomical parameters takes precedence over definitions borrowed from external traditions.

These vast cycles, whether of 1000 or 1008 *yugas*, provided the grand canvas upon which the more immediate calculations of planetary positions and calendrical science were painted. To make these immense timescales functional for practical astronomy, however, a precise starting point within this deep time is required. It is to this crucial task—the anchoring of his entire system to a specific moment in history—that Aryabhata turns next.

Notes:

- **Yugasahasraṃ kalpo brāhmo divaso bhavet sa vijñeyaḥ:** "A thousand yugas is a Kalpa, known as a Day of Brahmā." (Kalpa = 1000 Mahāyugas = 4.32 billion years, in this Puranic definition).
- **Rātris tasya tāvaty eva:** "His Night is just as long."
- **Prabhavo vināśaś cha:** "During the Day is creation, during the Night is dissolution."

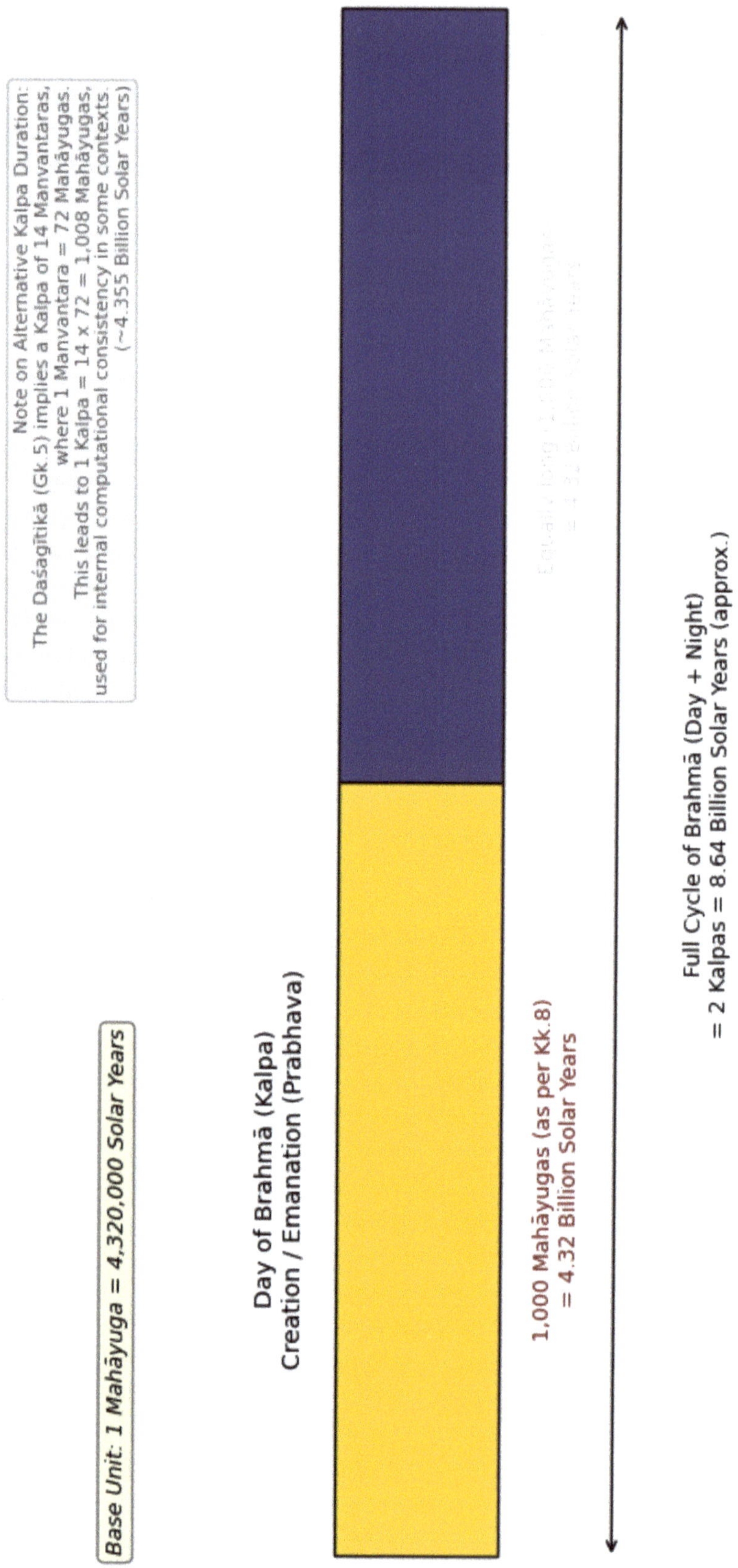

Figure 5.7: Kk.8: Aryabhata's Kalpa (Day of Brahmā). Illustrating the vast cosmological timescale defined in Kālakriyāpāda Verse 8.

5.7 The Starting Point: Epoch, Ahargana, and Calendrical Conventions (Kk .9-11)

Precise astronomical calculations demand a universally agreed-upon starting point in time – an epoch. Aryabhata's system is firmly anchored to such a moment within the grand cosmic cycles. All subsequent computations of planetary positions rely on accurately determining the number of civil days elapsed since this epoch. This elapsed day count, known as the ahargaṇa (literally, "heap of days"), is the master key that unlocks the state of the heavens for any given date. Verses Kk.9-11 of the Kālakriyāpāda first set the broader context of time units (Kk.9), then detail the critical calculation of the ahargaṇa (Kk.10), and finally specify key calendrical conventions (Kk.11) essential for these computations.

5.7.1 The Epoch: Dawn of the Kali Yuga (Foundation for Kk.9-11)

As established in the *Gītikāpāda* (Gk.4, Gk.5), the fundamental reference point for all calculations in the Aryabhatiya is the commencement of the current Kali Yuga. Based on the cosmic time elapsed as detailed in Gk.5, this epoch corresponds to sunrise on Friday, February 18, 3102 BCE (Julian calendar), at Aryabhata's prime meridian of Laṅkā. (While Gk.4 mentions "Wednesday" [budhāhni] in relation to the epoch, the computational start derived from elapsed yugas points to Friday. This discrepancy is a well-known point of scholarly discussion, with Friday being the operationally accepted day for the epoch in the Aryabhatian tradition). The subsequent verses operate from this defined starting point.

5.7.2 Broader Cyclical Divisions and Time Measures (Kk.9)

Before detailing the calculation of elapsed days, Verse Kk.9 establishes a broader context by listing various standard time units and their hierarchical relationships, situating Aryabhata's specific computational methods within the wider landscape of Indian time-reckoning.

Sanskrit:

युगवर्षमासार्धं तिथयो नाड्यो विनाडिकाः क्रमशः ।

षष्ट्या षष्ट्याभ्यस्ता गुरुरिति कालप्रमाणानि ॥ ९ ॥

IAST:

yugavarṣamāsārdhaṃ tithayo nāḍyo vināḍikāḥ kramaśaḥ |

ṣaṣṭyā ṣaṣṭyābhyastā gururiti kālapramāṇāni || 9 ||

Translation:

"Yuga, year, month, half-month (pakṣa), tithis, nāḍīs, [and] vināḍikās – successively (kramaśaḥ) [these units are related by being] multiplied sixty by sixty (ṣaṣṭyā ṣaṣṭyābhyastā). These (gururiti, literally 'the great ones' or 'important ones') are the measures of time (kālapramāṇāni)."

Notes (for Kk.9):

- **Yugavarṣamāsārdhaṃ tithayo nāḍyo vināḍikāḥ kramaśaḥ:** "Yuga, year, month, half-month (pakṣa), tithis, nāḍīs, vināḍikās successively." (Lists various time units).
- **Ṣaṣṭyā ṣaṣṭyābhyastā:** "[Related by being] multiplied by sixty by sixty." (Highlights sexagesimal relationships).
- **Gururiti kālapramāṇāni:** "These are the important/great measures of time."

The Many Scales of Time

Before detailing the precise mechanics of his calendrical system, Aryabhata takes a moment to acknowledge the many different scales on which time was measured in ancient India. This verse lists a familiar hierarchy of time units, creating a bridge from vast, traditional concepts to the precise, computational units he will use for his astronomy.

He lists a spectrum of durations:

- ***Yuga*** (an age)
- **Year** (*varṣa*)
- **Month** (*māsa*)
- **Half-month** or fortnight (*pakṣa*)
- **Lunar day** (*tithi*)
- ***Nāḍī*** (a unit of 24 minutes)
- ***Vināḍikā*** (a unit of 24 seconds)

The phrase "multiplied sixty by sixty" emphasizes the sexagesimal (base-60) structure that governs the smaller, more precise units of this system, which are the building blocks of astronomical calculation.

Acknowledging a Wider World

By including traditional and sometimes variable units like the *yuga* in this general list, Aryabhata may be showing an awareness of the diverse time-reckoning systems prevalent in his era, such as those found in Jaina or other philosophical traditions.

However, having acknowledged this broader context, his purpose becomes clear. He is about to pivot from this general list to the specific, rigorously defined units needed for his own scientific system. This verse serves as a final, panoramic view of the landscape of time before he zooms in to construct the precise, mathematical clockwork that will power his predictive astronomy.

5.7.3 Calculating the Ahargana – The Heap of Days (Kk.10)

The crucial input for finding planetary positions on any specific date is the ahargaṇa. This is the total number of civil (sāvana) days that have passed from the defined epoch (sunrise at Laṅkā, start of Kali Yuga) up to the sunrise of the desired date. Accurately determining the ahargaṇa was a fundamental task, and Verse Kk.10 provides the rules for this vital calculation.

Sanskrit:

अधिमासकावमानां शेषं वर्षाग्रमासगुणं क्रमशः ।

द्वादशषष्टिघ्नोनं षष्टिद्वयहीनवर्षमासाः स्युः ॥ १० ॥

IAST:

adhimāsakāvamānāṃ śeṣaṃ varṣāgramāsaguṇaṃ kramaśaḥ |

dvādaśaṣaṣṭighnonaṃ ṣaṣṭidvayahīnavarṣamāsāḥ syuḥ || 10 ||

(Due to its extreme condensation, a literal word-for-word translation is less useful than an algorithmic interpretation based on commentaries.)

The Day-Count: Calculating the *Ahargaṇa*

How do you find a planet's precise position for a specific date in the past or future? First, you need to know exactly how many days have passed between now and then. This is the single most important calculation in practical Indian astronomy: finding the ***ahargaṇa***, or "the heap of days."

> **The *ahargaṇa* is the total number of civil (solar) days that have elapsed from a fixed starting point (the beginning of the *Kali Yuga*) up to the sunrise of the target date. Think of it as the odometer on a great cosmic time machine. If you know this number, you can calculate the position of any celestial body.**

The Recipe for the *Ahargaṇa*

This highly condensed verse provides the complete, step-by-step recipe for calculating this crucial number. It is a meticulous process of accounting for the differences between the solar, lunar, and civil calendars.

Given a target date (e.g., 1000 years, 2 months, and 15 days after the start of the epoch), the procedure is as follows:

1. **Start with the Solar Calendar:** Calculate the total number of solar months that have passed.

2. **Add the Intercalary Months:** Using the long-term rates derived from the *Mahāyuga*, calculate how many extra lunar months (*adhimāsas*) have been added to the calendar during this time. Add these to your total to get the true number of lunar months.

3. **Convert to Lunar Days (*Tithis*):** Convert this total number of lunar months into lunar days, or *tithis*, and add the *tithis* from the current, incomplete month. This gives you a grand total of all the *tithis* that have passed.

4. **Subtract the Omitted Days:** A *tithi* is slightly shorter than a civil day. To keep the calendars aligned, some *tithis* are periodically omitted. Using the appropriate long-term rate, calculate the total number of these "omitted days" (*avamas*) and subtract them from your grand total.

The final result is the **ahargaṇa**—the precise, unified count of civil days elapsed since the epoch.

Putting the *Ahargaṇa* to Work

Once an astronomer had this "master number," the rest was a matter of simple proportion. To find the mean position of Mars, for example, they would use the Rule of Three:

Mean Position = (Total Revolutions of Mars per *Mahāyuga* / Total Civil Days per *Mahāyuga*) × *Ahargaṇa*

This would give the average position of the planet. This mean position was the essential starting point upon which the various corrections for anomaly (*manda* and *śīghra*) were applied to find the planet's true, observable position in the sky.

The *ahargaṇa* calculation is the heart of Aryabhata's predictive system. It is the crucial bridge that connects the abstract, long-term cycles of the cosmos to any specific, real-world date.

5.7.4 Calendrical Conventions – The Śuklādi System (Kk.11)

To ensure that calculations align with the calendar in use, Verse Kk.11 specifies Aryabhata's conventions for reckoning years and months.

Sanskrit:

चैत्रशुक्लादेर्मासा द्वादश वर्षाणि पञ्च वर्षाणि ।

युगभगणाद्यं तस्माद् दिनमासवर्षयुगादीनाम् ॥ ११ ॥

IAST:

chaitraśuklādermāsā dvādaśa varṣāṇi pañcha varṣāṇi |

yugabhagaṇādyaṃ tasmād dinamāsavarṣayugādīnām || 11 ||

The Starting Line: Defining the New Year

Every calendar needs a clearly defined starting point—an official New Year's Day. In this verse, Aryabhata establishes the precise moment that marks the beginning of the month and the year in his system.

The *Chaitra-śuklādi* System

He declares that the year begins on the **first day of the bright fortnight (*śukla pakṣa*) of the lunar month of Chaitra** (which corresponds to March-April in the Gregorian calendar). This is a logical and observationally grounded choice. A new lunar month begins with the New Moon, when the Moon is in conjunction with the Sun. The "first day" of the month is the day after this conjunction, when the first sliver of the waxing crescent moon becomes visible in the evening sky. The month of Chaitra is significant because it contains the vernal equinox, a natural and powerful marker of the Sun's return to the northern hemisphere.

This *Chaitra-śuklādi* system, therefore, anchors the start of the year to two observable celestial events: the first visible crescent moon and the arrival of spring.

The Five-Year Cycle

The verse also makes an intriguing reference to a period of "five years." This likely refers to a smaller, five-year *yuga* cycle that was used in some older traditions (like the *Vedāṅga Jyotiṣa*) for specific calendrical or ritual purposes. By mentioning it, Aryabhata shows his awareness of these other systems, but he makes it clear that his own primary astronomical calculations are based on the grand, long-term cycles of the *Mahāyuga*.

The Foundation of Timekeeping

With these final definitions, the temporal foundation of Aryabhata's system is complete. He has established:

- The grand cosmic cycles of the *Mahāyuga*.
- The precise units of time, from years down to breaths.
- The "master key" calculation of the *ahargaṇa* (the day-count).
- The official starting point of the calendar year.

This comprehensive framework allowed any practitioner of his system to pinpoint any moment in time relative to the great cosmic clock. With this temporal coordinate fixed, they could then begin the real work of astronomy: calculating the precise positions of the celestial bodies for that exact instant.

Notes (for Kk.11):

- **Chaitraśuklādermāsā dvādaśa varṣāṇi:** "Months [begin] from the first day of the bright fortnight of Chaitra; twelve [such] months [make] a year." (Establishes the Chaitra-śuklādi system for the start of months and the year).
- **Pañca varṣāṇi:** "Five years." (Likely refers to a smaller 5-year ritualistic yuga, distinct from the astronomical Mahāyuga).
- **Yugabhagaṇādyaṃ tasmād dinamāsavarṣayugādīnām:** "From the yuga-revolutions [etc.] is the beginning of days, months, years, yugas." (Reckoning of time periods is anchored to the epoch and the defined revolutions/parameters of the Mahāyuga).

5.8 Calculating Mean Positions and Orbital Scale (Kk.12-16)

Once the *ahargaṇa* (elapsed days since the epoch) is determined, Aryabhata outlines the next step: calculating the mean position (*madhyama graha*) of each celestial body for that day. This is achieved using the Rule of Three, leveraging the total revolutions per *mahāyuga* (from *Gītikāpāda* 3–4) and the total civil days in a *mahāyuga*. Verses Kk.12–16 of the *Kālakriyāpāda* then provide context for the physical scale of planetary orbits and the conceptual framework for transitioning from mean to true (observable) positions, introducing the critical concept of anomalies (*kendra*).

5.8.1 The Cosmic Scale and Celestial Motion

Sanskrit:

भचक्रं परिधिगुणं षष्ट्याप्ता योजनैः समन्तात् तु ।

दिवसेनेकेन समं प्रवहेण सदागतिर् नियता ॥ १२ ॥

IAST:

bhachakraṃ paridhiguṇaṃ ṣaṣṭyāptā yojanaiḥ samantāt tu|

divasenekena samaṃ pravaheṇa sadāgatir niyatā || 12 ||

Translation: The celestial sphere's circumference (*bhachakraṃ*), measured in *yojanas* and derived via a factor related to sixty (per *Gītikāpāda* 6), is traversed uniformly in one sidereal day by the constant motion of the *pravaha* (cosmic force/driving principle).

The Great Celestial Wheel

Having defined the units of time, Aryabhata now turns to the immense scale of the cosmos itself. In this verse, he describes the celestial sphere as a vast wheel, establishing a direct and powerful relationship between its size and its speed.

The Speed of the Heavens

Aryabhata states that the great "wheel of the asterisms"—the celestial sphere—travels a specific number of *yojanas* (a unit of distance) in the time it takes to draw one breath (*prāṇa*).

This establishes a **universal linear speed** for the daily rotation of the heavens. While the angular speed is constant (one minute of arc per breath), the linear speed depends on the size of the circle being measured. A point on the celestial equator moves at an incredible velocity, while a point near the celestial pole moves much more slowly, tracing a smaller circle in the same amount of time.

A Unified System of Space and Time

This concept is a cornerstone of Aryabhata's system. It is another expression of the elegant unity between space and time that he established with the *prāṇa-kalā* equivalence.

- **Time** is measured in breaths (*prāṇas*).
- **Angular Space** is measured in minutes of arc (*kalās*).
- **Linear Space** is measured in *yojanas*.

By linking all three—stating that in one unit of time (a *prāṇa*), the heavens turn by one unit of angle (a *kalā*) and travel a specific distance (in *yojanas*)—he creates a fully integrated and computationally seamless model of the cosmos.

This verse doesn't calculate the position of a single planet. Instead, it does something much more profound: it sets the universal stage upon which all planetary motion occurs, defining the fundamental relationship between the size of the universe and the rhythm of its daily rotation.

Notes:

- **Bhachakraṃ paridhiguṇaṃ ṣaṣṭyāptā yojanaiḥ:** "Celestial sphere's circumference (bhachakra), measured in yojanas (derived via Gk.6 with a factor of sixty)."
- **Divasenekena samaṃ pravaheṇa sadāgatir niyatā:** "Traversed uniformly in one sidereal day by the constant motion of the pravaha (cosmic wind/driving force)."

Figure 5.8: Conceptual diagram illustrating the scale of the celestial sphere or Kakṣyā, which Aryabhata measured as approximately 3.3 billion Yojanas (Gītikāpāda Verse 6 / Gk.6). In Aryabhata's system, this immense sphere completes a full rotation in one Sidereal Day, which is equivalent to 21,600 Prāṇas (Kālakriyāpāda Verse 12 / Kk.12, Gk.11). This establishes a relationship between linear size and daily angular motion.

5.8.2 Orbital Sizes and Mean Motions: Quantifying Celestial Paths (Kk.13 -14)

Having established the grand scale of the cosmos (Kk.12), Aryabhata now turns to defining the dimensions of individual planetary orbits and connecting their physical sizes to their observed motions. Verses Kk.13 and Kk.14 are pivotal in this regard, bridging the gap between the abstract angular speeds derived from Mahāyuga revolution counts and the tangible linear measures of celestial paths in yojanas.

Kk.13: Linear Measure of Daily Orbital Motion

This verse first establishes the principle that a planet's daily angular travel corresponds to a specific linear distance along its orbit.

Sanskrit:

कक्ष्यामण्डलभागो ग्रहस्य भगणान्तरस्य मध्यस्य ।

योजनसंख्या तस्य तु प्रमाणम् उक्तं स्वशास्त्रे तु ॥ १३ ॥

IAST:

kakṣyāmaṇḍalabhāgo grahasya bhagaṇāntarasya madhyasya |

yojanasaṃkhyā tasya tu pramāṇam uktaṃ svaśāstre tu || 13 ||

Translation:

"The segment of a planet's orbit (kakṣyāmaṇḍalabhāgaḥ) corresponding to its mean daily motion (madhyasya [bhagaṇāntarasya]) has its measure in yojanas (yojanasaṃkhyā), the standard for which (pramāṇam) is stated in this treatise (svaśāstre)."

From Angular Speed to Physical Distance

How far does a planet actually travel in one day? We can measure its movement in the sky as an angle—a certain number of degrees or minutes of arc. But in this verse, Aryabhata makes a crucial assertion: this abstract angular motion corresponds to a real, physical distance, measured in *yojanas*.

The Racetrack Analogy

Imagine two runners on a circular racetrack.

- One is on the inner lane.

- The other is on the outer lane.

If they both run a quarter of a lap (an angle of 90 degrees), they have covered the same *angular* distance. However, the runner on the outer lane has had to run a much greater *linear* distance to do so.

A Universe of Defined Orbits

Aryabhata applies this same logic to the planets. He states that a planet's average daily motion—measured in minutes of arc—has a specific, quantifiable equivalent in *yojanas*.

This implies a critical feature of his model: **every planet's orbit has a defined physical size**. While they all complete their revolutions within the same celestial sphere, they travel on different "lanes" of this great cosmic racetrack.

The verse explicitly states that the standard for these sizes is "stated in this treatise," linking this concept directly back to the immense circumference of the celestial sphere (the *kakṣyā*) that he defined in the *Gītikāpāda*. This grand celestial sphere provides the ultimate scale against which the size of each planet's individual orbit is measured.

This verse forges a vital link between the abstract geometry of angles and the physical reality of distance, transforming the planets from mere points of light into bodies moving through a structured, measurable, and vast cosmos.

Kk.14: Determining Individual Orbital Circumferences and Lunar Distances

Building on Kk.13, this verse explains how the specific size of each planet's orbit is determined and makes a crucial statement about the Moon's distance.

Sanskrit:

ग्रहभगणभाजितः स्यात् कक्षामानं ग्रहाणां तु ।

योजनकर्णाः क्रमशः शशिनां मध्यस्फुटौ स्फुटौ च तथा ॥ १४ ॥

IAST:

grahabhagaṇabhājitaḥ syāt kakṣāmānaṃ grahāṇāṃ tu |

yojanakarṇāḥ kramaśaḥ śaśināṃ madhyasphuṭau sphuṭau cha tathā || 14 ||

Translation:

"A planet's orbital circumference (kakṣāmānam) is obtained by dividing the [standard] celestial sphere's circumference by that planet's [number of] revolutions [in a Mahāyuga] (grahabhagaṇabhājitaḥ). The Moon's distances (yojanakarṇāḥ), both mean (madhya) and true (sphuṭa), are determined similarly [or by related principles]."

Building the Solar System to Scale

Having established that planets travel real, physical distances, Aryabhata now provides the recipe for calculating the size of each planet's individual orbit. These two verses work together to transform the abstract revolution counts of the *Mahāyuga* into a structured, dimensioned cosmos.

The Master Blueprint: The Great Celestial Sphere

The starting point for this calculation is the immense size of the great celestial sphere itself, whose circumference Aryabhata defined in the *Gītikāpāda*. This serves as the master blueprint for the entire solar system.

The Recipe for an Orbit

To find the circumference of any specific planet's orbit, Aryabhata gives a simple and logical rule:

> **Divide the circumference of the great celestial sphere by the total number of revolutions that planet makes in a *Mahāyuga*.**

This elegant method scales the entire cosmos down to define the size of each individual planet's orbital path. A planet that moves quickly and completes many revolutions (like the Moon) will have a smaller orbit, closer to the Earth. A planet that moves slowly and completes few revolutions (like Saturn) will have a vast orbit, much farther away.

This ensures a perfect, mathematical coherence between a planet's speed and the physical size of its path.

A Crucial Clue: The Moon's Changing Distance

Within this section, Aryabhata includes a subtle but profoundly important statement about the Moon. He notes that one can derive both its **mean distance** and its **true, variable distance** (*sphuṭa karṇa*).

This is a critical observation. By explicitly acknowledging that the Moon's distance from the Earth is not constant but *variable*, Aryabhata is signaling that a simple, circular orbit at a fixed distance cannot fully explain what we see.

The Moon's apparent size and speed change throughout the month. These observable facts require a more sophisticated model to explain them. This single phrase—"true, variable distance"—is the first hint of the need for the **epicyclic models** that he will use later. It is a brilliant piece of foreshadowing, preparing the ground for the complex but necessary corrections (the *manda* and *śīghra* equations) that are required to reconcile the simple, mean motions of the planets with their true, observed positions.

These verses, therefore, do more than just give the heavens a physical scale. They demonstrate the mind of a master theorist at work, building a coherent system while simultaneously acknowledging the observational complexities that will require even more ingenious solutions.

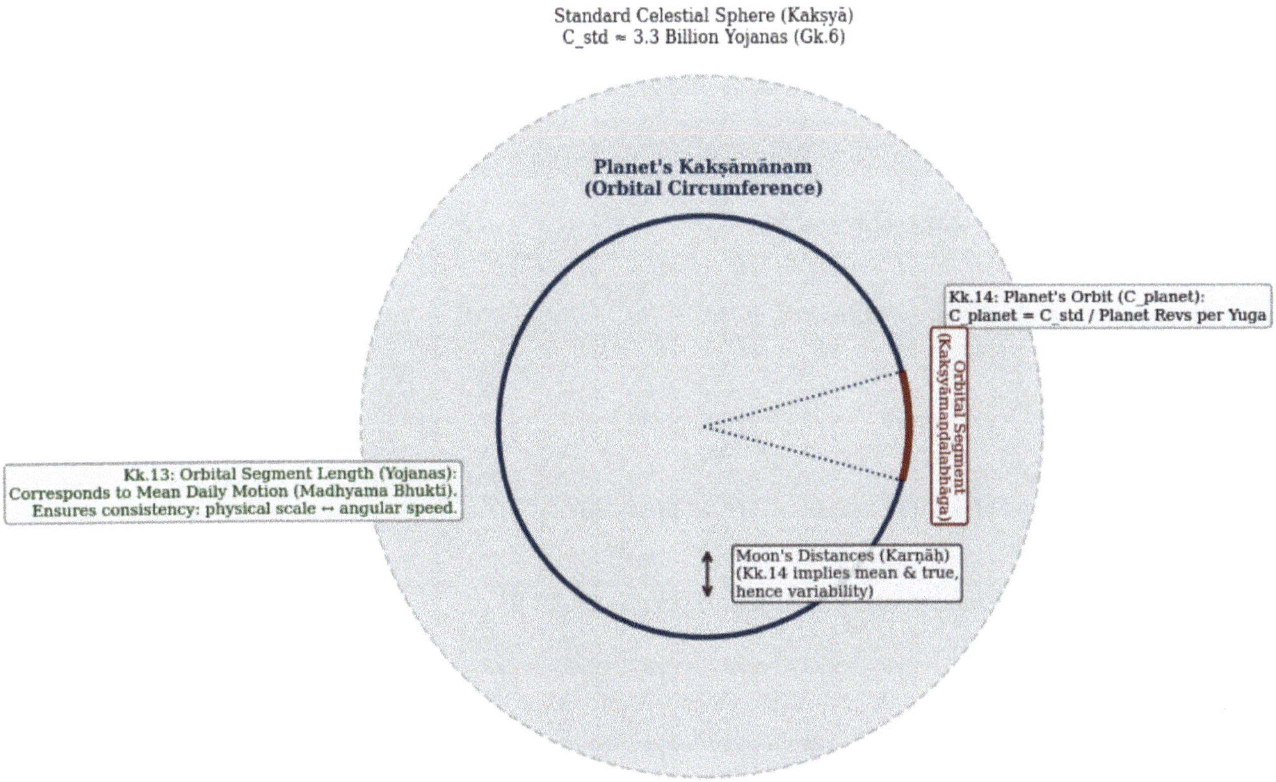

Figure 5.9: Defining Orbital Scales & Lunar Distances in Aryabhata's System (Kk.13-14). Shows planetary orbital circumferences (Kakṣāmānam, blue) determined by scaling the Standard Celestial Sphere (yellow dashed, Gk.6) with Yuga revolutions. The daily orbital segment (Kakṣyāmaṇḍalabhāga, red arc) corresponds to linear distance, aligning orbit size with mean motion, and acknowledges the Moon's variable mean/true distances (Karṇāḥ), foreshadowing orbital corrections

5.8.3 Geocentric Orbital Order

Sanskrit:

शशिरविकुजगुरुकोणा भगणाः क्रमशः स्थिता उपर्युपरि ।

तेषां चान्त्यसमीपे भ्रमति सदा भपञ्जरः स्वगत्या ॥ १५ ॥

IAST:

śaśiravikujagurukoṇā bhagaṇāḥ kramaśaḥ sthitā uparyupari|

teṣāṃ cāntyasamīpe bhramati sadā bhapañjaraḥ svagatyā || 15 ||

Translation: "The orbits (*bhagaṇāḥ*, contextually orbital spheres) of the Moon, Sun, Mars, Jupiter, and Saturn are arranged successively one above another (*kramaśaḥ sthitā uparyupari*). Near the outermost of these, the sphere of fixed stars (*bhapañjaraḥ*) revolves with its own motion [or: stands as the limit]."

Notes:

- **Śaśiravikujagurukoṇā bhagaṇāḥ kramaśaḥ sthitā uparyupari:** "Orbits of Moon, Sun, Mars, Jupiter, Saturn are arranged successively one above another." (Geocentric order from Earth out).
- **Teṣāṃ cāntyasamīpe bhramati sadā bhapañjaraḥ svagatyā:** "Near the outermost, the sphere of stars (bhapañjara) revolves with its own motion (or is fixed)."

The Order of the Heavens

Having defined the sizes of the planetary orbits, Aryabhata now arranges them in their proper order, creating a clear, nested model of the geocentric cosmos. The sequence is logical and intuitive, ordered by speed and, by implication, by increasing distance from the central Earth.

The journey outwards from the Earth proceeds as follows:

1. **The Moon:** The fastest moving body, with the smallest orbit.

2. **The Sun:** The next in order, defining the fundamental rhythm of the year.

3. **Mars**

4. **Jupiter**

5. **Saturn:** The slowest of the visible planets, with the largest orbit.

Finally, enclosing this entire planetary system is the vast **sphere of the fixed stars** (*bhapañjara*), which forms the ultimate boundary of the cosmos.

The Puzzle of Mercury and Venus

It is noteworthy that two familiar planets, Mercury and Venus, are not included in this simple, concentric list. This is not an oversight but a reflection of the complexity of their motion as seen from Earth.

The movements of these two inner planets are intricately tied to the position of the Sun. They are never seen far from it in the sky, appearing either as morning or evening stars. Because of this unique behavior, Aryabhata handles their positions not with a simple, independent orbit in this list, but through a more complex computational model known as the *śīghra* (fast) correction, which brilliantly accounts for their relationship with the Sun.

A Framework for Correction

This verse provides the fundamental, structural map of the solar system. By establishing this clear order of nested orbits, Aryabhata creates the necessary framework upon which the sophisticated epicyclic corrections—the adjustments needed to calculate a planet's true position—will be applied.

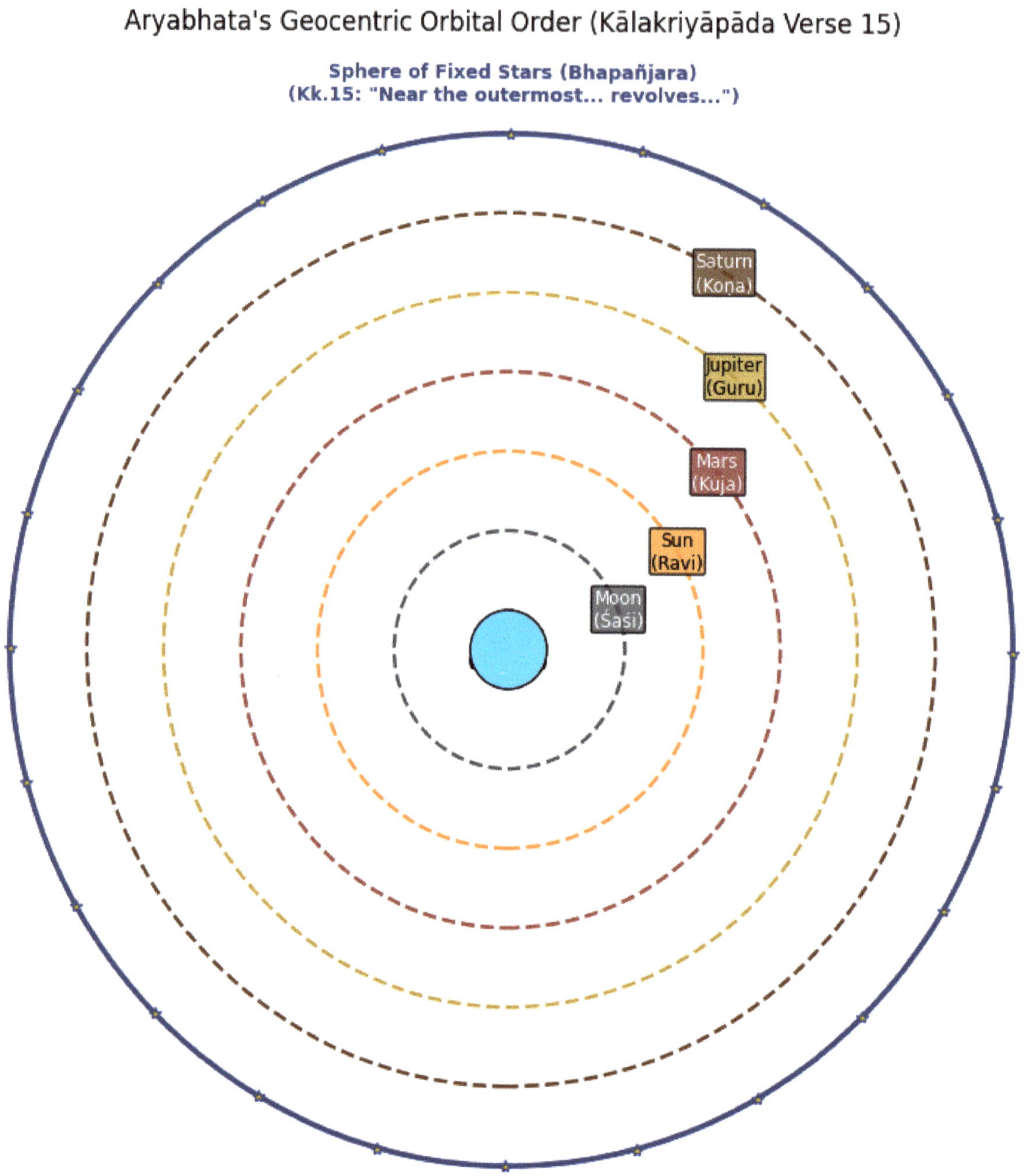

Figure 5.10: Aryabhata's Geocentric Orbital Order (Kk.15). Illustrates Earth at the center with concentric orbits arranged 'one above another' (uparyupari) for Moon, Sun, Mars, Jupiter, and Saturn as described in Kk.15, culminating in the outermost Sphere of Fixed Stars (Bhapañjara). (Note: Kk.15 does not place Mercury/Venus in this simple sequence).

5.8.4 Introducing Anomalies (Kendra)

Sanskrit:

वृत्तम् अपक्रमयुक्तं समवृत्तं प्राग्विलग्नयोर् मध्ये ।

मध्योच्चपातभगणाः केन्द्रान्त्यफलानि जानीयात् ॥ १६ ॥

IAST:

vṛttam apakramayuktaṃ samavṛttaṃ prāgvilagnayor madhye|

madhyocchapātabhagaṇāḥ kendrāntyaphalāni jānīyāt || 16 ||

Translation: The epicycle (*vṛttam*), associated with [correcting for] deviation (*apakramayuktaṁ*), is uniform [in its own motion]. From mean positions, apogees (*Uccha*), and node revolutions, one derives the anomalies (*kendras*) and final corrections (*antyaphalāni*, interpreted as encompassing *kendra* and *phala*).

Notes:

- **Vṛttam apakramayuktaṁ samavṛttaṃ prāgvilagnayor madhye:** "The epicycle (vṛtta), associated with correcting deviation (apakrama), is uniform [in its own motion]..."
- **Madhyocchapātabhagaṇāḥ kendrāntyaphalāni jānīyāt:** "From mean positions, apogees (Uccha), and node revolutions, one derives anomalies (kendra) and final corrections (antyaphala)."
- **Kendra:** Anomaly angle (Mean Longitude - Uccha Longitude).

The Key to Correction: The Anomaly (*Kendra*)

A planet's average position is easy to calculate, but it's not where the planet is actually seen in the sky. To get from the average to the true position, we need to apply a series of corrections. This verse introduces the single most important input for all of those corrections: the **kendra**, or **anomaly**.

The *kendra* is simply the **angular distance between a planet's average position and its relevant point of anomaly**.

There are two types of *kendra*, corresponding to the two types of corrections:

1. **The *Manda Kendra*:** This is the angle between the planet's average position and its apogee (the point of its slowest motion). It is the key to correcting for the planet's changing speed in its own orbit.
2. **The *Śīghra Kendra*:** This is the angle between the planet's average position and a "fast point" linked to the Sun. It is the key to correcting for effects caused by viewing the planet from a moving Earth, including retrograde motion.

Once an astronomer calculates the *kendra*, it becomes the primary input for the trigonometric engine of corrections. Using the table of sines (*jyā*), the *kendra* is transformed into the precise amount of adjustment needed to find the planet's true, observable position.

Summary: Building the Physical Cosmos

The preceding block of verses (Kk. 12-16) represents a crucial bridge in Aryabhata's system. They work together to translate the abstract, time-based data into the physical and conceptual architecture of the universe.

In this short but profound passage, Aryabhata has:

1. Established a **unified system of space and time**, linking the speed of the heavens to its physical size.

2. Provided the method to calculate the **physical scale of each planet's orbit** in *yojanas*.

3. Arranged these orbits into a clear **geocentric hierarchy**.

4. Finally, introduced the **kendra**, the key that unlocks the all-important corrections for anomaly.

This systematic progression is the hallmark of Aryabhata's genius. He meticulously builds his model from the ground up, moving from the simple rhythm of mean motions to a sophisticated, dimensioned cosmos capable of predicting the true, complex, and beautiful dance of the planets.

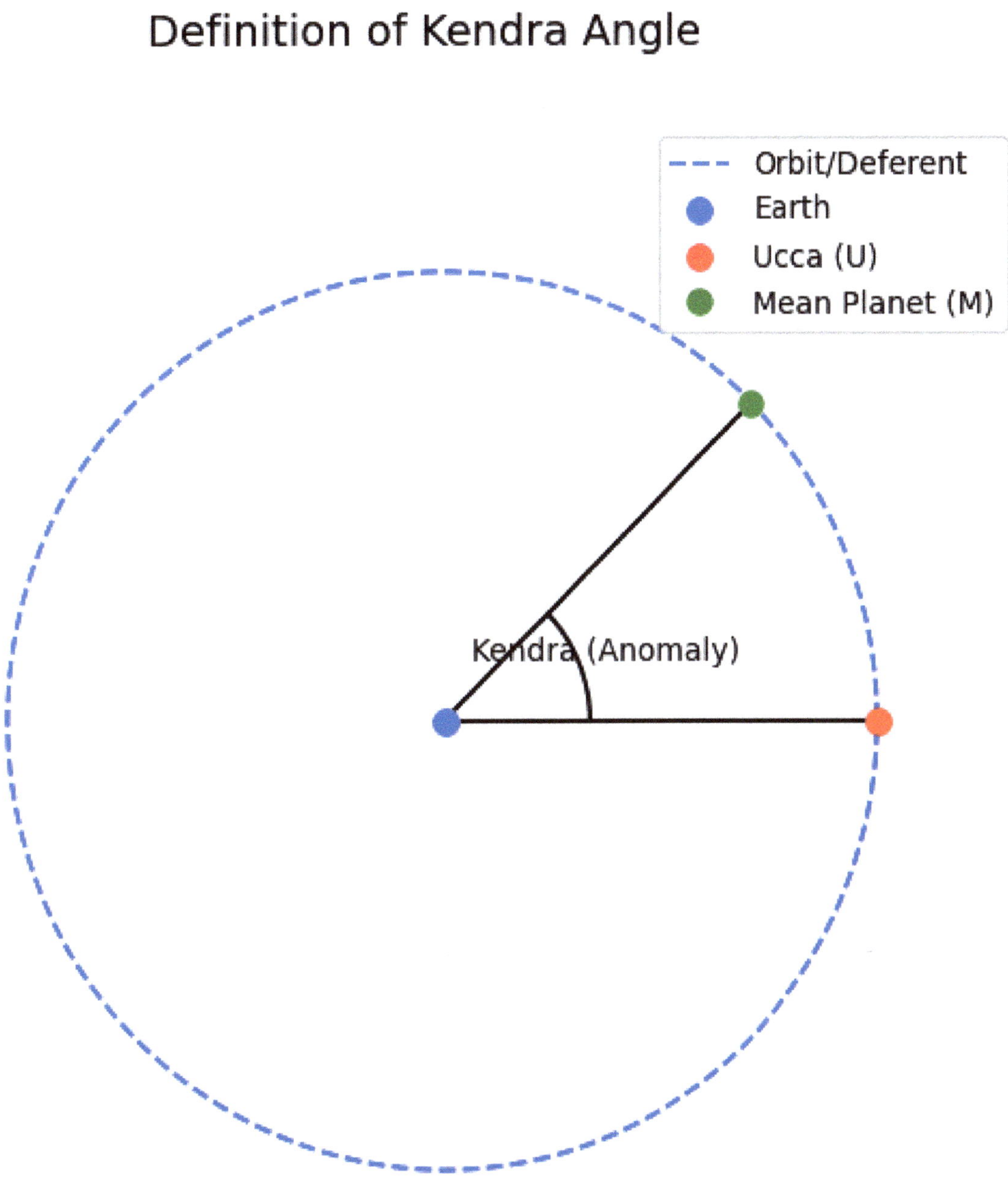

Figure 5.11: Diagram illustrating the definition of the Kendra (Anomaly) angle in Aryabhata's system. The Kendra is the angular distance between the planet's mean position (M) and its reference point (Uccha, U, or Śīghroccha), as seen from Earth (E). This angle is the fundamental input for calculating the Manda and Śīghra corrections to determine the planet's true position.

5.9 Wheels Upon Wheels: Correcting the Mean Motions (Kk.17-21)

The planets do not trace simple, steady circles across the sky. They speed up, slow down, and sometimes, in a startling display, even appear to move backward for a time against the stars—a phenomenon known as **retrograde motion**.

How can a geocentric model, with the Earth at the center, possibly account for this complex and irregular dance?

In this section of the *Kālakriyāpāda*, Aryabhata unveils his ingenious solution: a sophisticated system of corrections designed to transform a planet's simple, calculated **mean position** into its complex, observable **true position**. This is accomplished through two primary adjustments: the *manda* (slow) and *śīghra* (fast) corrections.

The Two Great Corrections

1. **The Slow Correction (*Manda*):** This first correction accounts for a planet's variable speed in its own orbit. Just as a race car on an oval track slows down on the far turn and speeds up on the straightaway, a planet's speed changes as it moves through its orbit. The *manda* correction adjusts for this, accounting for the fact that the planet moves faster when it is closer to the Earth and slower when it is farther away.

2. **The Fast Correction (*Śīghra*):** This second, more complex correction accounts for phenomena caused by viewing the planets from a moving Earth. It brillianty models the planet's appearance relative to the Sun, including the bounded back-and-forth motion of Mercury and Venus and, most importantly, the characteristic backward loops of **retrograde motion** seen for all the planets.

The Geometric Model: Wheels Upon Wheels

While Aryabhata's verses provide the precise mathematical recipes for these corrections, the geometric model that produces these effects is a beautiful system of **epicycles**, often described as "wheels upon wheels."

Imagine a large wheel (the **deferent**) rotating around the Earth. Now, imagine a smaller wheel (the **epicycle**) attached to the rim of the large one, spinning independently. The planet itself is located on the rim of this smaller wheel.

From our vantage point on Earth, the combined motion of these two wheels—the deferent carrying the epicycle, and the epicycle carrying the planet—can produce all the complex motions we see in the sky. It can make a planet appear to slow down, speed up, and even trace a backward loop.

Aryabhata's *manda* and *śīghra* corrections are the precise computational tools that translate this elegant "wheels upon wheels" geometry into calculable, predictive mathematics. They are the engine that allows his system to move beyond simple averages and accurately model the true, beautiful, and complex dance of the planets.

5.9.1 Defining Epicycle Sizes

Sanskrit:

वृत्तफलानि प्रथमात् त्रैराशिकेन यथाक्रमं योज्याः ।

मन्दशीघ्रफलं च यथास्थानं तद्विकल्पितं च ॥ १७ ॥

IAST:

vṛttaphalāni prathamāt trairāśikena yathākramaṃ yojyāḥ|

mandaśīghraphalaṃ cha yathāsthānaṃ tadvikalpitaṃ cha || 17 ||

Translation: "The results/effects of the [epicyclic] circles (*vṛttaphalāni*), starting from the first, are to be determined by the Rule of Three (*trairāśikena*) in order. The *manda* and *śīghra* corrections (*phalaṃ*) are [then applied] according to their respective positions and as specifically determined."

Notes:

- **Vṛttaphalāni prathamāt trairāśikena yathākramaṃ yojyāḥ:** "Results/effects (sizes/radii) of epicyclic circles (vṛtta) are determined by Rule of Three (trairāśika) from the first [maximum correction]."
- **Mandaśīghraphalaṃ cha yathāsthānaṃ tadvikalpitaṃ cha:** "Manda and śīghra corrections (phala) are [applied] according to position and as determined."

Tuning the Wheels: Sizing the Epicycles

The epicyclic model of "wheels upon wheels" is elegant, but how large should each of these wheels be? In this verse, Aryabhata establishes a fundamental principle: the size of a planet's epicycle is not arbitrary. It is precisely determined by the amount of correction needed.

The Logic of Proportionality

The core idea is simple and logical:

> **A larger observed anomaly requires a proportionally larger epicycle to model it.**

Imagine a planet that deviates only slightly from its average path. To account for this small wobble, you only need a small epicycle. But for a planet that shows a large variation in speed or a wide retrograde loop, you need a much larger epicycle to generate that significant effect.

The verse states that the size of each epicycle is to be derived using the **Rule of Three**. The known input is the *maximum observed deviation* for a given planet, and the output is the required size of the epicycle needed to produce that deviation.

Building a Model That Matches Reality

This verse is crucial because it grounds the entire epicyclic model in empirical data. The sizes of the "wheels" are not chosen at random; they are meticulously "tuned" to match the observed realities of the planets' motions.

This principle ensures that the geometric model is not just a theoretical construct, but a powerful predictive tool capable of accurately replicating the speeds, slowdowns, and backward loops that we actually see in the night sky. It is the first step in building a machine that is not only elegant in theory but also precise in practice.

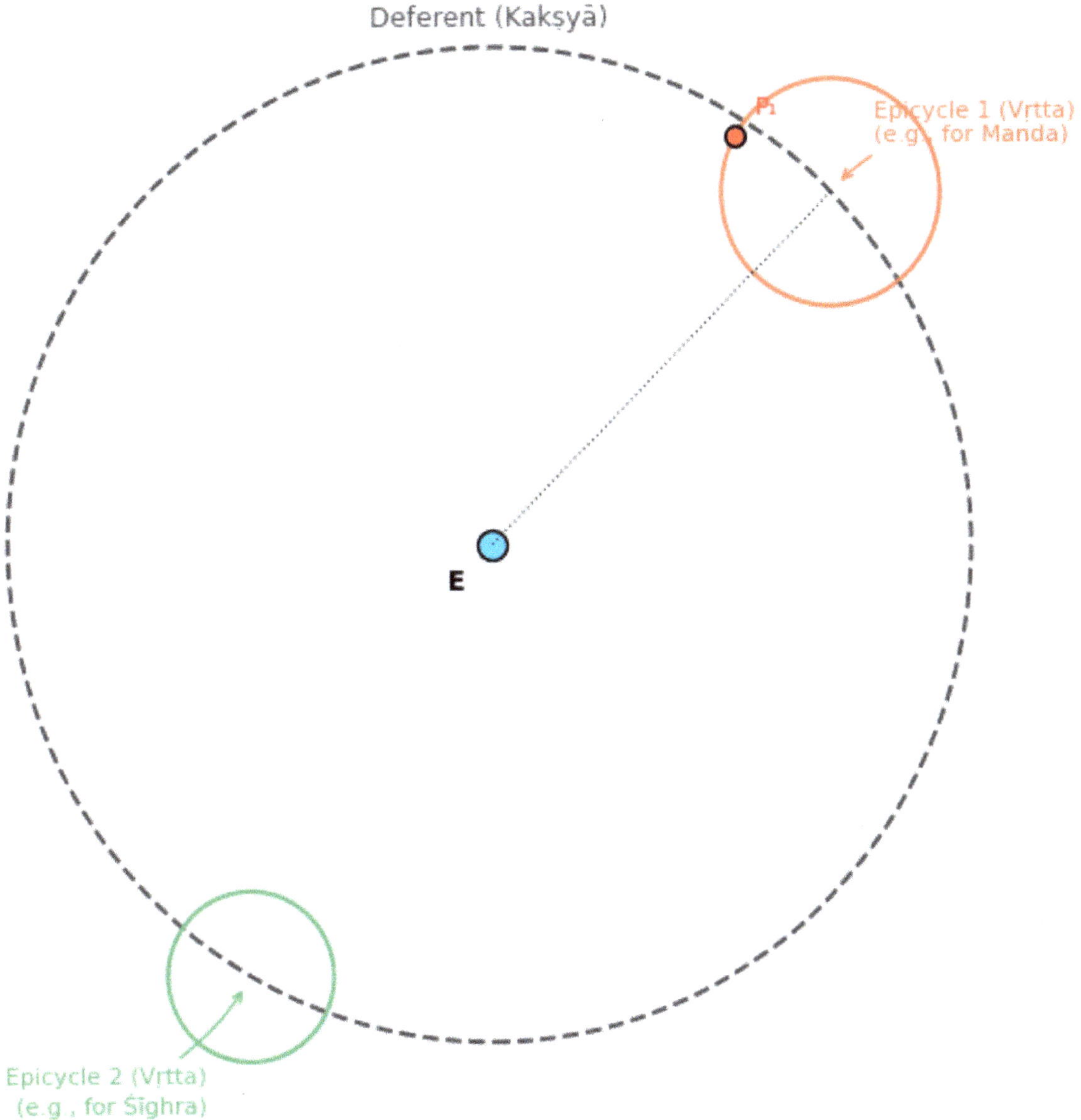

Figure 5.12: Kk.17 – Defining Epicycle Radii (Vṛttaphalāni) Proportionally. Illustrates how Aryabhata (Kk.17) sizes Manda and Śīghra epicycles (coral/green Vṛtta on grey dashed Kakṣyā) proportionally to their maximum angular corrections (Max Phala), effectively tuning the model to observed planetary motions.

5.9.2 Calculating the Bhujāphala (Sine Component of Correction)

Sanskrit:

भुजाफलं अर्धज्यया हृतं यात् त्रैराशिकेन संनादति ।

मन्दशीघ्रकृते च यथावद् योज्यं फलं तत् ॥ १८ ॥

IAST:

bhujāphalaṃ ardhajyayā hṛtaṃ yāt trairāśikena saṃnādati|

mandaśīghrakṛte cha yathāvad yojyaṃ phalaṃ tat || 18 ||

Translation: "The *bhujāphala*, derived from the R·sine of the anomaly (*ardhajyayā*, i.e., *kendra-jyā*), is obtained using the Rule of Three (*trairāśikena*). This resulting correction (*phalaṃ*) for the *manda* or *śīghra* [anomaly] is then applied appropriately."

Notes:

- **Bhujāphalaṃ ardhajyayā hṛtaṃ yāt trairāśikena saṃnādati:** "Bhujāphala, derived from R·sine of anomaly (kendra-jyā), is obtained by Rule of Three."
- **Mandaśīghrakṛte cha yathāvad yojyaṃ phalaṃ tat:** "This resulting correction for manda/śīghra is applied appropriately."
- **Bhujāphala:** Sine component of epicyclic correction, proportional to R·sin(kendra) and epicycle size.

The First Gear of Correction: The *Bhujāphala*

Once the anomaly angle (*kendra*) is known, the first step in calculating the required correction is to find a value called the ***bhujāphala***, which literally means the "arm-result." This is the first of two crucial components that make up the full correction.

The Calculation

The *bhujāphala* is found using a two-step process:

1. **Find the Sine:** First, the astronomer uses the sine table (*jyā*) to find the R·sine of the *kendra* angle.
2. **Scale by the Epicycle's Size:** This sine value is then scaled proportionally based on the size of the specific epicycle being used (as determined in the previous verse). A larger epicycle will yield a larger *bhujāphala*.

The Geometric Meaning: The Perpendicular Displacement

What does this "arm-result" actually represent in the "wheels upon wheels" model?

Imagine a line drawn from the Earth to the center of a planet's epicycle. Now, imagine a second line drawn from the center of the epicycle out to the planet itself. These two lines form an angle—the *kendra*.

The ***bhujāphala* represents the component of the planet's displacement that is perpendicular to the line of sight from Earth to the epicycle's center.**

Think of it like this: if the planet is at the "top" or "bottom" of its epicycle (from our perspective), its displacement is entirely perpendicular, and the *bhujāphala* is at its maximum. If the planet is on the line of sight (closest or farthest from us on its epicycle), its perpendicular displacement is zero, and the *bhujāphala* is also zero.

The *bhujāphala* is a critical intermediate value. It is the first gear in the intricate machine of epicyclic correction. It works in tandem with its partner, the *koṭiphala* (the cosine component), to determine the final, precise adjustment needed to find the planet's true position.

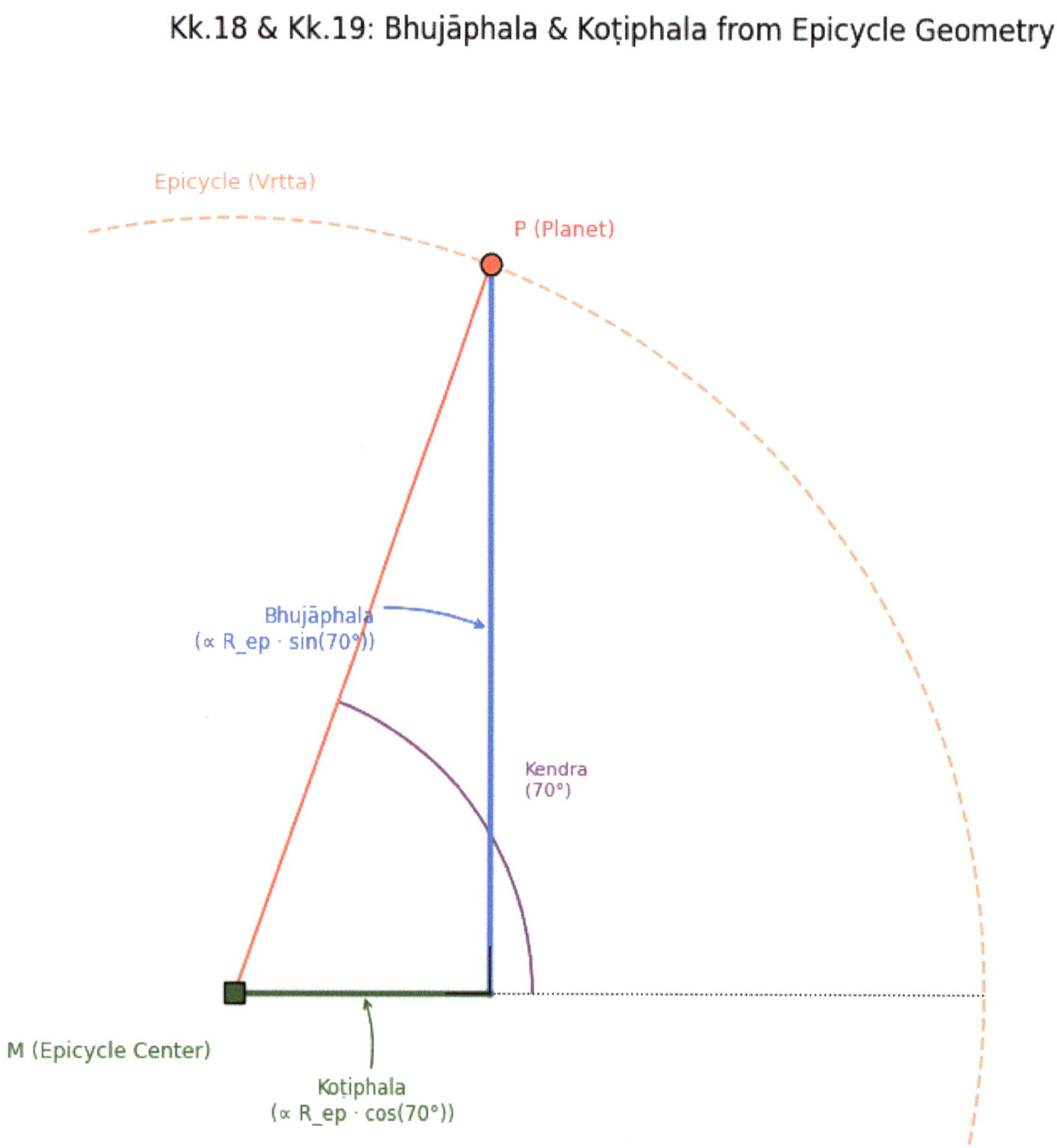

Figure 5.13: Bhujāphala and Koṭiphala in Aryabhata's Epicyclic Model (Kk.18-19). The planet (P) on its epicycle is positioned by the Kendra angle. The Bhujāphala (blue, sine component) and Koṭiphala (green, cosine component) represent its orthogonal displacements. These are derived via sine tables, scaled by epicycle size, and are crucial for calculating the final correction (Phala) and true distance (Karṇa).

5.9.3 Calculating the Kotiphala (Cosine Component of Correction)

Sanskrit:

कोटिफलं अर्धज्यया हृतं यात् त्रैराशिकेन संनादति ।

यथास्थानं मन्दशीघ्रकृते च योज्यं तत् फलं ॥ १९ ॥

IAST:

koṭiphalaṃ ardhajyayā hṛtaṃ yāt trairāśikena saṃnādat |

yathāsthānaṃ mandaśīghrakṛte cha yojyaṃ tat phalaṃ || 19 ||

Translation: "The *koṭiphala*, derived from the R·cosine of the anomaly (*ardhajyayā*, i.e., *kendra-koṭijyā*), is obtained using the Rule of Three. This correction (*phalaṃ*) for the *manda* or *śīghra* [anomaly] is applied according to its position."

Notes:

- **Koṭiphalaṃ ardhajyayā hṛtaṃ yāt trairāśikena saṃnādati:** "Koṭiphala, derived from R·cosine of anomaly (kendra-koṭijyā), is obtained by Rule of Three."
- **Yathāsthānaṃ mandaśīghrakṛte cha yojyaṃ tat phalaṃ:** "Applied according to position for manda/śīghra."
- **Koṭiphala:** Cosine component of epicyclic correction, proportional to R·cos(kendra) and epicycle size.

The Second Gear of Correction: The *Koṭiphala*

After calculating the *bhujāphala* ("arm-result"), the next step is to find its essential partner: the ***koṭiphala***, or "perpendicular-result." This is the second component needed to fully describe the planet's displacement on its epicycle.

The Calculation

The *koṭiphala* is calculated in a similar way to its counterpart. It is found by taking the **R·cosine** of the anomaly angle (*kendra*) and scaling it by the size of the epicycle. (In practice, the R·cosine was found using the same sine table, by looking up the sine of the complementary angle, 90° - *kendra*).

The Geometric Meaning: The Forward/Backward Displacement

If the *bhujāphala* represents the planet's *side-to-side* displacement, the ***koṭiphala* represents its *forward-and-backward* displacement along the line of sight from Earth.**

Let's return to our "wheels upon wheels" model:

- When the planet is at the "top" or "bottom" of its epicycle, its forward/backward displacement is zero, and the *koṭiphala* is also zero.

- When the planet is at the point on its epicycle closest to or farthest from Earth, its displacement is entirely along the line of sight, and the *koṭiphala* is at its maximum.

Two Pieces of a Puzzle

The *bhujāphala* and the *koṭiphala* are like the two sides of a right-angled triangle. Together, they perfectly describe the planet's position on its epicycle relative to the epicycle's center.

With these two fundamental components calculated, the astronomer now has all the information needed to determine two crucial final values: the planet's true, variable distance from Earth (*karṇa*), and the final angular correction that must be applied to its mean position.

5.9.4 Applying the Manda Correction

Sanskrit:

मन्दात् प्रतिलोमगत् यात् मन्दकेन्द्रं तद्वशात् फलं च।

यथावद् योज्यं मध्यात् स्फुरति स्फुटं तत् ॥ २० ॥

IAST:

mandāt pratilomogat yāt mandakendraṃ tadvāśāt phalaṃ cha|

yathāvad yojyaṃ madhyāt sphurati sphuṭaṃ tat || 20 ||

Translation:

"From the manda [apogee], moving in the opposite direction [to the planet's mean motion, to find the angle], the mandakendra is obtained, and from its value, the correction (phalaṃ) [is found]. Applied correctly to the mean position (madhyāt), the [more nearly] true position (sphuṭaṃ) then shines forth."

Notes:

- **Mandāt pratilomogat yāt mandakendraṃ tadvāśāt phalaṃ cha:** "From manda [apogee], moving oppositely, the mandakendra is obtained; from it, the correction (mandaphala)."
- **Yathāvad yojyaṃ madhyāt sphurati sphuṭaṃ tat:** "Applied correctly to mean position, the [more nearly] true position shines forth."
- **Mandakendra:** Mean Longitude - Mandoccha Longitude.

The First Refinement: Applying the Slow Correction

With the tools of the *bhujāphala* and *koṭiphala* in hand, Aryabhata now details the procedure for applying the first of the two great corrections: the **manda correction**, which accounts for the planet's variable speed in its orbit.

Step 1: Find the *Mandakendra*

The first step is to calculate the **Mandakendra**, or "slow anomaly angle." This is the angular distance between the planet's average position and its *Mandoccha*—the apogee, or point of its slowest motion.

The verse specifies that this angle is to be measured *pratilomagat*, or "against the grain"—meaning it is measured from the apogee backwards to the planet's average position. This *Mandakendra* is the key that tells us *where* the planet is in its cycle of speeding up and slowing down.

Step 2: Calculate the Correction Angle (*Mandaphala*)

Next, this *Mandakendra* angle is used as the input to calculate the **Mandaphala**, the actual angular correction needed. This is where the *bhujāphala* and *koṭiphala* come into play. Through a series of trigonometric steps, these components are used to determine the precise angular shift caused by the planet's position on its *manda* epicycle.

Step 3: Apply the Correction

Finally, this calculated correction angle (*Mandaphala*) is applied to the planet's original average position. Depending on which half of its anomaly cycle the planet is in, the correction is either added to or subtracted from the mean longitude.

The verse poetically states that after this adjustment, the "true position shines forth" (*sphuṭaṃ prakāśate*). While this is only the *first* of two major corrections, it represents a huge leap in accuracy. By accounting for the planet's own orbital anomaly, the astronomer moves from a simple average to a much more realistic and refined position, one step closer to the true, observable location of the planet in the sky.

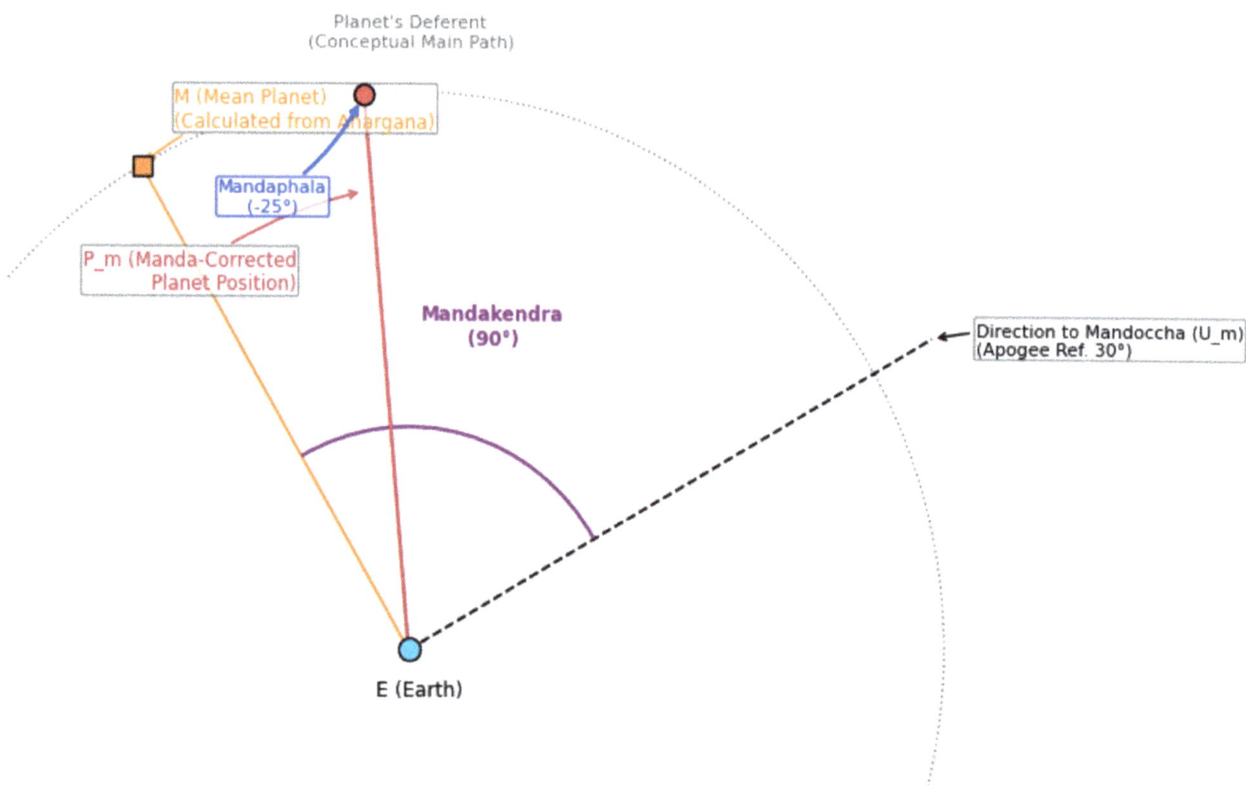

Figure 5.14: Applying the Manda Correction (Kk.20). Illustrates refining a planet's Mean Longitude (M, orange). The Mandakendra (purple arc) is the angle from the Mandoccha (U_m, apogee) to M. The derived Mandaphala (blue arc, e.g., -25°) is then applied to M, resulting in the Manda-Corrected Longitude (P_m, red), which accounts for variable orbital speed.

5.9.5 Applying the Śīghra Correction

Sanskrit:

शीघ्राद् अनुलोमगत् यात् शीघ्रकेन्द्रं तद्वशात् फलं च ।

यथावद् योज्यं स्फुटात् स्फुरति स्फुटं तत् ॥ २१ ॥

IAST:

śīghrād anulomogat yāt śīghrakendraṃ tadvāśāt phalaṃ cha |

yathāvad yojyaṃ sphuṭāt sphurati sphuṭaṃ tat || 21 ||

Translation:

"From the śīghra [fast point], moving in the same direction [as its motion, to find the angle], the śīghrakendra is obtained, and from its value, the correction (phalaṃ) [is found]. Applied correctly to the [previously corrected] position (sphuṭāt), the [even more] true position (sphuṭataraṃ / sphuṭaṃ) then shines forth."

Notes:

- **Śīghrād anulomogat yāt śīghrakendraṃ tadvāśāt phalaṃ cha:** "From śīghra [fast point], moving in same direction, śīghrakendra is obtained; from it, the correction (śīghraphala)."
- **Yathāvad yojyaṃ sphuṭāt sphurati sphuṭaṃ tat:** "Applied correctly to the [manda-corrected] position, the true position shines forth."
- **Śīghrakendra:** (Manda-corrected Longitude) - Śīghroccha Longitude.

The Second Refinement: Applying the Fast Correction

After correcting for the planet's own orbital speed, the astronomer must now apply the second great correction: the **śīghra correction**. This adjustment accounts for the complex effects caused by viewing the planet from a moving Earth, including the phenomenon of retrograde motion.

The process is a near-perfect echo of the *manda* correction, but it uses a different set of parameters.

Step 1: Find the *Śīghrakendra*

First, the astronomer calculates the **Śīghrakendra**, or "fast anomaly angle." This is the angular distance between the planet's position (which has already been corrected once by the *manda* process) and its *Śīghroccha*—a "fast point" that is computationally linked to the position of the mean Sun.

The verse specifies that this angle is measured *anulomagat*, or "with the grain"—meaning it is measured in the natural, forward direction of the zodiac. This *Śīghrakendra* is the key that tells us where the planet is in its synodic cycle relative to the Sun.

Step 2: Calculate the Correction Angle (*Śīghraphala*)

Next, this *Śīghrakendra* angle is used as the input to calculate the **Śīghraphala**, the angular correction needed to account for the "fast" anomaly. Just as before, this involves a series of trigonometric steps using the *bhujāphala* and *koṭiphala*, but this time they are scaled by the size of the *śīghra* epicycle.

Step 3: Apply the Correction to Find the Final True Position

Finally, this *Śīghraphala* is applied to the once-corrected longitude of the planet. The result, as the verse states, "becomes truer" (*sphuṭataraṃ bhavati*). This is the planet's final, doubly-corrected **true geocentric longitude**. This is the position that should match what an observer actually sees in the night sky.

The Heart of the Planetary Model: A Summary

This sequence of five verses (Kk. 17-21) forms the computational heart of Aryabhata's entire system for predicting planetary positions. In this masterful passage, he has integrated:

- **Geometric Concepts:** A "wheels upon wheels" model with precisely sized epicycles.
- **Trigonometric Precision:** The use of sine (*jyā*) functions to calculate the exact components of the needed corrections.
- **Systematic Refinement:** A clear, step-by-step process of first applying the *manda* correction and then the *śīghra* correction to transform a simple average position into a precise, observable reality.

This methodical approach allowed Aryabhata's system to account for all the complex and beautiful irregularities of the planets' paths, including their changing speeds and their strange backward loops. The mathematical depth and systematic rigor of this model underscore the profound achievement of the *Aryabhatiya*, a framework that would influence the course of Indian astronomy for a thousand years.

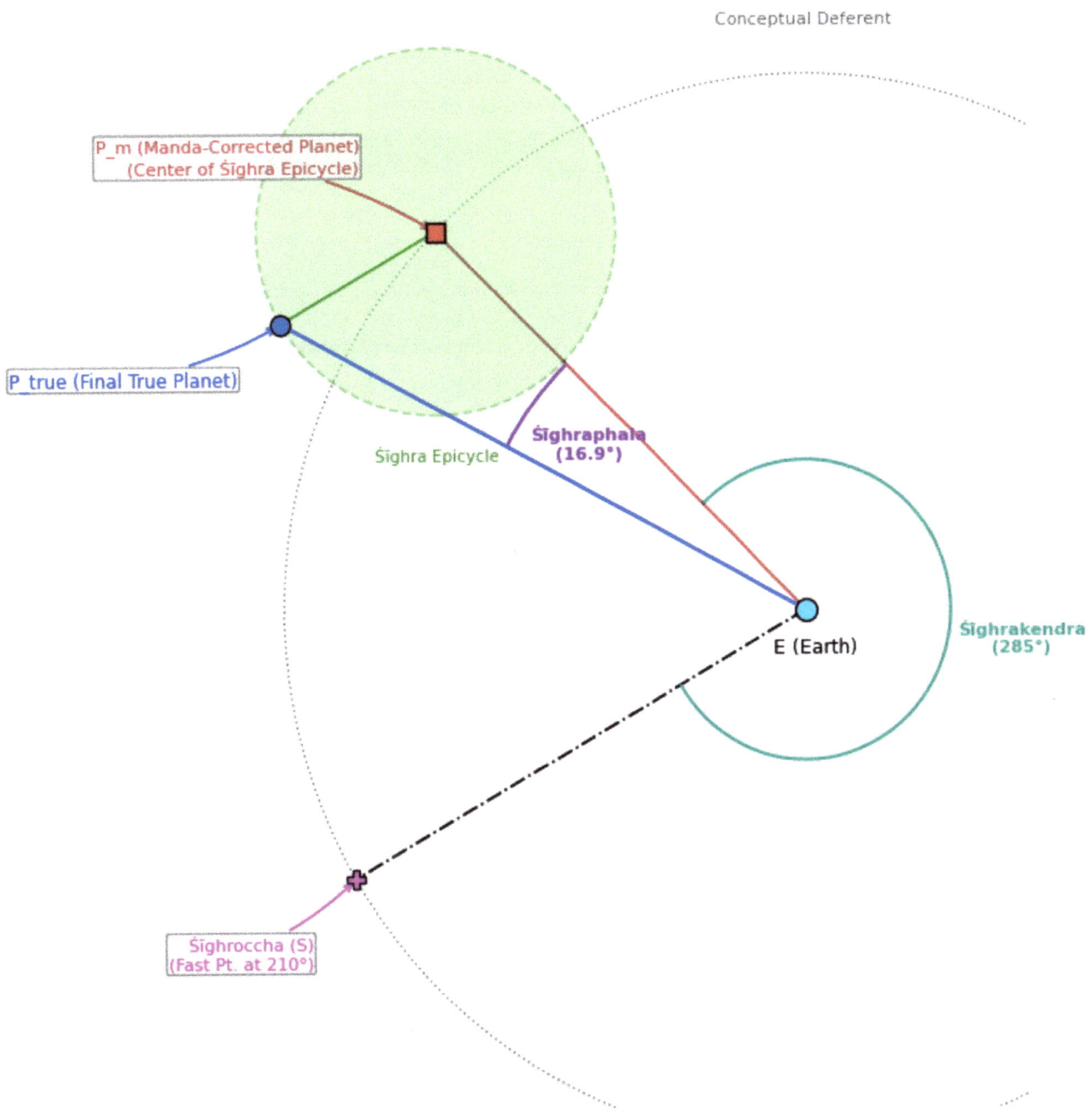

Figure 5.15: Śīghra Correction via Epicycle (Kk.21). Illustrates the Manda-Corrected Planet (P_m) as the deferent point for the Śīghra epicycle (green dashed). The Śīghrakendra (angle P_m-E-S) determines the True Planet's (P_true) position. Applying the resulting Śīghraphala (violet arc, angle P_mEP_true) to P_m's longitude gives the final True Longitude, explaining synodic phenomena.

5.10 The Final Polish: Iteration and Refinement (Kk.22-24)

The journey from a planet's average position to its true, observable position is not quite complete. There is one final, subtle layer of complexity that must be addressed to achieve the highest level of accuracy. The following verses detail this final polishing step—a process of **iteration and refinement**.

The Challenge: A "Chicken-and-Egg" Problem

The central challenge lies in a fundamental interdependence, a kind of astronomical "chicken-and-egg" problem:

1. The **correction angle** we need to apply depends on the planet's **true distance** from the Earth. (Just as a person waving their hand seems to move less if they are far away, a planet on its epicycle creates a smaller angular shift when it is farther from us).

2. But, the planet's **true distance** depends on its exact position on its epicycle.

3. And, its **position on the epicycle** is determined by the **anomaly angle**—the very thing we are trying to use to find the correction!

In short: **the size of the correction depends on a distance, which in turn depends on the correction.**

The Solution: A Process of Iteration

This circular dependency cannot be solved with a single, direct calculation. Instead, it requires a brilliant and practical technique: **iteration**.

Iteration is a process of successive approximation. It's like focusing a camera lens:

- You start with a good first guess (a "rough focus").
- You apply a correction based on that guess.
- This gives you a slightly better, more refined position.
- You then re-calculate the correction based on this *new* position.
- You repeat this process, with each step bringing the result closer and closer to the true, perfectly focused answer.

These next verses outline the steps for this iterative process. They provide the methods for refining the anomaly angles, calculating the true distance (*karṇa*), and determining the precise final correction, ensuring that the predicted position is not just an approximation, but a highly accurate reflection of the planet's true place in the heavens. This iterative approach is a hallmark of the numerical power and sophistication of Aryabhata's system.

5.10.1 Iteration with Half-Corrections for Anomaly Refinement

Sanskrit:

मन्दशीघ्रफलार्धानि केन्द्रयोः संनादति यथाक्रमम् ।

पुनः फलं तत् योज्यं स्फुटं च तद्वशात् ॥ २२ ॥

IAST:

mandaśīghraphalārdhāni kendrayoḥ saṃnādati yathākramam|

punaḥ phalaṃ tat yojyaṃ sphuṭaṃ cha tadvāśāt || 22 ||

Translation: "Halves of the *manda* and *śīghra* corrections (*phalārdhāni*) are to be applied successively to their respective anomalies (*kendrayoḥ*) in order. Then, the [recalculated] correction (*phalaṃ*) is to be applied again, and from that, the true [position] (*sphuṭaṃ*) [is obtained]."

Notes:

- **Mandaśīghraphalārdhāni kendrayoḥ saṃnādati yathākramam:** "Halves of manda and śīghra corrections (phalārdha) are applied successively to their respective anomalies (kendra)."
- **Punaḥ phalaṃ tat yojyaṃ sphuṭaṃ cha tadvāśāt:** "Then, the [recalculated] correction is applied again, and from that, the true [position]."

The Half-Step Method: A Smoother Path to a True Position

How does the iterative process of "focusing" on a planet's true position actually work? This verse describes a clever and practical technique at the heart of the method.

After making the first, rough calculation of the *manda* and *śīghra* correction angles, Aryabhata gives a simple instruction:

> **Take half of the calculated correction and apply it back to the planet's average position.**

This step creates a new, slightly improved anomaly angle (*kendra*). This refined angle is then used to recalculate a more accurate correction, and the process is repeated.

The "Half-Step" Advantage

Why use only half the correction? This is a brilliant technique for ensuring that the iterative process works smoothly. Imagine focusing an old camera lens. If you turn the ring too far, you will overshoot the perfect focus and have to come back. You might end up twisting back and forth, "hunting" for the sharpest image.

By applying only *half* the calculated correction at each step, Aryabhata's method avoids this problem. It takes a smaller, more careful step, ensuring that the calculation **converges smoothly and stably** towards the correct answer without overshooting it. It is a practical and efficient way to guarantee that the iterative "focusing" will quickly land on the true, sharpest position of the planet.

This refined anomaly angle, achieved through one or more of these half-step iterations, becomes the final input for calculating the definitive correction needed to find the planet's true place in the sky.

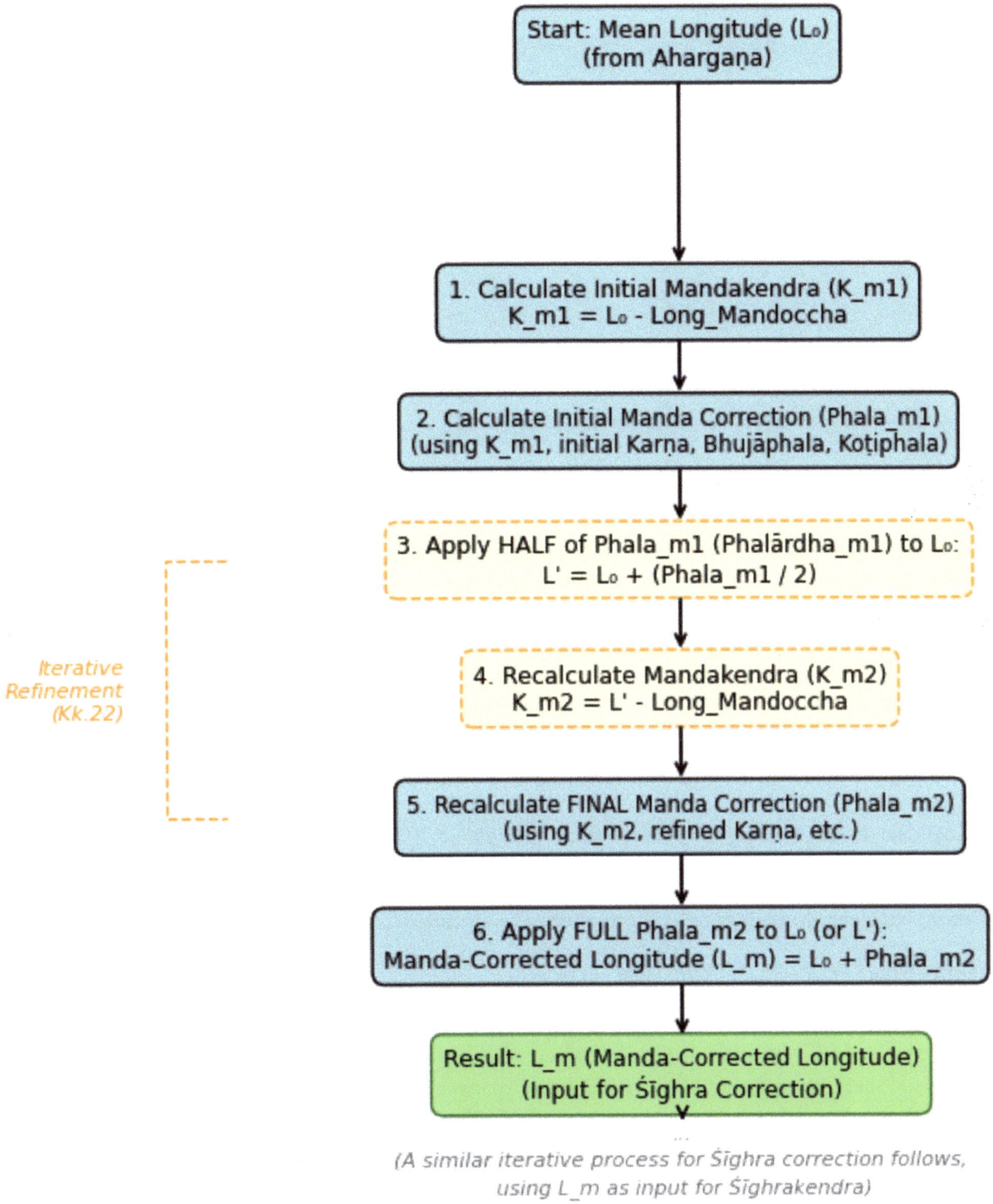

Figure 5.16: Kk.22: Iterative Refinement of Anomaly (Kendra) using Half-Corrections (Phalārdha) – Conceptual Flow for Manda Correction.

5.10.2 Calculating the Precise Correction Angle

Sanskrit:

कर्णेन भुजाफलहृता यात् फलज्या त्रैराशिकेन च ।

यथावद् योज्यं स्फुटं तत् मन्दशीघ्रकृते च ॥ २३ ॥

IAST:

karṇena bhujāphalahṛtā yāt phalajyā trairāśikena cha|

yathāvad yojyaṃ sphuṭaṃ tat mandaśīghrakṛte cha || 23 ||

Notes:

- **Karṇena bhujāphalahṛtā yāt phalajyā trairāśikena cha:** "Bhujāphala divided by true distance (karṇa) yields R·sine of correction angle (phalajyā), by Rule of Three."
- **Yathāvad yojyaṃ sphuṭaṃ tat mandaśīghrakṛte cha:** "Applied correctly for manda/śīghra to get true [position]."
- **Phalajyā:** R·sin(phala) = (R × bhujāphala) / karṇa. (Phala is the true angular correction).
- **Karṇa:** True distance to the planet (hypotenuse in epicyclic geometry).

The Final Calculation: Finding the True Correction Angle

After refining the anomaly angle through iteration, the final step is to calculate the precise correction angle (*phala*) that must be applied to the planet's longitude. This verse provides the crucial formula for this last, critical calculation.

The Underlying Geometry: The View from Earth

Imagine you are standing on Earth (E) watching a planet (P) move along its epicycle, whose center is M. This forms a triangle: E-M-P. The *bhujāphala* (calculated in Kk.18) represents the "height" of this triangle—the planet's perpendicular displacement from the line of sight EM. The *karṇa* is the hypotenuse of this triangle—the planet's true, instantaneous distance (EP) from you on Earth. The correction angle (*phala*) is the small angle at your eye (at E) created by this displacement. Basic trigonometry tells us that the sine of this angle is simply the ratio of the opposite side (the *bhujāphala*) to the hypotenuse (the *karṇa*).

The Formula: A Matter of Perspective

Aryabhata's verse captures this geometric insight perfectly. He states that the R·sine of the correction angle (*phalajyā*) is found by a simple proportion:

R·sin(Correction Angle) = (R × *Bhujāphala*) / *Karṇa*

This formula brilliantly accounts for the effects of perspective:

- If the planet is **closer** to Earth (a smaller *karṇa*), the same displacement on its epicycle will create a **larger** apparent correction angle.
- If the planet is **farther** from Earth (a larger *karṇa*), the same displacement will create a **smaller** apparent correction angle.

The Final Step

Once this R-sine value (*phalajyā*) is calculated, the astronomer simply uses the sine table in reverse to find the actual angle of the correction (*phala*). This angle is then added to or subtracted from the planet's longitude to arrive at its final, true, and observable position in the sky.

The Implied Calculation of True Distance (*Karṇa*)

This entire process relies on knowing the planet's true, variable distance from Earth—the *karṇa*. Although the verse focuses on *using* the *karṇa*, it implies that it has already been calculated. This is where the *koṭiphala* (the cosine component of the displacement) comes into play. By using both the *bhujāphala* and the *koṭiphala*, the astronomer can solve the E-M-P triangle and find the length of the *karṇa* with precision.

This intricate interplay—where the correction depends on a distance that is itself part of the calculation—is the hallmark of the sophistication of Aryabhata's system. It ensures that the final predicted position of a planet is not just a rough average, but a highly accurate result, rigorously corrected for distance and perspective.

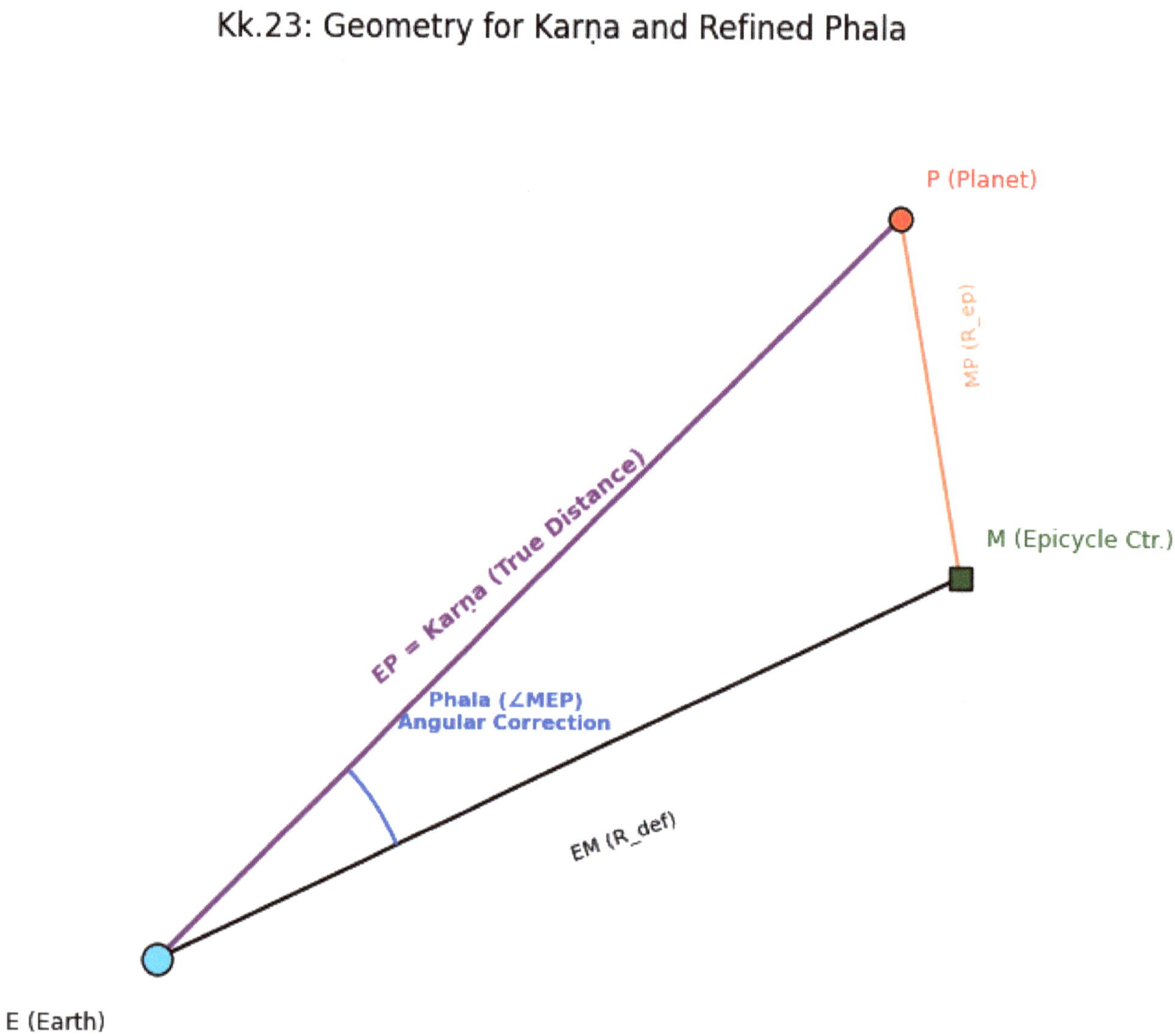

Figure 5.17: Geometric Basis for True Distance (Karṇa) & Correction (Phala) (Kk.23). Shows Karṇa (EP) derived from triangle E-M-P components (EM, MP, Kendra). R·sin(Phala) is then (Trijyā × linear Bhujāphala) / linear Karṇa, yielding an accurate Phala (angle MEP) that considers varying distance.

5.10.3 Calculating the True Daily Motion (*Sphuta Bhukti*)

Sanskrit:

स्फुटमध्यभुक्तिः केन्द्रफलवशात् यात् त्रैराशिकेन च ।

मन्दशीघ्रकृते यथावद् योज्यं तत् फलं ॥ २४ ॥

IAST:

sphuṭamadhyabhuktiḥ kendraphalavaśāt yāt trairāśikena cha|

mandaśīghrakṛte yathāvad yojyaṃ tat phalaṃ || 24 ||

Translation: "The true daily motion (*sphuṭamadhyabhuktiḥ*) is obtained from [corrections based on] the effect of the anomaly (*kendraphalavaśāt*) by means of the Rule of Three. For the *manda* and *śīghra* [anomalies], that correction (*phalaṃ*) is to be applied appropriately to yield the result."

Notes:

- **Sphuṭamadhyabhuktiḥ kendraphalavaśāt yāt trairāśikena cha:** "True daily motion (sphuṭa bhukti) is obtained from [corrections based on] the effect of the anomaly (kendra) by Rule of Three."
- **Mandaśīghrakṛte yathāvad yojyaṃ tat phalaṃ:** "For manda/śīghra, that correction is applied appropriately."

The Final Step: Calculating the True Daily Motion

Knowing a planet's precise position today is only half the battle. To predict where it will be tomorrow, an astronomer must also know its **true daily motion** (*sphuṭa bhukti*)—its actual, variable speed on that specific day. This verse provides the method for this final, crucial calculation.

From Average Speed to True Speed

A planet's *average* daily motion is a simple, constant value. But its *true* daily motion changes constantly as it speeds up and slows down in its orbit.

The process for finding this true speed mirrors the process for finding the true position. It involves taking the planet's **mean daily motion** and adjusting it with corrections derived from the same *manda* and *śīghra* anomalies.

The Logic of the Correction

The logic is intuitive:

- When a planet is moving from its slowest point (apogee) towards its fastest point (perigee), its true daily motion will be *greater* than its average speed. The correction will be additive.

- When it is moving from its fastest point back towards its slowest, its true daily motion will be *less* than its average speed. The correction will be subtractive.

The verse indicates that the size of this speed correction is found using the **Rule of Three**. The calculation takes into account the planet's current position in its anomaly cycle (*kendra*) and its true distance from Earth (*karṇa*) to determine the precise adjustment needed.

The Key to Prediction

The calculation of the true daily motion is the final, essential piece of the predictive puzzle. With this value, the astronomer can accurately extrapolate the planet's position to the next day, the day after that, and so on.

It is the engine that allows the system to be truly predictive, transforming a static snapshot of the heavens into a dynamic, forward-looking model capable of charting the future paths of the planets with remarkable accuracy.

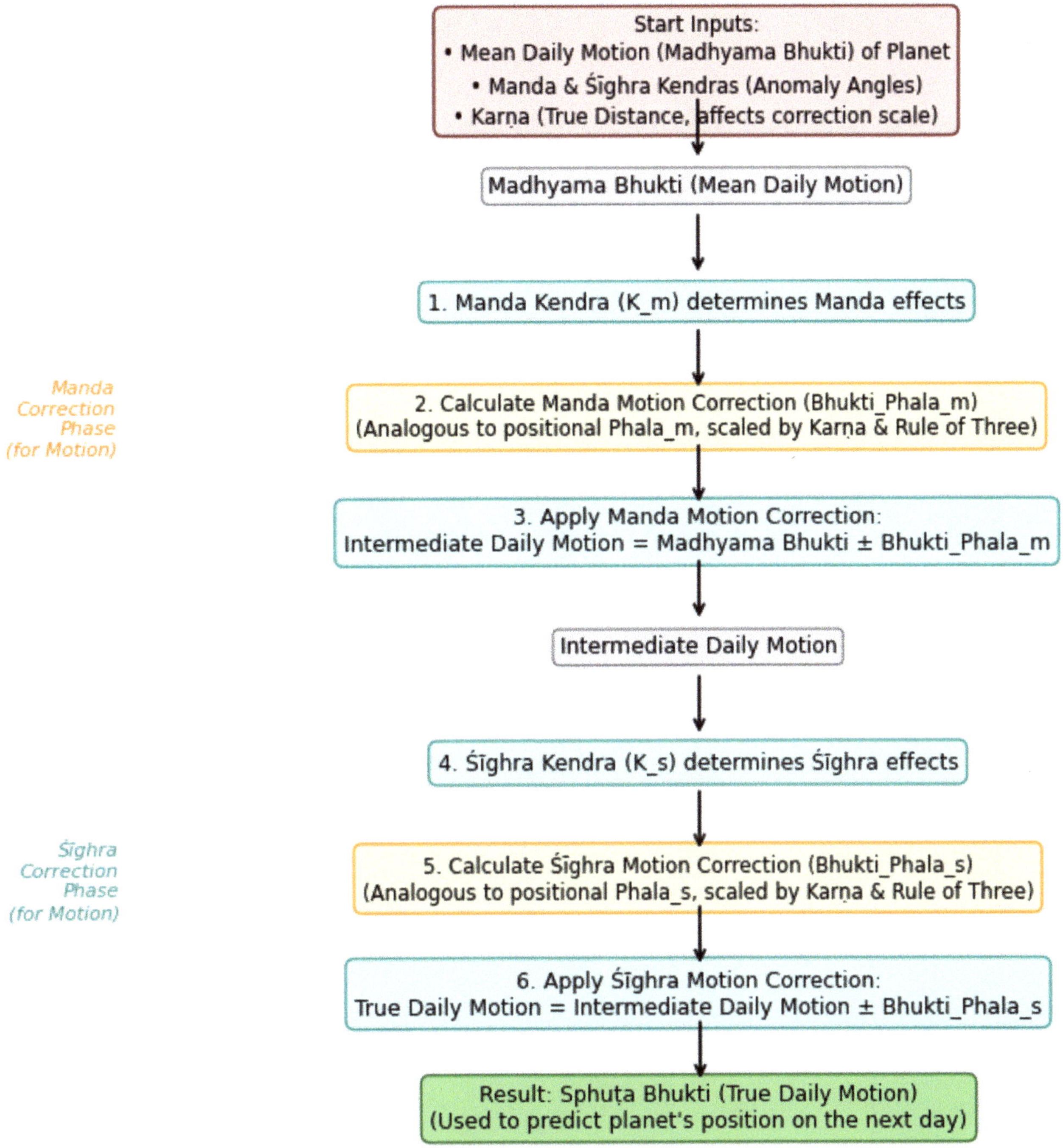

Figure 5.18: Kk.24: Conceptual Flow for Calculating Sphuṭa Bhukti (True Daily Motion).

5.10.4 From Average to Reality: The Complete System of Correction

The final verses of this section represent the culmination of Aryabhata's planetary theory. They reveal the intricate process of transforming a planet's simple, calculated **mean motion**—its average speed over vast cosmic ages—into a precise prediction of its **true motion**, the complex, variable speed we actually observe in the sky.

The Challenge: A Puzzle of Interdependence

The core challenge Aryabhata solves is a sophisticated "chicken-and-egg" problem. The angular correction needed to find a planet's true position depends on its true distance from the Earth. But that true distance itself depends on the very position we are trying to find.

The Solution: A Masterpiece of Iteration and Perspective

Aryabhata's solution is a masterful blend of geometry and computational prowess, built on two key principles:

1. **Iterative Refinement:** He resolves the circular dependency through a process of **iteration**, or successive approximation. By applying a "half-correction," recalculating, and repeating, his method allows an astronomer to zero in on the true position with increasing accuracy, like focusing a lens until the image is perfectly sharp.
2. **The Importance of Perspective (*Karṇa*):** The system meticulously calculates and uses the planet's true, variable distance from Earth (*karṇa*). This ensures that the final correction is perfectly scaled to the observer's perspective, correctly accounting for the fact that a planet's movement on its epicycle will appear smaller when it is farther away and larger when it is closer.

The Final Output: True Position and True Speed

This intricate process yields two essential results:

- The **true position** (*sphuṭa graha*): The planet's precise, observable longitude for that day.
- The **true daily motion** (*sphuṭa bhukti*): The planet's actual, non-uniform speed for that day.

This second result is what makes the system truly predictive. Knowing not just *where* a planet is, but *how fast* it is currently moving, allows an astronomer to accurately forecast its position for the days and weeks to come.

A Monument of Mathematical Rigor

This complete system of correction—blending iterative techniques, perspective-based adjustments, and the calculation of both true position and true speed—underscores the profound mathematical genius of the *Aryabhatiya*. It is a sophisticated, self-consistent, and remarkably accurate model that masterfully integrated geometry, trigonometry, and advanced computational methods. This framework not only provided a robust engine for astronomical prediction in its own time but also profoundly influenced the trajectory of Indian astronomy for a thousand years.

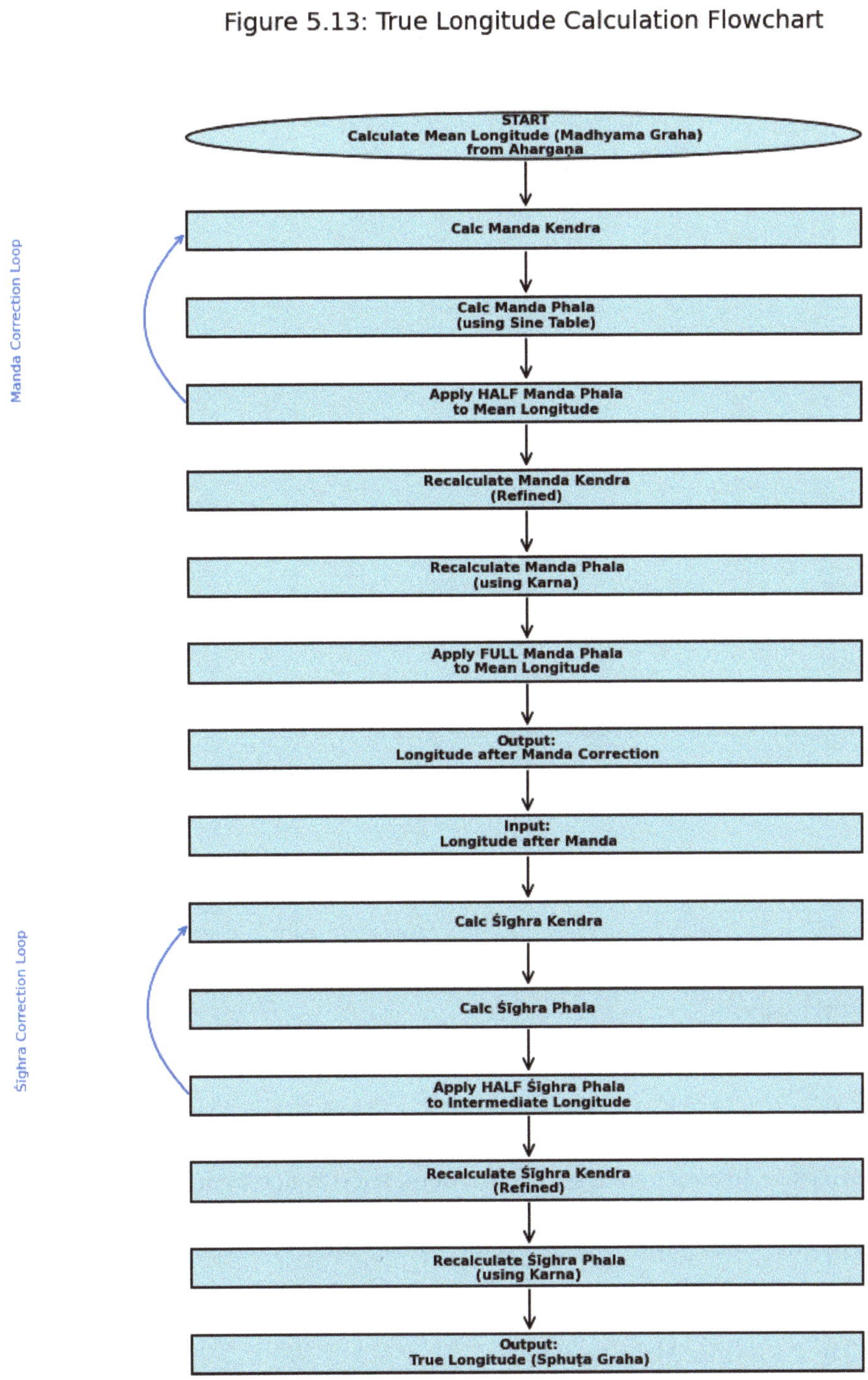

Figure 5.19: Aryabhata's Iterative Path to True Longitude (Sphuṭa Graha) (Kk.17-24). Flowchart illustrating the transformation from Mean Longitude (via Ahargaṇa) through iterative Manda (variable speed) and Śīghra (synodic) corrections, utilizing anomaly angles (Kendra) and true distance (Karṇa), to arrive at the True Longitude.

5.11 Summary of Planetary Motion Calculation (Kk.25)

The final verse of the Kālakriyāpāda, Kk.25, serves as a concise summation, reiterating the core principles and factors involved in the intricate process of determining a planet's true observed motion from its calculated mean motion.

Sanskrit:

एवं ग्रहमध्यमचारान् रविमध्योच्चपातपरिहीनान् ।

कृत्वा भूयः कुर्यात् क्षेत्रात् कालाद् गतिं च स्फुटम् ॥ २५ ॥

IAST:

evaṃ grahamadhyamachārān ravimadhyocchapātaparīhīnān|

kṛtvā bhūyaḥ kuryāt kṣetrāt kālād gatiṃ cha sphuṭam || 25 ||

Translation:

"Thus, having first determined the mean motions/positions of the planets (grahamadhyamachārān), [then] accounting for [their positions relative to, or deviations involving,] the mean Sun (ravimadhya), apogees (Uccha), and nodes (pāta), one should subsequently ascertain the true motion/position (gatiṃ cha sphuṭam) from the geometry (kṣetrāt) [of the model] and from time (kālāt) [elapsed]."

Notes:

- **Evaṃ grahamadhyamachārān ravimadhyocchapātaparīhīnān kṛtvā:** "Thus, having determined mean motions/positions of planets, [then accounting for relations to] mean Sun, apogees (Uccha), and nodes (pāta)."
- **Bhūyaḥ kuryāt kṣetrāt kālād gatiṃ cha sphuṭam:** "Subsequently ascertain the true motion/position from geometry (kṣetra) and time (kāla)."

The Grand Procedure: The Path from Mean to True

In this final, masterful verse, Aryabhata steps back to give us a bird's-eye view of his entire astronomical system. He encapsulates the grand procedure, the systematic path an astronomer must follow to travel from a simple, calculated average to the complex, observable truth.

The process is a logical, multi-stage journey:

1. **Start with the Mean:** The journey always begins with the **mean position** of a planet. This is calculated from the elapsed time (*ahargaṇa*) and the great cosmic cycles of the *Mahāyuga*.
2. **Use the Key Reference Points:** Next, this mean position is compared to the key reference points that define

the anomalies:

- The **Mean Sun**, which is essential for the *śīghra* (fast) correction.
- The planet's **Apogee** (*Mandoccha*), which is essential for the *manda* (slow) correction.
- The planet's **Nodes** (*pāta*), which are crucial for calculating latitude and predicting eclipses.

3. **Apply the Geometric Engine:** Finally, the true position is derived "from the geometry" (*kṣetrāt*) and "from time" (*kālāt*). This refers to the application of the powerful epicyclic model—the "wheels upon wheels"—which uses the anomaly angles derived from time to produce the precise geometric corrections needed.

In essence, Aryabhata reaffirms that the ultimate goal of his science is to derive the **true, observable position** (*sphuṭa*) from the **theoretical mean position** (*madhyama*) through a systematic interplay of time, geometry, and precise calculation.

Situating Aryabhata's Achievement

It is important to place Aryabhata's geocentric model in its proper historical context. For its era, it was a monumental achievement.

- **Remarkable Success:** His system achieved a significant degree of predictive accuracy. Its robust computational framework allowed astronomers to explain and forecast many complex celestial phenomena, including retrograde motion, with a precision that was unmatched in its time.
- **A Mathematical Model:** Like all geocentric systems, Aryabhata's model was a sophisticated **mathematical approximation** of physical reality, not a direct representation of the solar system's true mechanics. The parameters of the model—the sizes of the epicycles and the positions of the apogees—were meticulously "tuned" based on empirical observation to make the predictions match reality.
- **Inherent Limitations:** While powerful, the model could not perfectly account for every nuance of planetary motion. Its primary strength, and Aryabhata's focus, was its stunning ability to predict a planet's position along the ecliptic.

Aryabhata's great achievement was not in discovering the final, physical truth of the cosmos. It was in creating a mathematically coherent, computationally powerful, and predictively effective system that represented a giant leap forward in the scientific understanding of the heavens. His work profoundly shaped the course of Indian astronomy for a thousand years and stands as a timeless monument to the power of reason and mathematics.

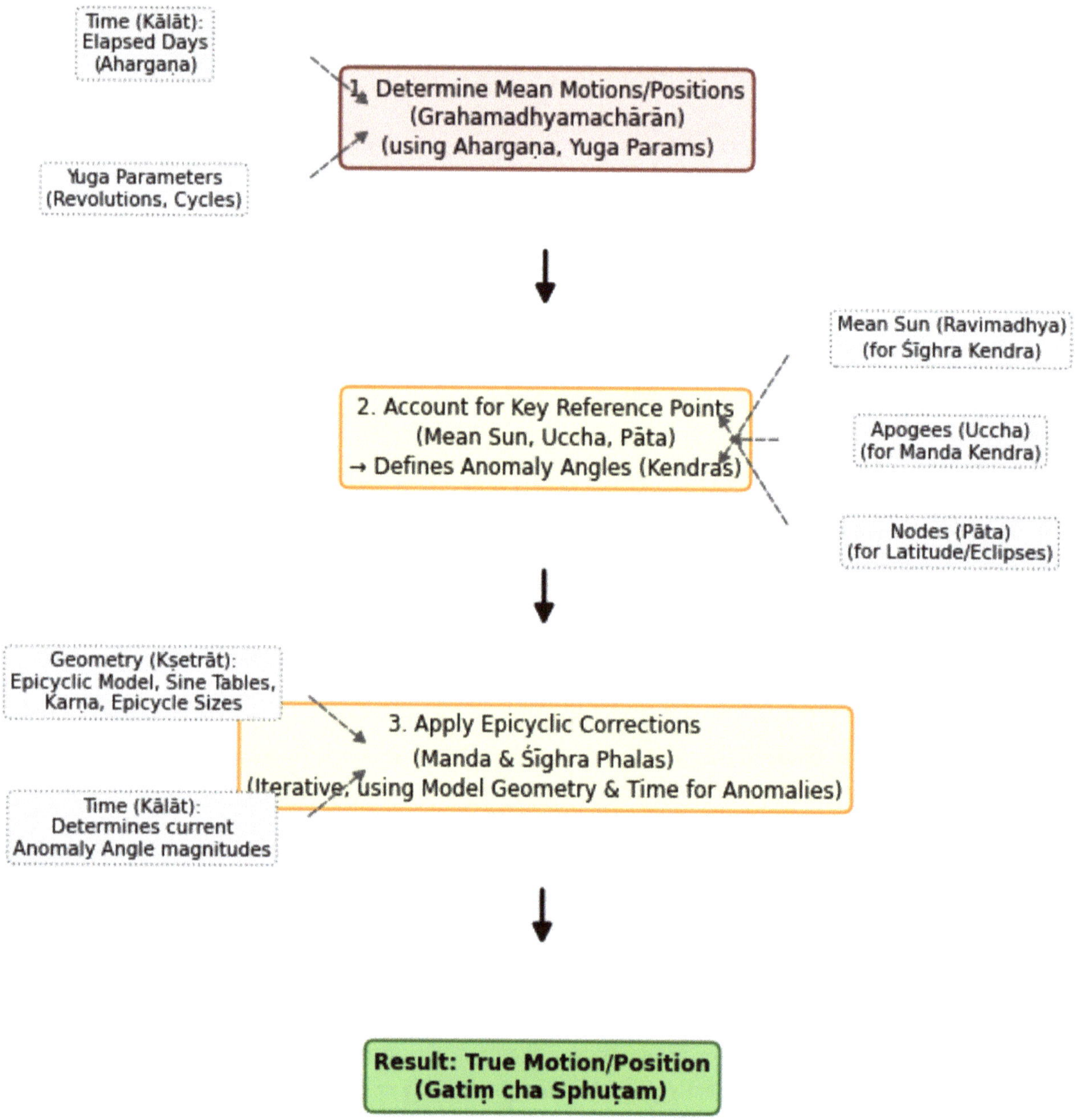

Figure 5.21: Kk.25: Summary of Aryabhata's Process for True Planetary Motion/Position.

5.12 Conclusion: The Temporal Engine of the Aryabhatiya

The *Kālakriyāpāda*, or "The Reckoning of Time," stands as the dynamic heart of the *Aryabhatiya*. In this pivotal chapter, Aryabhata has meticulously constructed the intricate clockwork that drives his entire astronomical system.

He began by defining the very fabric of time, from the smallest human quantum—the breath, or *prāṇa*—to the vast, cyclical ages of the cosmos. He then forged these units into a powerful computational engine, establishing the **ahargaṇa**, the "count of days," as the master key to unlocking the heavens. With this key in hand, he unveiled his sophisticated epicyclic model.

Through a systematic process of **manda** (slow) and **śīghra** (fast) corrections, he transformed simple, calculated averages into the complex, true motions of the planets. This remarkable system, blending iterative refinement with trigonometric precision, could account for the planets' changing speeds and their strange retrograde loops with an accuracy that was unparalleled for its era.

This potent machinery, anchored to the astronomical constants of the *Gītikāp-da* and wielding the mathematical tools of the *Gaṇitapāda*, does more than just predict. It provides the indispensable inputs for the final, grand articulation of Aryabhata's cosmic vision: the projection of these calculated positions onto the three-dimensional geometry of the celestial sphere.

The *Kālakriyāpāda*, therefore, is not merely about measuring time. It is about mastering it, providing the essential bridge from the abstract rhythm of numbers to the profound and culminating subject of the **Golapāda**—the Sphere.

***

Chapter Six

Mapping the Heavens – Golapāda

Aryabhata's Celestial Sphere: Earth's Rotation, Eclipse Science, and Cosmic Geometry

6.0 Introduction: A Revolution in Ancient Astronomy

Centuries before Copernicus proposed that the Earth revolved around the Sun, a young scholar in Gupta India quietly ignited his own astronomical revolution. His name was Aryabhata, and the culmination of his life's work is found in the fourth and final section of his masterpiece, the *Aryabhatiya*: the **Golapāda**, or "Chapter on the Sphere."

Composed around 499 CE, this remarkable chapter is where all the threads of Aryabhata's thought converge. It is the grand application of everything that came before. Here, he masterfully wields the mathematical toolkit forged in the *Gaṇitapāda* and the temporal framework of the *Kālakriyāpāda*, applying them to the astronomical constants laid down in the *Gītikāpāda*.

The *Golapāda* is a stunning fusion of mathematical rigor, empirical observation, and cosmological vision. Guided by the invaluable 7th-century commentary of Bhāskara I, we can see Aryabhata construct the *gola*—the celestial sphere—as a coherent geometric model. This was a profound shift in thinking. It transformed the heavens from a realm of mystical portents and myth into a rational, predictable, and, most importantly, **calculable** system.

The chapter unfolds with compelling logic, building a comprehensive picture of the cosmos, one piece at a time.

- It begins by defining the great celestial highways—the **ecliptic and the equator**—that form the backdrop for all heavenly motion.
- It then firmly places our world within this framework, establishing the **Earth as a sphere** at the center of the cosmos.
- From there, it explores the radical implications of this model, including the relativity of "down," the nature of time zones, and its most revolutionary idea: that the Earth itself **spins on its axis** once a day.
- With breathtaking intellectual dexterity, Aryabhata then integrates traditional cosmological symbols, like the great Mount Meru, by reinterpreting them as **rational, geometric features** of his new model.

He then puts this model to work, detailing the trigonometric tools needed to determine time and position from any latitude on Earth. The discussion culminates in his definitive, physical explanation for eclipses, which he describes not as the work of demons, but as the simple, predictable geometry of shadows. This rational foundation allows him to present sophisticated methods for predicting the precise timing, duration, and appearance of any eclipse.

The *Golapāda* is Aryabhata's masterwork, the final synthesis of his thought. As Bhāskara I aptly noted, it is like a physician's anatomical sketch or a grammarian's linguistic analysis—a simplified yet essential key for decoding the universe's complexity. It stands as a monument to scientific reason and marks a true revolution in ancient Indian thought.

6.1 Mapping the Ecliptic (Gola.1)

The Golapāda begins by defining the celestial 'highway' – the path the Sun appears to follow annually against the backdrop of stars. This verse establishes the fundamental geometry of the ecliptic (*apamaṇḍala*) and it's crucial tilt relative to the celestial equator, the relationship that governs Earth's seasons.

Sanskrit:

मेषादेः कन्यान्तं समम् उदग् अपमण्डलार्धम् अपयातम् ।

तौल्यादेर् मीनान्तं शेषार्धं दक्षिणेनैव ॥ १ ॥

IAST:

meṣādeḥ kanyāntaṃ samam udag apamaṇḍalārdham apayātam |

taulyāder mīnāntaṃ śeṣārdhaṃ dakṣiṇenaiva || 1 ||

Translation: "The ecliptic's half (*apamaṇḍalārdham*) stretching from Aries' beginning to Virgo's end (*meṣādeḥ kanyāntaṃ*), arcs equally north (*samam udag apayātam*); from Libra's beginning to Pisces' close (*taulyāder mīnāntaṃ*), the remaining half (*śeṣārdhaṃ*) bends equally south (*dakṣiṇenaiva*)."

The Tilted Path: Explaining the Seasons

Why do we have seasons? Why is the Sun higher in the sky in summer and lower in winter? This verse explains the fundamental geometric reason for this yearly cycle: the **tilt of the ecliptic**. The ecliptic is the great highway the Sun appears to travel along throughout the year. But this highway is not parallel to the celestial equator (the projection of Earth's equator into space). It is tilted.

Two Great Arcs: North and South

Aryabhata describes the ecliptic as being composed of two great, symmetrical 180-degree arcs.

- **The Northern Arc:** Starting from the vernal equinox (the beginning of Aries), the first half of the ecliptic, which includes the signs from Aries to Virgo, deviates to the **north** of the celestial equator. The Sun reaches its highest northern point at the summer solstice.

- **The Southern Arc:** Starting from the autumnal equinox (the beginning of Libra), the second half, from Libra to Pisces, mirrors this journey but to the **south** of the celestial equator. The Sun reaches its lowest southern point at the winter solstice.

The Crucial Symmetry of the Tilt

The verse emphasizes that these two arcs are perfectly symmetrical (*samam*). The maximum distance the Sun travels north of the celestial equator is exactly equal to the maximum distance it travels south. This angle of tilt is known as the **Obliquity of the Ecliptic**. Aryabhata had previously given its value as **24 degrees**.

The Cause of the Seasons

This simple, 24-degree tilt is the direct and sole cause of the Earth's seasons.

- When the Sun is on the **northern arc**, the Northern Hemisphere is tilted *towards* the Sun, receiving more direct sunlight for longer periods. This brings spring and summer.

- When the Sun is on the **southern arc**, the Northern Hemisphere is tilted *away* from the Sun, receiving less direct sunlight for shorter periods. This brings autumn and winter.

This elegant geometric description transforms the changing seasons from a mysterious natural phenomenon into a predictable and understandable consequence of the Earth's tilted path through the heavens.

Notes:

- ***Apamaṇḍalārdham:*** Half the ecliptic circle (180° arc).

- ***Meṣādeḥ kanyāntaṃ:*** From start of Aries (Vernal Equinox) to end of Virgo.

- ***Taulyāder mīnāntaṃ:*** From start of Libra (Autumnal Equinox) to end of Pisces.

- ***Samam:*** Equally, symmetrically. * *(Implicit reference to Gk.9):* Maximum deviation (Obliquity) = 24°.

- ***Udag/Dakṣiṇena:*** Northward/Southward (deviation from celestial equator).

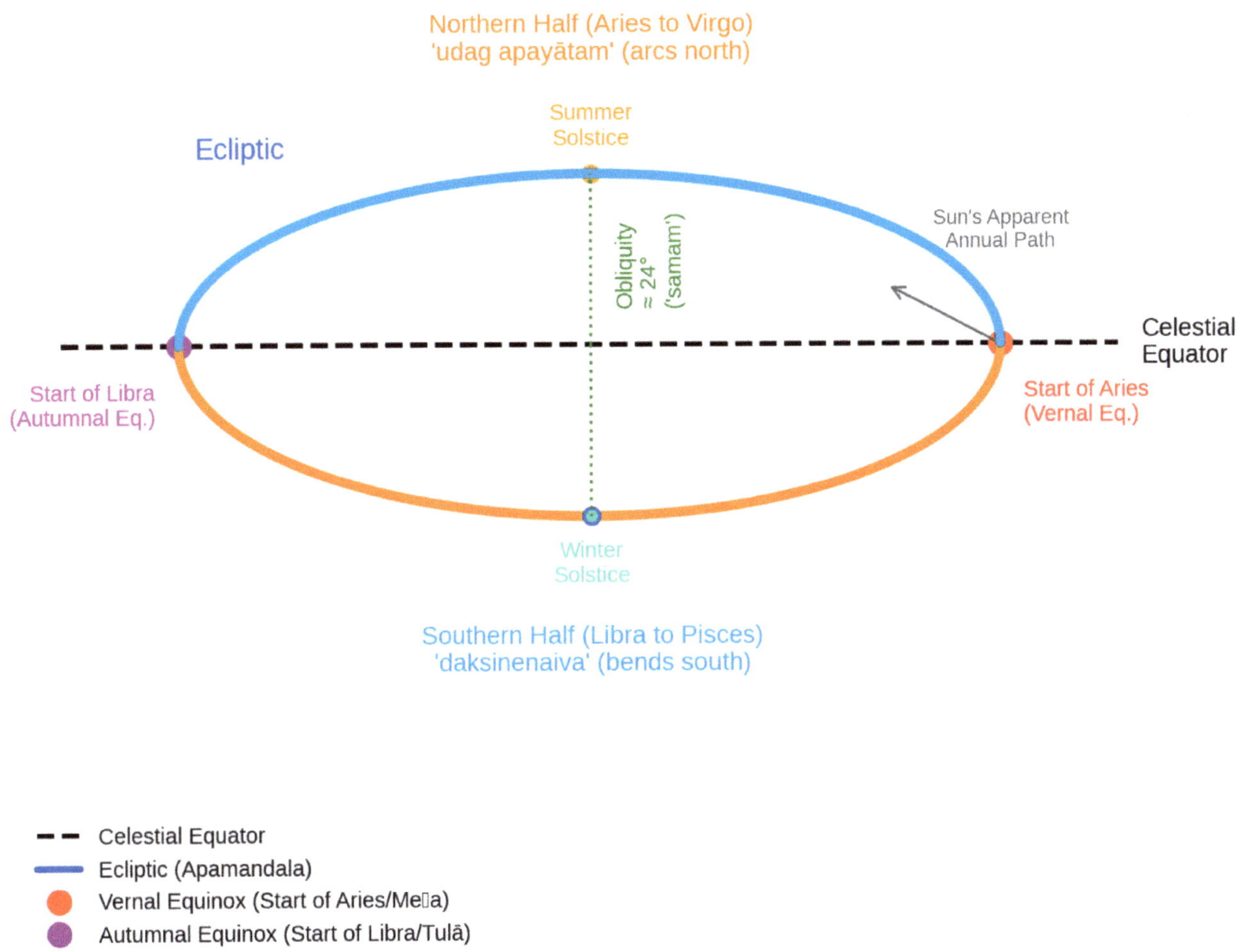

Figure 6.1a: The Ecliptic's Two Halves (Golapāda Verse 1). The diagram shows the Ecliptic tilted against the Celestial Equator. Its northern half (blue arc, Aries-Virgo) arcs 'equally north' (samam udag), and its southern half (orange arc, Libra-Pisces) bends 'equally south' (dakṣiṇenaiva), defining the solstices and equinoxes. This tilt (Obliquity ≈24°) causes seasons.

6.2 Motion Along and Around the Ecliptic (Gola.2-4)

Having defined the ecliptic as the central path, Aryabhata describes how other celestial bodies travel along or near it. He introduces the crucial concept of celestial latitude (deviation from the ecliptic) and sets practical limits for observing bodies when they are close to the glare of the Sun.

6.2.1 The Ecliptic Highway and Earth's Shadow

Sanskrit:

ताराग्रहेन्दूपाताः भ्रमन्ति अजस्रम् अपमण्डले अर्कश् च ।

अर्काच् च मण्डलार्धे भ्रमति हि तस्मिन् क्षितिच्छाया ॥ २ ॥

IAST:

tārāgrahendūpātāḥ bhramanti ajasraṃ apamaṇḍale arkaś cha |

arkāc cha maṇḍalārdhe bhramati hi tasmin kṣiticchāyā || 2 ||

Translation: "The star-planets (*tārāgraha* = Mars, Jupiter, Saturn, Mercury, Venus), Moon (*indu*), and nodes (*pātāḥ*) ceaselessly roam (*bhramanti ajasram*) on or near the ecliptic (*apamaṇḍale*), and so too does the Sun (*arkaś ca*); opposite the Sun (*arkāt ca*), at half the circle [180° away] (*maṇḍalārdhe*), Earth's shadow (*kṣiticchāyā*) moves within it [the ecliptic plane]."

The Celestial Highway: The Ecliptic and the Earth's Shadow

To map the movements of the heavens, astronomers needed a fundamental reference line—a great celestial highway upon which all motion could be measured. In this verse, Aryabhata defines this critical path: the **ecliptic**, or *apamaṇḍala*.

The Ecliptic: The Highway of the Planets

The ecliptic is the great circle that the Sun appears to trace through the stars over the course of a year. Aryabhata establishes this as the primary track for all celestial motion.

- The mean positions of the **Sun, the Moon, and all the planets** are calculated along this path.
- The **lunar nodes** (*pāta*)—the crucial points where the Moon's tilted orbit crosses the ecliptic—also slide along this same highway.

The ecliptic is, in essence, the master coordinate system for the entire solar system.

The Earth's Shadow: An Invisible Partner

Next, Aryabhata makes a simple but profound statement that is the key to his entire theory of lunar eclipses. He declares that the **Earth's shadow** (*kṣiticchāyā*) also moves along this same ecliptic highway. This invisible shadow is cast directly opposite the Sun, 180 degrees away. As the Sun moves east along the ecliptic, the Earth's shadow moves in perfect opposition, like an invisible anti-sun, silently tracing the same path on the opposite side of the sky.

The Foundation for Eclipse Theory

This elegant geometric insight lays the perfect foundation for explaining a lunar eclipse. A lunar eclipse occurs when the Full Moon, which is by definition on the opposite side of the sky from the Sun, passes through this invisible shadow. By establishing that both the Full Moon and the Earth's shadow travel along the same celestial highway, Aryabhata sets the stage perfectly. A lunar eclipse becomes a simple and predictable event: a celestial alignment that happens whenever the Moon, in its monthly journey, encounters the shadow at the exact same point on the ecliptic.

Notes:

- ***Apamaṇḍale:*** On/in the ecliptic plane.
- ***Tārāgrahendūpātāḥ:*** Planets, Moon, nodes.
- ***Ajasram:*** Ceaselessly, continuously.
- ***Maṇḍalārdhe:*** Half circle away (180° opposition).
- ***Kṣiticchāyā:*** Earth's shadow, essential for lunar eclipses.

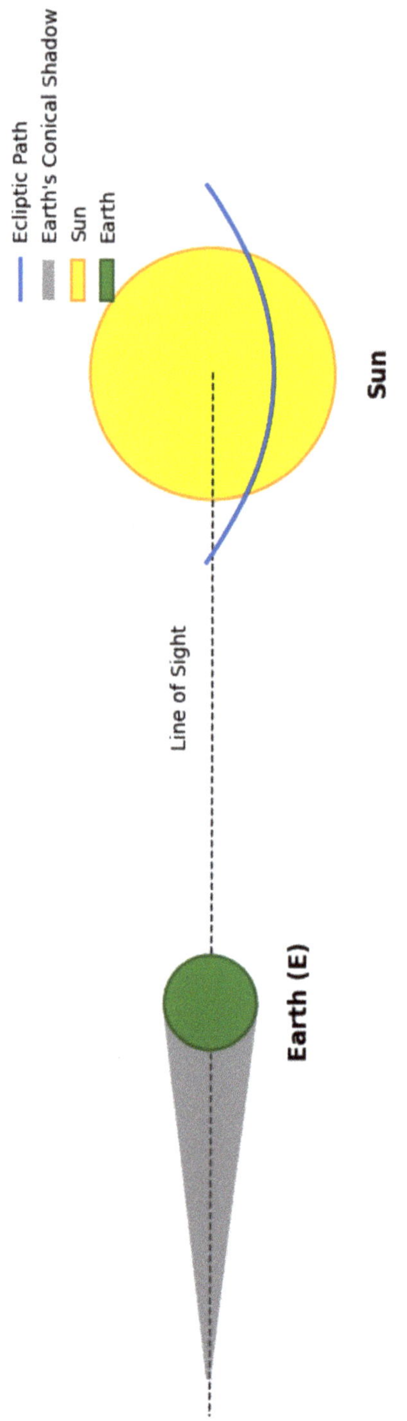

Figure 6.1b: Earth's Shadow (Bhūcchāyā) on the Ecliptic (Gola.2). Depicts the Earth's shadow cast 180° opposite the Sun and moving along the Ecliptic (Gola.2), which is the cause of lunar eclipses (Gola.37).

6.2.2 Latitude – Deviations from the Path

Sanskrit:

अपमण्डलस्य चन्द्रः पाताद् यात्य् उत्तरेण दक्षिणतः ।

कुजगुरुकोणाश् चैवं शीघ्रोच्चेनापि बुधशुक्रौ ॥ ३ ॥

IAST:

apamaṇḍalasya chandraḥ pātād yāty uttareṇa dakṣiṇataḥ |

kujagurukoṇāś chaivaṃ śīghrocchenāpi budhaśukrau || 3 ||

Translation: "The Moon (*chandraḥ*) strays north and south (*uttareṇa dakṣiṇataḥ*) of the ecliptic (*apamaṇḍalasya*) from its node (*pātād*); so do Mars (*kuja*), Jupiter (*guru*), and Saturn (*koṇa* = Saturn) likewise; Mercury (*budha*) and Venus (*śukra*) follow similarly via their ***Śīghroccha*** [reference points]."

Wandering off the Path: The Concept of Celestial Latitude

The ecliptic is the great central highway of the solar system, the path the Sun appears to follow through the stars. But the Moon and the planets are not perfect drivers; they do not stick to this main road. They regularly wander slightly above or below it. This north-south deviation from the ecliptic is called **celestial latitude**, or *vikṣepa*. It is a crucial concept for understanding the true positions of the celestial bodies.

The Moon's Tilted Path

The Moon's orbit is the most significant example. Its path is tilted relative to the ecliptic, causing it to swing north and south during its monthly journey. Aryabhata states that the Moon's maximum latitude is **4.5 degrees** north or south of the ecliptic. It only crosses the ecliptic—the point where its latitude is zero—at two specific points in its orbit. These are the **lunar nodes** (*pāta*).

The Planets' Deviations

The planets also have their own celestial latitudes, though generally smaller than the Moon's.

- **Mars, Jupiter, and Saturn:** Aryabhata assigns these outer planets a maximum latitude of **2.25 degrees** (exactly half the Moon's maximum).
- **Mercury and Venus:** The latitudes of these inner planets are more complex, as their apparent motion is tied to the position of the Sun. However, they also deviate from the ecliptic by a similar amount.

Why Celestial Latitude is Critical

Understanding and calculating celestial latitude is absolutely essential for predictive astronomy, especially for one of the most dramatic celestial events: **eclipses**.

For a solar or lunar eclipse to occur, two conditions must be met:

1. The Sun, Earth, and Moon must be aligned in longitude (a New or Full Moon).
2. The Moon must be at or very near one of its nodes, meaning its latitude must be close to zero.

If the Moon is too far north or south of the ecliptic during a New Moon, its shadow will miss the Earth entirely, and there will be no solar eclipse. If it is too far north or south during a Full Moon, it will pass above or below the Earth's shadow, and there will be no lunar eclipse. Celestial latitude, therefore, is the key that determines whether a perfect alignment in longitude will result in a spectacular eclipse or just another ordinary New or Full Moon.

Notes:

- ***Vikṣepa:*** Celestial latitude (N/S deviation from ecliptic). Maxima in Gk.9.
- ***Pātād:*** From the node (lunar nodes).
- ***Kuja/Guru/Koṇa:*** Mars/Jupiter/Saturn.
- ***Śīghroccha:*** Epicyclic point linked to mean Sun, used for inner planet motion and latitude calculations.

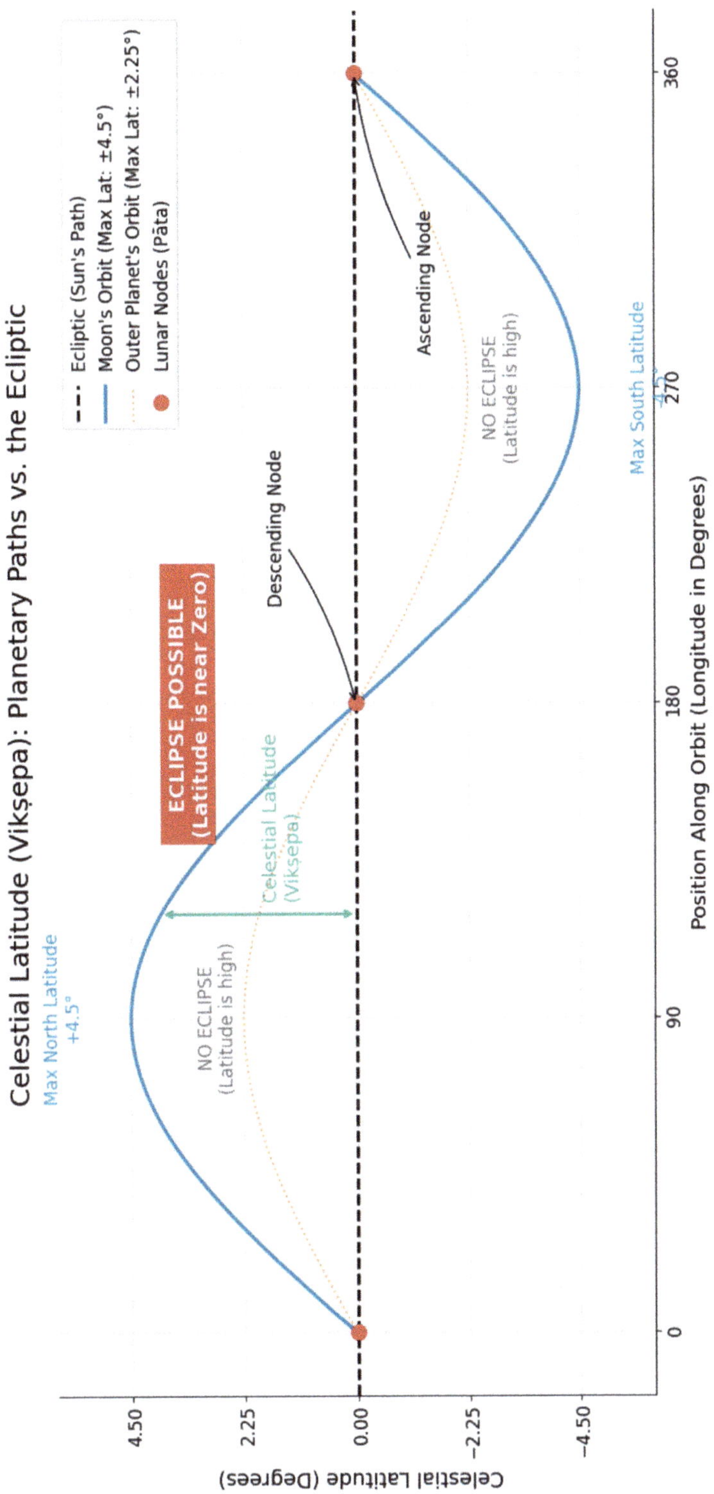

Figure 6.2: The Geometry of Celestial Latitude (Vikṣepa). This diagram illustrates how the Moon's orbit is tilted relative to the ecliptic. An eclipse is only possible when the Moon is at or near a node (pāta), where its latitude is zero.

6.2.3 Visibility Thresholds Near the Sun

Sanskrit:

चन्द्रः अंशैर् द्वादशभिर् अविक्षिप्तोऽर्कान्तरस्थितो दृष्यः ।

नवभिर् भृगुर् भृगोस् तैस् त्र्यधिकैर् द्व्यधिकैर् यथा श्लक्ष्णाः ॥ ४ ॥

IAST:

chandraḥ aṃśair dvādaśabhir avikṣipto'rkāntarasthito dṛśyaḥ |

navabhir bhṛgur bhṛgos tais tryadhikair dvyadhikaiḥ yathā ślakṣṇāḥ || 4 ||

Translation: "The Moon (*chandraḥ*) becomes visible (*dṛśyaḥ*) at twelve degrees (*dvādaśabhiḥ aṃśaiḥ*) of elongation from the Sun (*arkāntarasthito*) when without latitude (*avikṣiptaḥ* = on the ecliptic); Venus (*bhṛguḥ*) at nine degrees (*navabhiḥ*); from Venus [add increments successively]: by three more (*tryadhikaiḥ*) [Jupiter: 9+3=12°], by two more (*dvyadhikaiḥ*) [Mercury: 12+2=14°], by two more [Saturn: 14+2=16°], by two more [Mars: 16+2=18°], according to their faintness/subtlety (*yathā ślakṣṇāḥ*)."

Interpretation: This verse provides practical observational rules, defining the minimum angular separation from the Sun (**elongation, *arkāntara***) required for a celestial body to become visible to the naked eye in the twilight sky (either setting just after the Sun or rising just before it – known as **heliacal rising/setting**). These limits are crucial for calendrical purposes and basic observation.

1. **Observational Limits:** These values represent the threshold where the object emerges from the Sun's glare. They are given assuming the object has zero latitude (***avikṣiptaḥ***), meaning it lies directly on the ecliptic.

2. **Importance:**

    - Determining the first visibility of the thin crescent Moon after the New Moon is essential for starting the lunar month (***Cāndra māsa***) in many Indian calendars (especially ***śuklādi*** systems).

    - Predicting the first or last visibility of planets near the Sun (when they emerge from or disappear into conjunction).

3. **Values and Faintness:** The thresholds increase according to the object's faintness (***yathā ślakṣṇāḥ***), reflecting observational reality:

    - Moon: 12°

    - Venus (brightest): 9°

    - Jupiter: 12°

- Mercury: 14°
- Saturn: 16°
- Mars: 18°

4. **Type of Degree:** The "degrees" (***aṃśaiḥ***) here likely refer to ***kālāṃśa*** (time-degrees) or a similar measure related to the difference in rising/setting times near the horizon, rather than simple geometric elongation, as visibility depends heavily on atmospheric conditions and the angle of the ecliptic relative to the horizon.

Notes:

- ***Arkāntarasthito:*** Angular distance/elongation from the Sun.
- ***Avikṣiptaḥ:*** Without latitude (on the ecliptic).
- ***Aṃśaiḥ:*** Degrees (likely time-degrees, ***kālāṃśa***).
- ***Bhṛguḥ:*** Venus. Sequence: Moon(12°), Venus(9°), Jupiter(12°), Mercury(14°), Saturn(16°), Mars(18°).
- ***Yathā ślakṣṇāḥ:*** According to faintness/subtlety.

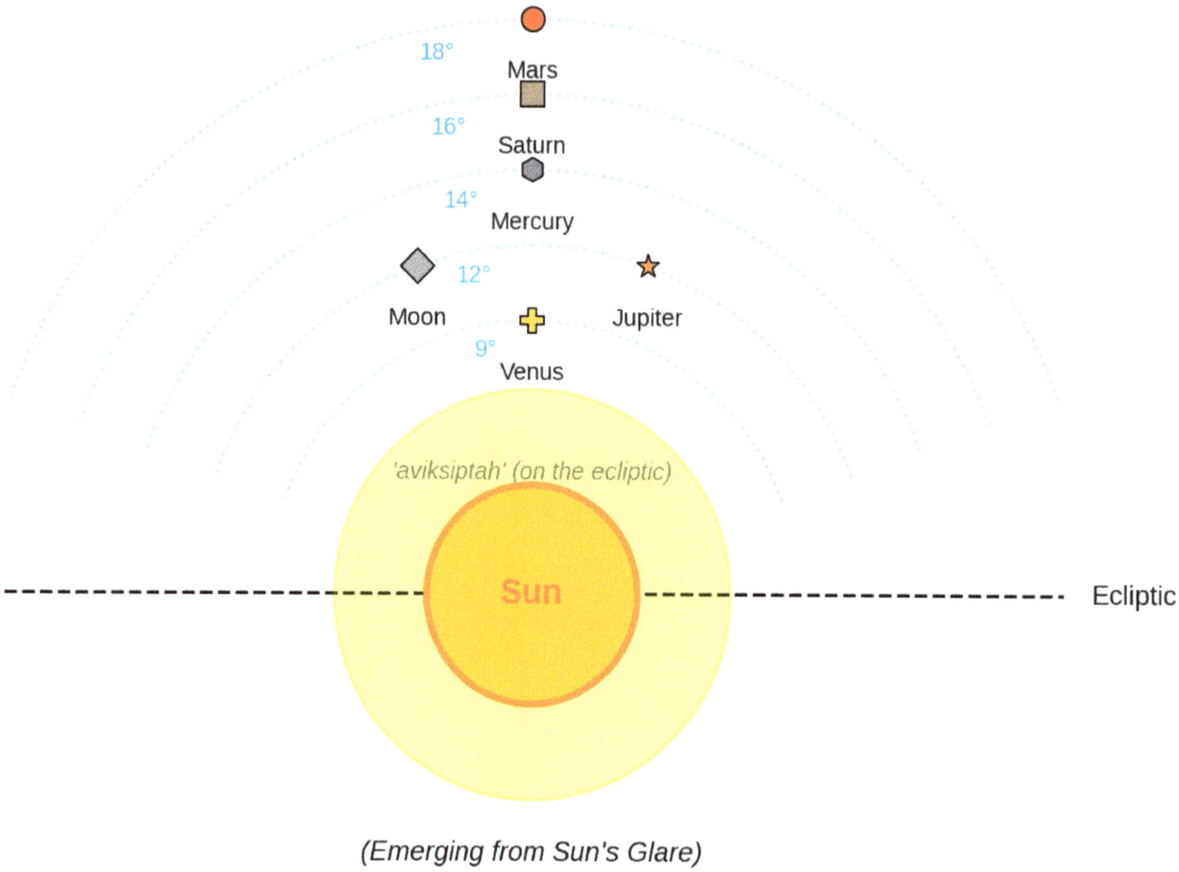

Figure 6.3a: Visibility Thresholds from the Sun (Gp.4). Illustrates the minimum angular elongation (arkāntara) for the Moon and planets to be seen when on the Ecliptic (avikṣiptaḥ), emerging from the Sun's glare. Values increase (9° to 18°) based on faintness (yathā ślakṣṇāḥ), as per Aryabhata.

6.3 Placing Earth in the Cosmos (Gola.5)

Having set the celestial stage and introduced the main actors, Aryabhata now explicitly positions the Earth within his model, definitively asserting its shape and central location within the framework of orbits.

6.3.1 Earth at the Center of Orbits

Sanskrit:

वृत्तभपञ्जरमध्ये कख्यापरिवेष्टितः खमध्यगतः ।

मृज्जलशिखिवायुमयो भूगोलः सर्वतः वृत्तः ॥ ५ ॥

IAST:

vṛttabhapañjaramadhye kakṣyāpariveṣṭitaḥ khamadhyagataḥ |

mṛjjalaśikhivāyumayo bhūgolaḥ sarvataḥ vṛttaḥ || 5 ||

Translation: "The Earth-sphere (***bhūgolaḥ***), made of earth, water, fire, and air (***mṛjjalaśikhivāyumayaḥ***), sits at the center of space (***khamadhyagataḥ***), encircled by the orbits (***kakṣyāpariveṣṭitaḥ***) [of planets], within the circular cage of stars (***vṛttabhapañjaramadhye***), [and is] round on all sides (***sarvato vṛttaḥ***)."

The Architecture of the Cosmos: A Central, Spherical Earth

In a few concise and powerful lines, Aryabhata lays out the fundamental architecture of his universe. This verse is not an argument but a declaration, establishing the five core principles upon which his entire astronomical system is built.

1. The Earth is a Sphere

Aryabhata begins with his most crucial assertion: the Earth is a sphere (*bhūgolaḥ*). He emphasizes this by adding that it is "round on all sides" (*sarvato vṛttaḥ*), leaving no room for doubt. This is the geometric bedrock of his system, essential for explaining everything from time zones to eclipses.

2. It is Made of the Elements

He grounds his model in the physical philosophy of his time, stating that the globe is composed of the classical elements: earth, water, fire, and air.

3. It is at the Center of Space

Next, he establishes the Earth's position in the cosmos. The Earth-sphere, he declares, is situated firmly "at the center of space" (*khamadhyagataḥ*). This confirms the geocentric nature of his model: the Earth is the stationary hub of the planetary system.

4. It is Encircled by Orbits

He then describes the motion of the heavens. The orbits of the Sun, Moon, and planets (*kakṣyā*) revolve around this central, unmoving Earth.

5. It is Contained Within the Stars

Finally, he places this entire system within the vast backdrop of the cosmos. The entire planetary system, with Earth at its heart, is enclosed within the sphere of the fixed stars, which he poetically calls the *bhapañjara*, or "cage of asterisms."

The Foundation is Laid

With these five clear and concise statements, Aryabhata establishes the fundamental blueprint of his universe. It is a cosmos with a spherical, stationary Earth at its absolute center, around which the Sun, Moon, and planets perform their celestial revolutions, all contained within the grand sphere of the stars. Upon this solid foundation, all the detailed explanations in the rest of the *Golapāda* will be built.

Notes:

- ***Bhūgolaḥ:*** Earth-sphere.
- ***Mṛj-jala-śikhi-vāyu-mayaḥ:*** Made of earth, water, fire, air.
- ***Khamadhyagataḥ:*** Situated at the center of space (geocentric).
- ***Kakṣyāpariveṣṭitaḥ:*** Encircled by the orbits.
- ***Vṛttabhapañjaramadhye:*** Within the circular cage of stars.
- ***Sarvato vṛttaḥ:*** Round on all sides, emphatically spherical.

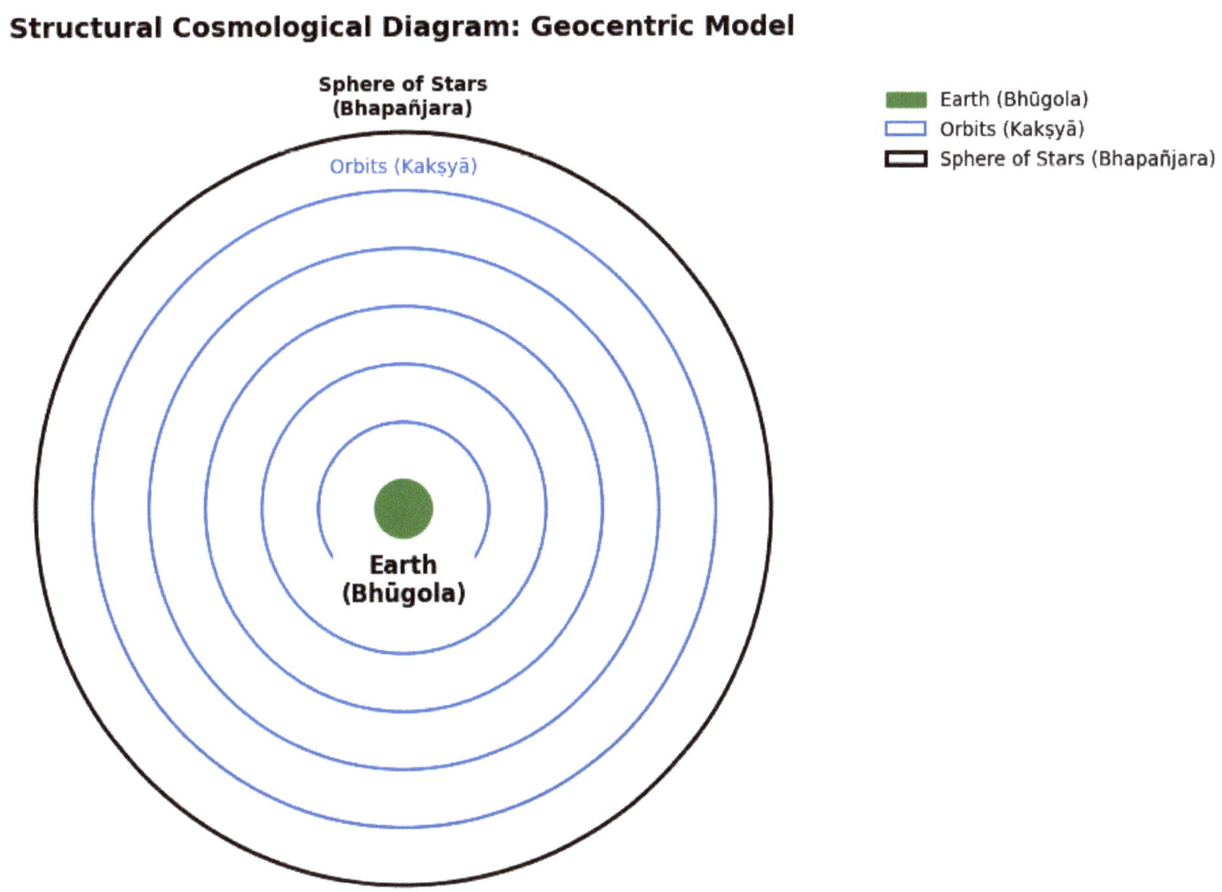

Figure 6.3b: Structural diagram of Aryabhata's geocentric cosmological model, as described in Golapāda Verse 5 (Gola.5). The Earth-sphere (Bhūgola) is situated at the center, encircled by the orbits (Kakṣyā) of the Sun, Moon, and planets, all contained within the outer sphere of fixed stars (Bhapañjara).

6.4 The Spherical Earth – A World Reimagined (Gola.6-7)

A spherical Earth was a radical idea compared to the flat-Earth cosmologies common in many traditions. Aryabhata not only asserts it but provides arguments and analogies to make this counter-intuitive concept comprehensible and plausible, tackling the conceptual challenges head-on.

6.4.1 Earth's Composition and Polar Features

Sanskrit:

भूगोलः सर्वतो वृत्तः पञ्चमहाभूतगुणसमन्वितः ।

मध्ये समन्तपर्वतः पर्यन्ते नरक वडवामुखं च ॥ ६ ॥

IAST:

bhūgolaḥ sarvato vṛttaḥ pañcamahābhūtaguṇasamanvitaḥ |

madhye samantaparvataḥ paryante naraka vaḍavāmukhaṃ cha || 6 ||

Translation: "The Earth-sphere (***bhūgolaḥ***) is round on all sides (***sarvato vṛttaḥ***), endowed with the qualities of the five great elements (***pañcamahābhūtaguṇasamanvitaḥ*** = earth, water, fire, air, ether). In the middle [North Pole region?] is the surrounding mountain [Meru?] (***madhye samantaparvataḥ***); at the edge/end [South Pole region?] are hell (***naraka***) and Baḍavāmukha."

The Blueprint of a Spherical World

Having declared the Earth to be a sphere, Aryabhata now begins to sketch the first features of his new world map. This verse lays the foundational blueprint, defining the Earth's nature and establishing its most important geographical and cosmological landmarks.

A Sphere of Five Elements

First, he reiterates the Earth's fundamental nature: it is "round on all sides" (*sarvato vṛttaḥ*), a perfect sphere. He also grounds it in traditional physics by stating that it is composed of the five great elements: earth, water, fire, air, and ether (*ākāśa*).

Mapping the Poles: North and South

Next, he maps the two most important regions onto this globe: the poles.

- **The North Pole:** He describes a "surrounding mountain in the middle" (*madhye samantaparvataḥ*). Later commentators like Bhaskara I explain that this is not a literal mountain on a flat plane. Instead, it is a rationalized, geometric representation of the **North Polar region**, a conceptual stand-in for the great Mount Meru of mythology. It is the pivot point of the northern world.

- **The South Pole:** He then describes the opposite pole, "at the boundary," associating it with *Naraka* (the underworld) and *Baḍavāmukha* (the mythical "Mare's Mouth," a fiery pit often associated with the far south). This maps the traditional underworld onto the **South Polar region**, the antipode to the known world.

The Natural Cause of Day and Night

By establishing a spherical Earth with a defined North and South Pole, Aryabhata lays the crucial groundwork for explaining day and night. If the Earth is a globe floating in space, and the Sun can only illuminate one half of it at a time, then the daily cycle of light and darkness becomes a simple, natural, and inevitable consequence of this geometry. The existence of day and night is no longer a mystery to be explained by myth, but a logical outcome of living on a spherical world.

Notes:

- ***Pañcamahābhūta:*** Five elements (earth, water, fire, air, ether).
- ***Samantaparvataḥ:*** Surrounding mountain (likely North Polar region/Meru).
- ***Paryante:*** At the boundary/end (likely South Polar region).
- ***Naraka/Baḍavāmukha:*** Associated with the South Pole.

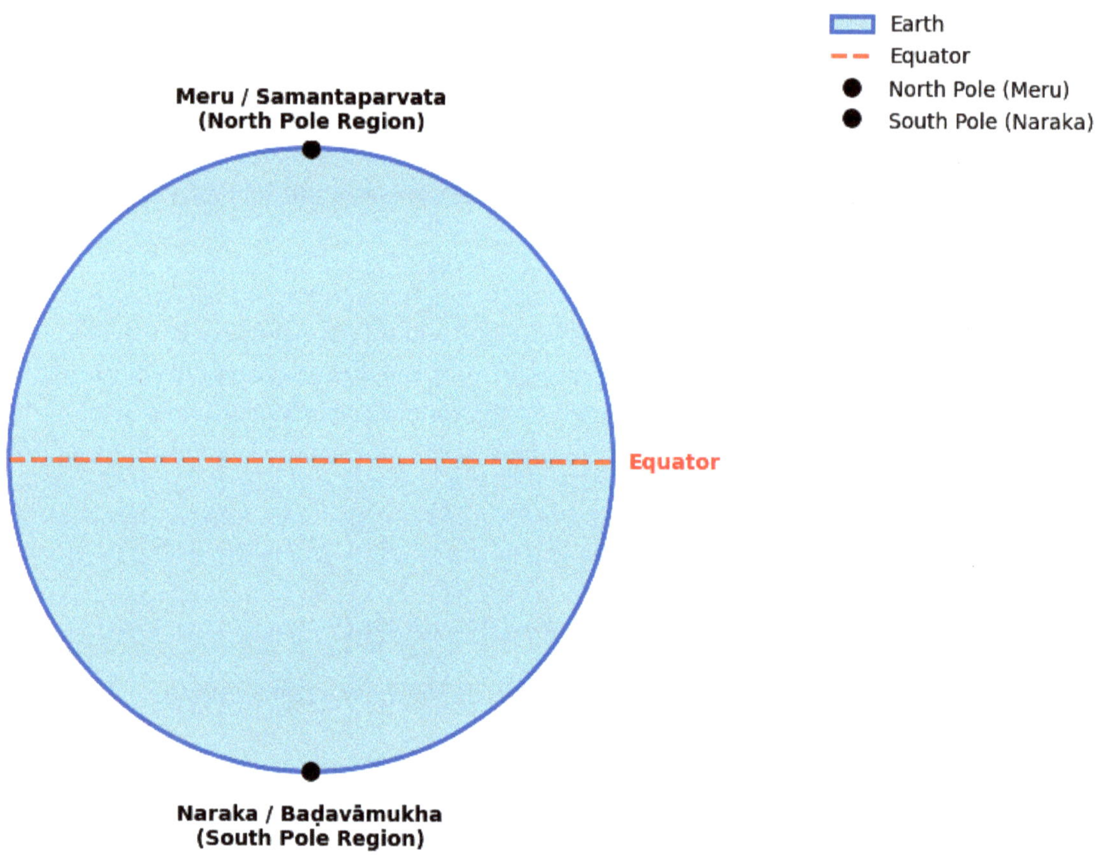

Figure 6.4: Diagram illustrating Aryabhata's spherical Earth (Bhūgola) and the mapping of traditional geographical and cosmological concepts onto its poles and Equator, as described in Golapāda Verse 6 (Gola.6) and elaborated by commentators. The North Pole region is associated with Meru, and the South Pole region with Naraka/Baḍavāmukha, fitting these concepts within a rational, spherical model.

6.4.2 The Analogy of the Kadamba Flower and Gravity

Sanskrit:

यद्वत् कदम्बपुष्पग्रन्थिः प्रचितः समन्ततः कुसुमैः ।

तद्वद् धि सर्वसत्त्वैर् जलजैः स्थलजैश् च भूगोलः ॥ ७ ॥

IAST:

yadvat kadambapuṣpagranthiḥ prachitaḥ samantataḥ kusumaiḥ |

tadvad dhi sarvasattvaiḥ jalajaiḥ sthalajaiś cha bhūgolaḥ || 7 ||

Translation: "Just as the spherical knot (*granthiḥ*) of a Kadamba flower (*kadambapuṣpa*) is covered all around (*samantataḥ*) with blossoms (*kusumaiḥ*), so indeed (*tadvad dhi*) the Earth-sphere (***bhūgolaḥ***) is covered with all beings (*sarvasattvaiḥ*), aquatic (*jalajaiḥ*) and terrestrial (*sthalajaiś cha*)."

The Kadamba Flower: A Spherical World in Miniature

If the Earth is a sphere, why don't the people on the "bottom" just fall off? This is perhaps the most common and powerful intuitive objection to the idea of a round world. In this famous verse, Aryabhata answers this puzzle not with a complex formula, but with a simple, elegant, and beautiful analogy from the natural world.

The Lesson of the Flower

Aryabhata asks us to consider the flower of the **Kadamba tree**. This tree produces perfectly spherical flower heads. From every point on the surface of this sphere, tiny individual blossoms emerge, each one pointing straight out, perpendicular to the surface.

The Analogy for a Spherical Earth

He then argues that our Earth is just like this Kadamba flower head.

- The spherical globe of the Earth is the central part of the flower.
- All living beings, scattered across the continents, are like the blossoms.

Just as the blossoms sprout naturally from every part of the flower's surface, so too do all creatures live all around the surface of the spherical Earth.

A Hint of Gravity: The Nature of "Down"

The true genius of this analogy is what it implies about the direction of **"down."** For each tiny blossom, "down" is the direction its roots point—*towards the center of the flower head.*

In the same way, for every person on Earth, "down" is not some absolute direction in space. It is simply the direction **towards the center of the Earth**.

This one simple image completely solves the "falling off" problem. If "down" always points to the center, then no one is ever "upside down." Everyone, no matter where they are on the globe, stands upright on their own local ground. This is a profound hint of a concept akin to gravity. Aryabhata is suggesting that the Earth possesses an intrinsic property—an attractive nature—that pulls all things on its surface towards its own center. While he does not define this as a universal force of attraction between all objects (Newton's great insight), he correctly identifies its most important local effect. He argues that this "pull" is what makes things fall, what gives them weight, and what keeps us firmly planted on the ground, no matter where we are on our spherical world.

Notes:

- ***Kadambapuṣpagranthiḥ:*** Spherical Kadamba flower head.
- ***Samantataḥ:*** All around, uniformly.
- ***Sarvasattvaiḥ:*** All living beings.
- ***Jalajaiḥ/Sthalajaiḥ:*** Aquatic/terrestrial.

Together, Gola.5-7 establishes Aryabhata's firm belief in a spherical Earth, providing descriptive mapping and a powerful analogy to make this foundational concept understandable and defensible.

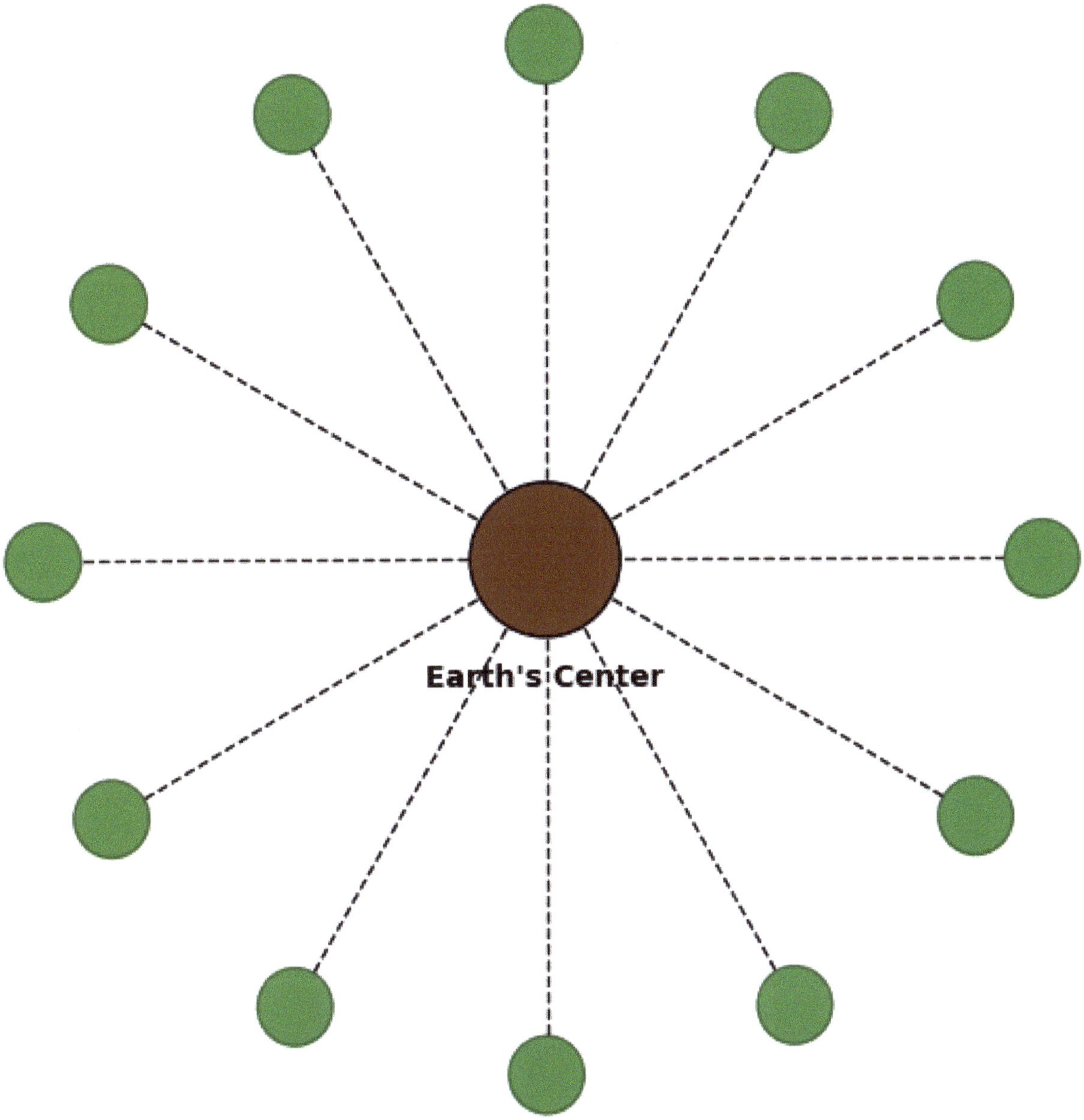

Figure 6.5: Illustration of the Kadamba flower analogy (Gola.7) used by Aryabhata to explain why beings can live all around a spherical Earth. Just as the flower's blossoms point radially outwards from the center, so too do inhabitants and objects on Earth orient towards its center, creating a sense of "down" relative to their location.

6.5 Earth's Rotation – A Radical Idea (Gola.9-10)

Here lies perhaps the most revolutionary and controversial assertion in the entire *Aryabhatiya*: the claim that the daily motion of the heavens is caused by the Earth itself spinning on its axis. Aryabhata presents his argument based on relative motion and contrasts it directly with the prevailing traditional explanation. This section marks a major conceptual leap in ancient astronomy.

6.5.1 The Relative Motion Argument for Earth's Rotation

Sanskrit:

अनुलोमगतिर् नौस्थः पश्यत्य् अचलं विलोमगं यद्वत् ।

अचलानि भानि तद्वत् समपश्चिमगानि लङ्कायाम् ॥ ९ ॥

IAST:

anulomagatir nausthaḥ paśyaty achalaṃ vilomagaṃ yadvat |

achalāni bhāni tadvat samapaścimagāni laṅkāyām || 9 ||

In this landmark verse, Aryabhata presents his revolutionary theory of Earth's axial rotation. He does so not through abstract declaration, but through a powerful and intuitive analogy grounded in the principle of relative motion.

Translation: "Just as a man on a boat moving forward sees a stationary object [on the bank] as moving backward, so too at Laṅkā, the stationary stars appear to move uniformly westward."

The analogy is simple yet profound: a person on a boat moving forward (*anulomagatir nausthaḥ*) perceives stationary objects on the bank (*achalam*) as if they are moving backward (*vilomagaṃ*).

Aryabhata then masterfully applies this everyday experience to the grandest scale imaginable. He argues that the apparent, uniform westward march of the stars (*samapaścimagāni*) is not a true motion of the heavens. Instead, it is an illusion, an apparent motion caused by the eastward rotation of the Earth itself, upon which the observer at Laṅkā resides.

The verse, therefore, makes three bold claims:

1. The stars are fundamentally stationary (*achalāni bhāni*) with respect to this daily motion.
2. Their apparent movement is perfectly uniform and westward.
3. The true cause of this perception is the eastward spin of the observer's terrestrial platform.

This is a profound kinematic insight, one of the most revolutionary in the history of ancient astronomy. It elegantly explains the fundamental diurnal motion—the daily rising and setting of all celestial bodies—as a direct consequence of the observer's own movement. This stands in sharp contrast to the prevailing models that required the entire, immense celestial sphere to revolve around a static Earth.

While Aryabhata's system remained geocentric in its planetary scheme—with the Sun, Moon, and planets still orbiting the Earth—attributing the primary daily motion to the Earth's own spin was a radical and brilliant simplification of the cosmos.

Notes:

- ***Anulomagatir:*** Forward motion (observer/Earth moving eastward).
- ***Vilomagaṃ:*** Backward (apparent) motion (bank/stars moving westward).
- ***Achalāni bhāni:*** Stationary stars (relative to diurnal motion).
- ***Samapaścimagāni:*** Uniformly westward moving (apparent motion).
- ***Laṅkāyām:*** At Laṅkā (prime meridian, but the principle is universal).

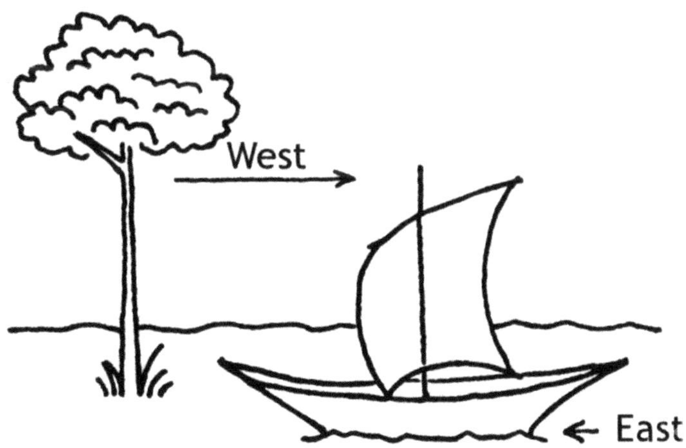

Figure 6.6: Illustration of Aryabhata's boat analogy (Gola.9). Just as a person in a boat moving eastward sees stationary objects on the shore appear to move westward, Aryabhata argued that the Earth's eastward rotation causes the apparent daily westward motion of the stars.

6.5.2 The Traditional View (*Pravaha* Wind)

Sanskrit:

उदयास्तमयनिमित्तं नित्यं प्रवहेण वायुना क्षिप्तः ।

लङ्कासमपश्चिमगो भपञ्जरः सग्रहो भ्रमति ॥ १० ॥

IAST:

udayāstamayanimittaṃ nityaṃ pravaheṇa vāyunā kṣiptaḥ |

laṅkāsamapaścimago bhapañjaraḥ sagraho bhramati || 10 ||

Translation: "As the eternal cause of rising and setting (*udayāstamayanimittaṃ*), propelled (*kṣiptaḥ*) by the ***pravaha*** wind (***pravaheṇa vāyunā***), the cage of stars (***bhapañjaraḥ***) along with the planets (***sagrahaḥ***) spins (*bhramati*), moving uniformly westward from Laṅkā (*laṅkāsamapaścimago*)."

The Prevailing View: The Great Cosmic Wind

Immediately after presenting his revolutionary theory of a spinning Earth, Aryabhata does something remarkable. In an act of intellectual honesty and shrewd rhetoric, he concisely states the prevailing, traditional explanation for the daily motion of the heavens.

The Theory of the *Pravaha* Wind

The traditional model, which stood in direct opposition to Aryabhata's, proposed that the Earth was stationary. In this view, the daily rising and setting of the stars was caused by a great, unseen cosmic current known as the ***pravaha vāyu***, or "provector wind." This eternal wind was thought to constantly push the entire celestial sphere—the great dome of the stars (*bhapañjara*) along with the planets embedded within it—in a uniform westward rotation around the unmoving Earth. This cosmic conveyor belt provided a complete, though very different, explanation for why the Sun, Moon, and stars appear to rise in the east and set in the west.

A Deliberate Contrast

By placing his own elegant theory of relative motion right before the traditional explanation, Aryabhata creates a powerful contrast. He implicitly asks the reader to choose between two models:

1. A simple, logical explanation based on a spinning Earth, analogous to a person in a moving boat.

2. A complex, traditional explanation requiring an unobservable cosmic wind to move the entire universe.

The Debate Ignited

Aryabhata's radical idea was not easily accepted. It ignited one of the most significant scientific debates in ancient India, a controversy that would last for centuries.

- **The Supporter:** Aryabhata's own commentator, Bhaskara I, fiercely defended the theory of Earth's rotation, relying heavily on the elegant "person in a boat" analogy.
- **The Critic:** The later, highly influential astronomer Brahmagupta (c. 628 CE) vehemently rejected the idea. He raised a series of powerful, common-sense objections that resonated widely.

Brahmagupta argued that if the Earth were spinning at a great speed, we should feel the effects. Why isn't there a constant, gale-force wind? Why aren't birds left behind in the sky when they fly? Why aren't tall buildings flung from the surface?

These were brilliant objections based on the physics of the time. Aryabhata had correctly described the *kinematics* (the geometry of relative motion), but without Newton's later concepts of gravity and inertia, he did not have the *dynamics* (the physics of forces) to answer these critiques. Despite the controversy, Aryabhata's proposal stands as a monumental conceptual achievement. It demonstrates a leap of logic and an ability to see the world from a new perspective that was centuries ahead of its time.

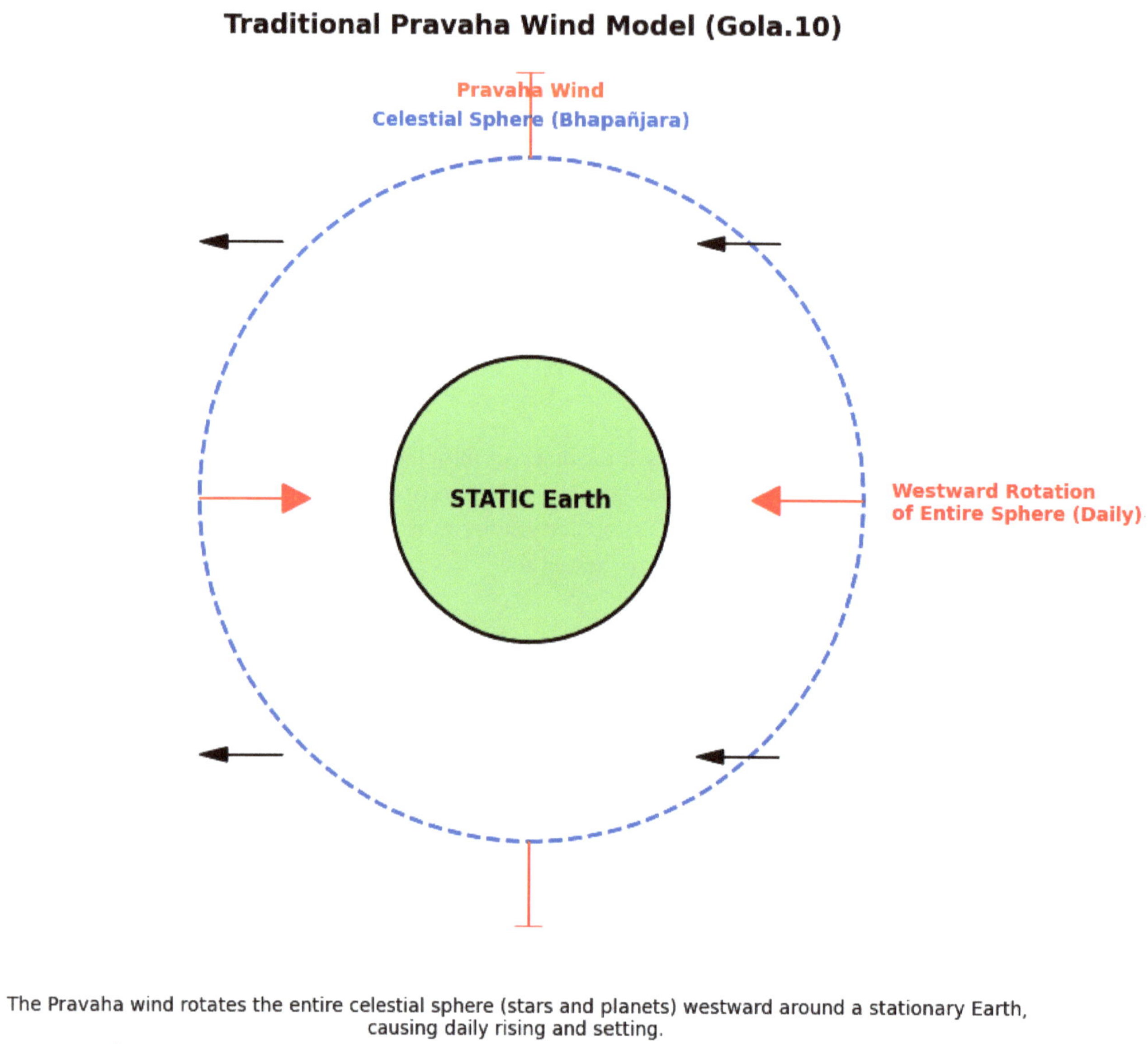

Figure 6.7: The Pravaha Wind Model for Diurnal Motion (Gola.10). Diagram illustrating the traditional explanation, presented by Aryabhata in Gola.10, where a cosmic Pravaha Wind rotates the Celestial Sphere (Bhapañjara, containing stars and planets) westward around a stationary Earth, causing daily rising and setting. (Aryabhata's own theory, Gola.9, attributed this apparent motion to Earth's rotation).

6.6 Cosmological Perspectives: Meru and Relativity (Gola.11-12)

Having proposed a spherical, rotating Earth, Aryabhata skillfully integrates elements of traditional Indian cosmology, like Mount Meru, into his new framework through rationalization. He also uses the spherical model to emphasize the relativity of perspective, a concept crucial for understanding a round world.

6.6.1 Meru Rationalized as the North Pole

Sanskrit:

मेरुर् योजनमात्रः प्रभाकरो हिमवता परिवृतः ।

नन्दनवनस्य मध्ये रत्नमयः सर्वतो वृत्तः ॥ ११ ॥

IAST:

merur yojanamātraḥ prabhākaro himavatā parivṛtaḥ |

nandanavanasya madhye ratnamayaḥ sarvato vṛttaḥ || 11 ||

Translation: “Meru, measuring only one yojana (***yojanamātraḥ***), radiant (***prabhākaro***), is perhaps encircled by the Himavat [range], and stands in the center of Nandana forest, jewel-like (***ratnamayaḥ***) and round on all sides (***sarvato vṛttaḥ***).”

Reimagining Meru: From Cosmic Mountain to Geometric Pole

In the ancient Puranic texts, Mount Meru was a central feature of the cosmos—an impossibly vast and golden mountain of immense height, the pivot of the universe. For a scientist like Aryabhata, who understood the Earth to be a sphere of a known size, this literal description was untenable. So what did he do? In this verse, he performs a brilliant act of scientific and cultural reinterpretation. He does not reject the powerful symbol of Meru; instead, he reimagines it to fit his new, rational model of the world.

A New Kind of Mountain

Aryabhata's Meru is dramatically different from the mountain of myth.

- **A Tiny Footprint:** Instead of being unimaginably large, he states that Meru measures only **one *yojana*** across (roughly 5 to 9 miles). This is a clear signal that he is not talking about a literal, continent-sized mountain.
- **Round and Radiant:** He describes it as being "round on all sides" (*sarvato vṛttaḥ*) and luminous (*prabhākaro*). This description is far more consistent with a geometric point or a polar region (perhaps alluding to the aurora) than a jagged, physical peak.

A Geometric Reinterpretation

Later commentators like Bhāskara I make the meaning explicit: Aryabhata's reimagined Meru is not a physical mountain at all. It is a **symbolic and geometric representation of the Earth's North Pole**—the northernmost point on the planet's axis of rotation. Its small, circular nature perfectly represents a geometric pole, and its location "in the center of the celestial garden" fits its role as the pivot point of the northern heavens.

A Bridge Between Worlds

This verse is a masterclass in intellectual diplomacy. Aryabhata shows profound respect for the traditional cosmology by retaining its most important symbols. However, he carefully redefines them, transforming them from literal, mythological features into rational, geometric concepts that align perfectly with his scientific model of a spherical, rotating Earth. He builds a bridge between the world of ancient myth and the world of empirical science, demonstrating a way to honor the past while paving the way for a new, more accurate understanding of the cosmos.

Notes:

- ***Yojanamātraḥ:*** Only one yojana (contrasting vast Puranic Meru).
- ***Prabhākaro:*** Radiant (symbolic? polar phenomena?).
- ***Sarvato vṛttaḥ:*** Round on all sides (fits pole/axis concept).

6.6.2 Polar Symmetry and Relativity of "Down"

Sanskrit:

स्वो मेरूर् स्थलमध्ये नरको वडवामुखं च जलमध्ये ।

अमरामरा मन्यन्ते परस्परम् अधःस्थितान् नियतम् ॥ १२ ॥

IAST:
svo merūr sthalamadhye narako vaḍavāmukhaṃ cha jalamadhye |

amarāmarā manyante parasparam adhaḥsthitān niyataṃ || 12 ||

Translation: "Heaven (*svaḥ*) and Meru are amidst land (*sthalamadhye*), [while] hell (*naraka*) and Baḍavāmukha [South Pole] are amidst water (*jalamadhye*); the gods (*amarā*) and the anti-gods/demons (*amarā / asurā*) constantly (*niyataṃ*) perceive each other (*parasparam*) as being situated below (*adhaḥsthitān*)."

The Great Puzzle of "Down": A Universe of Perspectives

If the Earth is a sphere, then what about the people on the other side? Wouldn't they be upside down? Wouldn't they fall off? This is a natural and ancient question, a major conceptual hurdle for anyone first encountering the idea of a round world. In this brilliant verse, Aryabhata provides a simple, elegant, and profoundly logical answer.

A Symmetrical World

First, he establishes a beautiful symmetry for the globe. He maps the traditional cosmological opposites onto the geographic poles:

- The **North Pole (*Meru*)** he associates with land and the gods (*amarā*) in their heavens.
- The **South Pole (*Baḍavāmukha*)** he associates with the great ocean and the anti-gods (*asurā*) in their underworld.

These two groups are *antipodal*—they are on the exact opposite sides of the Earth from each other.

The Relativity of "Down"

Now comes the crucial insight. Aryabhata states that the gods at the North Pole and the anti-gods at the South Pole **mutually perceive each other as being "below" them** (*adhaḥsthitān*).

This is a stunningly clear statement of the relativity of perspective.

- For the gods at the North Pole, "down" is the direction towards the South Pole.
- For the anti-gods at the South Pole, "down" is the direction towards the North Pole.

The profound implication is that **"down" is not a universal, absolute direction in space. "Down" is simply the direction towards the center of the Earth.** This insight solves the "upside-down" puzzle completely. Everyone on a spherical Earth, no matter their location, stands upright on their local ground, because "down" for them is simply the direction their feet are pointing. Just as he argued with the Kadamba flower analogy (Gola.7), where all the bristles point away from the center, Aryabhata shows that there is no universal "up" or "down"—only "away from the center" and "towards the center."

These verses show Aryabhata integrating traditional concepts like Meru geometrically and using the spherical Earth model to explain the relativity of perspective, thereby strengthening the conceptual foundations of his cosmology.

Notes:

- ***Svaḥ/Meru:*** Heaven/North Pole region (assoc. land, gods).
- ***Naraka/Baḍavāmukha:*** Hell/South Pole region (assoc. water, demons/asuras).
- ***Amarāmarā / Amarāsurā:*** Gods/Anti-gods (representing antipodal inhabitants).
- ***Adhaḥsthitān:*** Situated below (mutually perceived relative direction towards Earth's center).

6.7 Perspectives from a Spherical World (Gola.13-17)

Having established a spherical, rotating Earth with defined poles, Aryabhata now explores the direct observational consequences for someone located on this sphere. These verses demonstrate how longitude affects local time, how latitude dictates the view of the stars, and how the very perception of time cycles depends on the observer's location in the cosmos.

Sanskrit:

उदयो यः लङ्कायाम् सोऽस्तमयः सवितुर् एव सिद्धपुरे ।

मध्याह्नो यमकोट्याम् रोमकविषयेऽर्धरात्रः स्यात् ॥ १३ ॥

IAST:

udayo yaḥ laṅkāyāṃ so'stamayaḥ savitur eva siddhapure |

madhyāhno yamakoṭyāṃ romakaviṣaye'rdharātraḥ syāt || 13 ||

Translation: "What is sunrise in Laṅkā (***udayo yaḥ laṅkāyāṃ***) is indeed sunset for the Sun (***savituḥ***) in Siddhapura (***so'stamayaḥ siddhapure***); it is midday in Yamakoṭi (***madhyāhno yamakoṭyāṃ***); and it would be midnight in the Romaka region (***romakaviṣaye'rdharātraḥ***)."

A Journey Through Time: The Four Royal Cities

If the Earth is a spinning globe, then sunrise, noon, and midnight cannot happen for everyone at the same time. In this brilliant verse, Aryabhata demonstrates this profound consequence, effectively inventing the concept of time zones.

He asks us to imagine four "royal cities" located at equal distances around the Earth's equator, each separated by 90 degrees of longitude.

- **Laṅkā:** The prime meridian, the starting point (0° longitude).
- **Yamakoṭi:** A city 90 degrees to the east.
- **Romaka:** A city on the opposite side of the world, 180 degrees away.
- **Siddhapura:** A city 90 degrees to the west.

A Snapshot of a Spinning World

Aryabhata then takes a snapshot of the world at a single instant in time. He states that when the Sun is just beginning to rise in the east at the prime meridian city of **Laṅkā**:

- In **Yamakoṭi**, 90 degrees to the east, the Sun has already been up for six hours and is now at its highest point. It is **midday**.
- In **Romaka**, 180 degrees away on the other side of the planet, the Sun is just finishing its journey across the sky. It is **sunset**.
- In **Siddhapura**, 90 degrees to the west, the Sun is on the opposite side of the world and at its lowest point. It is **midnight**.

A Direct Proof of a Rotating Earth

This simple and elegant description is a direct and logical consequence of an eastward-spinning Earth. It clearly shows that for every 90 degrees of longitude you travel east, the local solar time advances by six hours. This is not just a description; it is a proof. It demonstrates the predictive power of Aryabhata's model. By treating the Earth as a rotating sphere, he could accurately deduce the simultaneous, yet different, experiences of time at various points across the globe. It is a foundational explanation of how time itself is tied to geography, a concept that remains central to our understanding of the world today.

Notes:

- ***Laṅkā:*** Prime meridian (0° Longitude).
- ***Yamakoṭi:*** 90° E.
- ***Romakaviṣaye:*** Romaka region, 180° from Laṅkā.
- ***Siddhapura:*** 90° W (or 270° E).

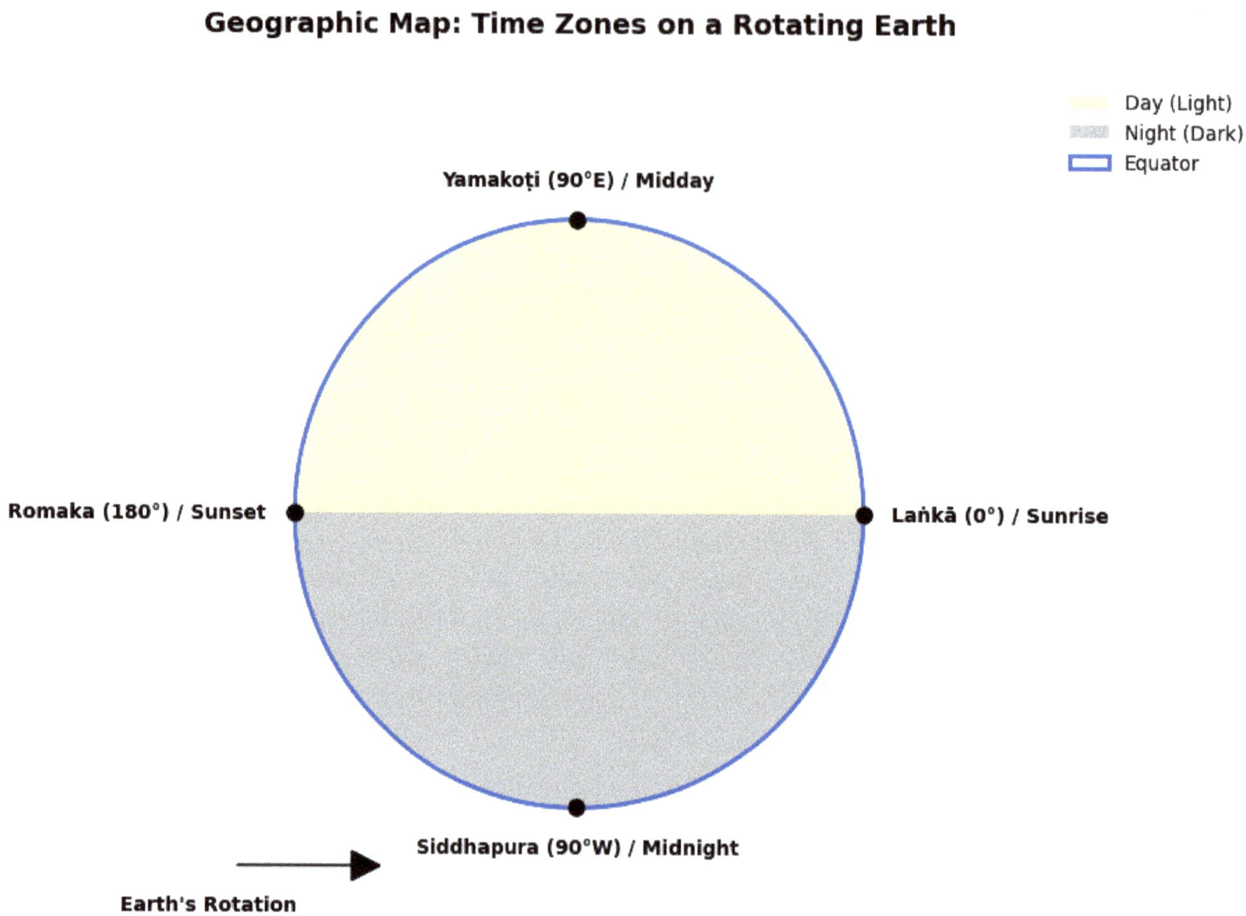

Figure 6.8: Aryabhata's Concept of Simultaneous Yet Different Times (Gola.13). Illustrates how, due to Earth's rotation, different locations on a spherical Earth experience distinct times of day (Sunrise, Midday, Sunset, Midnight) at the same instant, as shown for Laṅkā, Yamakoṭi, Romaka, and Siddhapura, separated by 90° longitude (6-hour difference).

6.7.1 Latitude and Mapping Key Locations

Sanskrit:

स्थलजलमध्याल् लङ्का भूकक्षाया भवेत् चतुर्भागे ।

उज्जयिनी लङ्कायास् तच्चतुरांशे समोत्तरतः ॥ १४ ॥

IAST:

sthalajalamadhyāl laṅkā bhūkakṣāyā bhavet chaturbhāge |

ujjayinī laṅkāyās tacchaturāṃśe samottarataḥ || 14 ||

Translation: "From the land-water midpoint (***sthalajalamadhyāl*** = North Pole), Laṅkā (Equator) lies at one-fourth of Earth's circumference (***bhūkakṣāyā chaturbhāge***). Ujjayinī stands due north (***samottarataḥ***) of Laṅkā by one-fourth of that quarter-span (***taccaturāṃśe***)."

Mapping the Earth: From the Pole to Ujjayinī

Having established that the Earth is a sphere, Aryabhata now begins to map its surface, defining the precise geographical locations of key reference points. This verse connects the abstract geometric model to the real-world places where astronomers lived and worked.

The process is a logical, step-by-step journey from the North Pole to the heart of Indian astronomy.

1. The Starting Point: The North Pole

Aryabhata begins at what commentators identify as the **North Pole**. He describes this as the "midpoint of land and water," a poetic reference consistent with the mythological geography where the great mountain Meru stood at the top of the world.

2. The First Landmark: The Equator

From the North Pole, he defines the position of the equator. He states that **Laṅkā**—the reference city located on the intersection of the equator and the prime meridian—is one-quarter of the Earth's total circumference away from the Pole.

This is a perfect geometric statement:

Distance from Pole to Equator = (1/4) × 360° = 90°

This correctly places the equator at a 90-degree angular distance from the North Pole.

3. The Center of Science: Ujjayinī

Finally, Aryabhata pinpoints the location of the most important center of ancient Indian astronomy: the city of **Ujjayinī**. He states that Ujjayinī is located due north of the equator by an amount equal to one-fourth of the total distance from the pole to the equator.

The calculation is simple and precise:

Latitude of Ujjayinī = (1/4) × 90° = 22.5° North

A Bridge Between Theory and Observation

This is a remarkable statement. The actual latitude of the modern city of Ujjain is approximately 23.15° N. Aryabhata's value of 22.5° is astoundingly close, demonstrating that his model was not just based on theory.

This value was almost certainly derived from careful, repeated observations made with a gnomon (a vertical stick used to measure the Sun's shadow), using the exact method he describes later in the text. This verse provides powerful evidence of Aryabhata's scientific approach, seamlessly integrating empirical data from real-world measurements into his elegant, spherical model of the Earth.

Notes:

- ***Sthalajalamadhyāl:*** Laṇd-water midpoint = North Pole (interpreted).
- ***Bhūkakṣā:*** Earth's circumference (360°).
- ***Chaturbhāge:*** One-fourth part = 90° arc (Pole to Equator).
- ***Tacchaturāṃśe:*** Quarter of that quarter-span = 90°/4 = 22.5°.
- ***Samottarataḥ:*** Due north. Establishes the latitude of Ujjayinī based on observation.

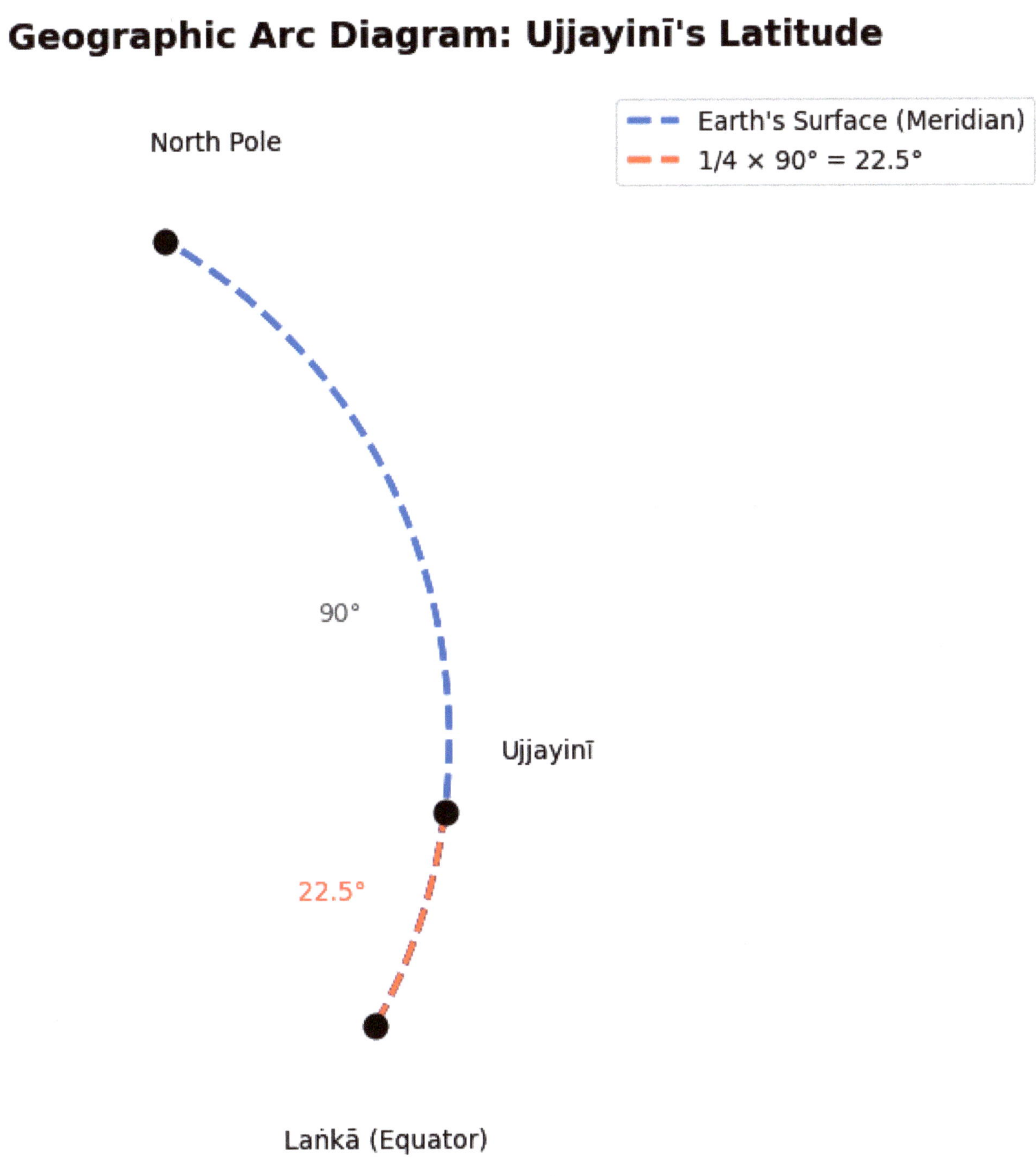

Figure 6.9: Ujjayinī's Latitude Calculation (Gola.14). Illustrates Ujjayinī as 1/4 the 90° arc (North Pole-Equator) north of Laṅkā (Equator), resulting in ≈22.5° N latitude, likely from observation.

6.7.2 The Limit of Vision (Horizon Dip) on a Spherical Earth

Sanskrit:

भूव्यासार्धेनोनं दृष्यं देशात् समाद् भगोलार्धम् ।

अर्धं भूमिच्छन्नं भूव्यासार्धाधिकं चैव ॥ १५ ॥

IAST:

bhūvyāsārdhenonaṃ dṛśyaṃ deśāt samād bhagolārdham |

ardhaṃ bhūmicchannaṃ bhūvyāsārdhādhikaṃ chaiva || 15 ||

Translation: "From level ground (***deśāt samāt***), slightly less than half the celestial sphere (***bhagolārdham***) is visible (***dṛśyaṃ***), diminished by [an amount related to] Earth's radius (***bhūvyāsārdhenonaṃ***); the [other] half, obscured by Earth (***bhūmicchannaṃ***), exceeds a hemisphere by that same amount (***bhūvyāsārdhādhikaṃ***)."

The Curved Horizon: A Consequence of a Spherical Earth

If you stand on the shore and look out at the sea, where exactly is the horizon? From our perspective, it seems to divide the world into two perfect halves: the sky above and the sea below. But in this verse, Aryabhata explains a subtle but crucial geometric fact that proves our world is a sphere.

The Dip of the Horizon

The verse states that the visible portion of the sky is always slightly *less* than a perfect hemisphere, while the portion hidden below our view is slightly *more* than a hemisphere.

Why is this? The answer lies in the curvature of the Earth.

> **Imagine you are a tiny ant standing on a giant beach ball. Your geometric horizon is a perfectly flat plane that touches the ball only at the point where you are standing. If the ball were flat, you would see a full 180 degrees of the sky. But because the ball is curved, your physical horizon—the line where the surface of the ball actually disappears from your view—is slightly *below* this geometric plane. The very curvature of the ball you are standing on hides a small sliver of the sky from your sight.**

This effect is known as the **dip of the horizon**. Aryabhata correctly states that the amount of this dip is directly related to the Earth's radius (*bhūvyāsārdha*).

Direct Evidence for a Round Earth

This is a powerful and elegant observation. It is not a complex astronomical theory but a simple, geometric consequence of living on a sphere. It provides direct, observable evidence for the Earth's curvature.

This subtle effect is critically important for precise astronomical observations made near the horizon. It means that to see a star rise, it must climb slightly above the "perfect" geometric horizon to clear the physical curvature of our own planet. This phenomenon, which Aryabhata calls the "shadow of vision" (*dṛkchāyā*), is a constant reminder that we are observers on a vast, round world.

Notes:

- ***Bhagolārdham:*** Half celestial sphere (180° arc).
- ***Bhūvyāsārdha:*** Earth's radius (its effect makes visible part < 180°).
- ***Dṛśyaṃ:*** Visible portion (< 180°).
- ***Bhūmicchannaṃ:*** Obscured portion (> 180°). Explains horizon dip.

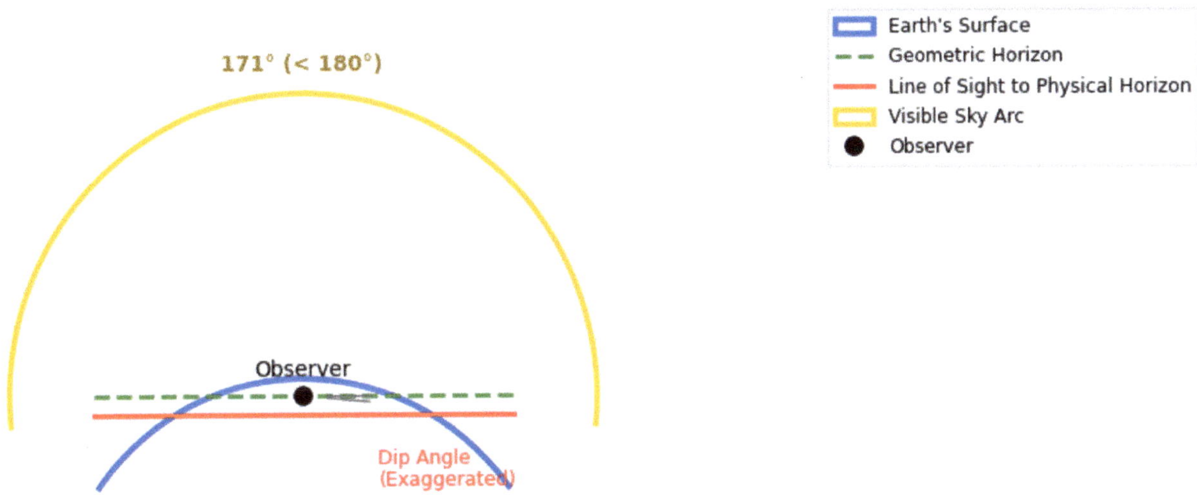

Figure 6.10: Horizon Dip (Gola.15). Due to Earth's curvature, the physical horizon (red) is below the geometric (dashed green), making the visible sky (yellow arc) < 180°. This effect (related to Dṛkchāyā, Gola.34) requires corrections for precise horizon observations. (Dip exaggerated).

6.7.3 Celestial Views from the Poles

Sanskrit:

देवाः पश्यन्ति भगोलार्धम् उदङ् मेरुसंस्थिताः सव्यम् ।

अर्धं त्व् अपसव्यगतं दक्षिणे वडवामुखे प्रेताः ॥ १६ ॥

IAST:

devāḥ paśyanti bhagolārdham udaṅ merusaṃsthitāḥ savyam |

ardhaṃ tv apasavyagataṃ dakṣiṇe vaḍavāmukhe pretāḥ || 16 ||

Translation: "Gods situated at northern Meru (***merusaṃsthitāḥ*** = North Pole observers) see half the celestial sphere (***bhagolārdham***) located to the north (***udaṅ***?) circling always above the horizon moving clockwise (***savyam***); spirits (***pretāḥ*** = South Pole observers) at southern Baḍavāmukha see the [other] half moving counter-clockwise (***apasavya-gataṃ***)."

A Carousel of Stars: The View from the Poles

What would the sky look like if you could stand at the very top or bottom of the world? In this verse, Aryabhata uses his model of a rotating Earth to describe the strange and beautiful celestial display that would greet an observer at the poles. This is not just a description; it is a powerful piece of evidence for his theory.

The View from the North Pole (*Meru*)

For an observer standing on the North Pole, the sky would be an unfamiliar and wondrous sight.

- The North Celestial Pole, the sky's unmoving pivot, would be directly overhead at the zenith.
- All visible stars would appear to trace perfect circles across the sky, parallel to the horizon. No star would ever rise or set; they would simply circle endlessly.

Most importantly, because the Earth is spinning eastward *underneath* the observer, the entire celestial sphere would appear to rotate in the opposite direction: **clockwise**. This rightward motion is what Aryabhata calls *savyam*.

The View from the South Pole (*Baḍavāmukha*)

For an observer at the South Pole—whom Aryabhata poetically associates with the *pretas* of the underworld—the view would be a perfect mirror image.

- The South Celestial Pole would be at the zenith.

- All stars in the southern celestial hemisphere would circle this point, never rising or setting.
- The apparent direction of rotation would be reversed: **counter-clockwise**. This leftward motion is what Aryabhata calls *apasavyam*.

The Logical Proof

This verse is a brilliant demonstration of the logical consistency of Aryabhata's system. He isn't just describing a random fact; he is making a specific, verifiable prediction based on his revolutionary model of an eastward-spinning Earth.

If the Earth spins east, then the view from the poles *must* be as he describes. By showing that his model accurately predicts these extreme cases, he strengthens his central argument and showcases the predictive power of his cosmology.

Notes:

- ***Merusaṃsthitāḥ:*** Situated at Meru (North Pole observer).
- ***Savyam:*** Clockwise / rightward motion (N. Hemi. pole view).
- ***Vaḍavāmukhe:*** At Baḍavāmukha (South Pole observer).
- ***Apasavyam:*** Counter-clockwise / leftward motion (S. Hemi. pole view).

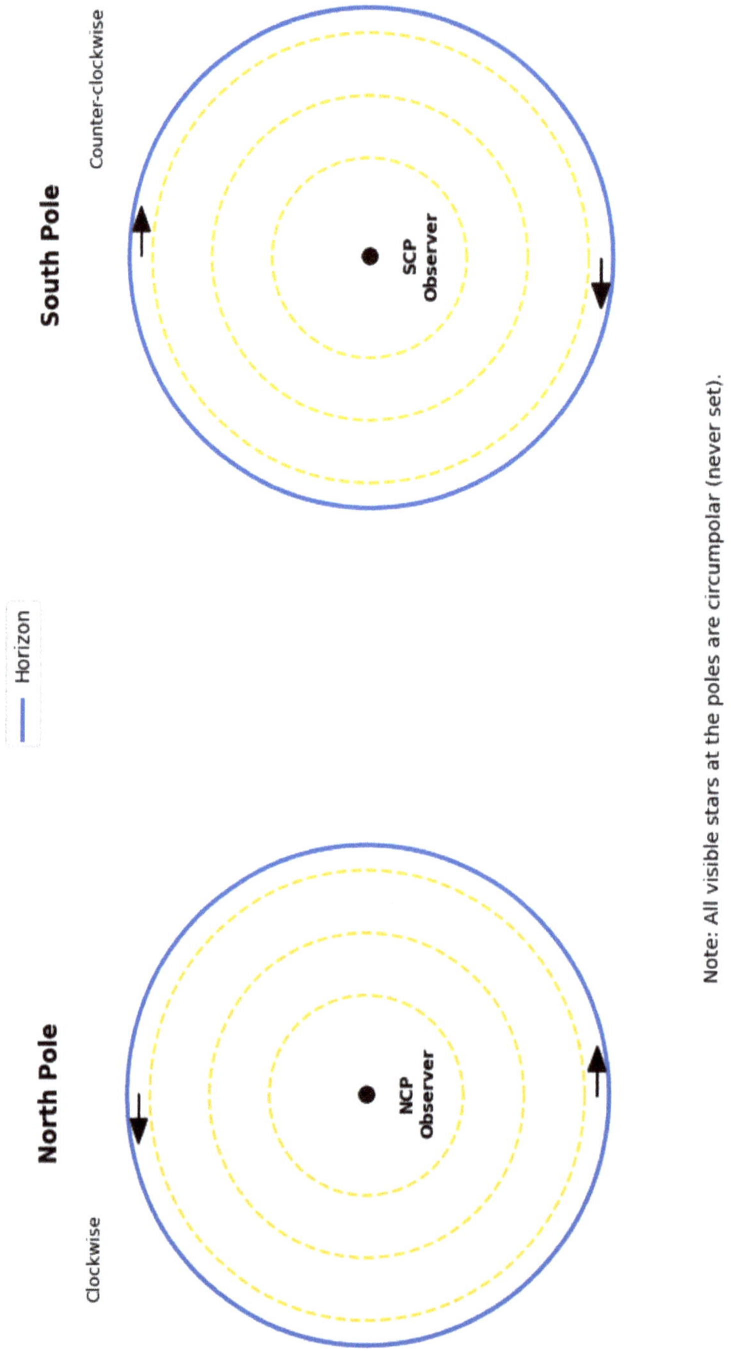

Figure 6.11: Stellar Motion at the Poles (Gola.16). Stars circle the Celestial Poles parallel to the horizon: clockwise at North Pole (left), counter-clockwise at South Pole (right). All are circumpolar. Described by Aryabhata (Gola.16) due to Earth's rotation (Gola.9).

6.7.4 Relativity of Temporal Cycles Based on Location

Sanskrit:

रविवर्षार्धं देवाः पश्यन्त्य् उदितं रविं तथा प्रेताः ।

शशिमासार्धं पितरः शशिगाः कुदिनार्धम् इह मनुजाः ॥ १७ ॥

IAST:

ravivarṣārdhaṃ devāḥ paśyanty uditaṃ raviṃ tathā pretāḥ |

śaśimāsārdhaṃ pitaraḥ śaśigāḥ kudinārdham iha manujāḥ || 17 ||

Translation: "Gods (***devāḥ*** = North Pole dwellers) see the risen Sun (***uditaṃ raviṃ***) for half a solar year (***ravivarṣārdhaṃ***), and likewise the spirits (***pretāḥ*** = South Pole dwellers); the ancestors dwelling on the Moon (***pitaraḥ śaśigāḥ***) [see it] for half a lunar month (***śaśimāsārdhaṃ***); humans here (***iha manujāḥ*** = on Earth away from poles) [see it] for half a civil day (***kudinārdham***)."

The Relativity of a Day

What is a "day"? The simple answer is the cycle of light and dark we experience from sunrise to sunset. But in this insightful verse, Aryabhata performs a brilliant thought experiment, showing that our experience of a "day" is not a universal constant but is entirely relative to our position in the cosmos.

He invites us to consider what a "day" would look like from three vastly different viewpoints.

1. The View from the Top of the World

First, Aryabhata considers an observer at the North or South Pole. For these observers, whom he poetically associates with the gods (*Devas*) and anti-gods (*Asuras* or *Pretas*), the experience of a day is completely different.

Due to the Earth's axial tilt, an observer at the North Pole sees the Sun remain continuously above the horizon for **six months**, followed by six months of continuous darkness. Their "day," therefore, is an entire Earth-year long, composed of one vast period of light and one vast period of darkness.

2. The View from the Moon

Next, he considers hypothetical observers on the Moon, whom he associates with the ancestral spirits (*Pitaras*). The Moon is tidally locked to the Earth, meaning it rotates on its axis at the same rate that it orbits the Earth.

The consequence is a lunar day that is incredibly long by our standards. An observer on the Moon would experience approximately **14.75 Earth-days** of continuous, harsh sunlight, followed by an equally long and frigid night. For a lunar inhabitant, a single cycle of day and night takes nearly a full Earth-month.

3. The View from "Here"

Finally, Aryabhata brings the perspective back to us—humans (*manujāḥ*) living "here" on the main, inhabited parts of the Earth. For us, a day is defined by the familiar cycle of roughly **12 hours** of daylight and 12 hours of darkness (*kudinārdham*).

Crucially, Aryabhata knows this is just an average. He understands that our day length varies with the seasons and our latitude, and he has already provided the precise mathematical tool—the *kṣitijyā* calculation (Gola.26)—to quantify this variation exactly.

Significance: A Universe of Perspectives

This verse is a profound demonstration of abstract, model-based thinking. Years before a similar verse on the relativity of "down" (Gola.12), Aryabhata applies the same logic to the concept of time. He uses his mental model of the solar system to deduce what an observer *would* see from different locations.

It shows an understanding that our perception of time is not absolute but is a direct consequence of our motion and location within a larger cosmic system. It is a remarkable insight that extends the principles of geometry and perspective from the domain of space to the very fabric of time itself.

Notes:

- ***Ravivarṣārdhaṃ:*** Half a solar year (polar "day" duration = 6 months).
- ***Śaśimāsārdhaṃ:*** Half a synodic lunar month (lunar "day" duration ≈ 14.75 Earth days).
- ***Kudinārdham:*** Half a civil day (≈ 12 hours, basic terrestrial average day length).
- ***Śaśigāḥ:*** Moon-dwellers (hypothetical observers). Demonstrates relativity of time perception.

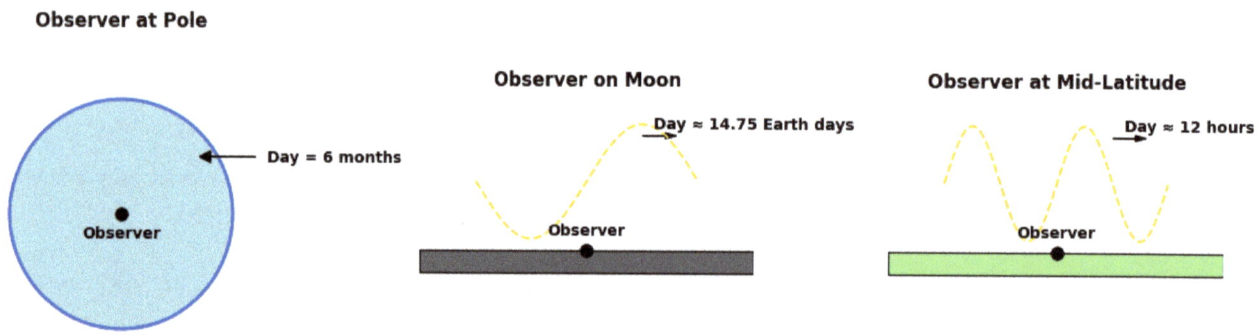

Figure 6.12: Relative 'Day' Lengths (Gola.17). Illustrates Aryabhata's concept of varied sunlight durations: ≈6 months at Earth's Poles, ≈14.75 Earth days on the tidally locked Moon, and ≈12 hours average at Earth's mid-latitudes.

6.8 The Celestial Sphere Model (Khagola) Details (Gola.18-25)

This section delves into the specific geometric structure of the ***khagola*** – the conceptual celestial sphere centered on the observer. Aryabhata defines the key reference circles and coordinate systems used for mapping and measuring positions in the sky from a specific location on the spherical Earth. This framework is essential for practical positional astronomy.

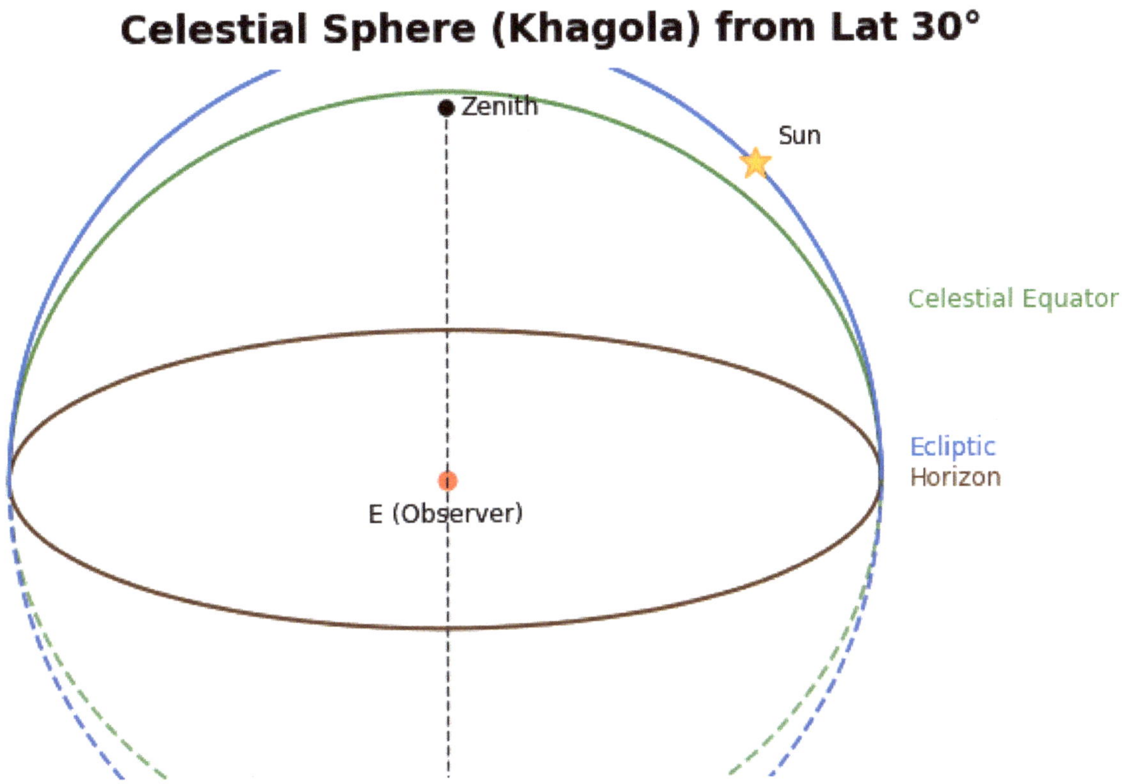

Figure 6.13: Simplified diagram of the celestial sphere (Khagola) from a geocentric perspective, illustrating key reference circles relative to an observer's location (E): the Horizon, the Celestial Equator, and the Ecliptic (Sun's path), showing their inclination relative to each other. This geometric model forms the basis for positional astronomy as detailed in the Golapāda.

6.8.1 Cardinal Circles of the Observer's Horizon

Sanskrit:

वृत्तं भजेत् समागं उदग् याम्यगं पश्चिमोदयम् चैव ।

प्राग् दक्षिणोत्तराशास्थम् ऊर्ध्वम् अधस् तत् प्रदेशस्थम् ॥ १८ ॥

IAST:

vṛttaṃ bhajet samāgaṃ udag yāmyagaṃ paścimodayam chaiva |

prāg dakṣiṇottarāśāstham ūrdhvam adhas tat pradeśastham || 18 ||

Translation: "One should divide the horizon circle (*vṛttaṃ*) by intersections (*samāgaṃ*) with lines going north (*udag*), south (*yāmyagam*), west-rising [=west point] (*paścimodayam*), and east (*prāg*), and also define circles passing through the south-north directions [=meridian] (*dakṣiṇottarāśāstham*), upward [to the zenith] (*ūrdhvam*), and downward [to the nadir] (*adhas*), all fixed for that location (*pradeśastham*)."

Your Personal Sky: The Horizon and its Landmarks

Before we can map the heavens, we must first understand our own personal view of the sky from where we stand. This verse lays out the fundamental framework of the observer's local world, establishing the key landmarks that define our window into the cosmos.

1. The Horizon: The Great Divider

The first and most fundamental reference is the **horizon**. This is the great circle that separates the visible sky above from the unseen sky below. It is the boundary of our world.

2. The Four Cardinal Points

The verse explains that on this horizon circle, there are four cardinal points:

- **North** (*udag*)
- **South** (*yāmyagam*)
- **East** (*prāg*)
- **West** (*paścimodayam*)

These four points anchor our view and provide the basic directions for all observations.

3. The Zenith and Nadir: The Poles of Our World

Next, Aryabhata defines the two "poles" of our local sky:

- **The Zenith:** The point directly overhead.
- **The Nadir:** The unseen point directly beneath our feet.

These two points define the vertical axis of our personal celestial sphere.

4. The Meridian: The Great North-South Line

Finally, the verse describes the most important line in the observer's sky: the **meridian**. This is a great circle that connects the key landmarks, running from the true north point on the horizon, up through the zenith, and down to the true south point.

The meridian serves two crucial functions:

- It defines the fundamental **north-south line** for the observer.
- It marks the line where every celestial object reaches its **highest point** in its daily journey across the sky.

The Foundation for Observation

Together, these elements—the horizon, the cardinal points, the zenith, and the meridian—create the local coordinate system. This framework is essential for measuring a star's position in terms that are meaningful to an observer on the ground, such as its **altitude** (height above the horizon) and its **azimuth** (compass direction). It is the personal stage upon which the universal drama of the heavens unfolds.

Notes:

- ***Vṛttaṃ:*** Circle (here, the Horizon).
- ***Samāgaṃ:*** Intersections (defining points).
- ***Paścimodayam:*** West-rising (identifies the West point).
- ***Ūrdhvam/Adhas:*** Zenith/Nadir.
- ***Dakṣiṇottarāśāstham:*** Situated in the South-North direction (defines the Meridian).
- ***Pradeśastham:*** Fixed for that specific location (local system).

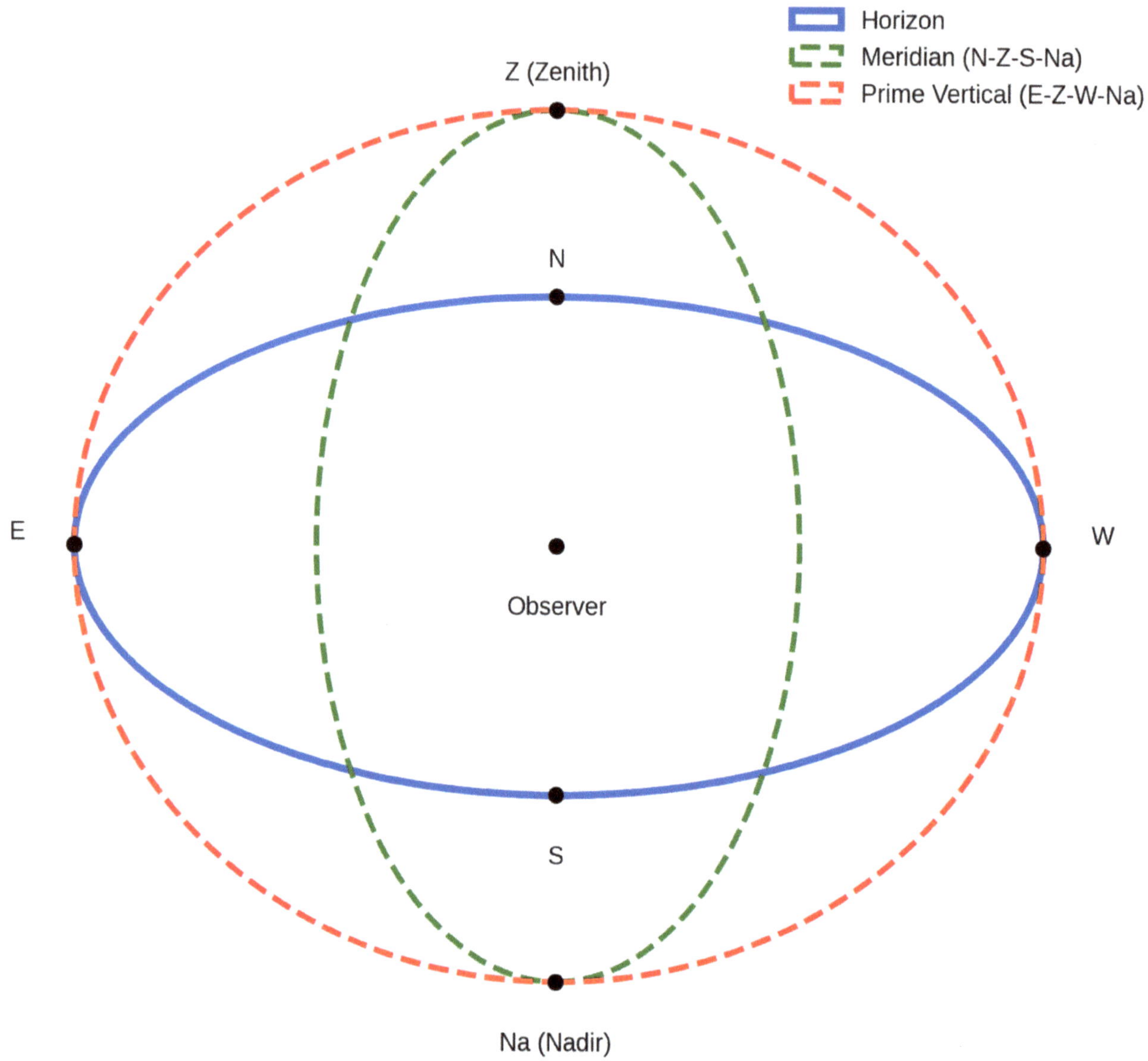

Figure 6.14a: The Observer's Local Coordinate System (Horizon System). This diagram illustrates the fundamental framework for local celestial observations, centered on the Observer. It is defined by: the Horizon (blue solid ellipse); the vertical axis connecting the Zenith (Z) overhead and the Nadir (Na) directly below; the Meridian (green dashed ellipse) passing through North (N), Z, South (S), and Na; and the Prime Vertical (red dashed ellipse) passing through East (E), Z, West (W), and Na. This system, introduced by Aryabhata in Golapāda Verses 18 and 20, is used for measuring a celestial object's altitude and azimuth from a specific location.

6.8.2 Declination Circle, Equatorial Horizon (*Unmandala*), and Daylight Variation

Sanskrit:

भगृहाणां गोलानाम् अपक्रममण्डलं समं वृत्तम् ।

उदग् याम्यस्थं तस्माद् उन्मण्डलं क्षयवृद्धी ॥ १९ ॥

IAST:

bhagrahāṇāṃ golānām apakramamaṇḍalaṃ samaṃ vṛttam |

udag yāmyasthaṃ tasmād unmaṇḍalaṃ kṣayavṛddhī || 19 ||

Translation: "The declination circle (***apakramamaṇḍalam***) of the spheres of stars and planets (*bhagrahāṇāṃ golānām*) is an equal circle (***samaṃ vṛttam*** = parallel to celestial equator), situated north or south (*udag yāmyasthaṃ*); from it [its geometry relative to the horizon] arises the ***unmaṇḍala*** [equatorial horizon / six o'clock line], connected with the decrease and increase (***kṣayavṛddhī***) [of daylight duration]."

The Geometry of Daylight: Paths and Variations

Why are summer days longer than winter days? This verse introduces the key geometric concepts that explain this fundamental rhythm of life. It defines the paths the stars follow and reveals why the time they spend above the horizon changes throughout the year.

1. The Daily Highway: The Declination Circle

First, the verse defines the path that every star and planet traces across the sky each day. This is the **Declination Circle** (*Apakramamaṇḍalam*).

Imagine the celestial equator as the great central highway of the heavens. A star's declination circle is its own personal, parallel lane on this celestial road system.

- A star with a high northern declination travels on a highway far to the north.
- A star with a southern declination travels on a highway far to the south.

Each day, a star completes one full lap on its designated declination circle.

2. The Reference Line: The Six O'Clock Line (*Unmaṇḍala*)

Next, Aryabhata defines a crucial reference line called the ***Unmaṇḍala***. This great circle passes through the true east and west points on the horizon and connects the north and south celestial poles.

It's often called the **"Six O'Clock Line"** for a simple reason: a star on the celestial equator (with zero declination) will cross this line exactly six hours after it rises. It acts as a standard, idealized "equatorial horizon," giving astronomers a fixed grid line against which to measure the real, observed motions.

3. The Result: The Increase and Decrease of Daylight

Finally, the verse explains the cause of the "increase and decrease" (*kṣayavṛddhī*) of daylight.

The length of the day is simply the amount of time the Sun spends above our local horizon. The puzzle is that the Sun's "daily highway" (its declination circle) is tilted with respect to our horizon.

Imagine two intersecting hula hoops:

- One hoop lies flat on the ground, representing your **local horizon**.
- The other hoop is the **Sun's daily path**, tilted at an angle.

The length of the arc of the Sun's path that is *above* your horizon hoop is the length of the day. Because the Sun's path is tilted, the portion above the horizon is generally not a perfect half-circle.

As the seasons change, the Sun's declination changes, meaning it moves to a different "daily highway." This changes the angle of intersection between the two hoops, and thus changes the length of the visible arc. This is what makes days longer in summer and shorter in winter.

This verse lays the conceptual groundwork. It establishes the geometric reason for the changing seasons, preparing the way for a later verse (Gola.26) that will provide the precise mathematical formula (*kṣitijyā*) to calculate this variation exactly.

Notes:

- ***Apakramamaṇḍalam:*** Declination circle (diurnal path parallel to equator).
- ***Samaṃ vṛttam:*** Equal circle (interpreted as parallel to equator).
- ***Unmaṇḍala:*** Equatorial horizon / Six o'clock line (passes E/W points and Celestial Poles).
- ***Kṣayavṛddhī:*** Decrease/increase (of daylight duration).
- ***NCP/SCP:*** North/South Celestial Pole

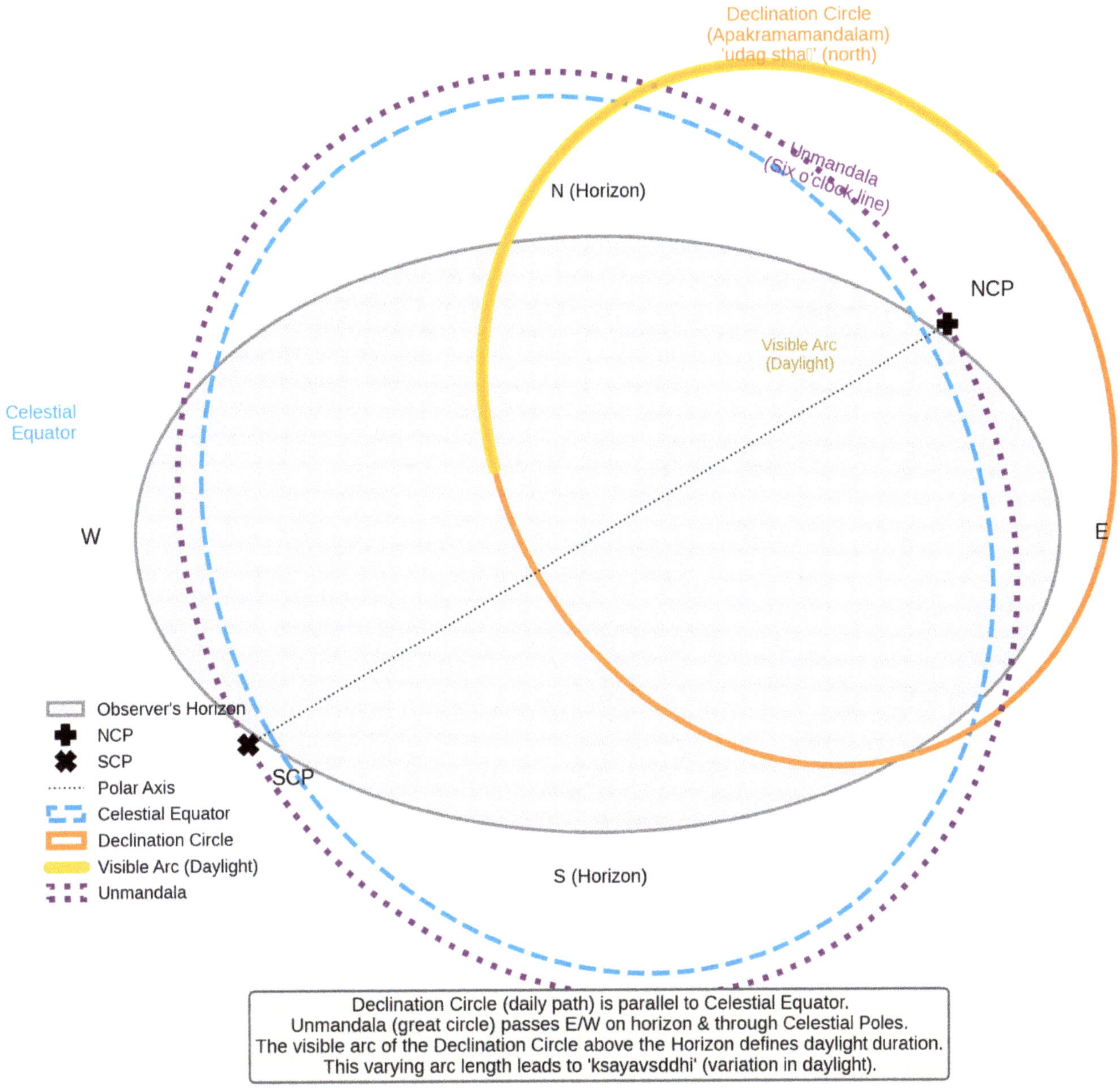

Figure 6.14b: Explaining Daylight Variation (Gola.19). Key celestial circles from Gp.19: the Horizon, Celestial Equator, a parallel Declination Circle (Apakramamaṇḍalam), and the Unmaṇḍala. The segment of the Declination Circle above the horizon (gold) determines daylight length. Its variation due to changing declination explains kṣayavṛddhī (increase/decrease of daylight).

6.8.3 Observer's Vertical and the Prime Vertical

Sanskrit:

ऊर्ध्वम् अधो द्रष्टुः स्याद् ग्रहयोगो द्रष्टृमध्यरेखैव ।

पूर्वापरदिक्स्थं तत् सममण्डलम् उच्यते बुधैः ॥ २० ॥

IAST:

ūrdhvam adho draṣṭuḥ syād grahayogo draṣṭṛmadhyarekhaiva |

pūrvāparadiksthaṃ tat samamaṇḍalam uchyate budhaiḥ || 20 ||

Translation: "The line connecting a planet to the observer (*grahayogaḥ*) passes above and below the observer (*ūrdhvam adho draṣṭuḥ syād*); the observer's central line (*draṣṭṛmadhyarekhaiva* = the vertical circle) passing through the east and west points (*pūrvāparadiksthaṃ*)—that is called the prime vertical (***samamaṇḍalam***) by the wise (*budhaiḥ*)."

The Great East-West Arc: The Prime Vertical

In addition to the meridian, which defines the north-south line in the sky, Aryabhata defines its crucial counterpart: the **Prime Vertical**, or *Samamaṇḍalam*.

Drawing the Line in the Sky

Imagine standing in an open field and pointing directly east. Now, trace a giant arc with your finger from that east point on the horizon, up through the point directly overhead (the zenith), and down to the true west point on the opposite horizon.

This great arc is the **Prime Vertical**.

A Fundamental Reference Circle

The Prime Vertical is a cornerstone of the observer's local sky.

- **Orientation:** It runs perfectly east-west, standing at a right angle to the north-south line of the meridian. Together, the meridian and the Prime Vertical divide the observer's sky into four cardinal quadrants.
- **Function:** It serves as a vital reference line. For ancient astronomers, noting the exact moment a star or planet crossed the Prime Vertical was a key observational technique. This data could be used to determine the time or to refine the calculation of a celestial body's position.

By defining this great east-west circle, Aryabhata completes the basic grid of the local sky, providing the essential framework needed for precise positional measurement and timekeeping.

Notes:

- ***Draṣṭuḥ:*** Of the observer.
- ***Draṣṭṛmadhyarekha:*** Observer's central line (vertical circle).
- ***Pūrvāparadikstham:*** Situated in the East-West direction.
- ***Samamaṇḍalam:*** Prime Vertical (East-Zenith-West-Nadir circle).
- ***Budhaiḥ:*** By the wise/learned.

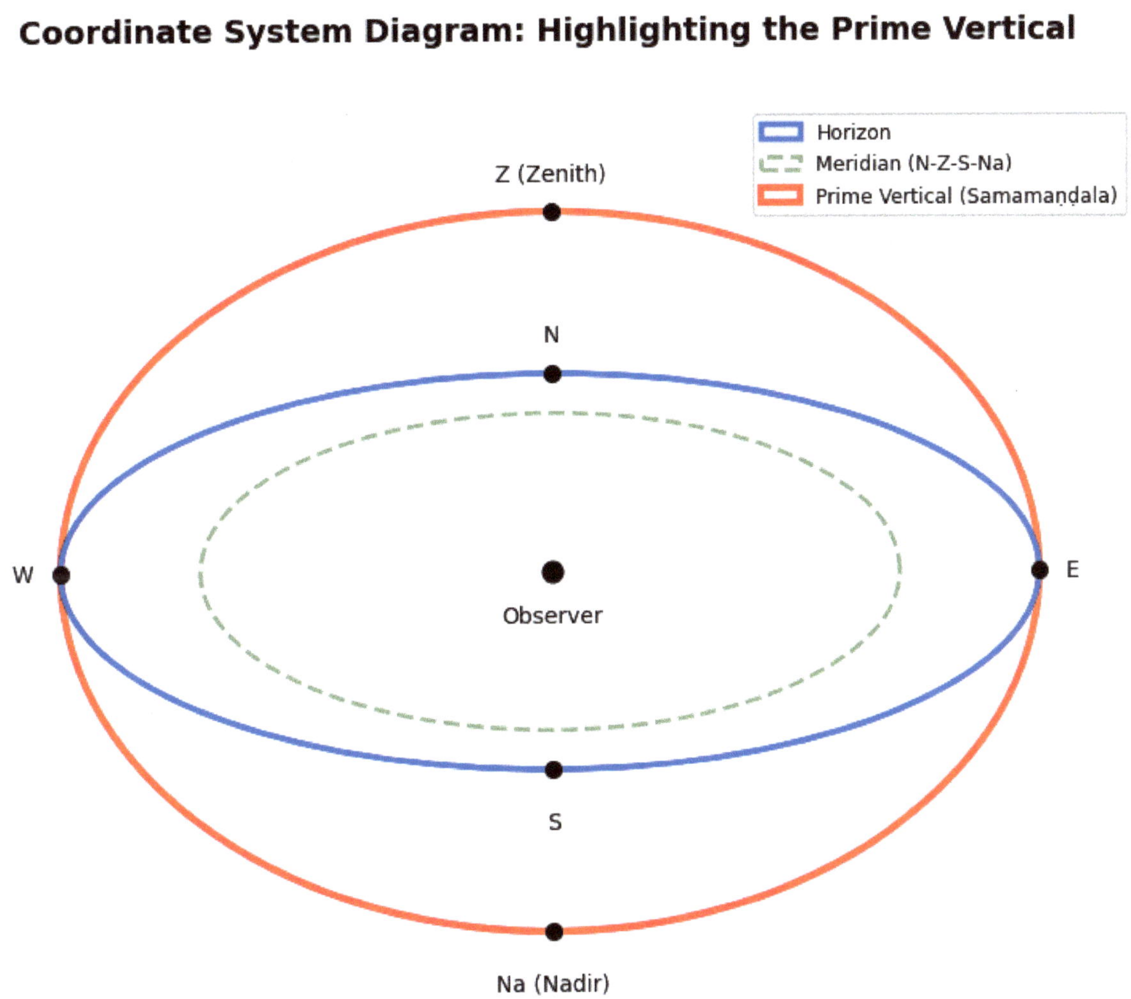

Figure 6.15: Aryabhata's Prime Vertical (Samamaṇḍala) (Gola.20). Illustrates the Prime Vertical (thick red curve), a great circle traversing the Zenith (Z), Nadir (Na), and the East/West points of the Horizon (blue ellipse), as per Gola.20. Perpendicular to the Meridian (dashed green), it is a fundamental reference for celestial observations and calculations.

6.8.4 Other Observer-Specific Vertical Circles

Sanskrit:

दृङ्मण्डलं च पूर्वं दिग्ज्या वृत्तं च दृग्गतिमार्गे ।

दृक्क्षेपमण्डलं स्याद् व्यासार्धे खग्रहयोगस्य ॥ २१ ॥

IAST:

dṛṅmaṇḍalaṃ cha pūrvaṃ digjyā vṛttaṃ cha dṛggatimārge |

dṛkkṣepamaṇḍalaṃ syād vyāsādhe khagrahayogasya || 21 ||

The Astronomer's Toolkit: Specialized Vertical Circles

Beyond the great, fixed circles that form the permanent structure of the sky (the horizon, meridian, and equator), Aryabhata defines a set of specialized circles. Think of these not as permanent highways on the celestial map, but as temporary lines an astronomer draws "on the fly" for a specific measurement or calculation. They are the essential tools for pinpointing a star's exact position and understanding observational effects.

1. The Altitude Circle (*Dṛṅmaṇḍala*)

When you look at a star, how do you measure its precise height above the horizon? You use the **Altitude Circle**.

This is a temporary vertical circle that you imagine stretching from the horizon, passing directly through the star you are observing, and continuing all the way to the point directly overhead (the zenith). Its sole purpose is to provide the perfect "celestial ruler" along which you can measure the star's altitude.

2. The Azimuth Circle (*Digjyā vṛtta*)

How do you measure a star's compass direction, or azimuth? For this, you use an **Azimuth Circle**.

The term *digjyā* refers to the sine of the azimuth, and these circles are the reference frames for that measurement. The most important of these is the **Prime Vertical**, the great circle that runs from the true east point on the horizon, up through the zenith, and down to the true west point. By referencing a star's position relative to this circle, an astronomer can determine its precise east-west direction.

3. The Parallax Circle (*Dṛkkṣepamaṇḍala*)

This is a more mysterious and highly specialized circle. Its name suggests a connection to parallax (*dṛkkṣepa*), the apparent shift in a star's position caused by the observer's location on Earth's surface.

While its exact function is a matter of scholarly discussion, it is likely a specialized reference circle—perhaps one drawn at a 90-degree angle to the Altitude Circle—used to help calculate the subtle, but crucial, effects of parallax.

Significance: A Sophisticated Geometric Toolkit

The definition of these circles is a testament to the sophistication of Aryabhata's system. They show a deep understanding of spherical geometry and its practical application to observation.

These are not just abstract concepts; they are the working tools of a practical astronomer. They provide the necessary geometric framework to:

- Pinpoint an object's precise coordinates (altitude and azimuth).
- Calculate subtle observational effects like parallax.
- Bridge the gap between a theoretical model and the reality of what is seen in the sky.

This toolkit demonstrates a system of thought that was both mathematically rigorous and grounded in the practical needs of observation.

Notes:

- ***Dṛṅmaṇḍalaṃ:*** Vertical circle through object and zenith (Altitude circle).
- ***Digjyā vṛttaṃ:*** Prime vertical or related azimuth reference circle.
- ***Dṛkkṣepamaṇḍalaṃ:*** Vertical circle possibly related to meridian transit or parallax calculation.
- ***Vyāsādha:*** Half-diameter = Radius = 90° angularly.

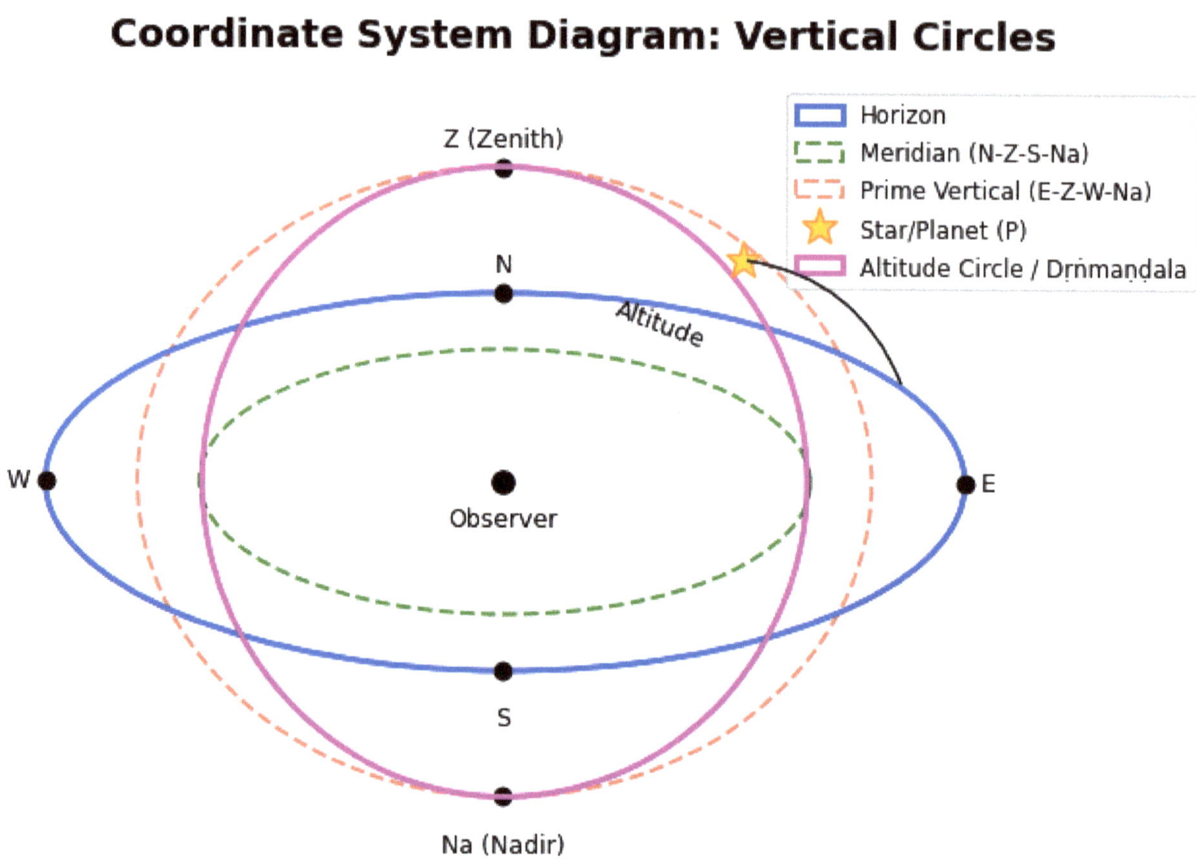

Figure 6.16: The Altitude Circle (Dṛṅmaṇḍala) (Gola.21). The Dṛṅmaṇḍala (magenta) passes through the Zenith and a celestial body (P); Altitude is measured along it from the Horizon. Prime Vertical and Meridian are shown for reference (Gola.21).

6.8.5 The Rotating Model Sphere (Armillary Sphere/Globe)

Sanskrit:

काष्ठमयं समवृत्तं समन्ततः समगुरुं लघुं गोलम् ।

पारतैलजलैस् तं भ्रामयेत् स्वधिया च कालसमम् ॥ २२ ॥

IAST:

kāṣṭhamayaṃ samavṛttaṃ samantataḥ samaguruṃ laghuṃ golam |

pāratailajalais taṃ bhrāmayet svadhiyā cha kālasamam || 22 ||

Translation: "A wooden sphere (***kāṣṭhamayaṃ golam***), perfectly round (***samavṛttaṃ***), uniformly dense all around (***samantataḥ samaguruṃ***) yet light (***laghuṃ***), should be rotated (***bhrāmayet***) by means of mercury, oil, or water (***pāratailajalaiḥ***), and by one's own intellect/skill (***svadhiyā***), in time with the [celestial] period (***kālasamam***)."

Interpretation: This fascinating verse describes the construction and use of a physical **celestial globe** or **armillary sphere** (***gola yantra***). Such models were essential teaching aids and potentially served as analog computational devices, bringing the abstract geometry to life.

1. **Construction:** It should be made of wood (***kāṣṭhamayam***), perfectly spherical (***samavṛttaṃ***), carefully balanced (***samagurum***, uniformly weighted so it rotates smoothly), and relatively light (***laghum***) for ease of handling.

2. **Operation:** The sphere should be rotated (***bhrāmayet***) to simulate the daily apparent motion of the heavens caused by Earth's rotation.

3. ***Pāratailajalaiḥ*** **("By means of mercury, oil, or water"):** This intriguing phrase likely refers to methods for achieving smooth, steady rotation. Possibilities include:

    - Floating the sphere in water for near-frictionless movement.
    - Using oil as a lubricant for bearings.
    - Using mercury in channels for balancing or perhaps even as part of a rudimentary driving mechanism (though this is highly speculative).
    - It might simply refer to methods of construction and finishing to ensure smooth surfaces.

4. ***Svadhiyā*** **("By one's own intellect/skill"):** Bhaskara I emphasizes this, indicating that skillful manual operation was key. The user needed the intelligence (*dhiyā*) to rotate it correctly.

5. ***Kālasamam* ("Synchronized with time"):** The rotation rate must match the actual diurnal period – one full rotation per sidereal day (or be scaled appropriately relative to a timekeeping device like a water clock).

Such a device allows students and practitioners to visualize complex concepts like rising and setting phenomena, meridian transits, the effects of latitude on celestial visibility, and the geometry of coordinate systems. It makes the abstract mathematical model tangible.

Notes:

- ***Kāṣṭhamayam:*** Made of wood.
- ***Samagurum:*** Uniformly weighted/dense, balanced.
- ***Pāratailajalaiḥ:*** With mercury, oil, water (for smooth rotation/balancing).
- ***Kālasamam:*** Synchronized with time (diurnal rate).
- ***Svadhiyā:*** By one's own intellect/skill (manual operation).

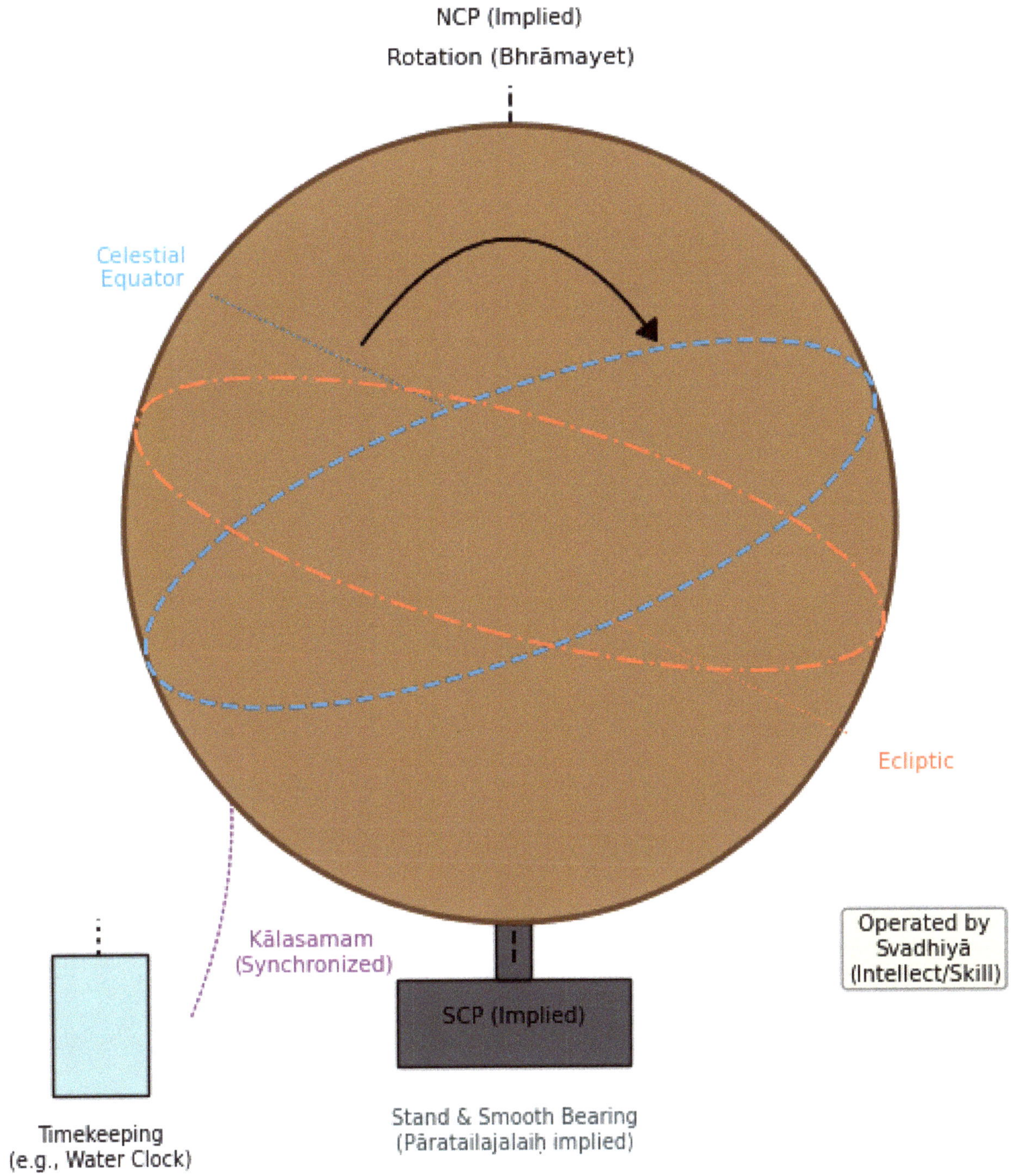

Figure 6.18: Gola.22 - The Rotating Model Sphere (Gola Yantra). This diagram depicts Aryabhata's description of a gola yantra, a physical model of the celestial sphere. The wooden sphere (kāṣṭhamayaṃ golam) is shown with illustrative celestial circles (Equator, Ecliptic) and an implied axis of rotation (NCP/SCP). It was to be rotated (bhrāmayet) by means of a smooth mechanism (pāratailajalaiḥ implied by the stand and bearing) and by the user's intellect (svadhiyā), synchronized with time (kālasamam) via a device like a water clock, to visualize celestial phenomena.

6.8.6 Visualizing Declination Circles on the Model

Sanskrit:

खगोलस्य तु मध्यस्थः स्ववृत्ते ग्रहो भ्रमण्डले ।

तदपक्रमवृत्तस्थः स्वकीये मण्डले सदा ॥ २३ ॥

IAST:

khagolasya tu madhyasthaḥ svavṛtte graho bhramaṇḍale |

tadapakramavṛttasthaḥ svakīye maṇḍale sadā || 23 ||

Translation: "A planet (*graho*), situated within the celestial sphere (*khagolasya madhyasthaḥ*) on its own diurnal circle (*svavṛtte*) within the sphere of asterisms (*bhramaṇḍale*), always rests on its declination circle (*tadapakramavṛttasthaḥ*) as it makes its own daily round (*svakīye maṇḍale*)."

The Daily Paths of the Stars

How do the stars and planets trace their paths across the sky each night? This verse explains the simple and elegant geometry of their daily motion, using the model of the rotating celestial sphere.

The Celestial Sphere: A Starry Globe

Imagine the entire night sky is a giant, hollow globe with the stars painted on its inner surface, and the Earth is at its center. This is the **celestial sphere**. As the Earth turns, this celestial sphere appears to rotate around us once every day, carrying the stars and planets with it.

Declination Circles: The Daily Highways

The verse explains that each celestial body follows its own specific circular path across this sphere. This daily path is its **circle of declination.**

Think of the celestial equator as the "equator" of this starry globe. A star's declination tells us how far north or south of this celestial equator it is.

- A star *on* the celestial equator will trace the celestial equator itself—the largest possible circle.
- A star *north* of the celestial equator will trace a smaller circle, parallel to the equator, in the northern celestial hemisphere.
- A star *south* of the celestial equator will trace a smaller circle in the southern hemisphere.

These parallel "highways" are the declination circles. A star's specific declination determines which of these highways it travels on every single day.

This simple concept is fundamental. It provides a clear, geometric way to visualize the daily apparent motion of every object in the sky, turning the complex dance of the heavens into a predictable and orderly system of nested circles.

Notes:

- ***Khagola:*** Celestial sphere (model).
- ***Svavṛtta / Svakīye maṇḍale:*** Own circle (diurnal path).
- ***Bhramaṇḍala:*** Sphere of asterisms.
- ***Apakramavṛtta:*** Declination circle (parallel to equator).

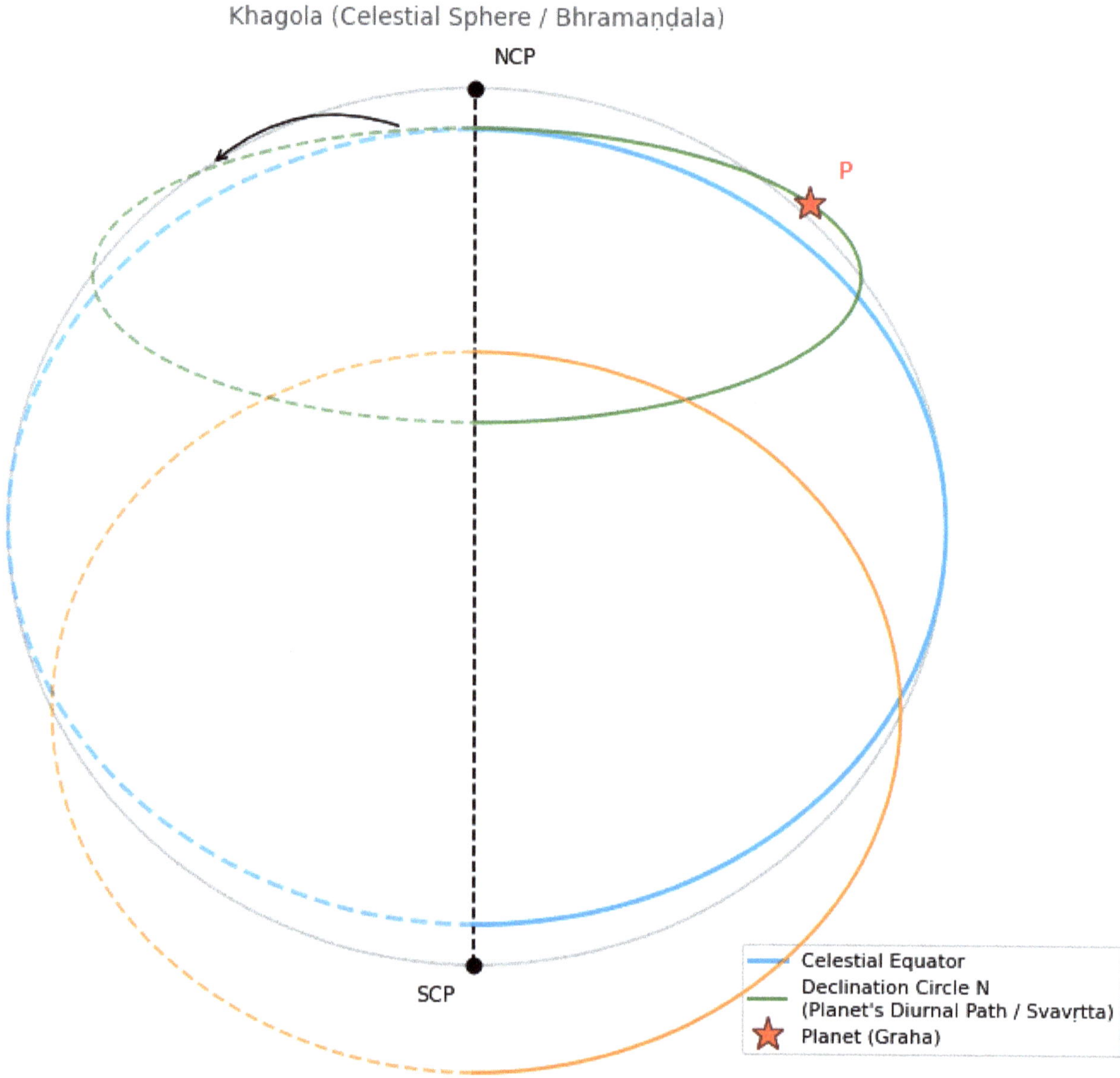

Figure 6.19: The Planet's Declination Circle (Svavṛtta) (Gola.23). A planet (P, red star) traces its daily path along its Declination Circle (Svavṛtta, green), parallel to the Celestial Equator (blue) and centered on the NCP-SCP axis. Orange circle shows another declination path.

6.8.7 Fixed Reference Points (Poles)

Sanskrit:

सूर्योदयास्तमयौ दक्षिणतः पश्यतां सदोदग् याम्यौ

विषुवत्स्थानाद् ऊर्ध्वं ग्रहाः खमध्यात् प्रवृत्ताः स्युः ॥ २४ ॥

IAST:

sūryodayāstamayau dakṣiṇataḥ paśyatāṃ sadodag yāmyau |

viṣuvatsthānād ūrdhvaṃ grahāḥ khamadhyāt pravṛttāḥ syuḥ || 24 ||

(Interpretation of this verse remains somewhat debated among scholars)

The Anchors of the Sky: Poles and Meridian

After defining the celestial grid, Aryabhata now turns to the sky as it appears to an observer on the ground. This verse establishes the two most important features of our personal view of the cosmos: the fixed points that act as anchors and the line that marks a star's daily peak.

1. The Unmoving Poles

The verse begins by noting that two points in the sky are "always north and south." These are the **North and South Celestial Poles**—the unmoving pivots around which the entire celestial sphere appears to turn each day.

For an observer in the Northern Hemisphere, the North Celestial Pole (marked very nearly by Polaris, the North Star) is a constant, fixed point in the sky. Its height above the horizon is a direct and powerful clue to your location on Earth:

The altitude of the visible celestial pole is exactly equal to your latitude.

These poles are the invariant anchors of the sky, providing a stable axis for the daily rotation of all celestial bodies.

2. The Meridian: The High Point of the Day

Next, the verse describes how planets cross the *khamadhya*—the "middle of the sky." This refers to the **meridian**, an imaginary line that runs from the due north point on your horizon, up through the point directly overhead (the zenith), and down to the due south point on the horizon.

When a star or planet crosses this line, it reaches its **culmination**, which is its highest point in its daily journey across the sky. The verse notes that the height of this culmination depends directly on the object's declination (its north-south "celestial latitude"). This provides a clear link between a star's fixed, universal coordinate and its predictable, daily behavior.

The View from the South

The verse includes a specific note "for southern observers." This is a crucial clarification. For someone in the Southern Hemisphere, the entire view is flipped:

- The **South Celestial Pole** is the visible, fixed anchor in the sky.
- Celestial bodies still culminate on the meridian, but they do so to the observer's *north*.

The Framework for Observation

This verse isn't about a complex calculation; it's about establishing the fundamental framework of the observable sky. By defining a fixed axis of rotation (the poles) and a defined line of maximum height (the meridian), Aryabhata lays the necessary groundwork upon which all other calculations of time and position can be built.

Notes:

- ***Dakṣiṇataḥ paśyatāṃ:*** For Southern observers (perspective might matter).
- ***Sadodag yāmyau:*** Always north and south (likely celestial poles).
- ***Viṣuvatsthāna:*** Plane of the celestial equator.
- ***Khamadhya:*** Mid-sky, the meridian.

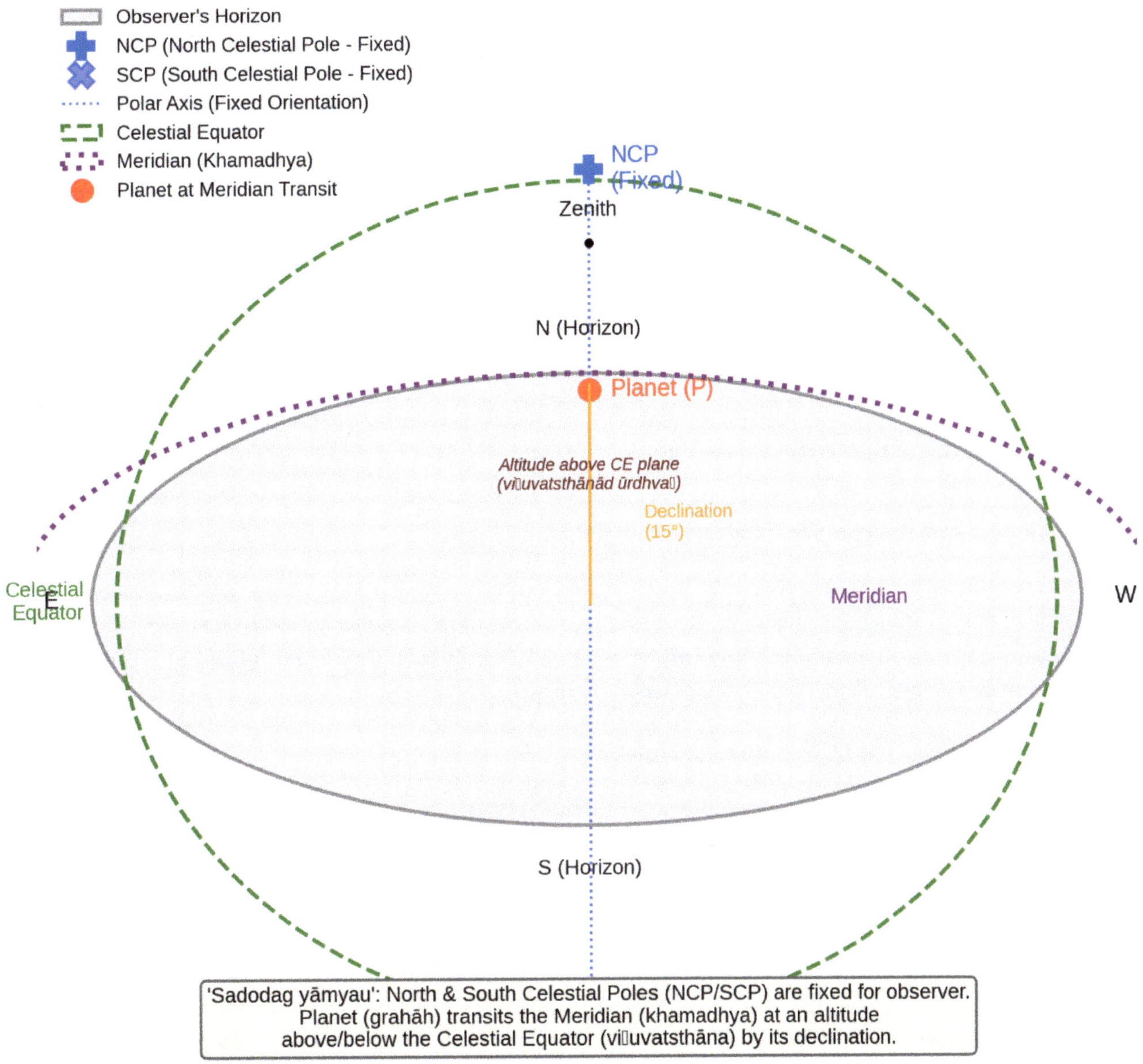

Figure 6.20: Aryabhata's Fixed References and Meridian Transit (Gola.24). As per Gp.24, the North & South Celestial Poles (NCP/SCP) are fixed (sadodag yāmyau). The diagram shows a planet (P) transiting the Meridian (khamadhya, purple dotted) at an altitude above/below the Celestial Equator (green dashed) by its declination, as indicated by visuvatsthānād ūrdhvam.

6.8.8 Measuring Declination Formally

Sanskrit:

भगोलस्यान्तरेणैव विषुवत्स्थानकख्यया ।

ग्रहाणाम् अपक्रमांशा ज्ञेयाः खगोलवृत्ततः ॥ २५ ॥

IAST:

bhagolasyāntareṇaiva viṣuvatsthānakakṣyayā |

grahāṇām apakramāṃśā jñeyāḥ khagolavṛttataḥ || 25 ||

Translation: "Solely through the interior [geometry] of the celestial sphere (***bhagolasyāntareṇa***), relative to the plane of the equator's path (***viṣuvatsthānakakṣyayā***), the degrees of declination (***apakramāṃśāḥ***) of planets are to be known [measured along] a celestial great circle (***khagolavṛttataḥ***)."

Mapping the Heavens: The Definition of Declination

To create a map of the heavens, astronomers needed a reliable grid system, just like the system of latitude and longitude used to map the Earth. This verse defines one of the two essential coordinates in that celestial grid: **declination**, or *apakrama*.

The Celestial Equator: The Starting Line

First, imagine extending the Earth's equator out into space, creating a vast ring among the stars. This is the **celestial equator**. It serves as the fundamental east-west starting line for the entire celestial coordinate system.

Declination: The "Latitude" of the Stars

Declination is simply a measure of how far north or south a celestial body is from this celestial equator. It is the celestial equivalent of latitude on Earth.

- A star with a **positive declination** is in the northern celestial hemisphere.
- A star with a **negative declination** is in the southern celestial hemisphere.
- A star with a declination of **zero** lies directly on the celestial equator.

The verse specifies that this angular distance is measured along a great circle that passes through the celestial poles and the star itself, much like a line of longitude on Earth runs from the North Pole to the South Pole.

A Universal Coordinate

Declination, when paired with a star's east-west coordinate (Right Ascension), gives it a fixed and universal address in the sky. Unlike coordinates that depend on an observer's location (like altitude and azimuth), a star's declination is the same for everyone on Earth, at all times.

This makes the equatorial coordinate system, which is built upon the concept of declination, the absolute foundation for all serious astronomical work. It is the system used to create star catalogs, calculate planetary positions, and build a timeless, objective map of the cosmos.

Notes:

- ***Bhagola:*** Celestial sphere.
- ***Viṣuvatsthānakakṣya:*** Plane of the celestial equator.
- ***Apakramāṃśāḥ:*** Degrees of declination (fundamental coordinate δ).
- ***Khagolavṛtta:*** Celestial great circle (specifically, an hour circle for measuring declination).

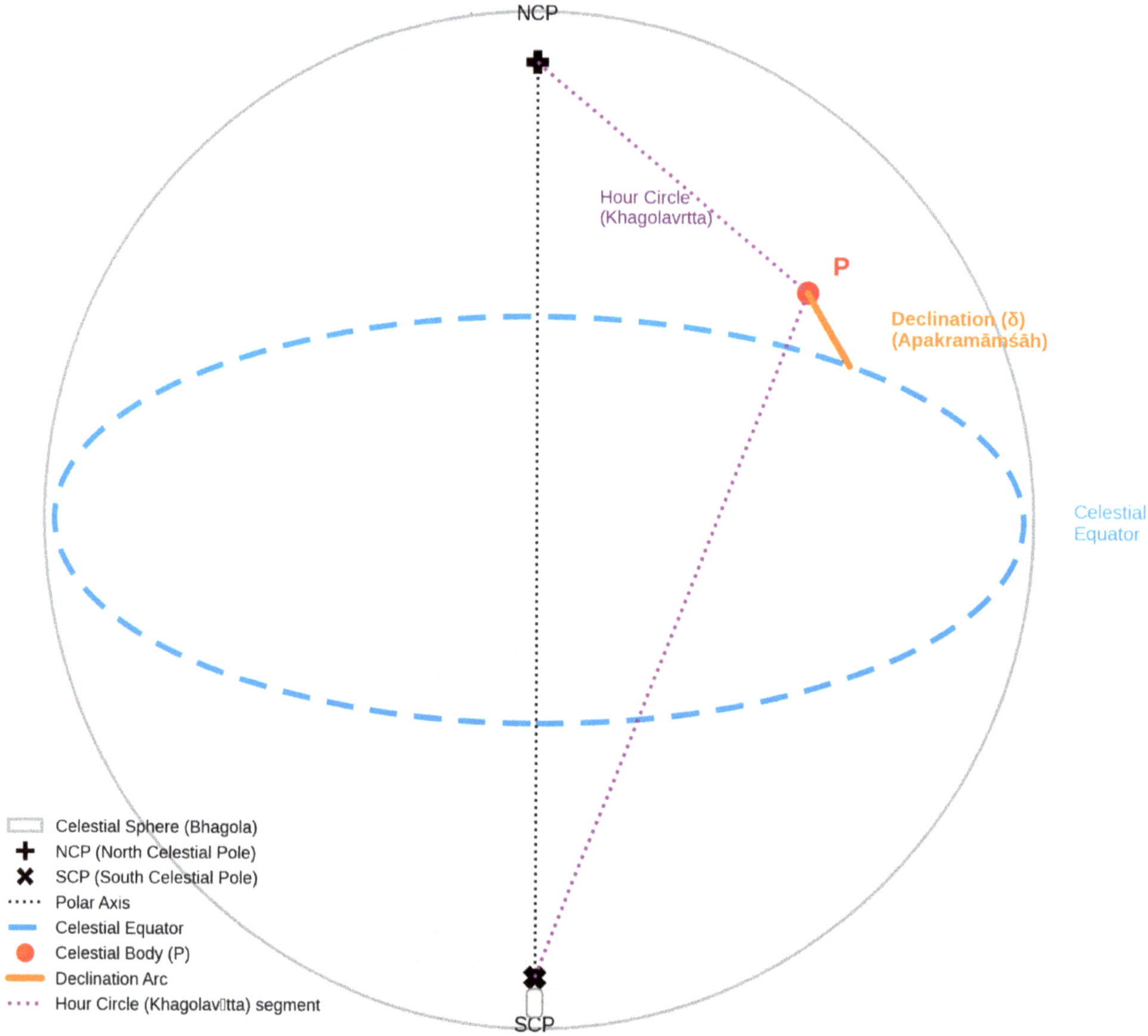

Figure 6.21: Defining Declination (Gola.25). Illustrates declination (δ, orange arc) as the angular distance of a celestial body (P) from the Celestial Equator (blue dashed), measured along its Hour Circle (magenta dotted) passing through the Celestial Poles (NCP/SCP).

6.9 Trigonometric Tools and Celestial Timing (Gola.26-34)

Having meticulously defined the geometry of the *khagola*, Aryabhata now demonstrates how to use his trigonometric tools (***jyā***, sine function) and relate them to observations (especially using the gnomon's shadow, ***chāyā***) to solve key problems in positional astronomy. This section bridges the geometric model with practical calculation and observation, addressing issues like the variation of daylight, unequal rising times of zodiac signs, determining altitude or time, finding directions, and accounting for observational effects like parallax.

6.9.1 Calculating Daylight Variation Sine (*Ksitijyā*)

Sanskrit:

इष्टापक्रमगुणिताम् अक्षज्याम् लम्बकेन हृत्वा या ।

स्वाहोरात्रे क्षितिज्या क्षयवृद्धिज्या दिननिशोः सा ॥ २६ ॥

IAST:

iṣṭāpakramaguṇitām akṣajyām lambakena hṛtvā yā |

svāhorātre kṣitijyā kṣayavṛddhijyā dinaniśoḥ sā || 26 ||

Translation: "Multiply the R·sine of the desired declination (iṣṭa-apakrama[-jyā]) by the R·sine of latitude (akṣajyām) and divide by the R·sine of co-latitude (lambakena hṛtvā = R·cos φ); the result (yā) is the kṣitijyā within the diurnal circle (svāhorātre). It is the R·sine corresponding to the increase or decrease (kṣayavṛddhijyā) of day and night duration (dinaniśoḥ sā)."

The Measure of Daylight: Unpacking the *Kṣitijyā*

Why are summer days long and winter days short? And why is this effect more extreme the farther you are from the equator? This verse answers this fundamental question by providing the mathematical key to calculating the precise length of daylight for any location on Earth, on any day of the year.

That key is a value called the ***Kṣitijyā***, which literally means the "Earth-sine" or "horizon-sine."

The Two Ingredients of Day Length

The length of daylight depends on the interplay between two crucial factors:

1. **The Time of Year (The Sun's Declination):** This is the Sun's north-south position relative to the celestial equator. In the Northern Hemisphere's summer, the Sun is far to the north (high declination); in winter, it is

far to the south (low declination).

2. **Your Location on Earth (Your Latitude):** This is your north-south position on the globe. Your latitude determines the angle at which you view the Sun's daily path across the sky.

The *Kṣitijyā* is a brilliant value that mathematically combines these two factors into a single, powerful number.

The Formula for the *Kṣitijyā*

Aryabhata's verse provides a direct formula for this "Earth-sine":

Kṣitijyā = [R·sin(Declination) × R·sin(Latitude)] / R·cos(Latitude)

Let's break down the components:

- **R·sin(Declination)** (*iṣṭa-apakrama-jyā*): Represents the Sun's seasonal position.
- **R·sin(Latitude)** (*akṣajyā*): Represents your location on Earth.
- **R·cos(Latitude)** (*lambaka*): The R·cosine of your latitude, which also helps define the geometry from your specific viewpoint.

From a Sine Value to a Time Correction

The *Kṣitijyā* itself is not the final answer; it is the crucial intermediate step. It is the raw material used to calculate a time correction called the ***chara***, which is the amount of time that needs to be added to or subtracted from a standard 6-hour half-day.

The process is a beautiful, multi-step calculation:

1. **Calculate the *Kṣitijyā*** using the formula above.
2. **Find the Angle:** Using his sine table in reverse, the astronomer converts the *Kṣitijyā* value back into an angle. This angle is known as the **Ascensional Difference.**
3. **Convert Angle to Time:** This angle is then converted into a time value, the *chara*. This is the "daylight correction time."
4. **Calculate the Length of Half the Day:** This time correction is then applied to the standard 6-hour half-day.
    - Length of half-day = 6 hours ± *chara* time
5. **Find the Total Daylight:** Simply double the result from Step 4 to get the total length of daylight for that day.

The *Kṣitijyā* is therefore the mathematical engine that drives the calendar. It connects the geometric position of the Sun in the heavens to the tangible, lived experience of the changing lengths of day and night, forming a cornerstone of all practical astronomy.

Notes:

- ***Iṣṭa-apakrama[-jyā]:*** R·sin(declination, δ).
- ***Akṣajyā:*** R·sin(latitude, φ).
- ***Lambaka:*** R·cos(latitude, φ).
- ***Kṣitijyā:*** R·sin(Ascensional Difference, A). Key for calculating daylight variation.
- ***Kṣayavṛddhijyā:*** Sine related to increase/decrease of day/night (synonym for ***kṣitijyā***).

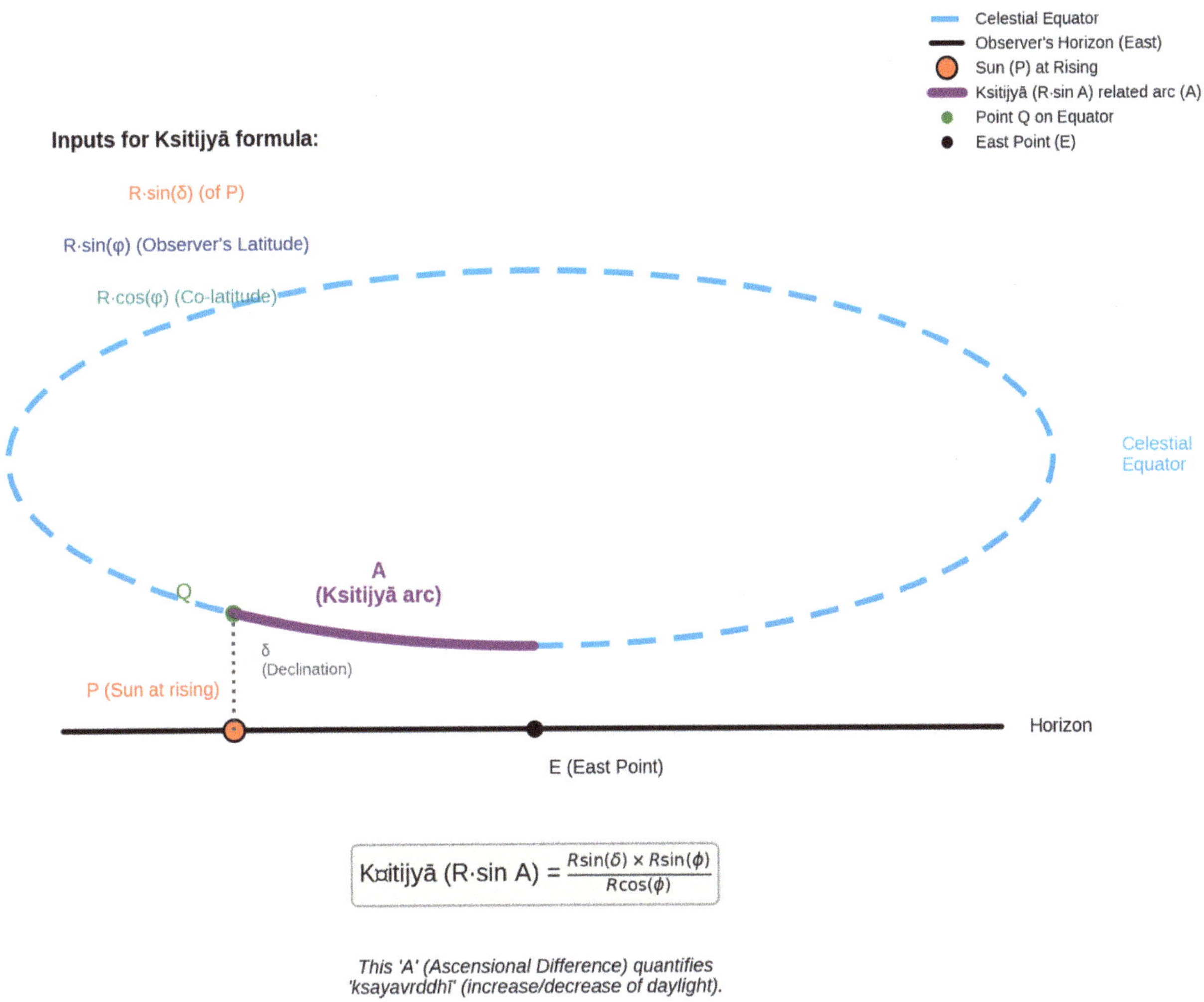

Figure 6.22: The Kṣitijyā Arc 'A' for Daylight Variation (Gola.26). Illustrates the Ascensional Difference arc 'A' (purple segment EQ on the Celestial Equator), between the East point (E) and point Q (equatorial point corresponding to the rising Sun P). Kṣitijyā (R·sin A), derived from this arc using inputs like declination (δ) and latitude (φ), quantifies daylight variation (kṣayavṛddhī).

6.9.2 Unequal Rising Times of Zodiac Signs (*chara* Application)

Sanskrit:

उदयति हि चक्रपादश् चरदलहीनेन दिवसपादेन ।

प्रथमोऽन्त्यश् चाथान्यौ तत्सहितेन क्रमोत्क्रमशः ॥ २७ ॥

IAST:

udayati hi chakrapādaś charadalahīnena divasapādena |

prathamo'ntyaś chāthānyau tatsahitena kramotkramaśaḥ || 27 ||

Translation: "Indeed, the first quadrant of the circle [ecliptic] (chakrapādaḥ) rises in a quarter-day (divasapādena) diminished by half the chara correction (charadalahīnena); so too the last quadrant (antyaḥ). The other two [middle quadrants] (anyau) rise in a quarter-day and increase by half chara (tatsahitena), in direct and reverse order (kramotkramaśaḥ)."

The Rising of the Signs: Calculating Oblique Ascension

If you watch the night sky from a northern city like London or New York, you will notice a strange and beautiful phenomenon: some zodiacal constellations seem to race over the horizon, while others rise slowly and majestically.

This verse answers a crucial question for ancient astronomers: **"At my specific latitude, how long does each sign of the zodiac take to rise?"** This is the problem of calculating **oblique ascension**.

The Tilted Racetrack Analogy

Imagine the celestial equator is a flat, horizontal racetrack. A 90-degree section of this track will always take exactly 6 hours to rise over the horizon, no matter where you are on Earth. This is our baseline.

Now, imagine the ecliptic—the path of the Sun and zodiac—is a second racetrack, tilted at a steep angle.

- The parts of the tilted track that are rising at a **shallow angle** to your horizon will appear to climb into the sky very quickly. These are the "signs of short ascension."
- The parts of the tilted track that are rising at a **steep, almost vertical angle** will take much longer to clear the horizon. These are the "signs of long ascension."

The Formula: Adjusting a Baseline

Aryabhata's method for calculating these rising times is elegant. He starts with a baseline and then applies a correction.

1. **The Baseline:** The starting point is the time it takes for a 90-degree quadrant of the celestial equator to rise, which is always a constant **6 hours** (a "quarter-day" or *divasapāda*).

2. **The Correction Factor (*Charadala*):** He then calculates a correction factor called the *charadala*. This value is derived from the *Kṣitijyā* (the "Earth-sine" discussed in Gola.26) and depends on both the observer's latitude and the Sun's declination (its north-south position). It perfectly quantifies the "steepness" of the ecliptic's angle at the horizon.

The Rules of Rising (for the Northern Hemisphere)

With these two components, the rising time for each quadrant of the zodiac can be found:

- **Spring and Winter Signs (Aries, Taurus, Gemini & Capricorn, Aquarius, Pisces):** These are the "signs of short ascension." They rise at a shallow angle and thus appear quickly.

    - **Rising Time = 6 hours – *Charadala***

- **Summer and Autumn Signs (Cancer, Leo, Virgo & Libra, Scorpio, Sagittarius):** These are the "signs of long ascension." They rise at a steep angle and take longer to appear.

    - **Rising Time = 6 hours + *Charadala***

The term *kramotkramaśaḥ* ("in direct and reverse order") in the verse points to the beautiful symmetry of this system. The correction for Aries is the same as for Pisces, the correction for Taurus is the same as for Aquarius, and so on, mirroring the Sun's symmetrical journey through the seasons.

Why This Was So Important

The ability to calculate oblique ascension was absolutely critical for one of the cornerstones of ancient astronomy and astrology: determining the **Ascendant (*Lagna*)**. The Ascendant is the precise degree of the zodiac rising on the eastern horizon at a specific moment, such as the time of a person's birth.

To find the Ascendant, one *must* know the rising times of the individual signs. Aryabhata's method provided the essential mathematical machinery to move from the idealized, uniform time of the celestial equator to the complex, latitude-dependent reality of how the zodiac actually appears in the sky.

Notes:

- ***Chakrapādaḥ:*** Quadrant of the ecliptic (90°).
- ***Divasapāda:*** Quarter-day (nominal rising time at equator = 15 *nāḍikās*).
- ***Charadala:*** Half the *chara* time correction (derived from ***kṣitijyā***).
- ***Prathama/Antya:*** First (Spring)/Last (Winter) quadrants.
- ***Kramotkramaśaḥ:*** In direct and reverse order (systematic application).

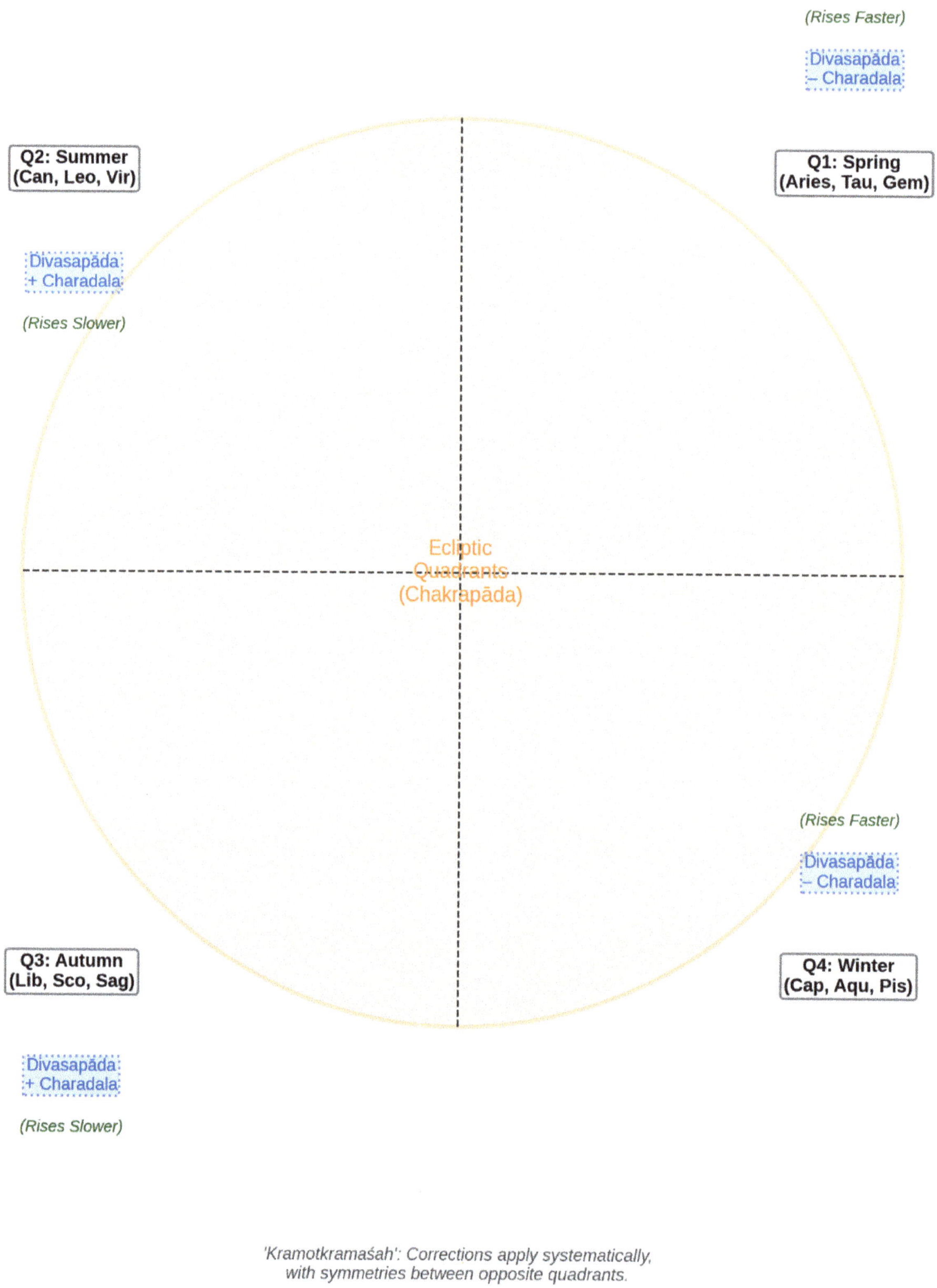

Figure 6.23: Oblique Ascension of Ecliptic Quadrants (Gola.27). Illustrates how the rising time for each 90° ecliptic quadrant (chakrapāda) is calculated by adjusting the baseline Divasapāda (quarter-day) with the Charadala correction. Quadrants 1 & 4 rise in (Divasapāda – Charadala), Quadrants 2 & 3 in (Divasapāda + Charadala), for a Northern Hemisphere observer.

6.9.3 Altitude Sine (Śanku) from Time (Hour Angle)

Sanskrit:

इष्टज्यावलम्बकगुणा त्रिज्याभक्ता स्ववृत्तगः शङ्कुः ।

स्वाहोरात्रे क्षित्युन्नतिः शङ्कोर् भगणार्धयोः एव ॥ २८ ॥

IAST:

iṣṭajyāvalambakaguṇā trijyābhaktā svavṛttagaḥ śaṅkuḥ |

svāhorātre kṣityunnatiḥ śaṅkor bhagaṇārdhayor eva || 28 ||

Translation: "Multiply the R·sine of the desired time [hour angle] (***iṣṭajyā***) by the R·sine of co-latitude (***avalambak-aguṇā*** = ***lambaka***) and divide by the radius (***trijyābhaktā***); the result is the gnomon-sine (***śaṅkuḥ***) of the object on its diurnal circle (***svavṛttagaḥ***). This elevation above the horizon (***kṣityunnatiḥ***) in its day-night arc (*svāhorātre*) is the *śaṅku*..."

Calculating a Star's Height in the Sky: The *Śaṅku*

If an astronomer knows their latitude and the "time" of a star—meaning how far it is in its daily arc past the high point of the meridian—how can they calculate its precise height above the horizon?

This verse provides the fundamental formula for finding the ***śaṅku***, which is the **R·sine of the star's altitude**. It's a direct way to convert celestial "time" into observable height.

The Formula: The Simple Case

Aryabhata first presents the formula for the simplest case: a star located on the celestial equator (where its declination is 0°). For such a star, the recipe is elegant and straightforward.

The *śaṅku* is found by multiplying the R·sine of the star's hour angle by the R·cosine of the observer's latitude, and then dividing by the standard radius.

Śaṅku **= [R·sin(Hour Angle) × R·cos(Latitude)] / Radius**

Let's break down the key components:

- **R·sin(Hour Angle)** (*iṣṭajyā*): This is the trigonometric measure of the "time" or angular distance the star has moved past the meridian.

- **R·cos(Latitude)** (*avalambaka*): This is the R·sine of the co-latitude, a crucial value that accounts for the observer's position on Earth.

Beyond the Simple Case: The General Formula

Of course, not all stars lie on the celestial equator. The true power of Aryabhata's system is that these same principles can be extended to find the altitude of *any* celestial body, at *any* time.

While the verse itself only states the simplified rule, the underlying spherical trigonometry known to this tradition provides a general formula that also incorporates the star's declination (δ), its north-south position relative to the celestial equator. The full formula is more complex, but it demonstrates the comprehensive nature of the system:

$$\mathbf{R\cdot sin(Altitude) = [R\cdot sin(\varphi)\times R\cdot sin(\delta) + R\cdot cos(\varphi)\times R\cdot cos(\delta)\times R\cdot cos(H)] / R^2}$$

Why This Calculation Was a Cornerstone

The ability to calculate the *śaṅku* was absolutely essential for a predictive astronomical system. It served three critical functions:

1. **Prediction:** It allowed an astronomer to predict the precise altitude (*kṣityunnatiḥ*) of a star or planet for any given moment.
2. **Verification:** The predicted altitude could be compared with the observed altitude, measured using an instrument like a gnomon (where shadow length is related to altitude), to verify the accuracy of the calculations or the instrument itself.
3. **Timekeeping:** By working the formula in reverse, an astronomer could observe a star's altitude and use that measurement to determine the exact time (the hour angle).

This formula, in its simple and general forms, created a powerful link between the abstract grid of the heavens and the visible, measurable reality of the sky.

Notes:

- ***Iṣṭajyā:*** R·sin(Hour Angle, H).
- ***Avalambaka/Lambaka:*** R·cos(latitude, φ).
- ***Trijyā:*** Radius (3438').
- ***Śaṅkuḥ:*** R·sin(altitude).
- ***Kṣityunnatiḥ:*** Elevation above horizon (altitude).* Verse gives the simplest case (δ=0); the general case requires a full spherical trig formula.

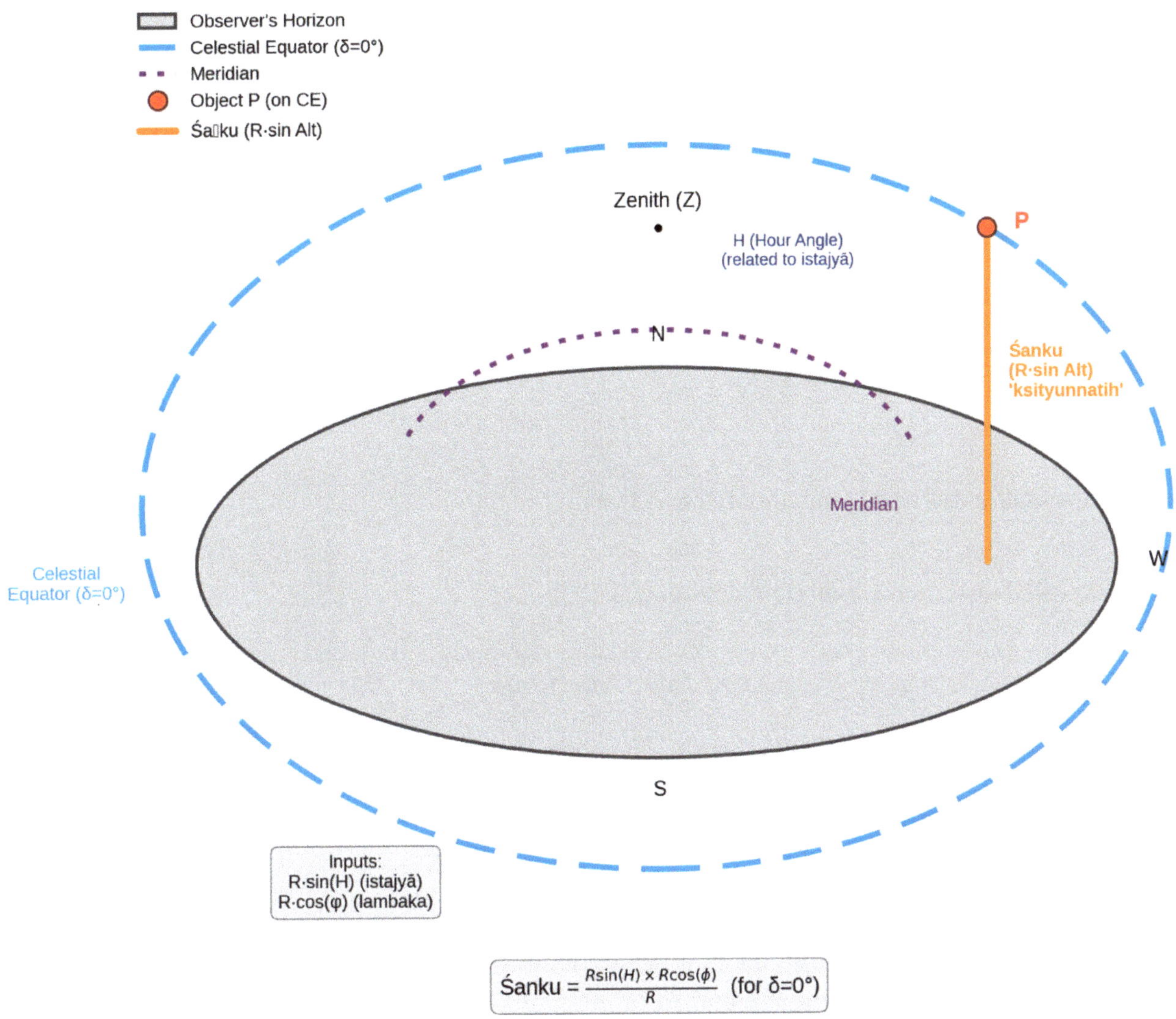

Figure 6.24: Calculating Śaṅku (R·sin Altitude) for an Object on the Celestial Equator (Gola.28). This diagram illustrates Aryabhata's method: the Śaṅku (orange vertical line, kṣityunnatiḥ) for an object (P) on the Celestial Equator (blue dashed) is derived from its Hour Angle (H, related to iṣṭajyā) from the Meridian (purple dotted) and the observer's co-latitude (lambaka), using the formula from Gp.28.

6.9.4 Calculating Shadow Components and Direction

Sanskrit:

शङ्क्वग्रं शङ्कुतलाद् इष्टा दिक् प्राग्गुणेन हृत्वा या ।

शङ्कोः स्वमूलतः सा याम्या सौम्या प्रदेशस्थः ॥ २९ ॥

IAST:

śaṅkvagraṃ śaṅkutalād iṣṭā dik prāgguṇena hṛtvā yā |

śaṅkoḥ svamūlataḥ sā yāmyā saumyā pradeśasthaḥ || 29 ||

(Interpretation requires relating terms to standard shadow calculations)

From the Heavens to the Ground: Predicting the Gnomon's Shadow

In previous verses, Aryabhata has shown how to calculate the Sun's precise position in the sky—its altitude and azimuth—for any given time and location. Now, he addresses the practical follow-up question:

> **"If I know where the Sun is in the sky, how can I predict the exact length and direction of the shadow cast by my gnomon?"**

This verse provides the method to connect the abstract, calculated position of the Sun to a concrete, observable measurement on the ground.

The Solution: A Bridge of Trigonometry

The solution lies in using trigonometry to project the Sun's position down to the ground plane. The gnomon (the vertical stick) and its shadow form a right-angled triangle, where the sunbeam is the hypotenuse.

To fully describe the shadow, we need two pieces of information: its total length and its orientation. Aryabhata's method involves calculating the shadow's components along two perpendicular axes:

1. **The North-South Component:** This tells us how much of the shadow's length points north or south along the meridian line.

2. **The East-West Component:** This tells us how much of the shadow's length points east or west.

By knowing the Sun's altitude and azimuth (its compass direction), a series of trigonometric calculations can reveal the precise lengths of these two components. Once these are known, the total length and exact direction of the shadow can be found easily using the Pythagorean theorem and basic trigonometry.

The Two-Way Street of Calculation

This method creates a powerful two-way street between observation and calculation:

- **Prediction:** An astronomer can use the calculated position of the Sun to predict exactly where a gnomon's shadow will fall at any given moment.
- **Measurement:** Conversely, by precisely measuring the length and direction of an existing shadow, an astronomer can work backwards to determine the exact time of day and verify the cardinal directions with great accuracy.

This verse is a perfect example of applied science. It demonstrates how the theoretical framework of spherical astronomy can be used to make precise, verifiable predictions about the world around us, turning the gnomon and its shadow into a sophisticated scientific instrument.

Notes:

- ***Śaṅkvagraṃ/Śaṅkutala:*** Gnomon tip/base (gives height G).
- ***Iṣṭā dik / Prāgguṇa:*** Terms related to trigonometric functions of azimuth/E-W displacement.
- ***Yāmyā/Saumyā:*** Southern/Northern component of shadow length.
- ***Pradeśasthaḥ:*** Specific to the observer's location (latitude dependent).

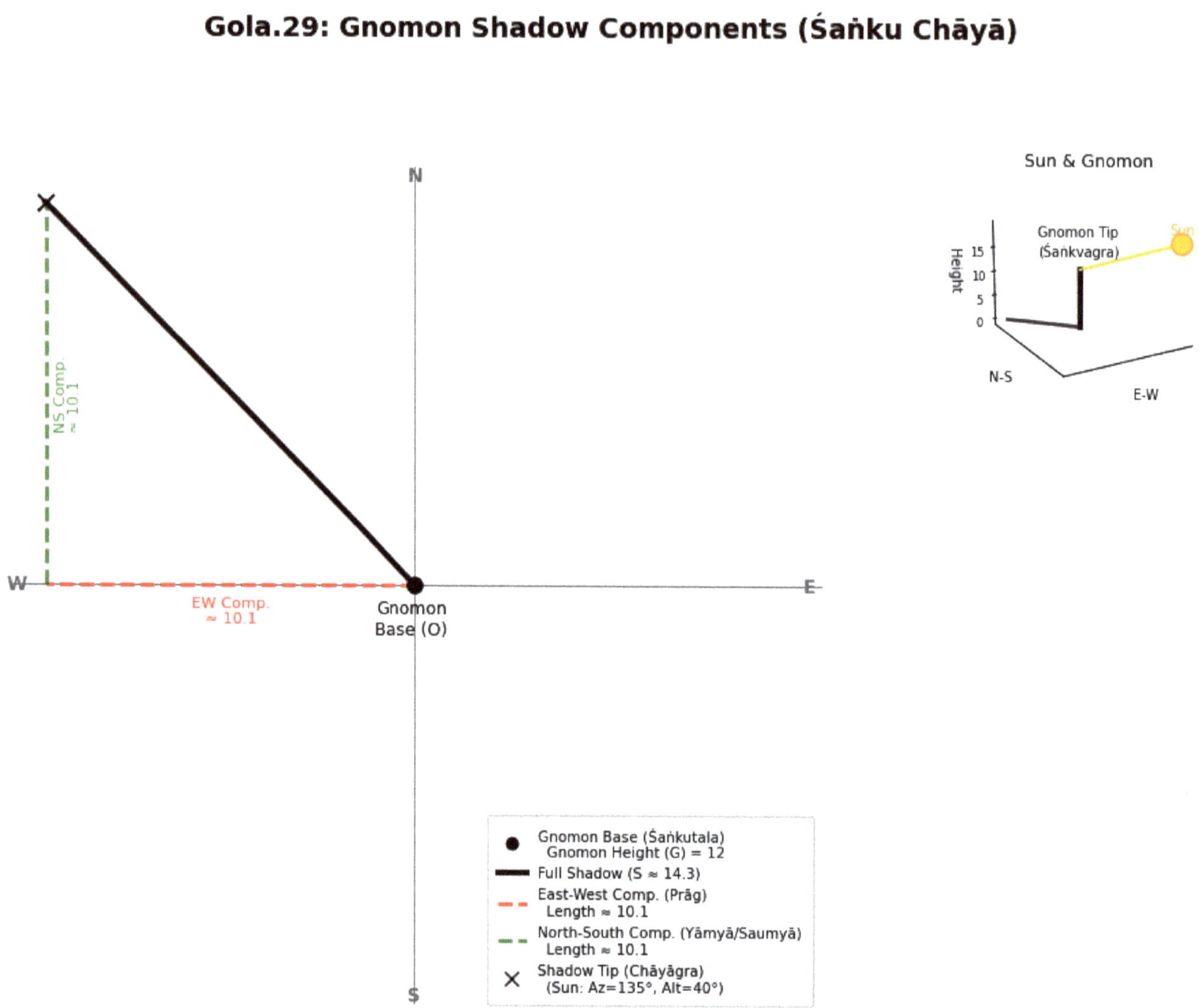

Figure 6.25: Gnomon shadow components (Śaṅku Chāyā) as described in Gola.29. The main diagram (left) shows a top-down view: the full shadow cast by the gnomon (thick black line) from its base (O) to the shadow tip (X). This shadow is resolved into its East-West (Prāg) component (red dashed) and North-South (Yāmyā/Saumyā) component (green dashed). The 3D inset (top right) illustrates the gnomon (Śaṅku), its tip (Śaṅkvagra), and the sunbeam that creates the shadow on the ground.

6.9.5 Calculating the Sun's Amplitude Sine (*Arkāgrā*)

Where exactly on the horizon does the Sun rise or set on a given day? It's rarely due East or due West, except on the equinoxes (and even then, with caveats). Golapāda Verse 30 provides Aryabhata's method for calculating the **Arkāgrā**, a term representing the R·sine of the Sun's **amplitude**—its angular distance along the horizon from the true East point at sunrise, or from the true West point at sunset.

Sanskrit:

अपक्रमगुणितां त्रिज्याम् अवलम्बकहृताम् अवाप्नोति ।

अर्काग्रा स्वाहोरात्रार्धे विषुवद् दिने न हि सदा भवति ॥ ३० ॥

IAST:

apakramaguṇitāṃ trijyām avalambakahṛtām avāpnoti |

arkāgrā svāhorātrārdhe viṣuvad dine na hi sadā bhavati || 30 ||

Translation: ""Multiply the R·sine of declination (apakrama[-jyā]) by the radius (trijyām) and divide by the R·sine of co-latitude (avalambakahṛtām = lambaka); one obtains (avāpnoti) the Sun's amplitude sine (arkāgrā) at sunrise/sunset [literally, 'at the half-day-night arc,' i.e., on the horizon] (svāhorātrārdhe). On the equinox day (viṣuvad dine), this is indeed not always [zero] (na hi sadā bhavati)."

The Sun's Journey Along the Horizon: *Arkāgrā*

We know the Sun rises in the east and sets in the west, but it rarely rises *exactly* due east or sets *exactly* due west. In the summer, it rises to the north of east, and in the winter, to the south of east.

This verse answers a fundamental observational question: **"How far north or south of true east will the Sun rise today?"**

The answer is a value called the **amplitude**, which is the angular distance along the horizon from the true east point to the sunrise point. The R·sine of this angle is what Aryabhata calls the ***Arkāgrā***.

The Formula for the Sun's Amplitude

Aryabhata provides a simple and elegant formula to calculate the *Arkāgrā*. It depends on just two factors: the Sun's position in its yearly cycle and your own position on Earth.

Arkāgrā **= [R·sin(Sun's Declination) × Radius] / R·cos(Observer's Latitude)**

Let's break down the components:

- **R·sin(Sun's Declination)**: This measures how far north or south the Sun is from the celestial equator on a given day. This value changes throughout the year, being largest at the solstices and zero at the equinoxes.
- **R·cos(Observer's Latitude)**: This accounts for your location on Earth. An observer at the equator will see a different amplitude than an observer in a northern or southern city.
- **Radius (R)**: This is the standard radius (3438') used to keep the calculations consistent.

This formula was incredibly useful for practical purposes, such as using the sunrise point to accurately determine the cardinal directions—a vital step for orienting instruments or sacred structures.

A Fascinating Note on the Equinox

The verse concludes with a subtle but brilliant observational note. Theoretically, on the day of an equinox, the Sun's declination is zero. Plugging this into the formula gives an *Arkāgrā* of zero, which means the Sun *should* rise exactly due east.

However, Aryabhata adds that this "is indeed not always so."

Why would he say this? Commentators like Bhāskara I explain that Aryabhata was acknowledging a real-world complication: **atmospheric refraction**. The Earth's atmosphere bends light, making the Sun appear slightly higher in the sky than it truly is, especially near the horizon.

This means we might see the Sun's disc appear just before its geometric center has reached the true east point, causing a very slight deviation. This small, parenthetical remark reveals the mind of a true scientist—one who understands that theoretical models must ultimately answer to what can be observed in reality. It is a nod to the complexities of nature that lie beyond the perfect world of geometry.

Notes:

- ***Apakrama[-jyā]:*** R·sin(declination, δ).
- ***Trijyā:*** Radius (3438').
- ***Avalambaka/Lambaka:*** R·cos(latitude, φ).
- ***Arkāgrā:*** R·sin(amplitude) (Sun's rising/setting deviation from E/W).
- ***Svāhorātrārdhe:*** At sunrise/sunset.
- ***Viṣuvad dine:*** On the equinox day. Note about non-zero effect might hint at refraction.

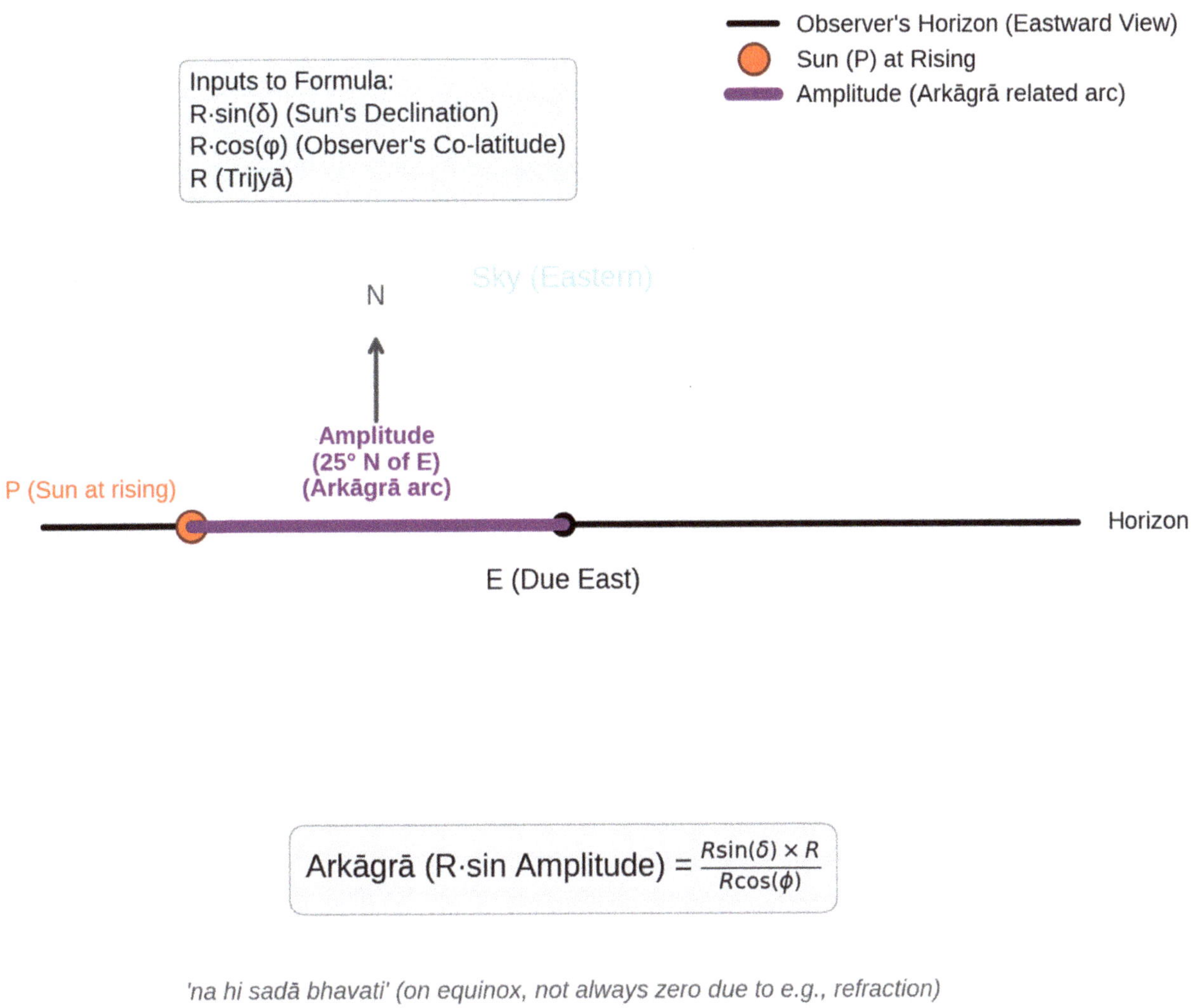

Figure 6.26: Sun's Amplitude (Arkāgrā) at Sunrise (Gola.30). This diagram illustrates the Sun's (P) amplitude at rising—its angular distance along the horizon from the true East point (E). The Arkāgrā (R·sine of this amplitude, represented by the purple arc EP) is calculated using the Sun's declination (δ), observer's co-latitude (related to R·cos φ), and Trijyā (R), as per Gp.30. The verse also notes that on the equinox, the amplitude is 'not always zero' (na hi sadā bhavati).

6.9.6 Finding Latitude from Equinox Noon Shadow

Sanskrit:

विषुवद्दिवसातपः पार्श्वे मध्याह्ने शङ्कुछाययोः युतिः ।

अक्षज्या भवति या सा मण्डलमध्ये ग्रहाः समाः ॥ ३१ ॥

IAST:

viṣuvaddivasātapaḥ pārśve madhyāhne śaṅkuchāyayoḥ yutiḥ |

akṣajyā bhavati yā sā maṇḍalamadhye grahāḥ samāḥ || 31 ||

Translation (based on common interpretation leading to the sine formula): "On the equinox day (viṣuvaddivase), at local noon (madhyāhne), the R-sine of latitude (akṣajyā) is obtained from the gnomon (śaṅkoḥ) and its shadow (chāyāyāḥ): [it is] the shadow length multiplied by the radius (trijyā) and divided by the hypotenuse (their 'combination' or yutiḥ, meaning $\sqrt{(G^2+S^2)}$, termed śruti or karṇa). At the equator (maṇḍalamadhye), celestial bodies rise vertically (grahāḥ samāḥ)."

Finding Your Place in the World: The Gnomon and Latitude

How can you determine your precise location on the Earth's surface using nothing more than a stick and its shadow? In this verse, Aryabhata answers this fundamental question, providing a simple and brilliant method for finding your geographical latitude.

The Observation: A Special Day and a Special Time

To perform this measurement, you need two specific conditions:

1. **The Equinox:** The observation must be made on the day of a spring or autumn equinox. On this day, the Sun is positioned directly over the Earth's equator.
2. **Local Noon:** The measurement must be taken at the exact moment of local noon, when the Sun is at its highest point in the sky for that day.

At this special time, you measure the length of the shadow cast by a vertical stick (a **gnomon**) of a known height.

The Elegant Geometry of the Solution

The brilliance of this method lies in its simple and beautiful geometry.

- Because it's the equinox, the Sun is on the celestial equator.

- Because it's local noon, the Sun is on your local meridian.

In this specific configuration, a remarkable thing happens: **the Sun's angular distance from the point directly overhead (the zenith) is exactly equal to your latitude (φ).**

This creates a perfect right-angled triangle on the ground, formed by:

- The vertical gnomon (height **G**).
- Its horizontal shadow (length **S**).
- The ray of sunlight connecting the top of the gnomon to the tip of the shadow.

The angle at the top of the gnomon, between the gnomon itself and the sunbeam, is your latitude, φ.

Calculating the R-sine of Latitude (*Akṣajyā*)

With this triangle, finding the R-sine of the latitude (*akṣajyā*) becomes a straightforward two-step process.

1. **Find the Hypotenuse (*Karṇa*):** First, calculate the length of the sunbeam from the top of the gnomon to the tip of its shadow using the Pythagorean theorem.
    - *Karṇa* = $\sqrt{(G^2 + S^2)}$
2. **Calculate the R-sine:** The R-sine of latitude is then the ratio of the shadow's length (the side opposite the angle φ) to the hypotenuse, multiplied by the standard radius (R).
    - *Akṣajyā* (R-sin φ) = (R × S) / *Karṇa*

This elegant method provides a direct, observational path to finding one of the most critical parameters in all of astronomy. An observer's latitude is essential for nearly every other calculation, from predicting the length of daylight to determining the precise timing of local celestial events.

A Note on the Equator

The verse concludes with an observation about the equator: "in the middle of the sphere, the celestial bodies are equal." This means that for an observer at the Earth's equator (where latitude is 0°), all celestial bodies rise and set perpendicular to the horizon. Their paths are cut perfectly in half by the horizon, leading to days and nights of equal length year-round (ignoring minor effects). This serves as a "zero-point" confirmation of the geometric principles at play.

Notes:

- ***Viṣuvaddivasa:*** Equinox day (Sun's declination δ=0).
- ***Madhyāhne:*** At midday (local noon).
- ***Śaṅku:*** Gnomon height (G).
- ***Chāyā:*** Noon shadow length (S).
- ***Yutiḥ:*** Combination/Calculation yielding ***akṣajyā***.

- ***Akṣajyā:*** R·sin(latitude, φ). Key practical method for finding latitude.
- ***Maṇḍalamadhye:*** At the equator (φ=0).

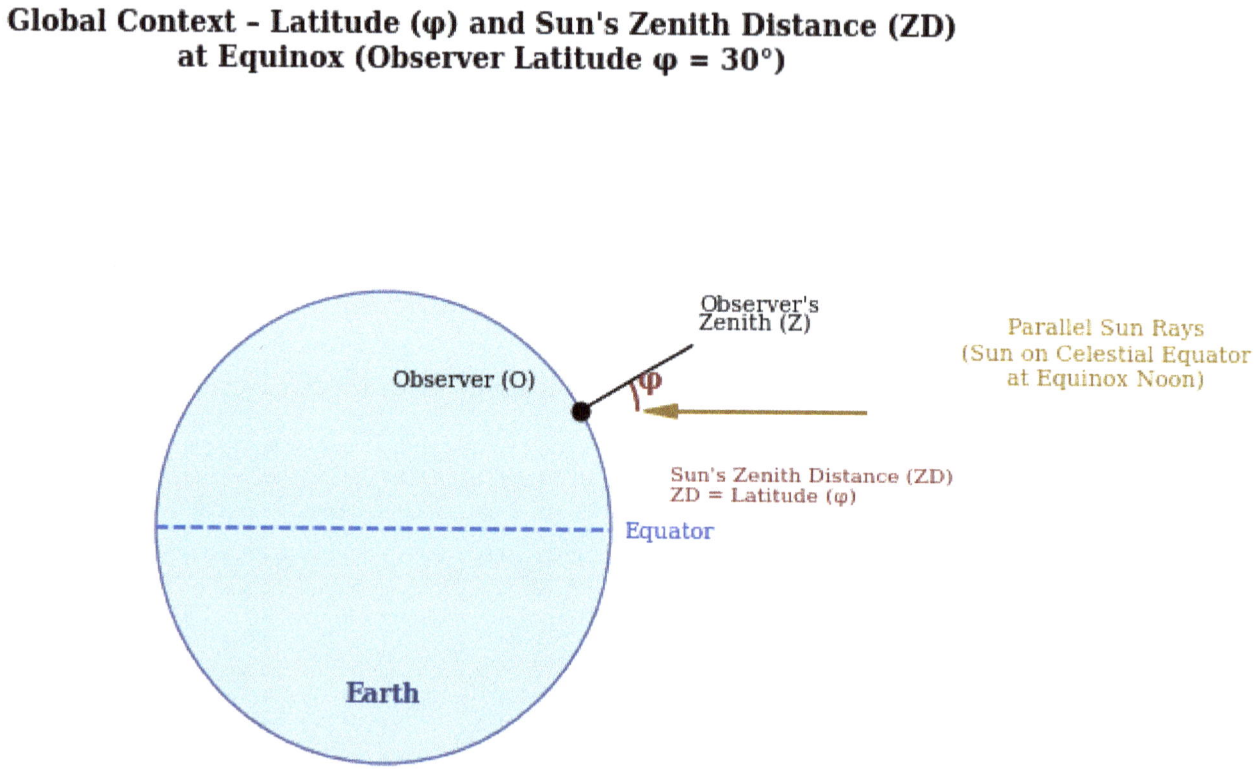

Figure 6.27a: Global context for determining latitude (φ). At equinox noon, the Sun's Zenith Distance (ZD) for an observer equals their latitude, as parallel Sun rays are aligned with Earth's equatorial plane.

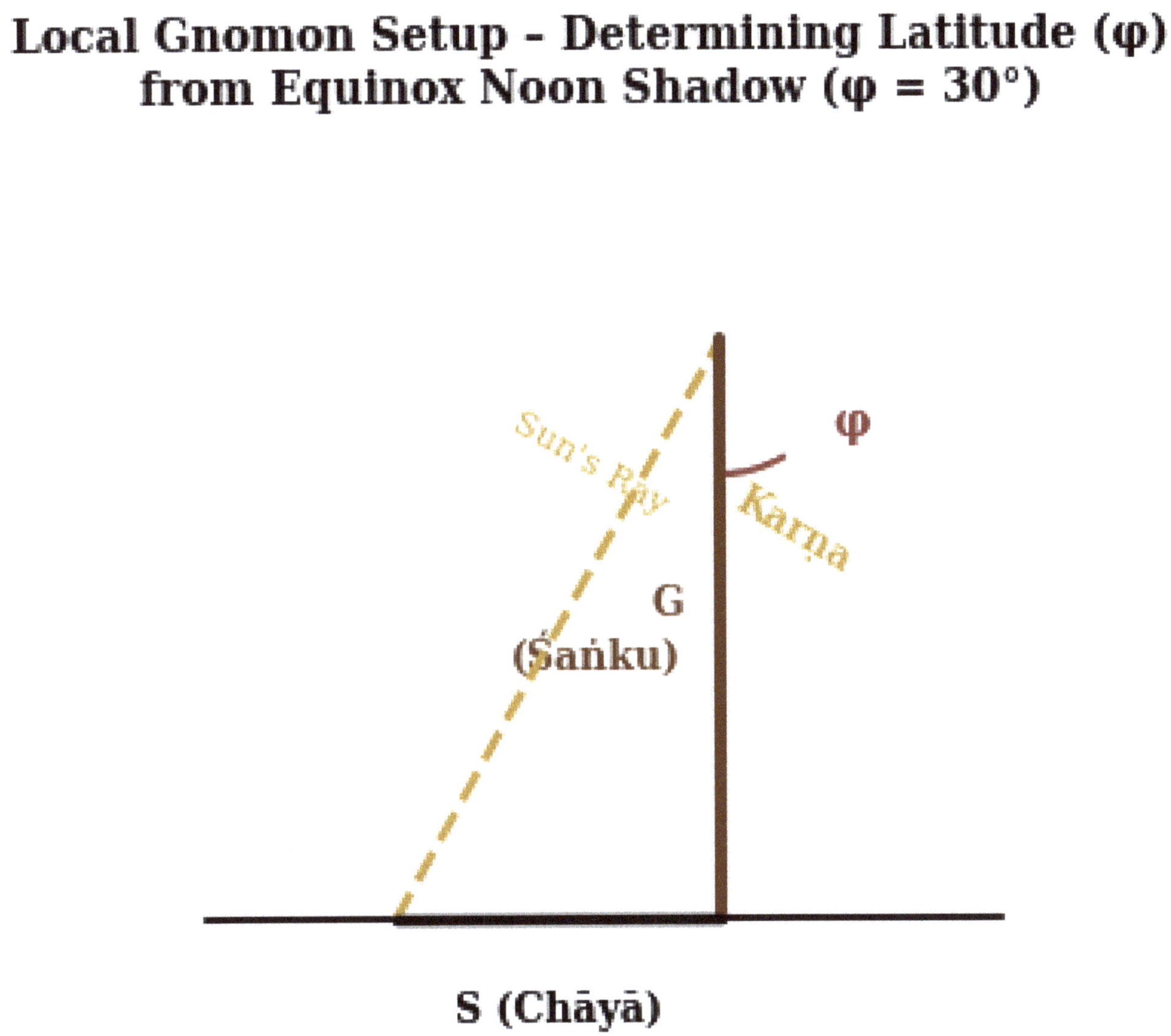

Figure 6.27b: Local gnomon setup for latitude determination at equinox noon. The gnomon (G) and its noon shadow (S) form a right triangle with the Sun's ray (Karṇa), allowing calculation of latitude (φ).

6.9.7 Relating Altitude, Zenith Distance, and Parallax Sines

Sanskrit:

शङ्कुर् यत्र गुणकारस् त्रिज्या भाज्यो भवेत् सदा तस्मिन् ।

व्यासार्धं दृक्क्षेपज्या स्याद् दृग्गतिज्या तु कर्णगुणा ॥ ३२ ॥

IAST:

śaṅkur yatra guṇakāras trijyā bhājyo bhavet sadā tasmin |

vyāsārdhaṃ dṛkṣepajyā syād dṛggatijyā tu karṇaguṇā || 32 ||

Translation (Interpretive, based on common understanding of these terms in parallax context): "The R·sine of zenith distance (dṛkṣepajyā) is [related to/complementary to] the R·sine of altitude (śaṅku, their 'combination' or yutiḥ in the sense of forming a right angle in the sky). From the difference [between geocentric and topocentric positions], the R·sine of parallax (dṛggatijyā) is known. This parallax sine (dṛggatijyā), modified by factors related to the true distance (karṇaguṇā), is applied to the planet's [calculated geocentric] position on the celestial sphere (grahagolāt)."

The Parallax Puzzle: Meet the Key Players

Before providing the recipe for the parallax correction, Aryabhata first introduces the three key trigonometric quantities that are the "main players" in the calculation. This verse acts as a cast list, defining the essential components needed to understand and solve the parallax puzzle.

1. The *Śaṅku*: The Measure of Height

The first player is the ***Śaṅku***. This is the **R·sine of the object's altitude**—its angular height above the horizon. In simple terms, the *Śaṅku* is a direct measure of how high the celestial body is in the sky. A larger *Śaṅku* means the object is higher.

2. The *Dṛkṣepajyā*: The Measure of Overhead Distance

The second player is the ***Dṛkṣepajyā***. This is the **R·sine of the object's zenith distance**—its angular distance from the point directly overhead (the zenith).

The *Śaṅku* and the *Dṛkṣepajyā* are a complementary pair. They are two sides of the same coin, locked together by the geometry of a right-angled triangle. Knowing one immediately tells you the other. Together, they perfectly define the vertical position of an object in the sky from the observer's perspective.

3. The *Dṛggatijyā*: The Measure of the Parallax Shift

The final and most important player is the ***Dṛggatijyā***. This is the **R·sine of the parallax angle itself**. It represents the magnitude of the apparent shift we are trying to calculate. The verse states that this value arises "from the difference" (*bhedāt*), referring to the difference between the object's true (geocentric) position and its apparent (topocentric) position.

The Crucial Role of Distance

Crucially, Aryabhata notes that the *Dṛggatijyā* (the parallax effect) is dependent on the object's true distance from the Earth's center, which he refers to with the term *karṇa* (the hypotenuse or true distance).

This is a key insight:

> **The parallax shift is inversely proportional to the object's distance. Closer objects have a larger parallax, while distant objects have a smaller one.**

This is why the parallax effect is so significant for the nearby Moon but almost negligible for the distant stars.

Setting the Stage

By defining these three key R·sines, Aryabhata sets the stage for the calculation to come. He has identified the necessary inputs: a measure of the object's altitude (*Śaṅku / Dṛkṣepajyā*) and a measure of its distance (related to the *Dṛggatijyā*). With these players introduced, he is now ready to provide the formula that shows how they interact to solve the parallax puzzle.

Notes:

- ***Śaṅkuḥ:*** R·sin(Altitude).
- ***Trijyā:*** Radius (R = 3438′).
- ***Vyāsārdhaṃ:*** Radius (R).
- ***Dṛkṣepajyā:*** R·sin(Zenith Distance) = R·cos(Altitude).
- ***Dṛggatijyā:*** R·sin(Parallax angle).
- ***Karṇa:*** True geocentric distance (relevant for parallax magnitude).

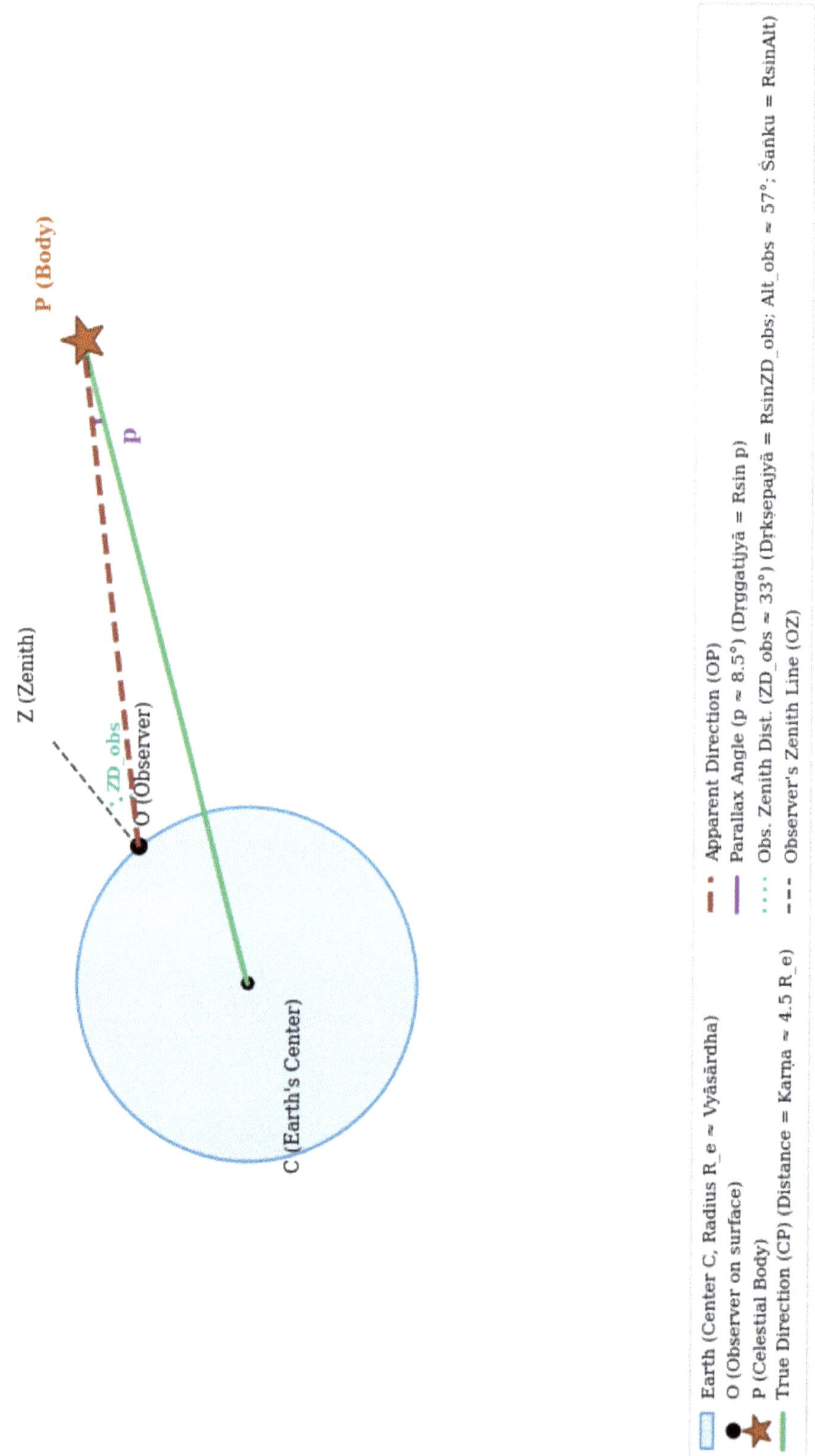

Figure 6.28: Gola.32 - Geometry of Parallax (Lambana). Depicts the apparent shift in a celestial body's (P) position: the Parallax Angle (p) between its geocentric (from C) and topocentric (from O) directions. Dṛggatijyā = R·sin(p). The diagram also shows observed Zenith Distance (ZD_obs, Dṛkṣepajyā) and related Altitude (Śaṅku). Parallax magnitude is influenced by the true distance Karṇa (CP).

6.9.8 Calculating Parallax Correction

Sanskrit:

दृग्गोलाद् ग्रहार्धोनात् स्वदृक्क्षेपहृताद् दृशः

तद्गुणा दृग्गतिज्या व्यासार्धभक्ता फलं लभ्दम् ॥ ३३ ॥

IAST:

dṛggolād grahārdhonāt svadṛkkṣepahṛtād dṛśaḥ |

tadguṇā dṛggatijyā vyāsārdhabhaktā phalaṃ labdham || 33 ||

(Interpretation requires understanding standard parallax formulae)

Translation (Interpretive, based on common understanding): "The R-sine of the observed zenith distance (svadṛkkṣepaḥ); the R-sine of horizontal parallax (dṛggatijyā) multiplied by this (tadguṇā), and then divided by the radius (vyāsārdhabhaktā), yields [the R-sine of] the parallax correction (phalam), after being diminished by half the [apparent angular size of the] planet (grahārdhonāt). This is the procedure for parallax (lambanavidhau) from the terrestrial sphere (bhūgolāt)."

The Parallax Effect: A Shift in Perspective

Imagine holding your thumb out at arm's length and closing one eye. Look at where your thumb appears against the distant background. Now, switch eyes—close the first eye and open the other. Your thumb will appear to jump to a different position.

This apparent shift is called **parallax**. It happens because your two eyes view your thumb from slightly different angles.

The same thing happens when we observe the Moon or a planet. An astronomer on the surface of the Earth sees the Moon from a slightly different angle than an imaginary observer at the Earth's center would. This difference in viewpoint causes the Moon to appear slightly lower in the sky than its "true" geometric position.

This verse from the *Golapāda* provides a precise mathematical recipe to calculate and correct for this vital effect.

The Key Ingredients

Aryabhata's method requires two key inputs, which he works with as R-sines:

1. **The Observed Zenith Distance:** This is the angular distance from the point directly overhead (the zenith) down to the Moon. The lower the Moon is in the sky, the larger this angle becomes.

2. **The Horizontal Parallax:** This is the *maximum possible parallax effect*, which occurs when the Moon is right on the horizon. This value depends on the Moon's distance from Earth—the closer the Moon, the larger the horizontal parallax.

Aryabhata's Recipe for the Correction

With these two ingredients, Aryabhata gives a simple proportional formula to find the parallax correction for any altitude:

Parallax Correction ≈ (Horizontal Parallax × R·sin(Observed Zenith Distance)) / R

The logic is brilliant and intuitive:

- The correction is at its maximum when the Moon is on the horizon (when the zenith distance is 90° and its R·sine is R).
- The correction is zero when the Moon is directly overhead at the zenith (when the zenith distance is 0°).
- For any position in between, the correction is a fraction of the maximum, determined by the R·sine of the zenith distance.

Once this correction is calculated, it is **added** to the observed altitude to find the "true" altitude as seen from the Earth's center.

An Observational Refinement: The Edge vs. The Center

Aryabhata includes a crucial instruction for practical observation. The Sun and Moon are not points of light; they are visible discs. An observation might be made of the Moon's *upper edge* or *lower edge* (limb), not its center.

He instructs the astronomer to first adjust for this by adding or subtracting the Moon's apparent radius (*grahārdhonāt*, "diminished by half the planet") to get the altitude of the Moon's center *before* applying the parallax correction. This ensures that the calculation is as precise as the observation itself.

Why Parallax is Essential

This correction is not a minor tweak; it is one of the most important in all of astronomy, especially for the Moon. The Moon is so close to Earth that its parallax effect can be as large as one degree—about twice its own apparent diameter.

Without an accurate parallax correction, it would be impossible to:

- **Predict Solar Eclipses:** A solar eclipse is a "cosmic bullseye" where the Sun, Moon, and a location on Earth must align perfectly. The Moon's parallax dramatically shifts its apparent position, so correcting for it is essential to predict where and when an eclipse will be visible.
- **Create Accurate Lunar Tables:** Any chart of the Moon's position must account for parallax to be useful.

Aryabhata's method provides a computationally efficient and geometrically sound rule for a correction that is fundamental to predictive astronomy, demonstrating a sophisticated understanding of observation on a spherical Earth.

Notes:

- ***Dṛggola:*** Observed celestial position framework.
- ***Grahārdha:*** Apparent semi-diameter (possible refinement).
- ***Svadṛkkṣepa[jyā]:*** R·sin(Observed Zenith Distance).
- ***Dṛggatijyā:*** R·sin(Horizontal Parallax angle).
- ***Vyāsārdha:*** Radius (R = 3438′).
- ***Phalaṃ:*** Correction result (parallax in altitude).

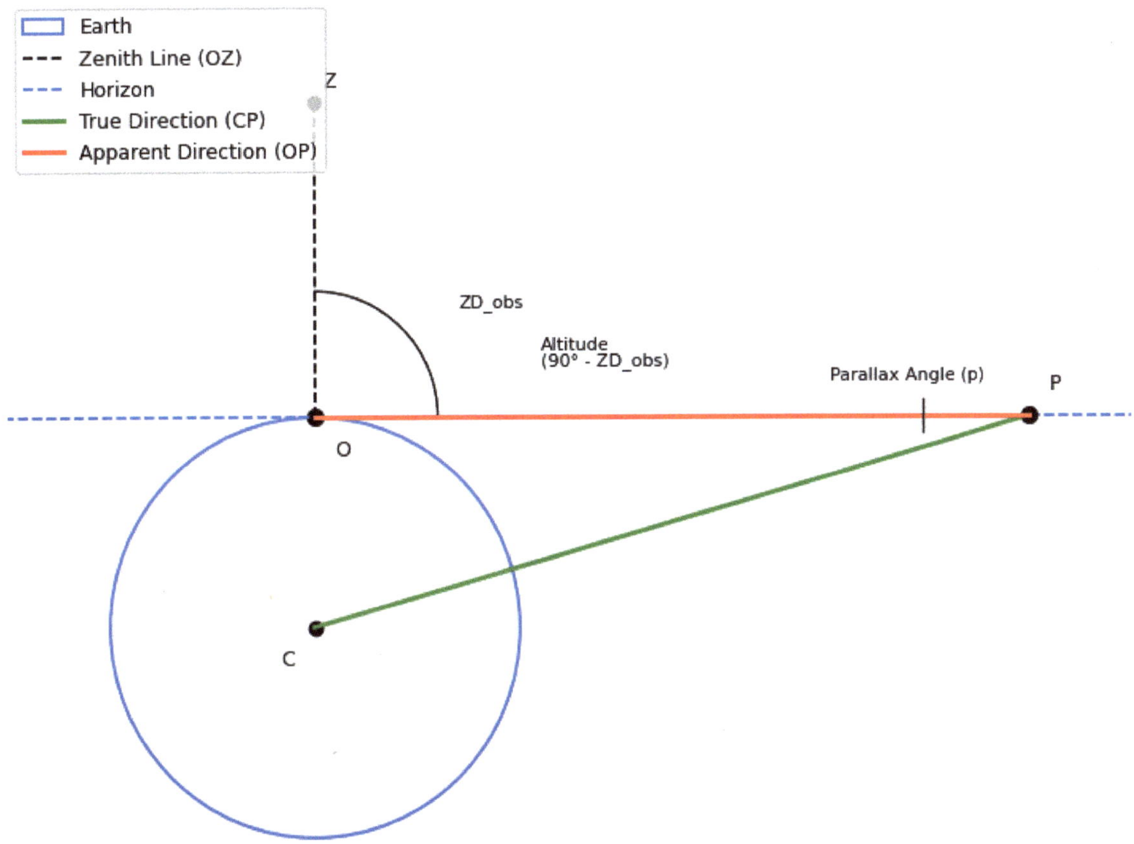

Figure 6.29: Geometry of Parallax (Gola.32-33). Illustrates an observer (O) on Earth (center C) viewing celestial body (P). The Parallax Angle (p) at P is the difference between the true geocentric direction (CP, green) and the apparent topocentric direction (OP, red). Shows observed Zenith Distance (ZD_obs, angle ZOP) and Altitude, with parallax causing P to appear lower than its geocentric direction would indicate from O.

6.9.9 Horizon Dip and Refinements (*Drkchāyā*)

Sanskrit:

भूव्यासार्धोनेन दृग्गोलेन ग्रहा दृष्याः समाः

तस्माद् दृक्छाँया स्याद् ग्रहपरिमाणाच् च संशुद्धा ॥ ३४ ॥

IAST:

bhūvyāsārdhonena dṛggolena grahā dṛśyāḥ samāḥ |

tasmād dṛkchāyā syād grahaparimāṇāch cha saṃśuddhā || 34 ||

Translation (Interpretive): "The effect of the Earth's radius (bhūvyāsārdhakṛtiḥ) on the visible portion of the sky (dṛggolāt), from which celestial bodies become visible (grahā dṛśyāḥ), gives rise to the 'shadow of vision' (dṛkchāyā). This [effect] is to be refined/corrected (saṃśuddhā) based on the apparent size of the celestial body (grahaparimāṇāt)."

The Final Refinement: The Reality of the Horizon

After building a sophisticated geometric model of the heavens, Aryabhata addresses a final, crucial question: what other real-world factors affect the precise moment a planet appears or disappears at the horizon?

This verse provides the final layer of refinement, moving from a perfect, theoretical model to the practical realities of observation. It introduces two key concepts: the "shadow of vision" and a correction for the apparent size of celestial bodies.

1. The Earth's Own Shadow: Horizon Dip

The verse first alludes to an effect caused by the "action of the Earth's radius" on what we can see. This is a direct reference to the phenomenon of **horizon dip**.

Because we live on a curved Earth, our true horizon—the physical line where sea meets sky—is always slightly *below* the idealized geometric horizon. Our own planet's bulk creates a "shadow of vision" (*dṛkchāyā*), hiding a small sliver of the sky. This means that for a planet to become visible, it must rise slightly *above* the perfect geometric horizon to clear the Earth's physical curvature.

2. The Apparent Size of the Sun and Moon

Next, Aryabhata states that this horizon effect must be "refined" or "corrected" (*saṃśuddhā*) based on the "apparent size" (*grahaparimāṇāt*) of the celestial body. This is a critical adjustment, especially for the Sun and Moon, which are not tiny points of light but have visible discs.

Think about a sunrise. We say the Sun has "risen" the moment its **upper edge (limb)** first peeks over the horizon. At that instant, the *center* of the Sun is still hidden below the horizon. To make an accurate prediction, an astronomer must account for the Sun's apparent radius. This "refinement" ensures that the calculation matches the real-world definition of a sunrise or sunset.

A Hint of a Deeper Correction?

This instruction to "refine" the calculation may also have served as a practical way to account for another complex near-horizon effect: **atmospheric refraction**. The Earth's atmosphere acts like a lens, bending light and making celestial objects appear slightly *higher* in the sky than they actually are.

While Aryabhata does not offer a specific theory of refraction, this "refinement" clause could have been an instruction to apply an empirically derived correction—a small adjustment known from observation to make the final prediction match reality. It would have effectively bundled the effects of horizon dip, apparent size, and atmospheric refraction into a single, practical fine-tuning step.

Significance: From Theory to Observation

This verse showcases a deep awareness of the challenges of practical astronomy. It demonstrates that Aryabhata was not content with an idealized geometric model. He understood that to be truly useful, a theory must be refined to account for the messy details of real-world observation. It is a final testament to his goal of creating a system that was not just mathematically elegant, but also predictively accurate.

Notes:

- ***Bhūvyāsārdha:*** Earth's radius (its effect on visibility).
- ***Dṛggola:*** Observed celestial sphere/horizon.
- ***Dṛkchāyā:*** Horizon dip, possibly including empirical adjustment for refraction.
- ***Grahaparimāṇa:*** Apparent size of the celestial body.
- ***Saṃśuddhā:*** Refined/Corrected.

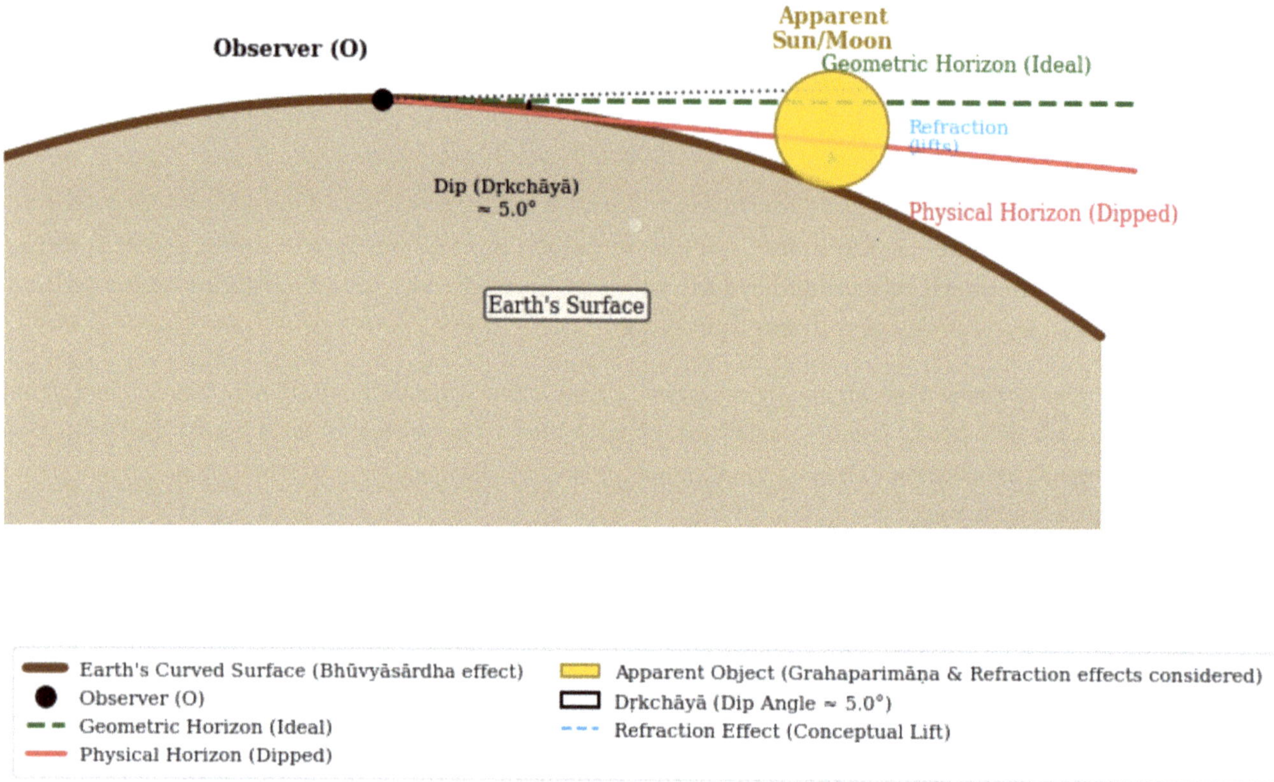

Figure 6.30: Horizon Effects at Sunrise/Sunset (Gola.34, effects exaggerated). Illustrates Horizon Dip (Dṛkchāyā) where the Physical Horizon (red) is below the Geometric (green dashed) due to Earth's curvature. Atmospheric refraction (lifting Sun/Moon, gold disk) and the object's apparent size (Grahaparimāṇa, e.g., limb rising) combine for the final observed rising/setting.

6.10 Positional Corrections & Eclipse Causation (Gola.35-37)

Before tackling the detailed mechanics of eclipses, Aryabhata introduces two further subtle corrections (***dṛkkarma***) needed to refine the calculated longitude of a celestial body, especially when it's near the horizon. He then makes his definitive statement on the true, physical cause of eclipses, marking a watershed moment in rational astronomy.

6.10.1 Latitude Visibility Correction - *Aksadrkkarma*

When a celestial body possessing celestial latitude (i.e., it is not precisely on the ecliptic) is observed near the horizon, its apparent ecliptic longitude can differ slightly from its calculated true ecliptic longitude. Golapāda Verse 35 introduces the **Akṣadṛkkarma**, a correction to account for this perspective effect, which arises from the interplay between the object's celestial latitude and the observer's terrestrial latitude.

Sanskrit:

विक्षेपगुणाक्षज्या लम्बकभक्ता भवेद् ऋणम् उदक्स्थे ।

उदये धनम् अस्तमये दक्षिणेगे धनं ऋणं चन्द्रे ॥ ३५ ॥

IAST:

vikṣepaguṇākṣajyā lambakabhaktā bhaved ṛṇam udaksthe |

udaye dhanam astamaye dakṣiṇage dhanaṃ ṛṇaṃ chandre || 35 ||

The Latitude Correction: *Akṣadṛkkarma*

The ecliptic is the great highway of the solar system, but planets, and especially the Moon, don't always stick perfectly to the road. They often wander slightly above or below it, a deviation known as **celestial latitude**.

This raises a critical question for an observer on Earth:

> **If a planet is, say, one degree *above* the ecliptic, how does my own position on Earth—my terrestrial latitude—affect where I *perceive* that planet to be along the ecliptic as it rises or sets?**

This is the puzzle solved by the **Akṣadṛkkarma**, or "Latitude-Visibility Correction."

The Cause: A Tale of Two Latitudes

To grasp this, imagine the ecliptic is a perfectly flat, transparent glass ring tilted in the sky. A planet is a drone flying one foot *above* this glass ring.

- If you stand directly under the center of the tilted ring (at the Earth's equator), the drone's shadow on the glass shows its "true" ecliptic longitude.
- But now, move to a position far to the side (e.g., London or another high terrestrial latitude). When you look at the drone as it just comes over your horizon, its position will seem to be shifted along the glass ring. From your angled viewpoint, its shadow appears to fall in a slightly different place.

This apparent shift—the *Akṣadṛkkarma*—is caused by the interplay of two things: the drone's height above the track (celestial latitude) and your own off-center viewing position (terrestrial latitude).

The Formula for the Correction

Aryabhata provides a formula that accounts for both factors. He states that the correction is found by multiplying the **R·sine of the planet's celestial latitude** by the **R·sine of the observer's terrestrial latitude**, and then dividing by the standard radius.

In modern terms, the correction (Δλ) is calculated as:

$$\mathbf{R \cdot \sin(\Delta\lambda) \approx [\ R \cdot \sin(\beta) \times R \cdot \sin(\varphi)\ ] / R}$$

Where:

- β (*vikṣepa*) is the planet's celestial latitude.
- φ (*akṣa*) is the observer's terrestrial latitude.
- **R** (*trijyā*) is the standard radius (3438').

The logic of the formula is clear: the correction is largest when *either* the planet is very far from the ecliptic (a large β) *or* the observer is very far from the Earth's equator (a large φ).

Applying the Correction: The Rules of Rising and Setting

After calculating the size of the correction, Aryabhata gives a set of rules for whether to add (*dhanam*) or subtract (*ṛṇam*) it. The rule depends on two conditions:

1. Whether the planet is **rising** (*udaye*) or **setting** (*astamaye*).
2. Whether its celestial latitude is **North** or **South**.

For an observer in the Northern Hemisphere, the rule for a planet North of the ecliptic is to *subtract* the correction at rising and *add* it at setting. The rules are reversed for a planet South of the ecliptic.

Why the Moon is a Special Case

The verse specifically mentions applying this correction to the "true Moon" (*tatsphuṭachandre*). This is because the Moon has a large celestial latitude (up to $\pm 5°$), making this correction significant and essential for accurate predictions of its position, especially at moonrise and moonset. For other planets with smaller latitudes, the effect is less pronounced but still important for precision work.

Notes:

- ***Vikṣepa:*** Celestial latitude (or its sine).
- ***Akṣajyā:*** R·sin(observer's latitude, φ).
- ***Lambaka:*** R·cos(observer's latitude, φ).
- ***Ṛṇam/Dhanam:*** Negative/Positive correction to longitude.
- ***Chandre:*** Especially important for the Moon.

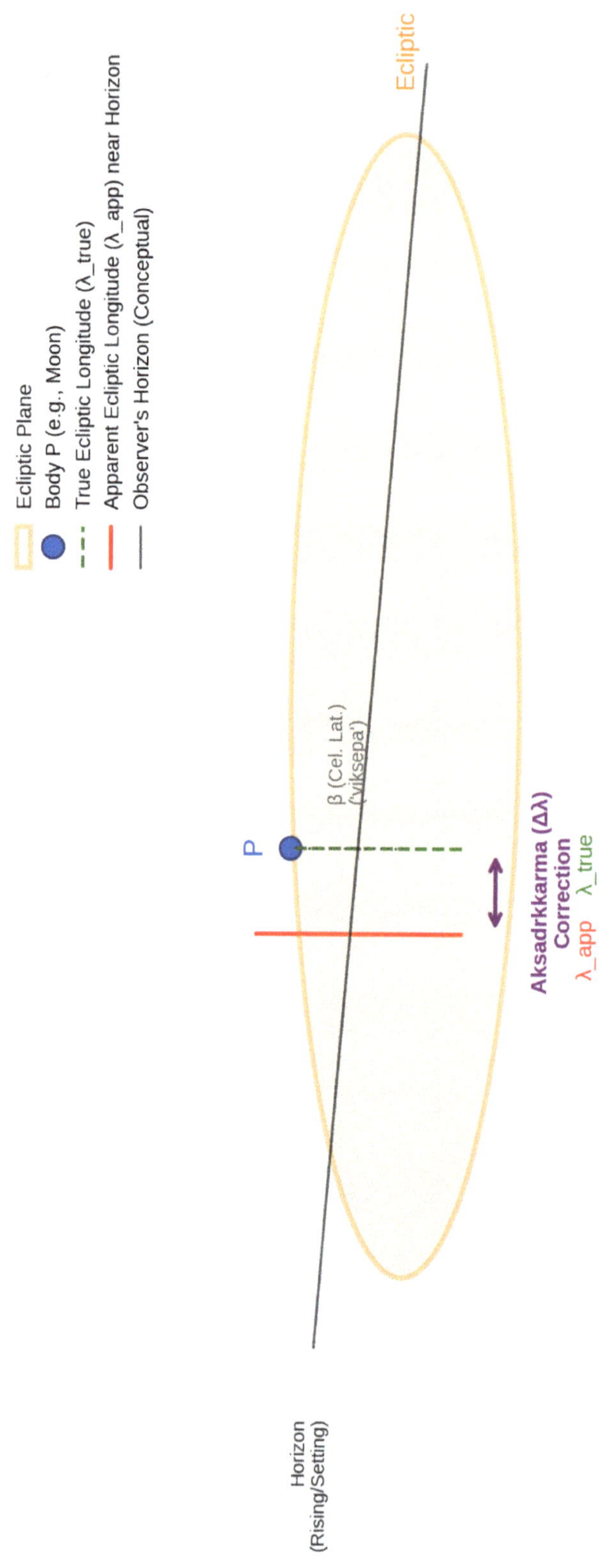

Figure 6.31: Akṣadṛkkarma – Longitude Correction near Horizon (Gola.35). Illustrates how Akṣadṛkkarma (Δλ) adjusts a body's true ecliptic longitude (λ_true) to its apparent longitude (λ_app) due to its celestial latitude (β) and observer's latitude, for precise horizon observations.

6.10.2 Ecliptic Inclination Correction- *Ayanadrkkarma*

Sanskrit:

विक्षेपापक्रमगुणा उत्क्रमज्या विस्तारार्धकृतिभक्ता ।

उदग् ऋणधनं उदगयने दक्षिणेगे धनं ऋणं याम्ये ॥ ३६ ॥

IAST:

vikṣepāpakramaguṇā utkramajyā vistarārdhakṛtibhaktā |

udagṛṇadhanaṃ udagayane dakṣiṇage dhanaṃ ṛṇaṃ yāmye || 36 ||

The Solstice Correction: *Ayanadṛkkarma*

Imagine a planet is moving perfectly along the ecliptic path, with no deviation. We know its true longitude. But when we observe this planet as it rises or sets near the horizon, a strange thing happens: its *observed* position seems to be slightly ahead of or behind its *calculated* position. Why?

This is the puzzle that the **Ayanadṛkkarma**, or "Solstice-Course Visibility Correction," is designed to solve. It addresses the visual distortion caused by the tilt of the ecliptic itself.

The Cause: A Matter of Perspective

To understand this, let's use a simple analogy.

> **Imagine a vast, circular racetrack that is tilted at a steep angle. You are standing on the ground watching the cars. A 100-foot section of the track at its highest point (the "solstice") will look much shorter to you than a 100-foot section on the side, closer to your eye level (the "equinox"). This is a trick of perspective. Even though the sections of track are equal, their projection onto your field of view is not.**

The ecliptic is this tilted racetrack. The horizon is your viewpoint. Because the ecliptic is tilted at about 24 degrees to the celestial equator, the way it intersects your local horizon changes constantly. This creates a perspective effect that "squashes" or "stretches" the appearance of longitude near the horizon. The *Ayanadṛkkarma* corrects for this visual distortion.

The Recipe for the Correction

Aryabhata provides a concise formula to calculate the size of this correction. The verse states that the correction is found by multiplying the **R·versed-sine of the planet's longitude from a solstice** by the **R·sine of the ecliptic's maximum declination** (i.e., the obliquity, 24°), and then dividing the result.

In modern trigonometric terms, this formula is equivalent to:

$$\textbf{Correction} \approx [\, \text{R·versin}(\lambda\_\text{sol}) \times \text{R·sin}(\varepsilon) \,] \,/\, \text{R}$$

Where:

- **R·versin(λ_sol)** is the R·versed-sine of the planet's longitude measured from the nearest solstice. This term quantifies *where* the planet is on the tilted path.
- **R·sin(ε)** is the R·sine of the ecliptic's obliquity ($\varepsilon \approx 24°$). This term accounts for *how tilted* the path is.
- **R** is the standard radius of the circle used for calculations (3438′).

The verse then gives a crucial sign rule for applying the correction: it is added (*dhanam*) when the planet is on its northern course (from the vernal equinox to the autumnal equinox) and subtracted (*ṛṇam*) when it is on its southern course.

A Tale of Two Corrections: *Ayana-* vs. *Akṣa-dṛkkarma*

It is crucial to understand how this correction works with the *Akṣadṛkkarma* mentioned in the previous verse. They are a sophisticated pair of corrections that account for two entirely different effects.

- ***Akṣadṛkkarma*** corrects for a planet being *off* the ecliptic (having its own celestial latitude). It answers the question: "How does a planet's height above or below the main path affect its observed position?"
- ***Ayanadṛkkarma*** corrects for the *ecliptic path itself being tilted*. It answers the question: "How does the tilt of the main path affect the observed position of a planet, even one that is perfectly on it?"

The Goal: Pinpoint Accuracy

By applying both of these "visibility corrections" (*dṛkkarmas*), Aryabhata could achieve an extraordinary level of precision. These adjustments were essential for accurately predicting the exact timing and location of phenomena that happen near the horizon, such as the rising and setting of planets, the precise moment of moonrise, and the calculation of the Ascendant (*Lagna*), a critical point for horoscopy.

Together, these two corrections showcase a profound mastery of spherical geometry and the complex interplay between the celestial grid and an observer's local perspective.

Notes:

- ***Vikṣepa:*** Celestial latitude (sine).
- ***Apakrama:*** Declination (sine) - used in calculation.

- ***Utkramajyā:*** R·versed sine (of longitude from nearest solstice).
- ***Vistarārdhakṛti:*** Square of radius (R^2).
- ***Udagayane/Yāmye:*** During northern/southern solar course (relative to solstices).

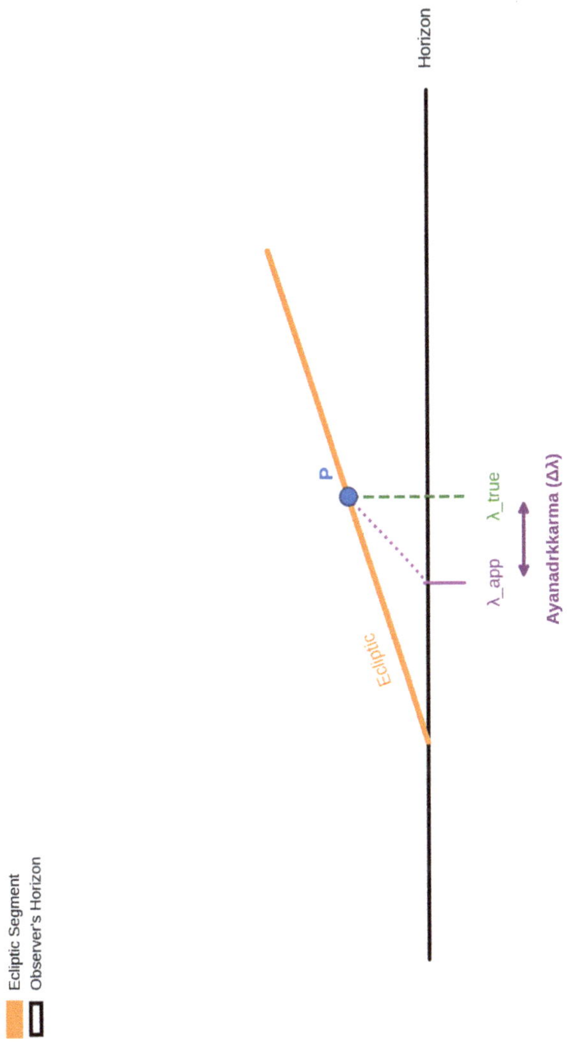

Figure 6.32: Ayanadṛkkarma - Longitude Correction (Gol a.36). This diagram shows the Ayanadṛkkarma (Δλ), which adjusts a body's (P) true ecliptic longitude (λ_true) to its apparent longitude (λ_app) when observed near the horizon. This correction is due to the ecliptic's tilt.

6.10.3 The True Cause of Eclipses - Shadow and Obscuration

Sanskrit:

चन्द्रो जलम् अर्कोऽग्निर् मृद् भूश् छाँयापि या तमस् तत् तु

छादयति शशी सूर्यं शशिनं महती च भूच्छाँया ॥ ३७ ॥

IAST:

chandro jalam arko'gnir mṛd bhūś chāyāpi yā tamaḥ tat tu |

chādayati śaśī sūryaṃ śaśinaṃ mahatī cha bhūcchāyā || 37 ||

Translation: "The Moon is water, the Sun is fire, Earth is earth, and the shadow (***chāyā***) indeed is darkness (***tamaḥ***). The Moon (***śaśī***) covers (***chādayati***) the Sun (***sūryaṃ***), and the great Earth-shadow (***mahatī bhūcchāyā***) covers the Moon (***śaśinaṃ***)."

Interpretation: This is one of the most significant verses in the *Aryabhatiya* and a landmark in the history of science. Aryabhata provides the correct, physical, and geometric explanation for both solar and lunar eclipses, **decisively rejecting prevalent mythological accounts** (e.g., the demon Rāhu swallowing the Sun or Moon).

1. **Physical Entities & Shadow:** While using conventional elemental associations (Moon=water, Sun=fire, Earth=earth), Aryabhata defines the crucial element, the shadow (***chāyā***), simply and physically as darkness (***tamaḥ***) – the absence of light.

2. **Solar Eclipse Cause:** A solar eclipse occurs when the physical body of the **Moon** (***śaśī***) passes between the Earth and the Sun, physically blocking (***chādayati***) the Sun's light (***sūryaṃ***) from reaching the observer.

3. **Lunar Eclipse Cause:** A lunar eclipse occurs when the **Moon** (***śaśinaṃ***) passes into the large (***mahatī***) conical **shadow cast by the Earth** (***bhūcchāyā***) into space on the side opposite the Sun.

Significance: This rational explanation fundamentally relies on:

- Light traveling in straight lines.
- The Earth is a sphere (***bhūgola***, Gola.5-7) capable of casting a substantial shadow.
- The Moon and Sun being opaque physical bodies at different distances from Earth.
- The Moon is orbiting the Earth and periodically aligning with the Sun (New Moon) or the Earth's shadow (Full Moon).

This marks a major conceptual shift towards a scientific understanding of celestial phenomena, replacing myth with a **causal explanation based on geometry and the physics of light and shadow**. It provides the necessary rational foundation for the detailed eclipse calculation methods presented in the following verses.

Notes:

- ***Chandraḥ jalam / Arko'gniḥ:*** Symbolic associations, secondary to the mechanism.
- ***Bhūcchāyā:*** Earth's shadow (a real, geometric entity).
- ***Tamaḥ:*** Darkness (the nature of the shadow).
- ***Chādayati:*** Covers/Obscures/Eclipses.

Gola.37: The True Cause of Eclipses - Shadow and Obscuration

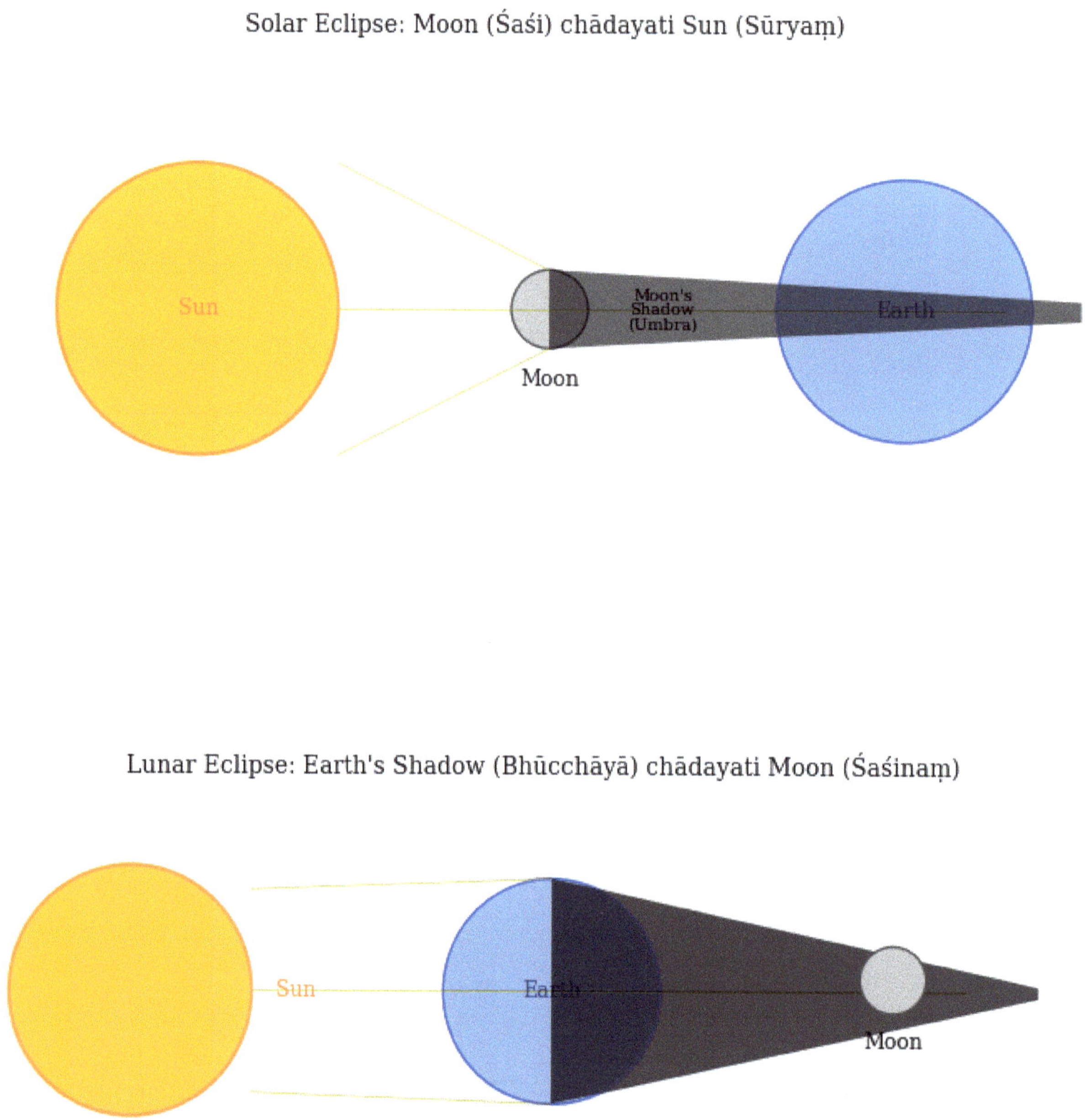

Figure 6.33: Aryabhata's Eclipse Mechanism (Gola.37). Solar Eclipse (Top): Moon's shadow on Earth obscures Sun. Lunar Eclipse (Bottom): Earth's shadow (Bhūcchāyā) on Moon obscures it.

6.11 Eclipse Calculations – Decoding the Shadows (Gola.38-48)

Building on the correct physical understanding of eclipses (Gola.37), Aryabhata now presents the sophisticated geometric and computational methodology required to predict their specific circumstances: exact timing, duration of partial and total phases, maximum magnitude (how much is covered), and visual orientation. This section exemplifies the practical power of his integrated mathematical and astronomical system.

6.11.1 Conditions and Timing of Mid-Eclipse

Sanskrit:

स्फुटशशिमासान्तेऽर्कं पातासन्नो यदा प्रविशतीन्दुः ।

भूच्छाँयां पक्षान्ते तदाधिकोनोनं ग्रहणमध्यम् ॥ ३८ ॥

IAST:

sphuṭaśaśimāsānte'rkaṃ pātāsanno yadā praviśatīnduḥ |

bhūcchāyāṃ pakṣānte tadādhikononaṃ grahaṇamadhyam || 38 ||

Translation: "When the Moon (*induḥ*), being near a node (***pātāsannaḥ***), comes to the Sun (*arkaṃ*) at the end of the true lunar month [new moon] (***sphuṭaśaśimāsānte***), or enters the Earth's shadow (*bhūcchāyāṃ*) at the fortnight's end [full moon] (***pakṣānte***), then addition or subtraction (***adhikononaṃ***) [is applied to find] the mid-eclipse time (***grahaṇamadhyam***)."

Problem: When can eclipses happen, and how do we find the time of maximum eclipse?

Solution: This verse sets the fundamental conditions and notes the need for timing corrections.

1. **Phase Requirement:** Eclipses are tied to specific lunar phases:
   - **Solar Eclipse:** Possible only at **New Moon** (***sphuṭaśaśimāsānte***), when Moon is between Earth and Sun.
   - **Lunar Eclipse:** Possible only at **Full Moon** (***pakṣānte***), when Moon enters Earth's shadow opposite the Sun.
2. **Nodal Proximity Requirement (*Pātāsannaḥ*):** Alignment in phase isn't enough. The Moon must also be **near one of its orbital nodes** (points where its path crosses the ecliptic). This ensures its celestial latitude (***vikṣepa***) is small enough for it to actually cover the Sun or enter the shadow. If the Moon is too far from a node at New/Full Moon, it passes above or below the Sun/shadow, and no eclipse occurs.
3. **Mid-Eclipse Timing (*Grahaṇamadhyam*):** The theoretically calculated time of exact New Moon (con-

junction) or Full Moon (opposition) is usually the starting point. However, this needs to be adjusted (***adhikononaṃ*** = addition or subtraction) to find the actual observed time of maximum eclipse. These corrections primarily account for **parallax** (especially parallax in longitude, *lambana*), which shifts the apparent time of conjunction/opposition for the observer on Earth's surface compared to the geocentric time.

Notes:

- ***Sfuṭaśaśimāsānte:*** End of true lunar month = New Moon (solar eclipse).
- ***Pakṣānte:*** End of fortnight = Full Moon (lunar eclipse).
- ***Pātāsannaḥ:*** Near a node (latitude condition crucial).
- ***Adhikononaṃ:*** Addition/subtraction (corrections like parallax needed for observed time).
- ***Grahaṇamadhyam:*** Middle of the eclipse (time of maximum phase).

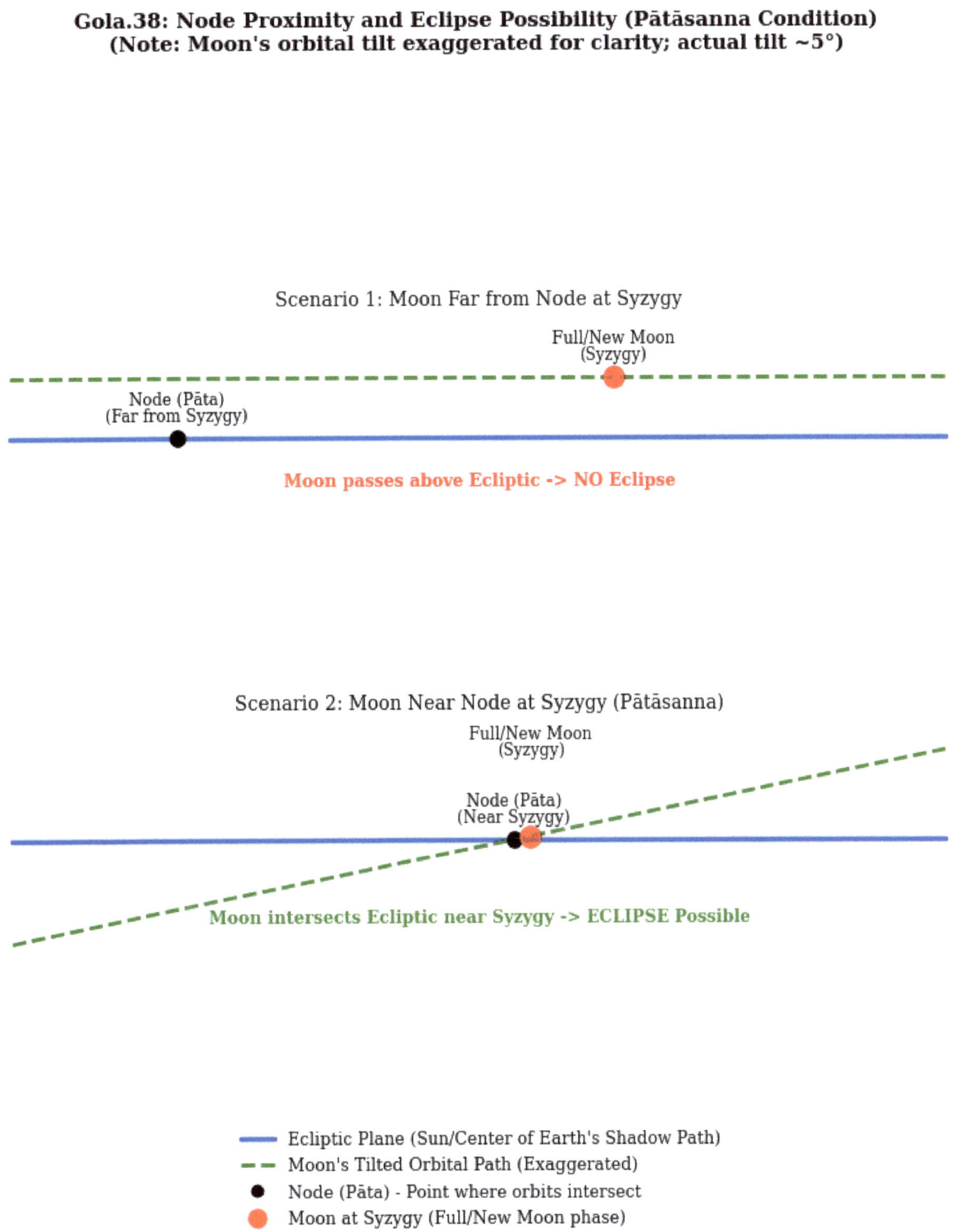

Figure 6.34: Node Proximity (Pātāsanna) for Eclipses (Gola.38). Illustrates why eclipses require the Moon to be near a Node (Pāta, orbit intersection) at Syzygy (New/Full Moon). Top: Moon far from Node at Syzygy = NO eclipse. Bottom: Moon near Node at Syzygy = ECLIPSE possible. (Moon's orbital tilt exaggerated).

6.11.2 Calculating the Length of Earth's Shadow Cone

Sanskrit:

रविभूविवरभक्तो रविग्रहयोः परिधियोगार्धम् ।

भूगोलविष्कम्भगुणं भूच्छाँयादीर्घत्वम् एतत् ॥ ३९ ॥

IAST:

ravibhūvivarabhakto ravigrahayoḥ paridhiyogārdham |

bhūgolaviṣkambhaguṇaṃ bhūcchāyādīrghatvam etat || 39 ||

(Interpretation relies on correcting the literal reading based on geometric necessity and commentary)

Problem: How far does the Earth's full shadow (umbra) extend into space?

Solution: This verse aims to calculate the length (***dīrghatva***) of Earth's umbral shadow cone (***bhūcchāyā***). While the literal reading of the calculation (*paridhiyogārdham*, "half sum of circumferences") is geometrically problematic for finding the shadow length, the underlying principle, clarified by commentators and geometric logic, uses **similar triangles** formed by the Sun, Earth, and the apex of the shadow cone.

Implied Formula (Similar Triangles):

Shadow Length (L) ≈ (Earth Diameter × Sun-Earth Distance) / (Sun Diameter - Earth Diameter)

L ≈ (D_E × Dist_SE) / (D_S - D_E)

Where:

- D_E = Earth's Diameter (1050 yojanas, Gk.7)
- D_S = Sun's Diameter (4410 yojanas, Gk.7)
- Dist_SE = Sun-Earth distance (*ravibhūvivara*, derived from mean motions/orbital sizes)

Knowing this shadow length is the first step in determining the shadow's properties at the Moon's distance for lunar eclipse calculations.

Notes:

- ***Bhūcchāyādīrghatvam:*** Length of Earth's shadow.
- Formula derived from similar triangles, correcting the literal text based on context.
- Uses diameters from Gk.7 and assumed/calculated distances.

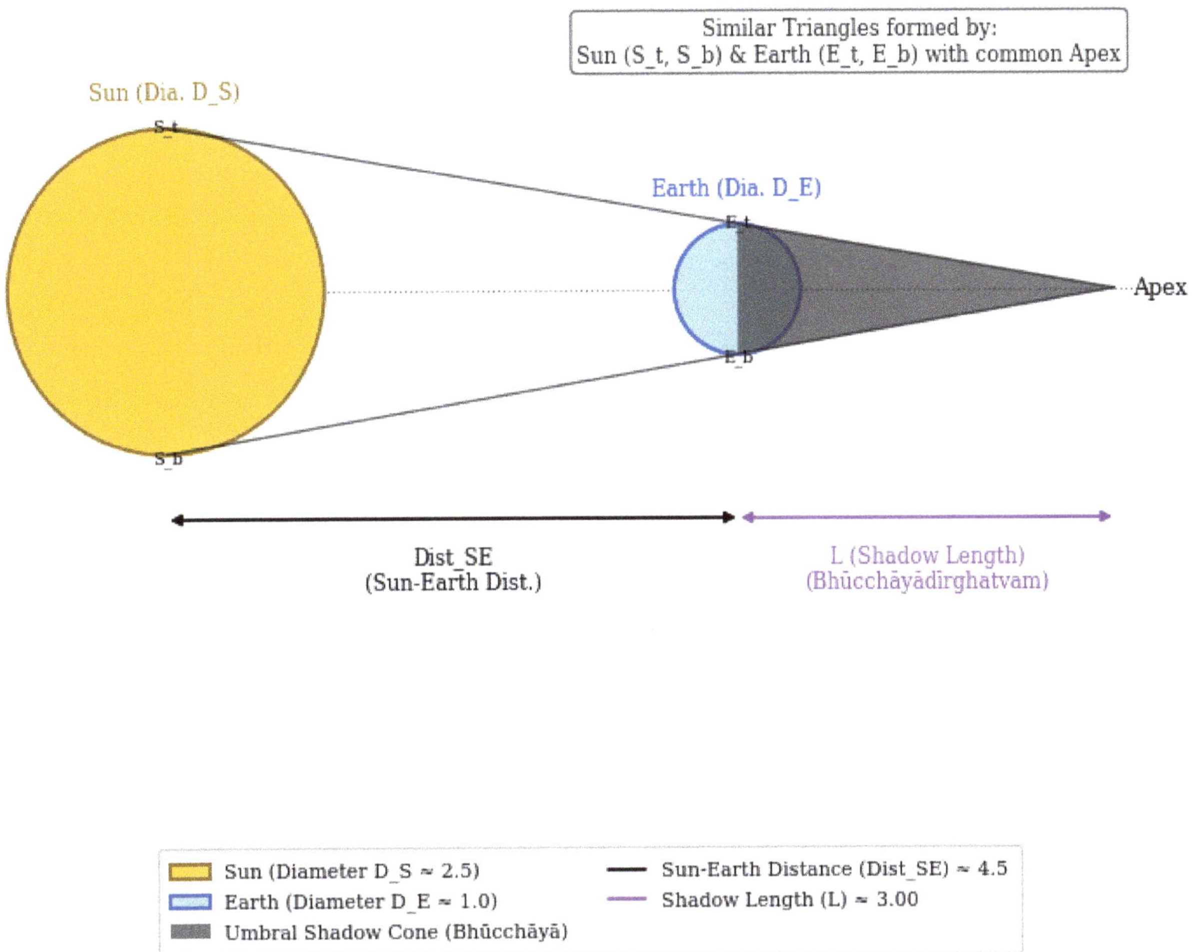

Figure 6.35: Geometric diagram illustrating the calculation of the length (L) of Earth's umbral shadow cone (Bhūcchāyādīrghatvam), as described in Golapāda Verse 39. The method relies on similar triangles formed by the Sun (diameter D_S), the Earth (diameter D_E), the Sun-Earth distance (Dist_SE), and the shadow extending to its Apex. Knowing this shadow length is foundational for lunar eclipse predictions.

6.11.3 Calculating the Shadow's Diameter at the Moon's Distance

Sanskrit:

छाँयाग्रहे विवरगते भूविवराच्छशिविवरकृतिभागे ।

भूव्यासगुणो लभ्दस्तमसो विष्कम्भ इन्दोः सः ॥ ४० ॥

IAST:

chāyāgrahe vivaragate bhūvivarāc chaśivivarakṛtibhāge |

bhūvyāsaguṇo labdhas tamaso viṣkambha indoḥ saḥ || 40 ||

Problem: How wide is the Earth's full shadow (umbra) where the Moon passes through it during a lunar eclipse?

Solution: This verse provides the method to calculate the diameter of Earth's umbral shadow (***tamaso viṣkambha*** = diameter of darkness) at the Moon's distance (***indoḥ***). This width is crucial because it determines whether a lunar eclipse will be total or partial, and for how long. The calculation again uses similar triangles, comparing the full shadow cone (length L from Gola.39) with the truncated cone at the Moon's distance (Dist_EM).

Formula (Similar Triangles):

Shadow Diameter at Moon (D_Sh) = Earth Diameter × (Shadow Length - Earth-Moon Distance) / Shadow Length

D_Sh = D_E × (L - Dist_EM) / L

Where:

- D_E = Earth's Diameter (1050 yojanas)
- L = Shadow Length (from Gola.39 calculation)
- Dist_EM = Earth-Moon distance (*śaśivivara*, mean or true distance)

Notes:

- ***chāyāgrahe vivaragate:*** "at the distance [of the Moon] during a shadow eclipse."

- ***bhūvivarāc chaśivivarakṛtibhāge:*** Refers to the term (L - Dist_EM) / L.
- ***tamaso viṣkambha indoḥ:*** The umbra's diameter at the Moon.
- ***bhūvyāsaguṇo:*** Multiplied by Earth's diameter.

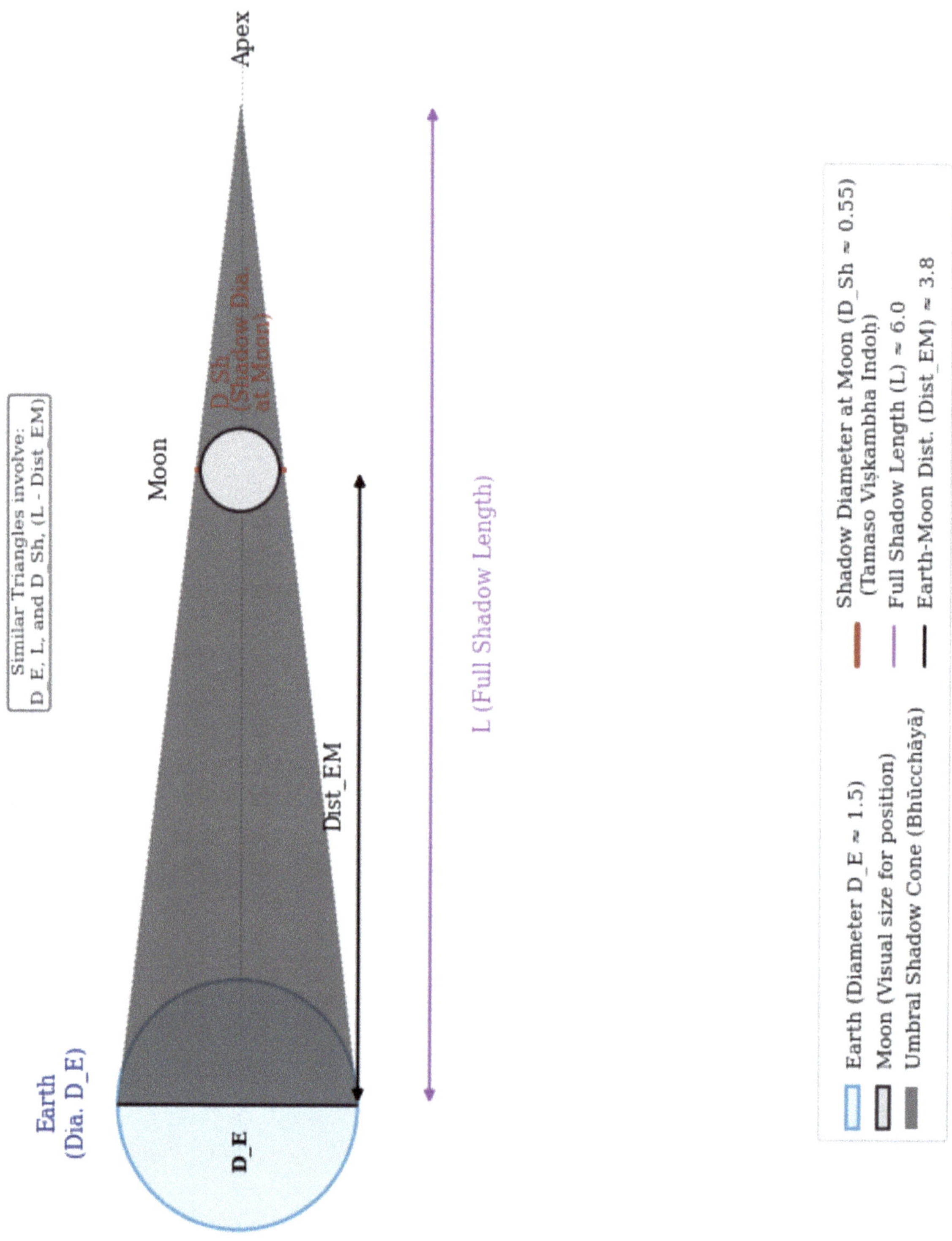

Figure 6.36: Earth's Umbral Shadow Diameter at Moon's Distance (Gola.40). Illustrates calculating the shadow diameter (D_Sh, Tamaso Viṣkambha Indoḥ) where the Moon intersects Earth's shadow cone (Bhūcchāyā). Gola.40's method uses similar triangles formed by Earth's diameter (D_E), full shadow length (L), D_Sh, and remaining shadow length (L - Dist_EM) from the Moon's distance (Dist_EM).

6.11.4 Radii for Eclipse Duration Calculations

Sanskrit:

ग्राहकमाण्डलयोर् योगोऽर्धं स्थितिः स्यात् तयोर् विशेषार्धम् ।

विमर्दः कालो ज्ञेयो ग्रहणस्य तु मध्यकालस्य ॥ ४१ ॥

IAST:

grāhakamaṇḍalayor yogaḥ ardhaṃ sthitiḥ syāt tayor viśeṣārdham |

vimardaḥ kālo jñeyo grahaṇasya tu madhyakālasya || 41 ||

Problem: How do we set up the geometry to calculate the duration of the partial and total phases of an eclipse?

Solution: This verse defines two crucial effective radii based on the sizes of the **eclipsing body (*grāhaka*)** and the **eclipsed body (*maṇḍala*** or ***grāhya*)**. These radii determine the geometric conditions for contact and totality/annularity.

1. ***Sthityardha*** **Radius (Radius for Partial Phase):**
   - Calculated as: **R_Sum = (Diameter_Eclipsing + Diameter_Eclipsed) / 2**
   - Represents: Half the sum of the diameters. This is the **maximum distance between the centers** of the two bodies for *any* part of the eclipse (first to last contact) to occur.
   - Determines: The total duration of the eclipse event (*sthiti*).
2. ***Vimardārdha*** **Radius (Radius for Total/Annular Phase):**
   - Calculated as: **R_Diff = |Diameter_Eclipsing - Diameter_Eclipsed| / 2**
   - Represents: Half the absolute difference of the diameters. This is the **maximum distance between the centers** for the eclipsed body to be *fully* immersed within the eclipsing body (or its shadow).
   - Determines: The duration of the total or annular phase (***vimardaḥ kālaḥ***). Totality/annularity only occurs if the eclipsing body appears larger than the eclipsed body (D_Grāhaka > D_Maṇḍala).

These radii are the key inputs for the Pythagorean calculations in the next two verses, which convert these geometric limits into time durations.

Eclipse Type	Grāhaka (Eclipsing)	Maṇḍala/Grāhya (Eclipsed)
Lunar	Earth's Shadow (D_Sh)	Moon (D_Moon)
Solar	Moon (D_Moon)	Sun (D_Sun)

Notes:

- ***grāhaka / maṇḍala:*** Eclipsing / eclipsed body.
- ***yogaḥ ardhaṃ:*** Half-sum of diameters → ***sthityardha*** radius (sets limit for any contact).
- ***viśeṣārdham:*** Half-difference of diameters → ***vimardārdha*** radius (sets limit for totality/annularity).
- ***vimardaḥ kālaḥ:*** Duration of the total/annular eclipse phase.
- ***madhyakālasya:*** Times are calculated relative to the eclipse midpoint.

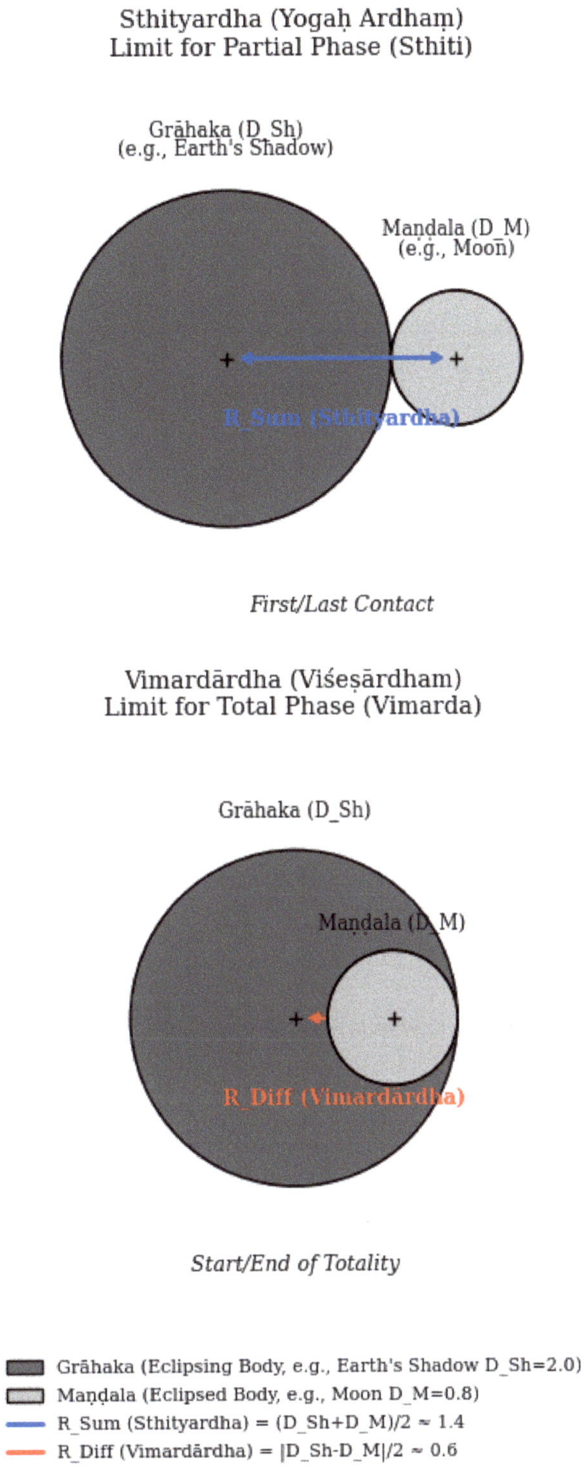

Figure 6.37: Eclipse Radii: Sthityardha (half sum of diameters) and Vimardārdha (half absolute difference of diameters), crucial for duration calculations (Gola.41). The former defines partial phase (Sthiti) limits, the latter total phase (Vimarda) limits (if eclipsing body is larger).

6.11.5 Calculating the Half-Duration of the Partial Eclipse (*Sthityardha*)

Sanskrit:

स्थित्यर्धवर्गद् विक्षेपवर्गं शोधयित्वा पदम् दिनः ।

षष्ट्या गुणितं भुक्त्यन्तरभक्तं स्थित्यर्धनाडिकाः स्युः ॥ ४२ ॥

IAST:

sthityardhavargād vikṣepavargaṃ śodhayitvā padam dinaḥ |

ṣaṣṭyā guṇitaṃ bhuktyantarabhaktaṃ sthityardhanāḍikāḥ syuḥ|| 42 ||

Problem: How long does the eclipse last from first contact to mid-eclipse (or mid-eclipse to last contact)?

Solution: This verse uses the Pythagorean theorem and the relative speed of the bodies to calculate the half-duration of the entire eclipse event (including partial phases) in time units (***nāḍikās***).

1. **Inputs:**
    - ***Sthityardha*** Radius (R_Sum from Gola.41).
    - Moon's celestial latitude (***vikṣepa***) relative to the ecliptic (and thus relative to the center of the Sun or Earth's shadow) at the time of mid-eclipse. This represents the minimum separation between the centers during the event.
    - Relative angular speed (***bhuktyantara***) between the Moon and the Sun (for solar eclipse) or the Moon and the center of the Earth's shadow (for lunar eclipse), measured in arc-minutes per *nāḍikā* (or per day, requiring conversion).
2. **Calculate Half-Path Length:** Imagine a right triangle where the hypotenuse is R_Sum, and one leg is the latitude ***vikṣepa***. The other leg represents the distance the Moon travels relative to the eclipsing body's center from the point of first contact to the point of closest approach (mid-eclipse).**Half-Path = $\sqrt{}$(R_Sum^2 - Vikṣepa^2)**(An eclipse occurs only if ***vikṣepa*** $\leq$ R_Sum).
3. **Convert Path to Time:** Divide the angular path length by the relative angular speed to get the time duration. The verse includes multiplication by 60, likely related to converting units (perhaps from degrees/day to minu tes/nāḍikā).**Half-Duration (Sthityardha Nāḍikās) = (Half-Path × 60) / Relative Speed (units adjusted)**

Notes:

- ***Sthityardhavargād:*** From the square of the ***sthityardha*** radius (R_Sum).

- ***Vikṣepavargaṃ:*** Square of Moon's latitude at mid-eclipse.
- ***Padam:*** Square root (gives the half-path length in angular units).
- ***Bhuktyantara:*** Difference in daily motions = relative angular speed.
- ***Sthityardhanāḍikāḥ:*** Half-duration of the entire eclipse (partial included) in ***nāḍikās***.

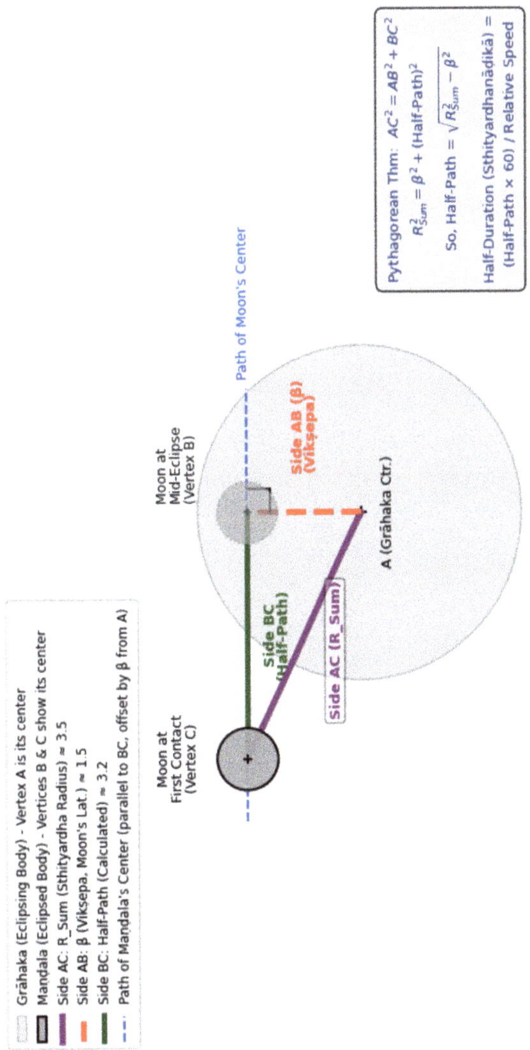

Figure 6.38: Determining Partial Eclipse Duration (Sthiti) via Half-Path (Gola.42). Shows the right triangle ABC for calculating the Moon's half-path (BC) during the partial phase. AC is the Sthityardha Radius (R_Sum, Grāhaka-Maṇḍala distance at contact), AB is the Moon's latitude (Vikṣepa, β) at mid-eclipse (B relative to A). BC, found via Pythagorean theorem, determines half the partial duration.

6.11.6 Calculating Half-Duration of Totality (*Vimardārdha*)

Sanskrit:

विमर्दार्धवर्गद् विक्षेपवर्गं विशोध्य मूलं यत् ।

षष्ट्या संगुण्य भुक्त्योर् विशेषेण विभाज्यः कालः ॥ ४३ ॥

IAST:

vimardārdhavargād vikṣepavargaṃ viśodhya mūlaṃ yat |

ṣaṣṭyā saṃguṇya bhuktyor viśeṣeṇa vibhājyaḥ kālaḥ || 43 ||

Problem: If the eclipse is total or annular, how long does that phase last (from start of totality to mid-eclipse, or mid-eclipse to end of totality)?

Solution: This uses the same Pythagorean logic as Gola.42, but applies it to the conditions for totality/annularity, using the ***vimardārdha*** radius (R_Diff).

1. **Inputs:**
    - ***Vimardārdha*** Radius (R_Diff from Gola.41).
    - Moon's latitude (***vikṣepa***) at mid-eclipse.
    - Relative angular speed (***bhuktyor viśeṣeṇa*** = ***bhuktyantara***).
2. **Calculate Half-Path Length (Totality):** This calculates the distance the Moon travels relative to the center while *fully* inside the boundary defined by R_Diff.
    a. **Half-Path_Totality = $\sqrt{}$(R_Diff^2 - Vikṣepa^2)**(Totality/annularity occurs only if ***vikṣepa*** < R_Diff). If ***vikṣepa*** ≥ R_Diff, this calculation yields an imaginary number or zero, meaning no totality.
3. **Convert Path to Time:** Divide the angular path length during totality by the relative angular speed.
    a. **Half-Duration Totality (Vimardārdha Nāḍikās) = (Half-Path_Totality × 60) / Relative Speed**

Notes:

- ***Vimardārdhavargād:*** From the square of the ***vimardārdha*** radius (R_Diff).
- ***Vikṣepavargaṃ:*** Square of the Moon's latitude at mid-eclipse.
- ***Mūlam:*** Square root (gives half-path length during total/annular phase).

- ***Bhuktyor viśeṣeṇa:*** Difference in motions = relative angular speed.
- ***Kālaḥ:*** Time (half-duration of total/annular phase in ***nāḍikās***).

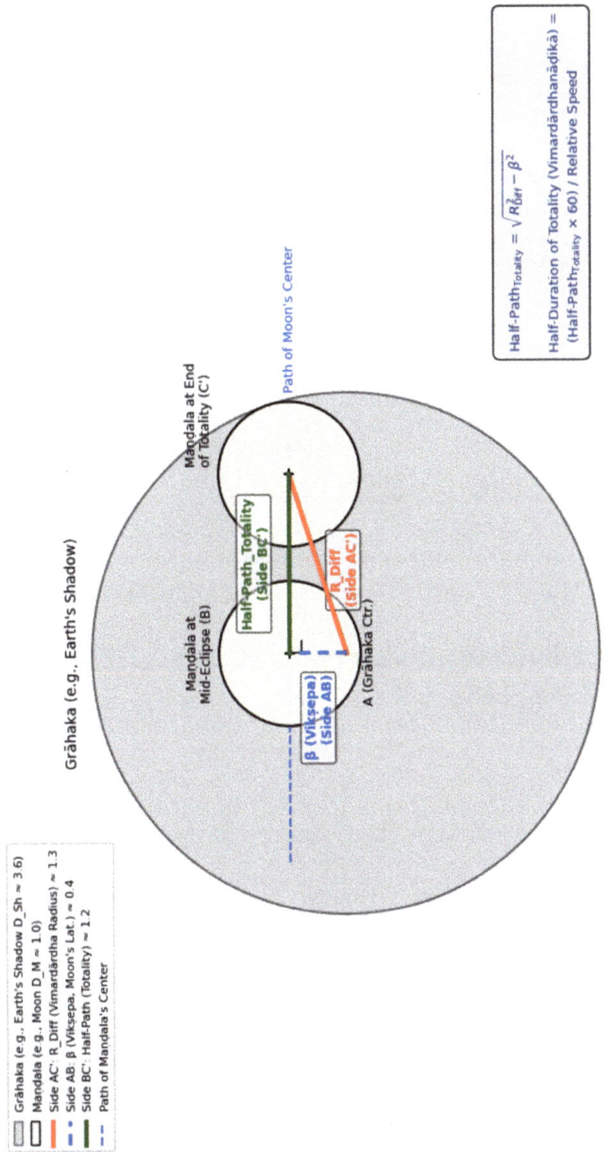

Figure 6.39: Totality Half-Path Geometry (Vimarda) (Gol a.43). Right triangle ABC' determines the Moon's half-path (BC') for totality duration. AC' = Vimardārdha Radius (R_Diff), AB = Moon's latitude (Vikṣepa, β) at mid-eclipse (B). BC' gives half the totality duration (Vimardārdhanāḍikā).

6.11.7 Measuring Eclipse Magnitude (*Grāsa*)

Sanskrit:

स्थित्यर्धाद् विमर्दार्धं शोधितं शेषम् अथ यत् ।

तेन ग्राहकमाण्डलं विभज्य लब्धो ग्रासोऽधिकः ॥ ४४ ॥

IAST:

sthityardhād vimardārdhaṃ śodhitaṃ śeṣam atha yat |

tena grāhakamaṇḍalaṃ vibhajya labdho grāso'dhikaḥ || 44 ||

Problem: How much of the Sun/Moon's diameter is covered at the moment of maximum eclipse?

Solution: This verse calculates the **eclipse magnitude (*grāsa*)**, representing the maximum fraction of the *eclipsed* body's diameter that is obscured by the *eclipsing* body.

1. **Inputs:**
    - Radius Sum (R_Sum from Gola.41) = (D_Grāhaka + D_Maṇḍala) / 2
    - Moon's latitude (***vikṣepa***) at mid-eclipse (minimum center separation).
    - Diameter of the eclipsed body (D_Maṇḍala or D_Grāhya).
2. **Calculate Maximum Obscuration Depth:** This is the linear amount by which the two disks overlap at maximum eclipse. It's the sum of their radii minus their minimum separation.**Max Obscuration = R_Sum - Vikṣepa**
3. **Calculate Magnitude (*Grāsa*):** The magnitude is the ratio of this maximum obscuration depth to the diameter of the body being eclipsed.**Grāsa = Max Obscuration / Diameter_Eclipsed**
    - Lunar Eclipse: Grāsa = (R_Sum - Vikṣepa) / D_Moon
    - Solar Eclipse: Grāsa = (R_Sum - Vikṣepa) / D_Sun

*(Verse Correction Note: The verse literally says "divide by the eclipsing body's disk" (grāhakamaṇḍalaṃ), but Bhaskara I and the geometric context clarify that the divisor must be the diameter of the eclipsed body (grāhyamaṇḍala) to get the conventional magnitude fraction).*A magnitude (grāsa) ≥ 1 indicates a total or annular eclipse.

Notes:

- ***sthityardhād vimardārdhaṃ śodhitaṃ śeṣam atha yat:*** The first line is obscurely phrased but the calculation relies on R_Sum and Vikṣepa.
- ***grāhakamaṇḍalaṃ vibhajya:*** Divisor corrected to eclipsed body's diameter by commentators.
- ***labdho grāso'dhikaḥ:*** The magnitude obtained (must be positive for eclipse).

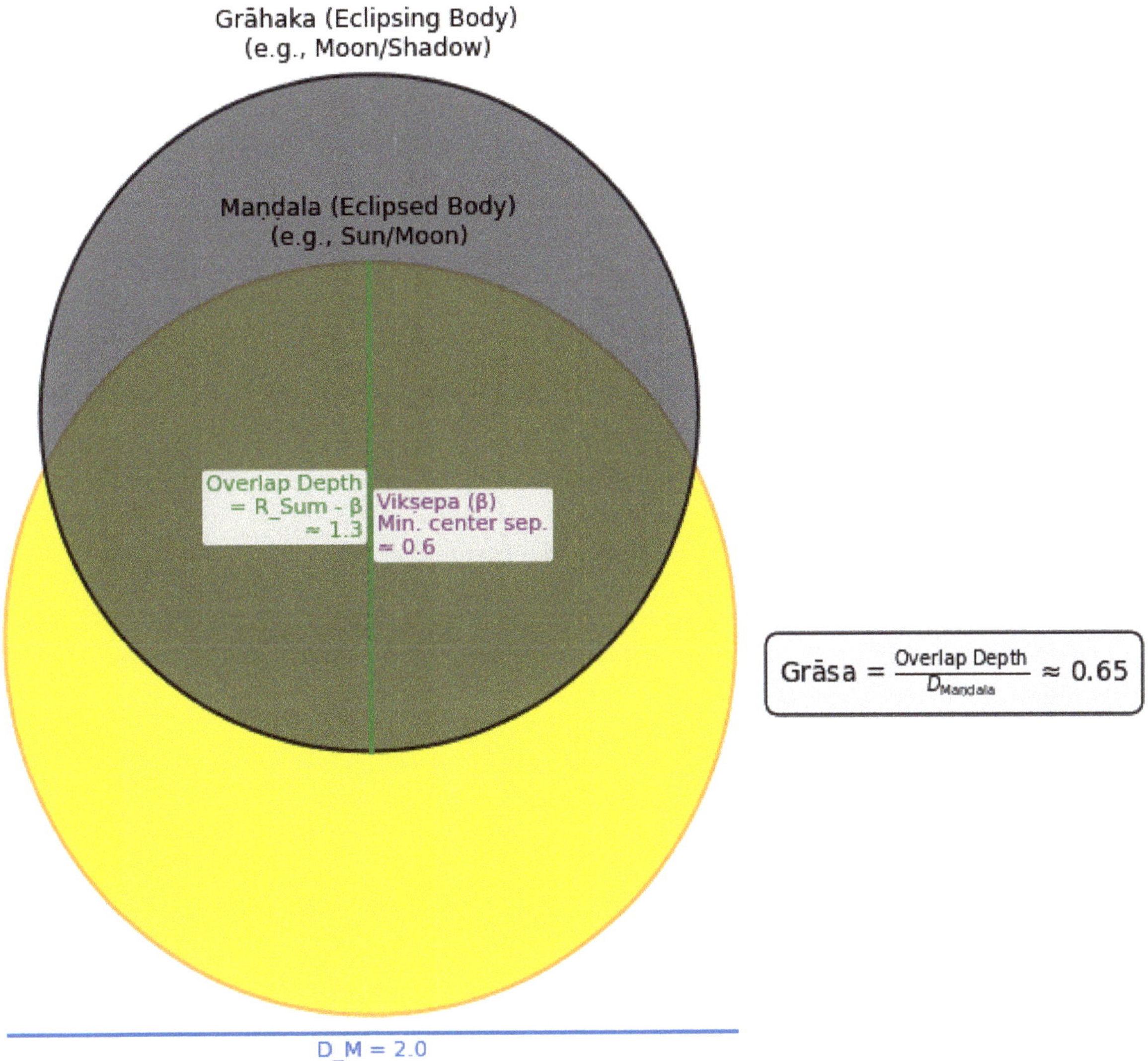

Figure 6.40: Eclipse Magnitude (Grāsa) Calculation (Gola.44). Shows the Maṇḍala (eclipsed) obscured by the Grāhaka. Grāsa equals (Max Obscuration Depth) divided by Diameter of Maṇḍala, where Max Obscuration equals (Radius of Grāhaka plus Radius of Maṇḍala) minus Vikṣepa (beta, min. center separation).

6.11.8 Angular Precision in Eclipse Viewing (*Valana*)

Sanskrit:

अक्षवलनम् अयनवलनं चैवं दृग्ग्रहणकाले ।

साध्यम् अनेन विधानाद् ग्रहपरिलेखस्य दर्शनम् ॥ ४५ ॥

IAST:

akṣavalanam ayanavalanaṃ chaivaṃ dṛggrahaṇakāle |

sādhyam anena vidhānād grahaparilekhasya darśanam || 45 ||

Problem: Where exactly on the Sun/Moon's disk (e.g., top-right, bottom-left) will the eclipse begin and end as seen from my location?

Solution: This verse introduces two crucial angular corrections collectively known as ***valana*** (= deflection or inclination). These are needed to determine the precise orientation of the eclipse contact points on the celestial disk from the observer's perspective. They account for the difference between the coordinate system fixed to the ecliptic (along which latitude is measured) and the observer's local coordinate system (oriented to their horizon or the celestial equator).

1. ***Akṣavalana*** **(Latitude Deflection):** This correction accounts for the apparent rotation of the observer's local sky framework (North point, Zenith) relative to the poles of the *ecliptic* due to the observer's terrestrial latitude (*akṣa*).

2. ***Ayanavalana*** **(Seasonal/Declination Deflection):** This correction accounts for the tilt of the *ecliptic* itself relative to the celestial equator (***ayana*** = Sun's course related to seasons/solstices). Its value depends on the specific point on the ecliptic where the eclipse is occurring.

3. **Calculation:** Both *valanas* are calculated using spherical trigonometry involving the observer's latitude, the ecliptic coordinates of the Sun/Moon at the time of eclipse, and the obliquity of the ecliptic. Their sum gives the total deflection angle.

4. **Purpose (*grahaparilekhasya darśanam*):** Computing the total *valana* allows the astronomer to create an accurate **eclipse diagram (*grahaparilekha*)**. This diagram shows the path of the eclipsing body across the eclipsed body's disk and precisely indicates the angle (relative to North or the Zenith) where the first contact, maximum eclipse, and last contact will occur on the limb. This demonstrates a high level of sophistication in predicting the visual appearance of an eclipse.

Notes:

- ***Akṣavalana:*** Deflection angle due to observer's latitude.

- ***Ayanavalana:*** Deflection angle due to ecliptic inclination/season.
- ***Dṛggrahaṇakāle:*** At the time of the observed eclipse.
- ***Grahaparilekha:*** Diagram/representation of the eclipse progression and orientation.

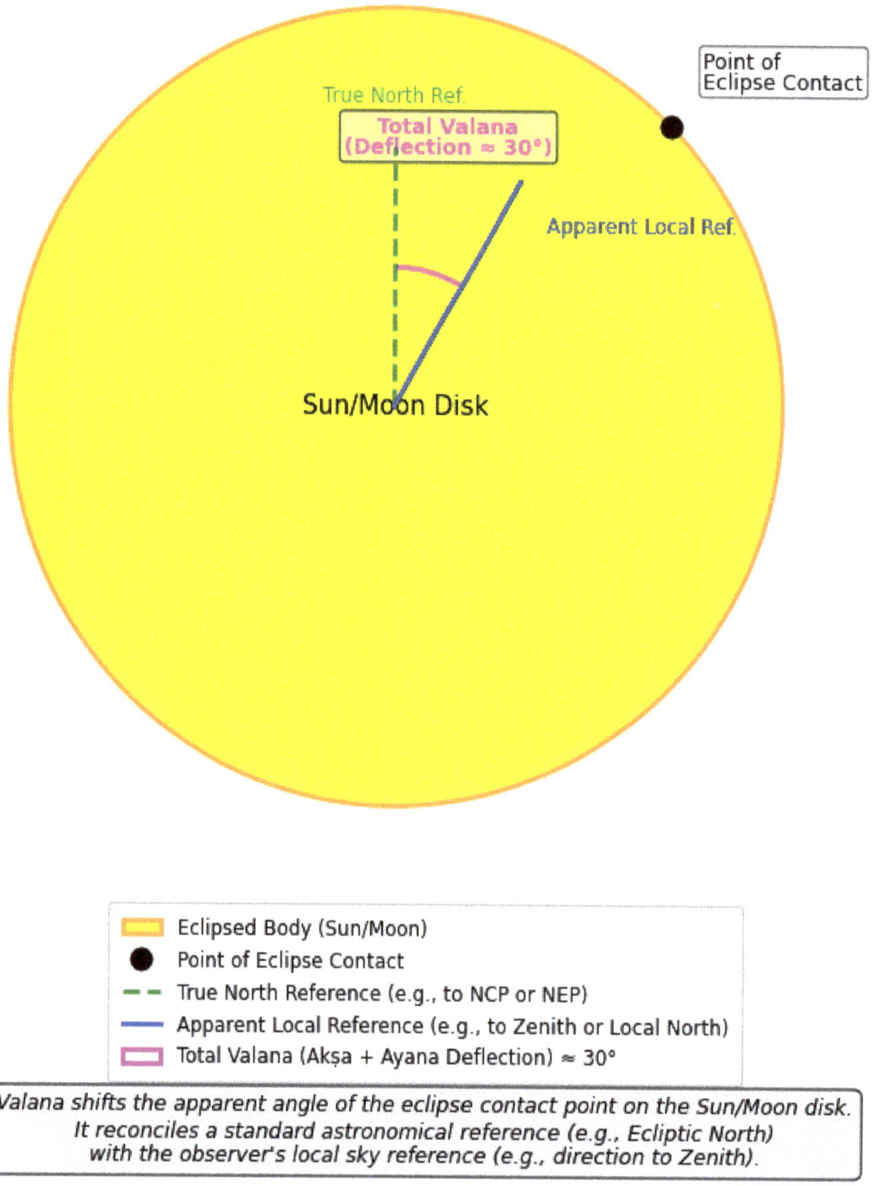

Figure 6.41: Total Valana – Angular Deflection at Eclipse Contact (Gola.45). Illustrates Total Valana (magenta arc), the angular difference between a standard astronomical reference (e.g., True North Ref. via NCP/NEP, green dashed) and the observer's apparent local sky reference (e.g., Apparent Local Ref. via Zenith/local North, blue). This Valana, a sum of Akṣavalana and Ayanavalana, effectively rotates the perceived angle of the eclipse contact point on the Sun/Moon disk.

6.11.9 The Colors of a Lunar Eclipse

Sanskrit:

प्रग्रहणान्ते धूम्रः खण्डग्रहणे शशी भवति कृष्णः ।

सर्वग्रासे कपिलः सकृष्णताम्रो तमोमध्ये ॥ ४६ ॥

IAST:

pragrahaṇānte dhūmraḥ khaṇḍagrahaṇe śaśī bhavati kṛṣṇaḥ |

sarvagrāse kapilaḥ sakṛṣṇatāmro tamomadhye || 46 ||

Observation: What does the Moon actually *look* like during a total lunar eclipse?

Description: This verse provides an accurate qualitative description of the changing colors and appearance of the Moon during the different phases of a total lunar eclipse, clearly based on careful observation rather than just geometric calculation.

1. ***Dhūmraḥ* (Smoky/Dim):** At the beginning and end of the eclipse (***pragrahaṇānte***), as the Moon enters and leaves the Earth's faint outer shadow (penumbra), its overall brightness dims slightly, giving it a "smoky" appearance.

2. ***Kṛṣṇaḥ* (Black/Dark):** As the Moon moves into the Earth's dark central shadow (umbra) during the partial phase (***khaṇḍagrahaṇe***), the eclipsed portion appears dark or black against the still-bright part.

3. ***Kapilaḥ* (Tawny/Brownish):** When the Moon is fully immersed in the umbra (***sarvagrāse***) but is near the edge of the shadow, it often doesn't vanish completely but appears a dull, brownish, or "tawny" color.

4. ***Sakṛṣṇatāmraḥ* (Dark Coppery-Red):** When the Moon passes deep into the central part of the Earth's shadow (***tamomadhye***), it often takes on the famous deep reddish or coppery hue – the characteristic color of a "blood moon."

Significance: This sequence is observationally accurate. The residual light and reddish color during totality are now understood to be caused by sunlight filtering and being refracted (bent) through the Earth's atmosphere all around its limb. The atmosphere scatters away blue light more effectively, allowing predominantly red light to pass through and weakly illuminate the Moon even when it's deep inside the geometric shadow. Aryabhata's description shows meticulous observation and perhaps an implicit qualitative understanding that *something* (like the atmosphere) affects the light passing into the shadow.

Notes:

- Describes visual phases: Penumbral (smoky), Partial Umbral (black), Total Umbral near edge (tawny), Total Umbral in core (dark copper-red).
- ***Sakṛṣṇatāmraḥ:*** Dark copper-red ("blood moon") - accurate observation of totality.

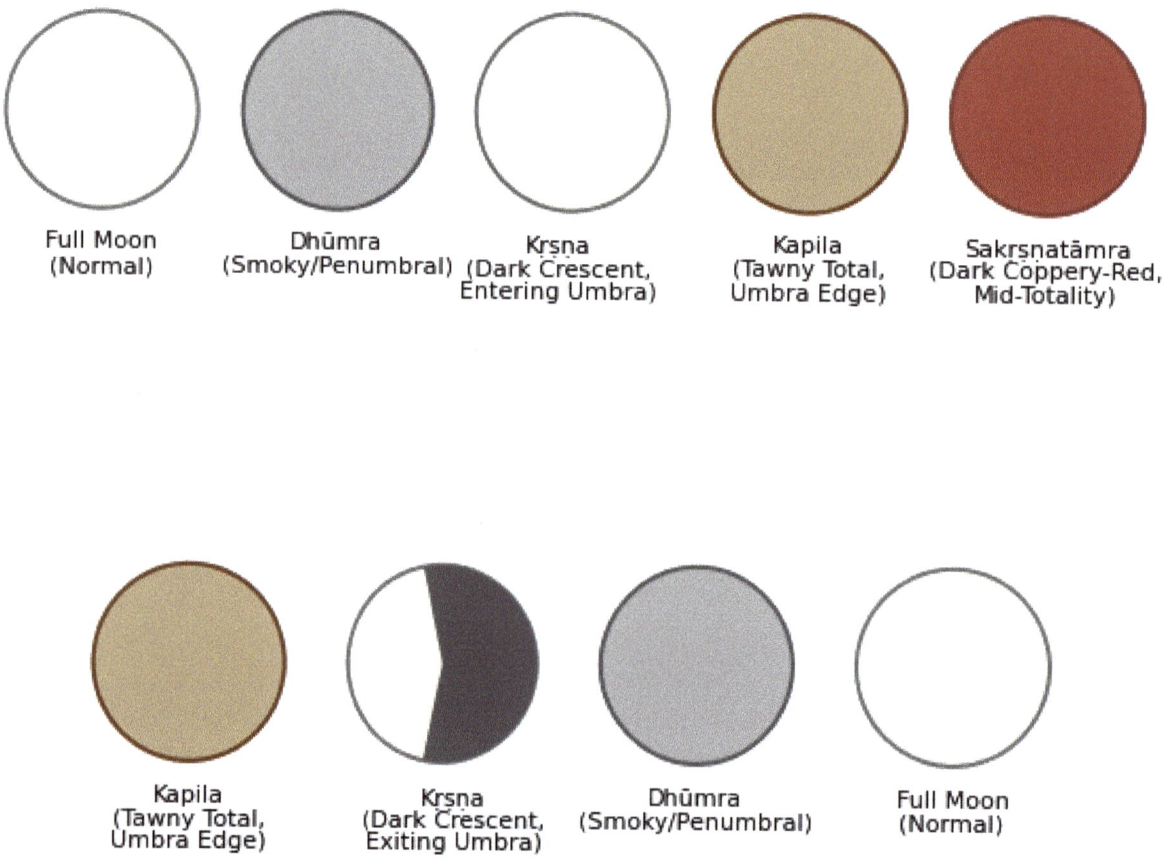

Figure 6.42: Gola.46: Lunar Eclipse Appearance Changes. This diagram illustrates the sequence of visual changes the Moon undergoes during a total lunar eclipse, corresponding to Aryabhata's descriptions in Golapāda Verse 46. The stages shown are: Full Moon (Normal), Dhūmra (Smoky/Penumbral), Kṛṣṇa (Dark Crescent as it enters/exits the umbra), Kapila (Tawny Total near umbra edge), and Sakṛṣṇatāmra (Dark Coppery-Red at mid-totality).

6.11.10 Why Earth's Shadow Cannot Eclipse the Sun

Sanskrit:

रविं न छादयति भूच्छाँया भूमौ स्थितो विदुर् विप्राः ।

अदृश्याख्यानात् तत् तु ग्रहणं रविणा समं भवति ॥ ४७ ॥

IAST:

raviṃ na chādayati bhūcchāyā bhūmau sthito vidur viprāḥ |

adṛśyākhyānāt tat tu grahaṇaṃ raviṇā samaṃ bhavati || 47 ||

Question: If Earth casts a shadow, why doesn't that shadow sometimes fall on the Sun and cause an eclipse *of* the Sun *by the Earth's shadow*?

Answer: Aryabhata addresses this logical point to reinforce his shadow theory.

1. **Shadow Direction:** He points out the basic geometry: the Earth's shadow (***bhūcchāyā***) is cast *away* from the Sun. An observer on Earth (***bhūmau sthitaḥ***) is located *between* the Sun and the direction of the shadow. Therefore, the shadow cannot possibly fall upon the Sun from the perspective of someone on Earth (***raviṃ na chādayati bhūcchāyā***).

2. **Known Impossibility (*Adṛśyākhyānāt*):** He states this is known to be impossible or unobservable ("from the statement/knowledge of its impossibility/invisibility").

3. **Contrast with Solar Eclipse:** He contrasts this impossible scenario with a real solar eclipse (***tat tu grahaṇaṃ***), which occurs when the eclipsing body (which must be the Moon, per Gola.37) is aligned *with* the Sun (***raviṇā samaṃ***) from the observer's viewpoint.

This verse clarifies the distinct geometric configurations and causal mechanisms for lunar versus solar eclipses, further solidifying the rationality of his shadow-based explanation.

Notes:

- Confirms Earth's shadow points away from the Sun relative to Earth observers.
- ***Adṛśyākhyānāt:*** "From the statement/knowledge of impossibility/invisibility."
- ***Raviṇā samaṃ:*** Aligned with the Sun (condition for solar eclipse by the Moon).

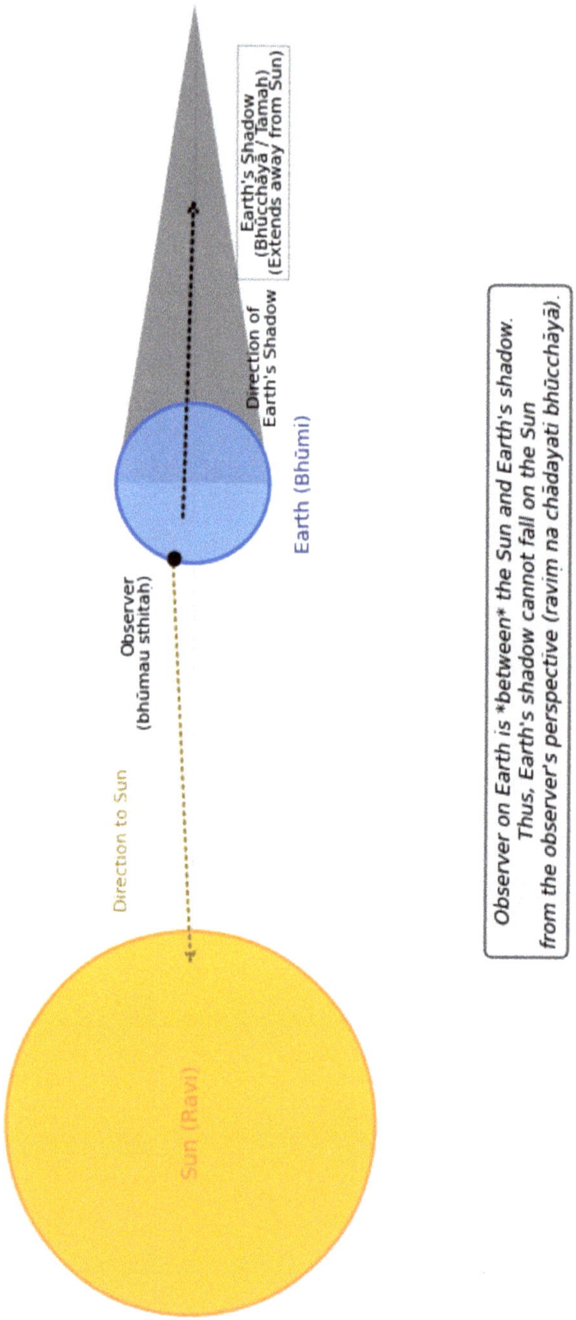

Figure 6.43: Diagram illustrating Aryabhata's explanation from Golapāda Verse 47 on why Earth's shadow (Bhūcchāyā) cannot eclipse the Sun (Ravi) for an observer on Earth. The Earth, positioned between the Sun and the shadow it casts, means an observer on Earth (bhūmau sthitaḥ) always sees the Sun in one direction and the region of Earth's shadow (tamaḥ) in the opposite direction. Therefore, the shadow cannot obscure the Sun from this perspective.

6.11.11 The Observational Threshold for Eclipse Visibility

Sanskrit:

अष्टभागो ग्रहणस्य न दृष्यो भवति भूतले सदा ।

ग्रहैर् मलिनीकृतस्य विमलस्य वा ग्रहमाण्डलस्य ॥ ४८ ॥

IAST:

aṣṭabhāgo grahaṇasya na dṛśyo bhavati bhūtale sadā |

grahair malinīkṛtasya vimalasya vā grahamaṇḍalasya || 48 ||

Question: Are very small eclipses always visible?

Answer: This verse states a practical, empirical limit for the naked-eye visibility of eclipses from Earth's surface (***bhūtale***).

1. **Visibility Threshold:** An eclipse where the maximum magnitude (***grāsa***) covers less than **one-eighth** (***aṣṭabhāgaḥ***) of the eclipsed body's diameter (i.e., magnitude < 0.125) is generally **not discernible** (***na dṛśyo bhavati sadā***) by an unaided observer.

2. **Application:** This limit applies both to perceiving the small darkened portion (***malinīkṛtasya***) creeping onto the disk during a very small partial solar eclipse, and to detecting the subtle overall dimming (***vimalasya***) of the Moon during the penumbral phases of a lunar eclipse or very slight partial umbral phases.

3. **Observational Realism:** This reflects a pragmatic understanding of the limitations of human vision in detecting faint contrasts or small changes in brightness without optical aid. It adds a layer of observational realism to the theoretical calculations – just because an eclipse is calculated to occur geometrically doesn't mean it will be noticeable.

Notes:

- ***Aṣṭabhāgaḥ:*** One-eighth part (magnitude threshold for naked-eye visibility ≈ 0.125).
- ***Na dṛśyaḥ:*** Not visible (to the unaided eye).
- Reflects practical limits for observing small magnitude eclipses or faint penumbral phases.

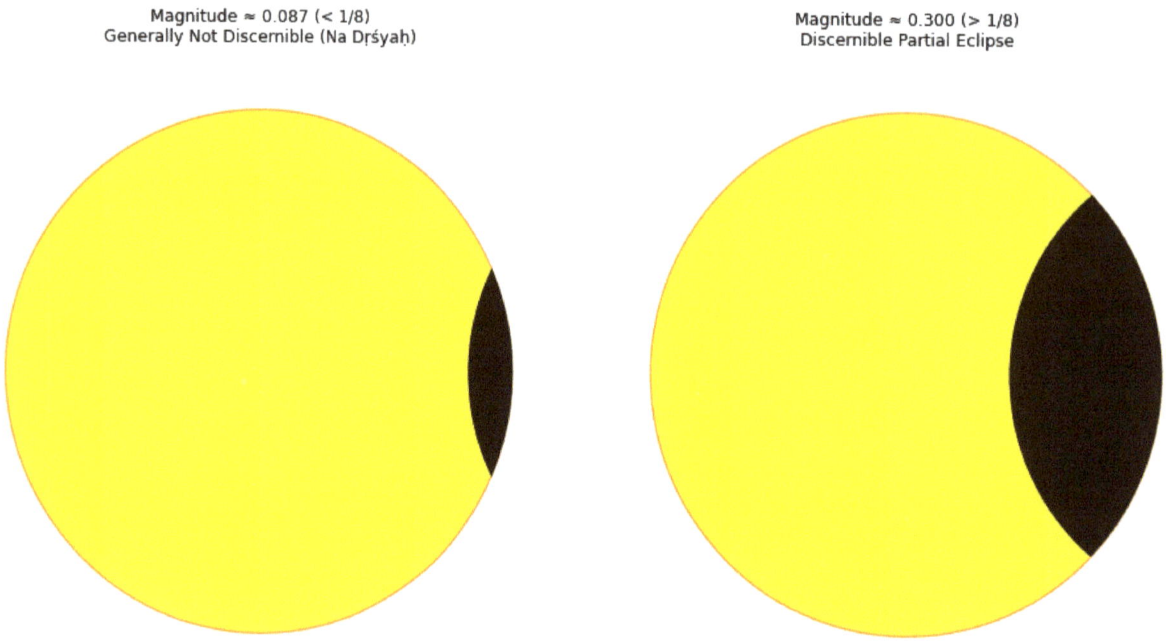

Figure 6.44: Diagram illustrating the observational threshold for eclipse visibility, as described by Aryabhata in Golapāda Verse 48. Left Panel: Shows an eclipse with a magnitude less than one-eighth (Aṣṭabhāgaḥ) of the celestial body's diameter. Such a small obscuration is generally not discernible (Na Dṛśyaḥ) to the unaided eye. Right Panel: Depicts a partial eclipse with a magnitude significantly greater than one-eighth, which would be clearly visible.

6.12 Knowledge, Intellect, and Legacy (Gola.49-50)

Concluding the Golapāda and the entire *Aryabhatiya*, Aryabhata steps back from specific calculations to reflect on the source and nature of the knowledge he has presented. These final verses blend reverence for tradition with confidence in his own intellectual contribution, and emphasize the profound importance of preserving this hard-won knowledge accurately.

6.12.1 Source of Knowledge – Tradition, Grace, and Intellect

Sanskrit:

सदसज्ज्ञानसमुद्रात् समुद्धृतं ब्रह्मणः प्रसादेन ।

सज्ज्ञानोत्तमं रत्नं मया निमग्नं स्वमतिनावा ॥ ४९ ॥

IAST:

sadasajjñānasamudrāt samuddhṛtaṃ brahmaṇaḥ prasādena |

sajjñānottamaṃ ratnaṃ mayā nimagnaṃ svamatināvā || 49 ||

Translation: "From the ocean of true and false knowledge (***sadasajjñānasamudrāt***), by the grace of Brahmā (***brahmaṇaḥ prasādena***), this supreme jewel of true knowledge (***sajjñānottamaṃ ratnam***), which was submerged (***nimagnam***), has been rescued by me using the boat of my own intellect (***svamatināvā***)."

Interpretation: In this deeply insightful verse, Aryabhata outlines the dual foundation of his work:

1. **Divine/Traditional Source:** He acknowledges the origin of true knowledge lies within the vast "ocean containing both true and false knowledge" (***sadasajjñānasamudrāt***), suggesting a complex inheritance of existing astronomical theories, some accurate, some flawed. He attributes the recovery of the "supreme jewel of true knowledge" (***sajjñānottamaṃ ratnam***) partly to "the grace of Brahmā" (***brahmaṇaḥ prasādena***). As interpreted by Bhaskara I and others, "Brahmā's grace" can signify alignment with the underlying cosmic truth (Brahman), adherence to the authentic scientific tradition (*śāstra-sampradāya*), or perhaps divine inspiration accompanying deep intellectual effort. He sees himself as recovering something valuable that had become obscured or "submerged" (***nimagnam***).

2. **Individual Intellect (*Svamati*):** Crucially, Aryabhata emphasizes his own active role and agency. He states this rescue was accomplished "by me using the **boat of my own intellect**" (***mayā ... svamatināvā***). He asserts the power and validity of his **own reasoning, mathematical analysis, and critical discernment** (***svamati***) as the essential vessel (*nāvā* = boat) required to navigate the confusing ocean of existing knowledge, separate truth from error, and bring forth a coherent, refined system.

This verse beautifully encapsulates a methodology that profoundly respects inherited wisdom while simultaneously championing the indispensable role of individual rational inquiry (*svamati*) in validating, correcting, and advancing scientific understanding.

Notes:

- ***Sadasajjñānasamudra:*** Ocean containing both true and false knowledge (complex inheritance).
- ***Brahmaṇaḥ prasādena:*** By Brahmā's grace (acknowledging divine/traditional source or cosmic truth).
- ***Sajjñānottamam ratnam:*** Supreme jewel of true knowledge.
- ***Nimagnam:*** Submerged/Lost/Obscured.
- ***Svamatināvā:*** By the boat of my own intellect (emphasizing power of individual reason/analysis).

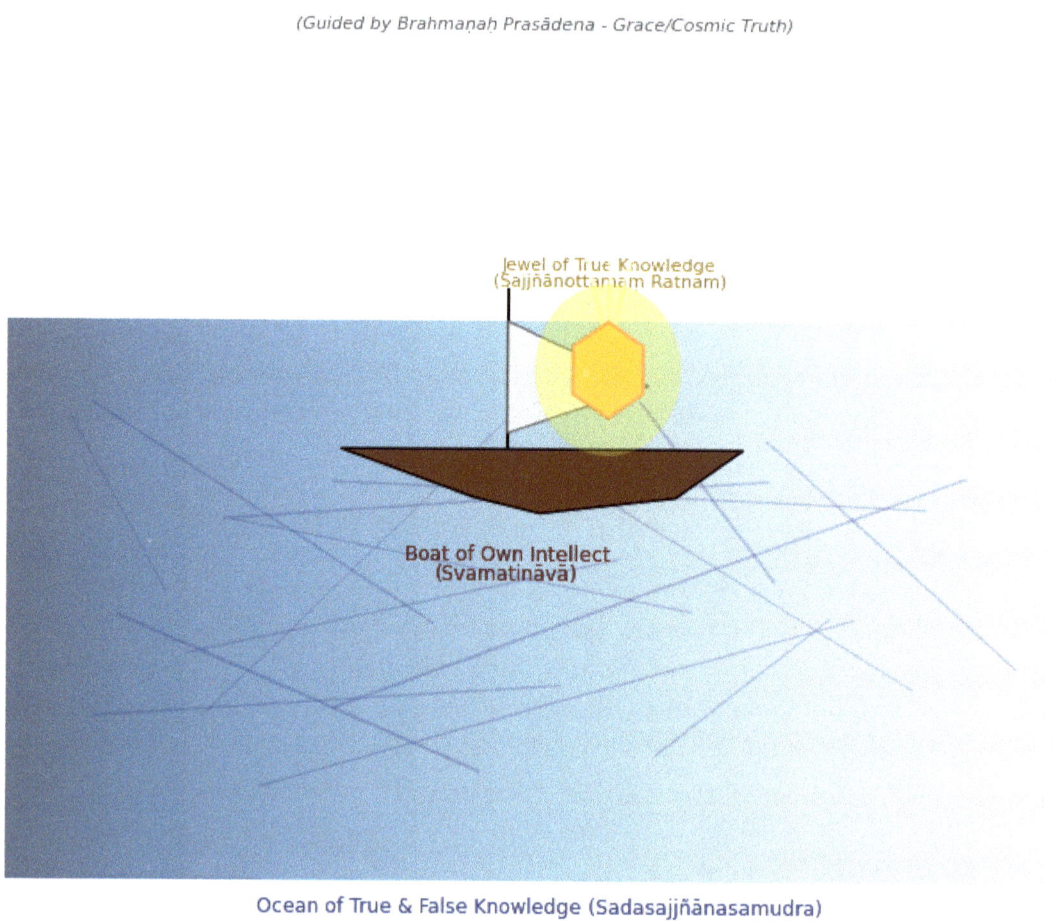

Figure 6.45: Aryabhata's Metaphor of Knowledge (Gola.49). The 'boat of intellect' (Svamatināvā) rescues the 'jewel of true knowledge' from the 'ocean of true/false knowledge,' guided by 'Brahmā's grace.'

6.12.2 The Charge of Preservation and the Work's Identity

Sanskrit:

आर्यभटीयं नाम्ना पूर्वं स्वायंभुवं सदा सत्यम् ।

सुकृतायुषोः प्रणाशं कुरुते प्रतिकञ्चुकं यः अस्य ॥ ५० ॥

IAST:

Aryabhatiyaṃ nāmnā pūrvaṃ svāyaṃbhuvaṃ sadā satyam |

sukṛtāyuṣoḥ praṇāśaṃ kurute pratikañchukaṃ yaḥ asya || 50 ||

Translation: "This [work], known from the beginning by the name ***Aryabhatiya*** (***Aryabhatiyaṃ nāmnā pūrvaṃ***), conforms to self-existent [eternal truth] (***svāyaṃbhuvaṃ***) and is always true (***sadā satyam***). He who creates a counterfeit/distortion (***pratikañchukaṃ***) of it causes the destruction (***praṇāśaṃ***) of his own merit (***sukṛta***) and lifespan (***āyuṣoḥ***)."

Interpretation: In this powerful concluding verse, Aryabhata asserts the validity and importance of his work, and issues a stern warning regarding its faithful preservation:

1. **Identity and Truth Claim:** He formally names his work the **Aryabhatiya**. He claims it conforms to ***svāyaṃbhuvaṃ*** – self-existent reality, inherent natural law, or fundamental truth – and is therefore ***sadā satyam*** ("always true," or eternally valid in its principles). This is a strong assertion that his mathematical and astronomical system accurately reflects the rational order of the cosmos he sought to model.

2. **Warning Against Distortion:** He delivers a potent injunction against anyone who creates a ***pratikañchukaṃ*** – literally a "counter-jacket," meaning a **counterfeit, plagiarism, malicious critique, or deliberate distortion** – of his treatise.

3. **Ethical Consequences:** Such intellectual dishonesty or corruption of knowledge is framed not just as a scholarly error but as a grave **ethical transgression** with severe personal repercussions: the destruction (***praṇāśaṃ***) of the perpetrator's accumulated religious or moral merit (***sukṛta***) and even a shortening of their lifespan (***āyuṣoḥ***).

This final statement underscores the immense value Aryabhata placed on the accurate preservation and transmission of the scientific knowledge he had painstakingly systematized. It secures the identity of the *Aryabhatiya* and places a strong ethical responsibility on future scholars to engage with his work honestly, accurately, and with intellectual integrity.

Notes:

- ***Aryabhatiyaṃ nāmnā:*** By the name Aryabhatiya (identifies the work).

- ***Svāyaṃbhuvaṃ:*** Self-existent / Conforming to eternal truth/reality.
- ***Sadā satyam:*** Always true (assertion of validity).
- ***Pratikañchukaṃ:*** Counterfeit / Distortion / Plagiarism / Malicious Critique.
- ***Sukṛtāyuṣoḥ praṇāśaṃ:*** Destruction of merit and lifespan (strong ethical warning).

6.13 Summary Table of Key Golapāda Formulas

Formula/Concept	Calculation	Variables	Verse Ref.	Notes
Kṣitijyā (Sine Asc. Diff.)	(R·sin(δ) × R·sin(φ)) / R·cos(φ)	δ=Declination, φ=Latitude	Gola.26	Basis for daylight variation.
chara (Rising Time Corr.)	Time derived from arcsin(Kṣitijyā / R)	Use Kṣitijyā. Applied as ± charadala to quarter-day.	Gola.27	Corrects zodiac sign rising times for latitude.
Śaṅku (Sine Alt, δ=0)	(R·sin(H) × R·cos(φ)) / R	H=Hour Angle, φ=Latitude	Gola.28	Basic formula for objects on the equator. General case needs a full formula.
Arkāgrā (Sine Amplitude)	(R·sin(δ) × R) / R·cos(φ)	δ=Declination, φ=Latitude	Gola.30	Sun's deviation N/S from E/W at rising/setting.
Akṣajyā (Sine Latitude)	R·sin(φ) = (R × S) / $\sqrt{(G^2 + S^2)}$	G=Gnomon Ht, S=Equinox Noon Shadow	Gola.31	Practical method to find latitude φ.
Dṛkṣepajyā (Sine Zenith Dist)	R·cos(Altitude)	Altitude	Gola.32	Complementary to Śaṅku (R·sin Alt).
Parallax Correction (Altitude)	Phala ≈ (Rsin(HP) × Rsin(Observed ZD)) / R	HP=Horizontal Parallax, ZD=Observed Zenith Distance	Gola.33	Refines observed altitude. HP implicit.
Dṛkchāyā (Horizon Dip)	Related to Earth Radius & Observer Height	Bhūvyāsārdha, Observer Height	Gola.34	Affects rising/setting times. May include refraction effects.
Akṣadṛkkarma (Long. Corr.)	Correction ≈ (Rsin(Vik) × Rsin(φ)) / Rcos(φ) (Sign depends)	Vik=Cel. Lat, φ=Observer Lat	Gola.35	Visibility correction to longitude due to observer latitude.
Ayanadṛkkarma (Long. Corr.)	Magnitude complex. (Sign depends on season & latitude)	Vik=Cel. Lat, δ=Declination, Long from Solstice	Gola.36	Visibility correction to longitude due to ecliptic tilt.

Formula/Concept	Calculation	Variables	Verse Ref.	Notes
Shadow Length (Umbra)	Length L ≈ (D_E × Dist_SE) / (D_S - D_E) (Interpreted)	D_E, D_S = Earth/Sun Diameters, Dist_SE = Sun-Earth Distance	Gola.39	Length of Earth's umbral shadow cone.
Shadow Width at Moon	Width D_Sh = D_E × (L - Dist_EM) / L	D_E=Earth Diam, L=Shadow Length, Dist_EM = Earth-Moon Distance	Gola.40	Diameter of umbra at Moon's distance.
Sthityardha Radius (Partial)	R_Sum = (D_Grāhaka + D_Maṇḍala) / 2	D=Diameters of Eclipsing/Eclipsed bodies	Gola.41	Geometric basis for partial phase duration.
Vimardārdha Radius (Total)	R_Diff = abs(D_Grāhaka - D_Maṇḍala) / 2	D=Diameters of Eclipsing/Eclipsed bodies	Gola.41	Geometric basis for total/annular phase duration.
Sthityardha (Time - Partial)	Time ≈ (√(R_Sum^2 - Vikṣepa^2) × 60) / Rel. Speed	R_Sum, Vikṣepa = Latitude, Rel. Speed	Gola.42	Half-duration of partial phase in nāḍikās.
Vimardārdha (Time - Total)	Time ≈ (√(R_Diff^2 - Vikṣepa^2) × 60) / Rel. Speed	R_Diff, Vikṣepa = Latitude, Rel. Speed	Gola.43	Half-duration of total/annular phase in nāḍikās.
Grāsa (Magnitude)	Magnitude = (R_Sum - Vikṣepa) / D_Maṇḍala	R_Sum, Vikṣepa, D_Maṇḍala = Diam of Eclipsed body	Gola.44	Max fraction covered. (Divisor interpretation crucial).
Valana (Orientation Corr.)	Total Valana = Akṣavalana + Ayanavalana	Components calc. based on lat, season, time	Gola.45	Corrects apparent angle of eclipse contact points.

6.14 Conclusion: A Compass for the Ages

The Golapāda stands as the brilliant capstone of Aryabhata's *Aryabhatiya*, demonstrating the power and coherence of his entire astronomical system. It masterfully applies the foundational parameters of the Daśagītikā, the mathematical machinery of the Gaṇitapāda, and the temporal-kinematic framework of the Kālakriyāpāda to construct a comprehensive, quantitative model of the cosmos. This model, centered on a **spherical, rotating Earth**—a truly revolutionary concept for its era—provided geometrically elegant explanations for a vast range of celestial phenomena. From the fundamental cause of seasons and the varying perspectives across the globe (longitude, latitude) to the complex apparent motions of planets and the precise prediction of eclipses based on shadow geometry (rejecting myth), Aryabhata charted the heavens with unprecedented rational clarity.

Illuminated by Bhaskara I's vital commentary, the Golapāda showcases a remarkable synthesis of:

- **Empirical Grounding:** Rooted in observation (gnomon measurements for latitude, visibility thresholds, eclipse appearances).
- **Mathematical Sophistication:** Applying geometry and sine-based trigonometry (*jyā*) to solve complex spatial problems and compute positional corrections (*dṛkkarma*, *valana*) and eclipse parameters.
- **Theoretical Insight:** Boldly proposing Earth's axial rotation based on relative motion and providing the correct physical cause for eclipses.
- **Critical Engagement with Tradition:** Skillfully integrating or rationalizing traditional concepts (like Meru) while prioritizing mathematical consistency and observation.
- **Ethical Commitment:** Concluding with a profound statement on the value of true knowledge and the responsibility for its accurate preservation (*svamati* vs. *pratikañcukaṃ*).

Aryabhata's Golapāda, particularly its radical theory of Earth's rotation, ignited centuries of scholarly debate within India but served as an indispensable reference point and computational foundation for subsequent generations of astronomers like Lalla, Śrīpati, and Bhaskara II. While distinct from Ptolemy's influential geometric system, Aryabhata's approach offered a powerful, computationally oriented model featuring groundbreaking insights.

The Golapāda stands as an enduring testament to the power of human reason, mathematical precision, and observational diligence to map the cosmos and understand our place within it. It demonstrates a pivotal moment in the global history of science, leaving a legacy whose mathematical echoes, transmitted across cultures and centuries, continue to resonate today. Aryabhata provided a compass crafted in 5th-century India, whose needle pointed towards a rational understanding of the heavens, guiding scientific inquiry for ages to come.

***

Chapter Seven

Two Systems of the World: Aryabhata and Ptolemy

A Comparison of the Cosmological and Mathematical Models

7.0 Introduction: Parallel Worlds, Divergent Paths

Having explored the intricacies of Aryabhata's mathematical and astronomical system as laid out in the *Aryabhatiya*, it is illuminating to place his work in comparative perspective with the other monumental achievement of ancient mathematical astronomy: the system of Claudius Ptolemy (fl. c. 150 CE) presented in his *Mathematical Syntaxis*, universally known as the *Almagest*. Though separated by over three centuries and developed in distinct cultural milieu—Aryabhata in Gupta India, Ptolemy in Roman Egypt (Alexandria)—both astronomers grappled with the same fundamental challenge: creating comprehensive, mathematically rigorous models capable of accurately predicting the complex motions of the celestial bodies, particularly the planets, as observed from Earth.

Both worked within a broadly geocentric framework and employed sophisticated geometric and computational tools. Yet, their approaches to structuring the cosmos, explaining planetary motion, and the specific mathematical techniques they favored reveal fundamental differences in methodology, underlying assumptions, and perhaps even scientific epistemology. Comparing these two giants allows us to appreciate the unique contributions of each tradition and the diverse ways ancient thinkers sought to "save the phenomena" – to create models that matched observation while adhering to certain philosophical or mathematical ideals. This chapter contrasts their approaches to Earth's role, planetary motion mechanisms, mathematical tools, and their respective historical legacies.

7.1 Model Structures Compared: Center, Motion, and Earth's Role

Both Aryabhata and Ptolemy placed the Earth at or very near the center of the cosmos. Their primary task was to reconcile the observed irregularities—planets speeding up, slowing down, and exhibiting retrograde motion—with the prevailing philosophical ideal (stronger perhaps in the Greek tradition but influential in India too) that celestial motions should fundamentally be based on uniform circular motion. Both used the concept of **epicycles** (small circles whose centers move on larger circles called **deferents**) as a key explanatory device. However, their specific model structures and treatment of Earth diverged significantly.

7.1.1 Earth's Role: Rotation vs. Static Stability

The most striking and conceptually profound difference lies in their conception of the Earth itself:

- **Ptolemy's Static Earth:** In the *Almagest* (Book I, Chapter 7), Ptolemy systematically argues against any motion of the Earth, including axial rotation. He employs arguments based on Aristotelian physics (heavy objects naturally seek the center and remain at rest), lack of observable effects (no constant strong wind from the west, projectiles falling vertically, clouds and birds keeping pace), and the perceived absurdity of the immense speed required for the Earth's surface to rotate daily. For Ptolemy, the Earth was demonstrably fixed and immobile at the center of the universe, and the apparent daily rotation of the entire celestial sphere was a physical reality.

- **Aryabhata's Rotating Earth:** As established in the Golapāda (Gola.9 / Chapter 6.5), Aryabhata proposed the radical hypothesis that the Earth **rotates on its axis** once daily from west to east (*prāk*). He presented this not with physical arguments addressing potential objections (like Ptolemy did against rotation), but as a **kinematic explanation** based on the principle of relative motion ("as a man in a boat moving forward sees stationary objects moving backward..."). This explanation elegantly accounted for the uniform diurnal rotation of the entire celestial sphere as an *apparent* motion caused by the observer's own movement. While Aryabhata's model remained geocentric regarding planetary orbits (the planets, Sun, and Moon still orbited the central Earth, Kk.15), his attribution of diurnal motion to Earth's rotation was a revolutionary departure from the prevailing consensus in both Indian and Greco-Roman traditions.

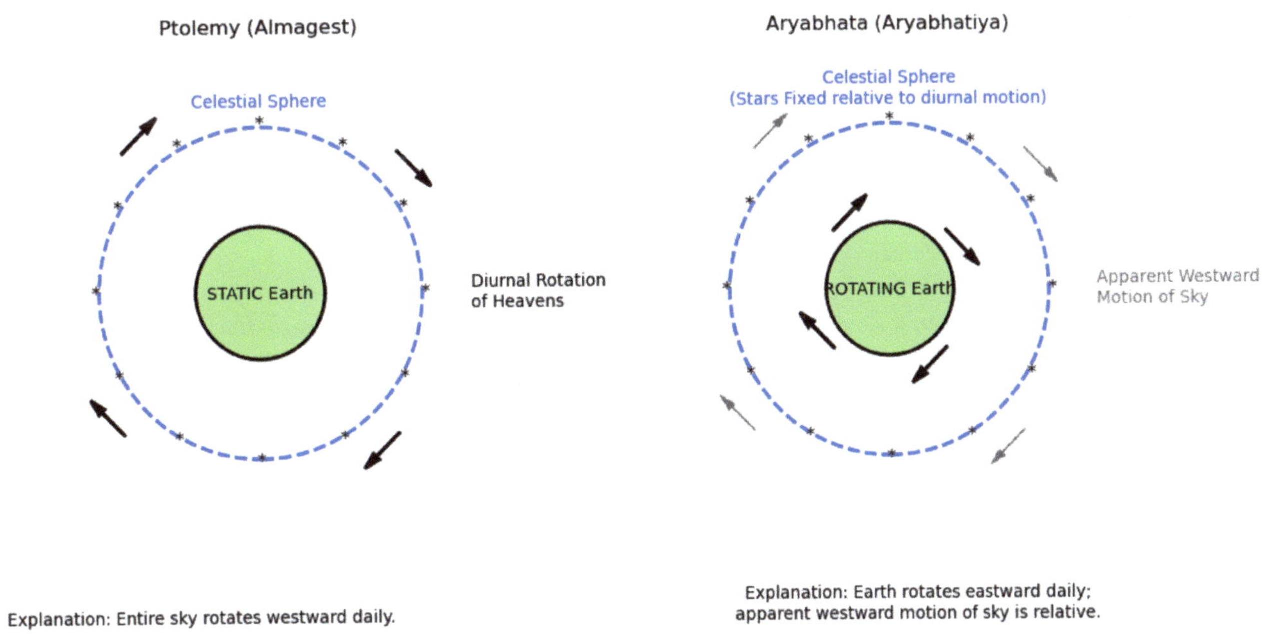

Figure 7.1: Comparative diagram illustrating the contrasting explanations for the apparent daily westward motion of the celestial sphere in the systems of Ptolemy and Aryabhata. Ptolemy's model (left) posits a Static Earth at the center, with the entire Celestial Sphere rotating westward around it. Aryabhata's model (right) proposes a Rotating Earth spinning eastward on its axis, which causes the apparent westward motion of the stars to be a relative phenomenon (as explained with the boat analogy, Gola.9).

7.1.2 Mechanisms for Planetary Motion: Equants vs. Computational Corrections

To account for the observed variations in planetary speeds (the "first inequality," related to orbital eccentricity) and the complexities of retrograde motion and varying elongation from the Sun (the "second inequality," related to synodic phenomena), both astronomers needed to refine the basic deferent-epicycle model. Their methods, however, were distinct:

- **Ptolemy's Geometric Machinery (Eccentric and Equant):** Ptolemy introduced two sophisticated geometric devices beyond the basic deferent and epicycle:
    - **Eccentric Deferent:** The center of the large deferent circle was offset from the position of the Earth. This helped account for variations in distance and speed simply by placing the Earth away from the geometric center of the orbit.
    - **The Equant:** This was Ptolemy's most ingenious and controversial device. It was a point located offset from the deferent's center, usually such that the deferent center lay midway between the Earth and the equant. The crucial rule was that the center of the epicycle moved along the deferent in such a way that it swept out equal angles in equal times *as viewed from the equant point*, not from the center of the deferent or the Earth (*Almagest*, Book IX, Chapter 5). This mathematical construct allowed Ptolemy to reproduce the observed non-uniform motion along the deferent (necessary for accurate predictions, especially for the superior planets) while preserving a semblance of regulated angular motion. However, it explicitly violated the principle of strict uniform circular motion around the deferent's *center*, a philosophical compromise that drew criticism from later Islamic astronomers who sought models adhering more strictly to combinations of uniform circular motions.
- **Aryabhata's Computational/Epicyclic Approach (Manda and Śīghra):** Aryabhata's system, as interpreted through his computational procedures (Kālakriyāpāda) and Bhaskara I's commentary, relied more heavily on applying numerically calculated corrections derived from trigonometric functions within the epicyclic framework. His model effectively used two primary corrections (often visualized as resulting from two epicycles, or one epicycle plus an eccentric, though the text focuses on the calculation):
    - **Manda Correction:** Calculated using the *mandakendra* (angle from the *Mandoccha*/apogee), this correction (*mandaphala*) adjusted the mean longitude to account for speed variations analogous to orbital eccentricity (slower near apogee, faster near perigee). Functionally, this achieves a similar outcome to Ptolemy's eccentric deferent.
    - **Śīghra Correction:** Calculated using the *śīghrakendra* (angle from the *Śīghroccha*/fast point, linked to the Sun), this correction (*śīghraphala*) accounted for synodic phenomena, including retrograde motion. Functionally, this achieves results similar to those produced by Ptolemy's combined use of deferent and epicycle, especially in accounting for the second inequality.

These corrections were calculated using Aryabhata's sine tables (*jyā*, Gk.10) and applied algorithmically, often iteratively (Kk.22-23), to the mean longitude. While the underlying geometry likely involved epicycles and potentially eccentrics, the emphasis in the text is on the computational procedure for finding the *phala* (correction). This approach, focusing on algorithmic steps and sine-based adjustments, mathematically achieved results similar to Ptolemy's geometric devices but avoided the explicit introduction of a potentially problematic construct like the equant.

*Note: A **synodic phenomenon** (or more commonly, a **synodic period**) refers to the time it takes for a celestial body to reappear at the **same point in its cycle of phases or configurations relative to the Sun, as observed from Earth** (or another specific observing point).*

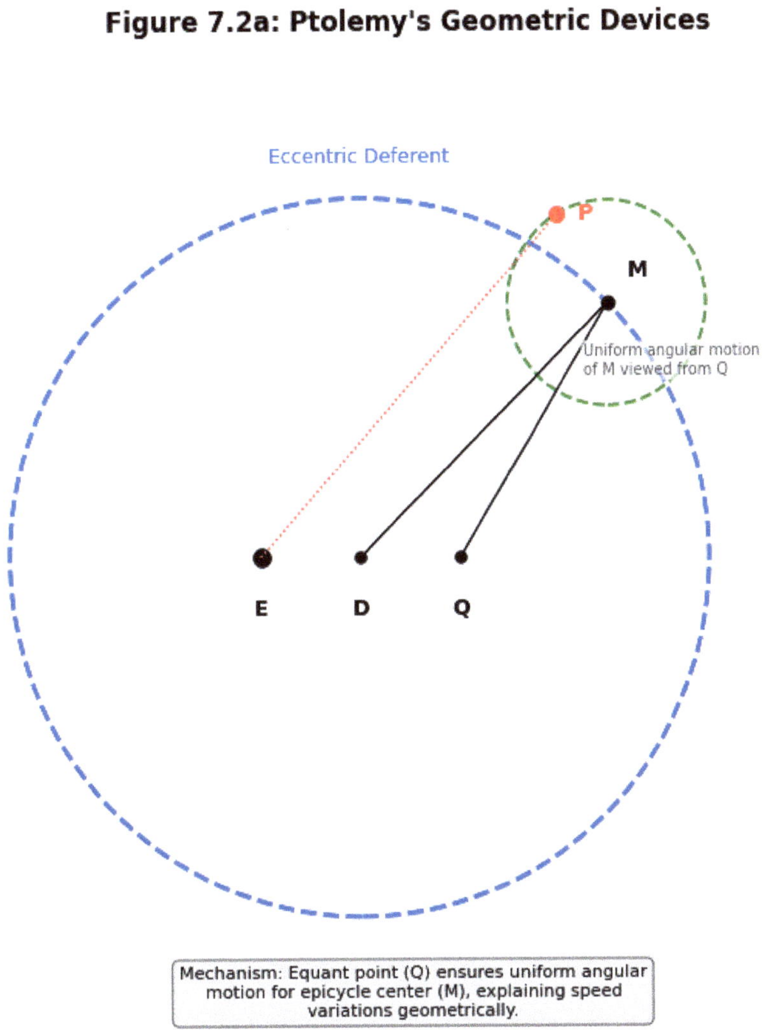

Figure 7.2a: Ptolemy's Geometric Devices for Planetary Anomalies. This diagram illustrates Ptolemy's model using an eccentric deferent and an equant point to explain variations in planetary speed. The Earth (E) is offset from the center of the deferent (D). The epicycle center (M), carrying the planet (P), moves along the deferent such that its angular motion is uniform as viewed from the Equant point (Q), not from the deferent's center or the Earth. This geometric construction aimed to replicate observed planetary motions.

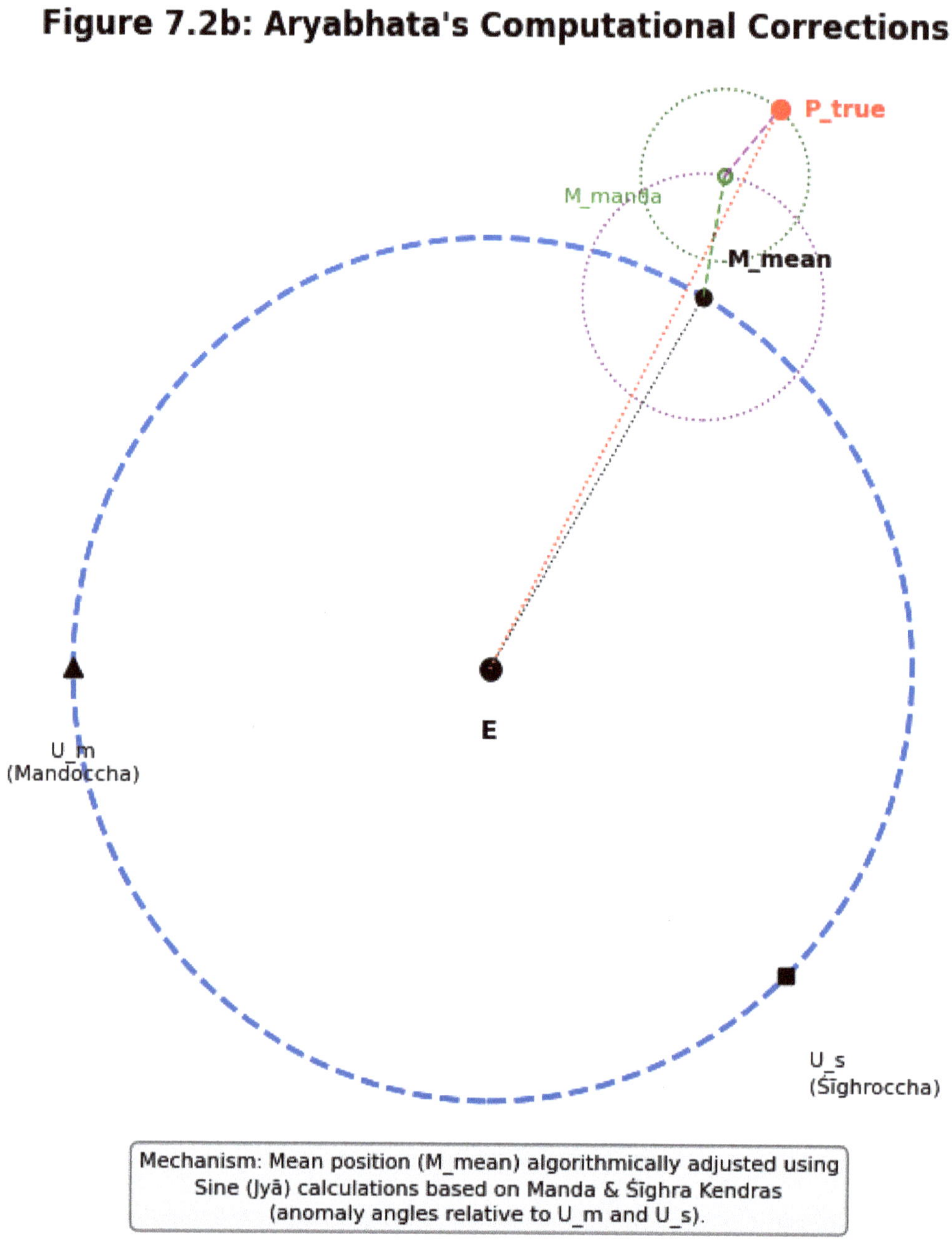

Figure 7.2b: Aryabhata's Computational Corrections for Planetary Anomalies. This diagram illustrates Aryabhata's approach, which, while likely involving underlying epicyclic geometry, emphasized algorithmic calculation of corrections. The mean position of the planet (M_mean) on its deferent is adjusted. The Mandoccha (U_m, apogee) and Śīghroccha (U_s, fast point) are key reference points. Corrections (phala), derived from Manda and Śīghra Kendras (anomaly angles relative to U_m and U_s) using Sine (jyā) tables, are applied to M_mean (conceptually moving it via M_manda) to determine the true position (P_true).

7.1.3 Divergent Approaches: Geometric Modeling vs. Computational Algorithm

This contrast highlights potentially different scientific priorities and epistemologies:

- **Ptolemy's Geometric Realism/Instrumentalism:** Ptolemy aimed to create a **geometric model**—a complex machine of nested circles, eccentrics, and equants—that visually and mathematically replicated the observed celestial motions ("saved the phenomena") with the highest possible accuracy. His *Almagest* is a masterpiece of geometric construction and demonstration. Whether he believed this machinery represented physical reality or was merely a predictive tool is debated, but the emphasis is clearly on the geometric structure.
- **Aryabhata's Computational/Algorithmic Focus:** Aryabhata appeared to prioritize **computational efficiency, mathematical consistency** (e.g., equal yugas), and **algorithmic elegance**. His system, deeply rooted in India's strong traditions of algebra (*kuṭṭākāra*, Gp.32-33) and numerical calculation (place value, root extraction, sine differences), provided a powerful set of procedures to accurately predict celestial positions. The epicycles and corrections might have been viewed more as indispensable mathematical tools for computation rather than necessarily representing literal physical wheels in the sky.

7.2 Mathematical Tools: Sines vs. Chords, Algorithms vs. Geometry

The contrasting cosmological structures of Aryabhata and Ptolemy were necessarily built upon and enabled by distinct mathematical toolkits, reflecting the different strengths and focuses of the Indian and Greco-Roman mathematical traditions.

7.2.1 Sine Tables and Algorithms: Aryabhata's Engine

Aryabhata's astronomy was intrinsically powered by his innovative use of the sine function (formulated as ***jyā*** or *ardha-jyā*, the length of the **half-chord** relative to a standard radius R=3438′) and sophisticated numerical algorithms:

- **Sine (** The cornerstone was his table of 24 R-Sine differences (Gk.10) covering a 90° quadrant at intervals of 3°45′. Crucially, he also provided the recursive, second-order finite difference algorithm (Gk.12) for constructing or verifying this table. This represented a significant achievement in numerical methods, allowing efficient and accurate computation of the trigonometric values essential for his planetary models.
- **Computational Procedures:** As seen extensively in the Kālakriyāpāda (Chapter 5), his system relied on step-by-step algorithms for calculating mean positions, anomalies (*kendras*), corrections (*phalas*) using the sine table and iterative refinements involving the *karṇa* (true distance), and ultimately the true positions (*sphuṭa graha*) and true daily motions (*sphuṭa bhukti*).
- **Algebraic Strength:** The Gaṇitapāda (Chapter 4) showcased advanced algorithms for square and cube roots (Gp.4-5) and, most notably, the *kuṭṭākāra* method for solving linear indeterminate equations (Gp.32-33), a powerful algebraic tool directly applicable to astronomical cycle problems (*ahargaṇa*).

7.2.2 Chord Tables and Geometry: Ptolemy's Foundation

Ptolemy's *Almagest*, conversely, exemplifies the power of classical Greek geometric methods applied rigorously to astronomy:

- **Chord Tables:** His primary trigonometric tool was the **chord function** (the length of the **full chord** subtending a given angle in a circle of fixed radius, typically R=60). In Book I, Chapter 11 of the *Almagest*, he meticulously details the geometric derivation and calculation of a comprehensive table of chords for angles from 0.5° to 180° in steps of 0.5°. These served a similar function to sine tables (as $\text{Chord}(2\theta) = 2R\sin\theta$) but were conceptually tied directly to geometric figures.
- **Geometric Constructions and Proofs:** As discussed previously, his model relied fundamentally on geometric constructs—deferents, epicycles, eccentrics, and the equant. His derivations and proofs throughout the *Almagest* extensively employed the **rule-based approach** and theorems of Euclidean geometry (e.g., properties of circles, triangles, similar figures, and theorems like Menelaus' theorem for spherical trigonometry, and his own Ptolemy's theorem for cyclic quadrilaterals used in deriving chord values).

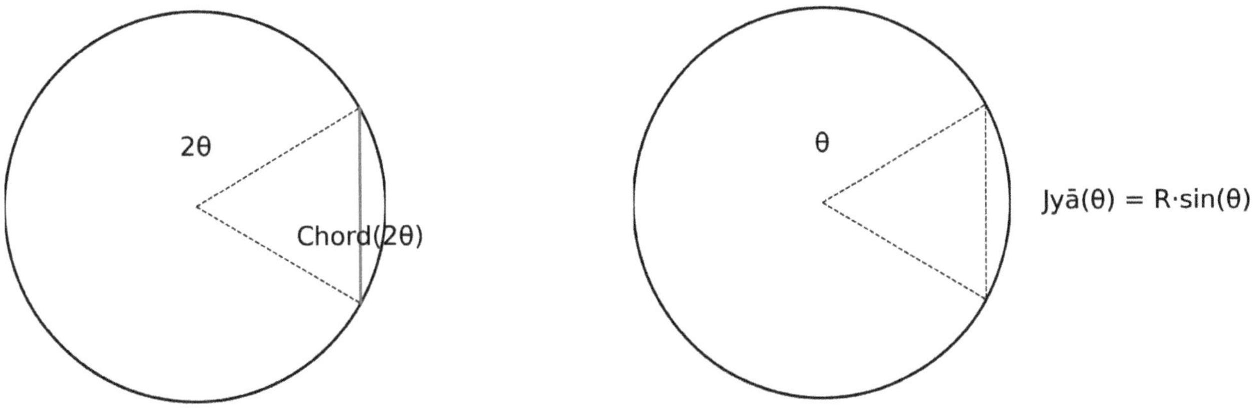

Figure 7.3: Comparison of the geometric definitions of the Greek Chord function and the Indian Sine (Jyā) function. Ptolemy's system (left) primarily used the length of the full chord subtending an angle (Chord(2θ)). Aryabhata's system (right) used the length of the half-chord (Jyā(θ) = R·sin(θ)), which is equivalent to the modern sine multiplied by the radius (R). This difference in definition reflects distinct mathematical traditions and approaches to trigonometry.

7.2.3 Comparing the Tools: Computation vs. Construction

The difference in mathematical toolkits highlights a potential paradigm distinction and differing strengths:

- **Focus:** Aryabhata's approach leaned more towards the **algebraic and computational**, seeking efficient numerical recipes and algorithms. Ptolemy's was fundamentally **geometric**, prioritizing accurate spatial representation and demonstration through rigorous constructions.
- **Trigonometric Function:** The Indian sine (***jyā*** = **Rsinθ**) proved mathematically more convenient and ultimately more powerful for the later development of analysis (calculus) than the Greek chord (Crd(2θ)). Furthermore, Aryabhata's difference-based method for generating the sine table was a distinctly numerical algorithm, whereas Ptolemy's chord table relied on geometric theorems.
- **Algebra vs. Geometry:** While Ptolemy used arithmetic extensively, his framework was predominantly geometric. Aryabhata's system integrated sophisticated algebraic techniques (like *kuṭṭākāra*) directly into astronomical problem-solving, reflecting the strength of Indian algebra.

7.3 Legacy and Influence

Both systems were monumental intellectual achievements that profoundly shaped the future course of astronomy, albeit through different pathways:

- **Ptolemy's Dominance:** The *Almagest*, with its comprehensive scope, detailed geometric models, extensive tables, and catalogue of stars, became the undisputed authority in astronomy for the Islamic world and medieval/Renaissance Europe. Its geometric models served as the essential foundation for astronomical thinking and practice for over 1400 years, acting as the baseline against which later astronomers (including Copernicus) reacted and developed their own theories.
- **Aryabhata's Computational Legacy:** While Aryabhata's specific cosmological model (especially Earth rotation) faced criticism and didn't achieve the same universal dominance as Ptolemy's, his **mathematical innovations** had a transformative, though often indirect, global impact. His sine function (via *jayb* to *sinus*), his computational algorithms, and the decimal place-value system he clearly articulated became fundamental tools for later science worldwide, transmitted largely via Islamic intermediaries. Within India, his work established a powerful computational (*gaṇita*-based) tradition in astronomy that flourished for centuries, inspiring critical engagement (Brahmagupta) and remarkable advancements (Kerala School).

Conclusion: Comparing Aryabhata and Ptolemy reveals not a simple contest of "right" versus "wrong," but two distinct, brilliant approaches to modeling the cosmos based on the tools and perspectives of their respective traditions. Ptolemy offered a geometric masterpiece that described the heavens with remarkable accuracy for centuries, profoundly shaping Western and Islamic astronomy through its comprehensive system. Aryabhata presented a computationally elegant system, incorporating radical ideas like Earth's rotation and powerful mathematical techniques like the sine function and algebraic algorithms; while his complete model saw less direct global adoption, the transmission of his mathematical tools fundamentally reshaped global numeracy and provided essential building blocks for later scientific development. Both stand as testaments to the power of human intellect to find mathematical order in the celestial sphere.

***

Chapter Eight

Ripples and Refinements: Brahmagupta's Critique and the Kerala Legacy

India's Scientific Dialogue: How Aryabhata's Legacy Was Tested, Transformed, and Advanced Within the Subcontinent

8.0 Introduction: The Dialogue Within India

While Aryabhata's *Aryabhatiya* laid a new foundation for mathematical astronomy in India, its influence was not monolithic or instantaneous. Like any significant scientific advance, it sparked debate, invited criticism, and inspired further refinement within the vibrant intellectual traditions of the subcontinent. This chapter traces the internal Indian dialogue surrounding Aryabhata's work, focusing on two crucial developments: the pointed critiques leveled by the influential 7th-century astronomer Brahmagupta, and the remarkable mathematical and astronomical advancements made centuries later by the Kerala School, which explicitly built upon and extended the Aryabhatan lineage. This internal trajectory reveals a dynamic process of engagement, correction, and innovation that characterized Indian science for over a millennium.

8.1 Brahmagupta's Critique: A Scholarly Clash (c. 628 CE)

No significant intellectual advance proceeds unchallenged, and Aryabhata's bold reforms soon invited scrutiny from another giant of Indian mathematics and astronomy: **Brahmagupta** (c. 598–c. 670 CE). Composing his magnum opus, the ***Brāhmasphuṭasiddhānta*** ("Correctly Established Doctrine of Brahma," BSS) around 628 CE, Brahmagupta made immense contributions himself, particularly in algebra (systematizing solutions to quadratic equations, Pell's equation, and indeterminate equations) and providing his own comprehensive astronomical system based on traditional frameworks but with refined parameters.

However, Brahmagupta explicitly and sometimes sharply criticized several key aspects of Aryabhata's work, particularly concerning cosmological principles and fundamental astronomical parameters. His critique, primarily found in Chapter XI (*Tantra-parīkṣā-adhyāya* - Examination of Systems) and elsewhere in the BSS, highlights the tension between Aryabhata's innovative, mathematically driven approach and the more established cosmological and astronomical views Brahmagupta represented.

8.1.1 Rejecting Earth's Rotation

Brahmagupta staunchly rejected Aryabhata's most radical departure: the theory of the Earth's axial rotation (Gola.9). Aligning himself with traditional static-Earth cosmologies (like the *Pravaha* wind model presented in Gola.10) and arguments perhaps influenced by Aristotelian physics (which may have reached India), he forcefully refuted the idea:

"If the earth rotates, then towers, walls, towns, etc., would fall down, and birds flying in the sky would not return to their nests when they fly towards the west. If the earth moves one *yojana* in a *prāṇa*, then arrows shot towards the east would fall at the place where the archer is, and clouds would move towards the west."*(Paraphrased interpretation of arguments in BSS XI, often cited or alluded to)*

Brahmagupta argued against axial rotation based on:

- **Scriptural/Traditional Authority:** Implicitly upholding the standard cosmological view.
- **Common Sense Perception:** The Earth feels stationary.
- **Lack of Observable Physical Effects:** He argued (incorrectly from a modern perspective, but logically from his framework) that rotation should produce observable effects like objects being flung off, disruption of flight paths relative to the ground, and constant strong winds – none of which were perceived.

For Brahmagupta, Aryabhata's proposal violated both established knowledge and empirical intuition. He preferred the traditional *Pravaha* wind explanation for diurnal motion.

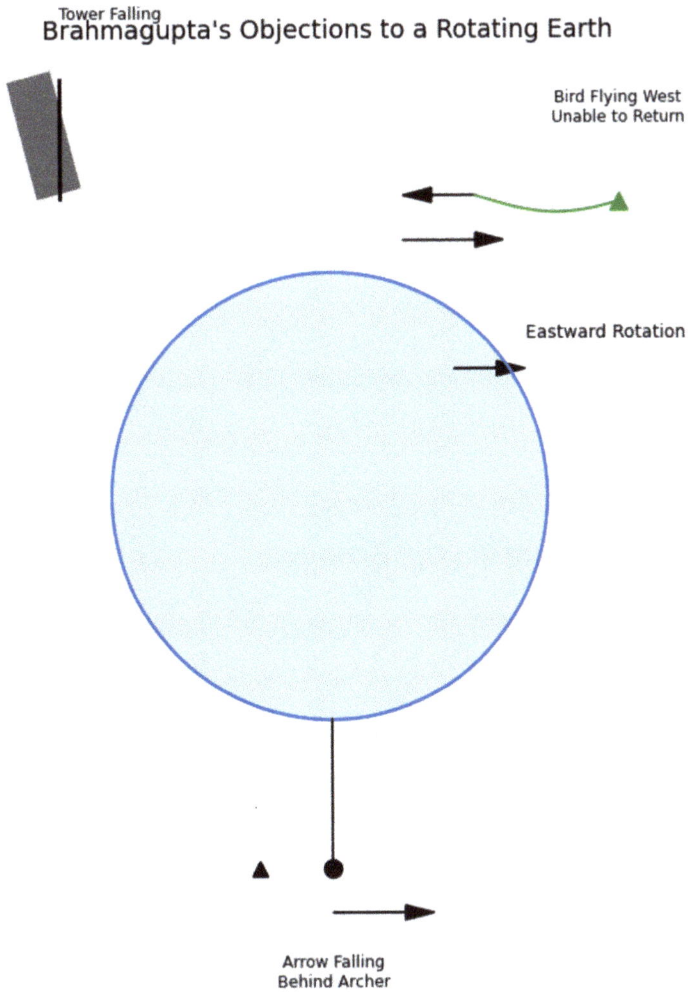

Figure 8.1: Diagram illustrating some of the physical objections raised by Brahmagupta and other traditional astronomers against Aryabhata's theory of Earth's axial rotation. They argued that if the Earth were spinning rapidly, phenomena like falling towers, birds flying westward, and projectile motion would be observably affected, which they claimed was not the case.

8.1.2 Battles Over Numbers: Parameters and π

The disagreements extended to core astronomical data and computational methods:

- **Lunar Revolutions:** Their values for the Moon's sidereal revolutions per *mahāyuga* differed slightly but significantly:
    - Aryabhata (Gk.3): 57,753,336

- Brahmagupta (BSS I.19): 57,753,300 (a difference of 36)This implies distinct underlying models or observational data for lunar motion, leading Brahmagupta to criticize Aryabhata's parameters.

- **Division:** Brahmagupta adhered to the traditional Puranic **unequal division** of the *mahāyuga* (4:3:2:1 ratio for Kṛta, Tretā, Dvāpara, Kali), rejecting Aryabhata's mathematically convenient **equal division** (*yugapādas*) championed by Bhaskara I.

- **Value of π:** While Aryabhata provided the highly accurate $\pi \approx 3.1416$ (Gp.10), Brahmagupta, in the mathematical sections of the BSS (BSS XII.40), primarily recommended using the less accurate but computationally simpler $\pi \approx \sqrt{10} \approx 3.1622$ for practical mensuration. While Brahmagupta was undoubtedly a highly capable mathematician likely aware of better approximations (perhaps including Aryabhata's), his choice reflects a different priority in this context – valuing ease of calculation for everyday geometric problems over theoretical precision.

8.1.3 A Productive Clash: Advancing the Dialogue

Despite the sometimes acerbic tone of his critique (common in scholarly debates of the era), Brahmagupta's engagement with Aryabhata's work was deep and ultimately productive for the development of Indian science.

- **Acknowledged Influence:** Brahmagupta clearly studied the *Aryabhatiya* closely. He adopted or refined several of its mathematical techniques, particularly in algebra (his work on indeterminate equations significantly built upon Aryabhata's *kuṭṭākāra*) and trigonometry (he provided his own methods and likely used sine concepts).

- **Stimulating Debate:** Their disagreements, particularly on Earth's rotation and fundamental parameters, fueled intense scholarly debate for centuries. This compelled subsequent astronomers to re-examine assumptions, justify their chosen parameters, perhaps refine observational techniques to settle disputes, and generally enhance the mathematical rigor and logical consistency of their arguments.

- **Establishing Traditions:** This clash helped solidify two major schools or computational traditions (*pakṣas*) within Indian astronomy: the ***Ārya-pakṣa*** (following Aryabhata, often using his parameters and midnight epoch, distinct from Aryabhata's own sunrise epoch but associated with his school) and the ***Brāhma-pakṣa*** (following Brahmagupta's parameters and system). Later astronomers often calculated results according to both systems or attempted syntheses.

Brahmagupta's critique represents a vital characteristic of a mature scientific tradition: foundational ideas are subject to intense scrutiny, empirical testing (within the limits of the time), and refinement through critical dialogue. His work ensured that Aryabhata's ideas were not accepted uncritically but became part of an ongoing scientific conversation that drove further progress.

8.2 The Kerala School: Calculus Precursors and Model Refinements (c. 14th–16th Centuries CE)

Centuries after the debates between Aryabhata's defenders and critics, a remarkable and somewhat isolated flourishing of mathematical and astronomical thought occurred in the southwestern region of Kerala. A lineage of brilliant scholars, consciously identifying with and extending the Aryabhatan tradition, developed highly sophisticated techniques involving **infinite series expansions** and iterative methods that contain clear **precursors to modern calculus**, significantly predating similar discoveries documented in Europe.

8.2.1 Mādhava and the Leap to Infinite Series

Mādhava of Sangamagrāma (c. 1340 – c. 1425) stands as the fountainhead of this movement. A brilliant mathematician-astronomer, he dramatically extended Aryabhata's trigonometry. Where Aryabhata had provided a table of 24 sine (*jyā*) values calculated using an ingenious finite difference method (Gp.12), Mādhava pioneered the use of **infinite series equivalent to the Taylor series expansions** for sine, cosine, and arctangent functions.

- **Transcending Finite Differences:** Aryabhata's second-order difference method yielded accurate sine values at discrete 3°45' intervals. Mādhava sought methods to calculate sine values for *any* angle with arbitrary precision.
- **Infinite Series Equivalents:** Though expressed in the geometric and algebraic language of medieval Kerala (often using recursive relations or verbal algorithms), his series captured the essence of infinite summation and convergence. For example, his series for sine, translated into modern notation, is:$\sin(x) \approx x - x^3/3! + x^5/5! - x^7/7! + \ldots$Similarly, he developed series for cosine and arctangent (the latter being equivalent to the Gregory-Leibniz series).

Unprecedented Precision for π: Using his arctangent series and sophisticated convergence acceleration techniques, Mādhava computed the value of π **to an astonishing accuracy of at least 11 decimal places** (3.141 59265359). This precision surpassed all known contemporary global approximations and remained unsurpassed outside India for nearly two centuries, highlighting the extraordinary conceptual leap made by the Kerala School.

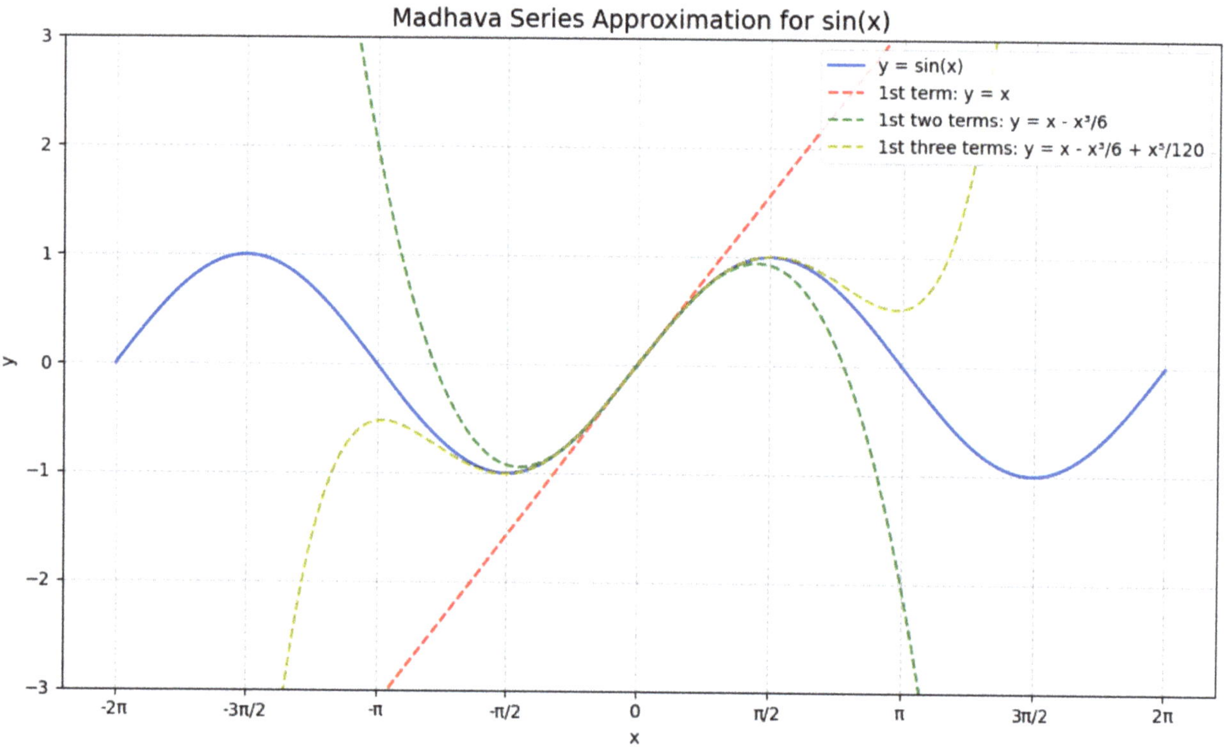

Figure 8.2: Graph illustrating how successive terms of the Madhava series (equivalent to the Taylor series for sine) approximate the sine function (blue curve). Mādhava of Sangamagrāma (c. 14th c.), building on the Aryabhatan tradition, developed infinite series expansions for trigonometric functions. Adding more terms (red, green, yellow dashed lines) provides progressively more accurate approximations of the sine value for a given angle.

8.2.2 The Guru-Śisya Paramparā

Mādhava founded an influential school or lineage (*guru-śiṣya paramparā*). His successors meticulously documented, refined, and extended these methods in seminal texts:

- **Parameśvara** (c. 1370–1460): Known for observational work and revisions of astronomical parameters.
- **Nīlakaṇṭha Somayājī** (c. 1444–1544): Author of the comprehensive astronomical treatise *Tantrasaṅgraha* and a detailed commentary (*bhāṣya*) on the *Aryabhatiya*. He explicitly acknowledged his debt to Mādhava and aimed to update Aryabhata's system.
- **Jyeṣṭhadeva** (c. 1500–1575): Author of the ***Yuktibhāṣā*** ("Rationale-Language" or "Exposition of Proofs"). This unique work, written in the local Malayalam language rather than Sanskrit, is particularly remarkable because it provides detailed logical rationales and **proofs (*yukti*)** for many of the infinite series and other theorems developed by Mādhava and the school, offering invaluable insight into their mathematical reasoning.

(Note on Calculus Precursors: It is crucial to understand the context. While these series are mathematically equivalent to Taylor series and demonstrate mastery of infinite processes and iterative approximation, they were derived and understood algorithmically and geometrically, focusing on numerical computation for astronomical and trigonometric problems. They lacked the formal analytic framework based on limits, derivatives, and integrals established by Newton and Leibniz centuries later. Hence, the term "precursor" accurately reflects both the profound insights achieved and the different conceptual foundations.) [Ref: Plofker, K. (2009). Mathematics in India. Princeton University Press; Joseph, G.G. (2011). The Crest of the Peacock. Princeton University Press.]

8.2.3 Fusion of Systems: Nilakantha's Geo-Heliocentric Model

This advanced mathematics was often applied to refine astronomical models. Nilakaṇṭha Somayājī, in his *Tantrasaṅgraha* and *Aryabhatiya Bhashya*, proposed a revised planetary model that represents a significant step towards heliocentric ideas, while still retaining a central Earth.

- **Nilakaṇṭha's Model:** In his system, the five naked-eye planets (Mercury, Venus, Mars, Jupiter, Saturn) orbit the Sun, which in turn orbits the stationary Earth. The Moon also orbits the Earth.
- **Computational Advantages:** This **geo-heliocentric model** offered a more accurate geometric explanation for the bounded elongation of inner planets and the retrograde motions of outer planets than purely geocentric epicyclic models (like Ptolemy's or even Aryabhata's original description). It allowed for more accurate calculations of planetary positions.
- **Independent Development:** Developed independently and significantly earlier than Tycho Brahe's similar model in late 16th-century Europe, Nīlakaṇṭha's system demonstrates a skillful blending of traditional geocentric frameworks with innovative mathematical and observational refinements driven by the desire for greater accuracy.

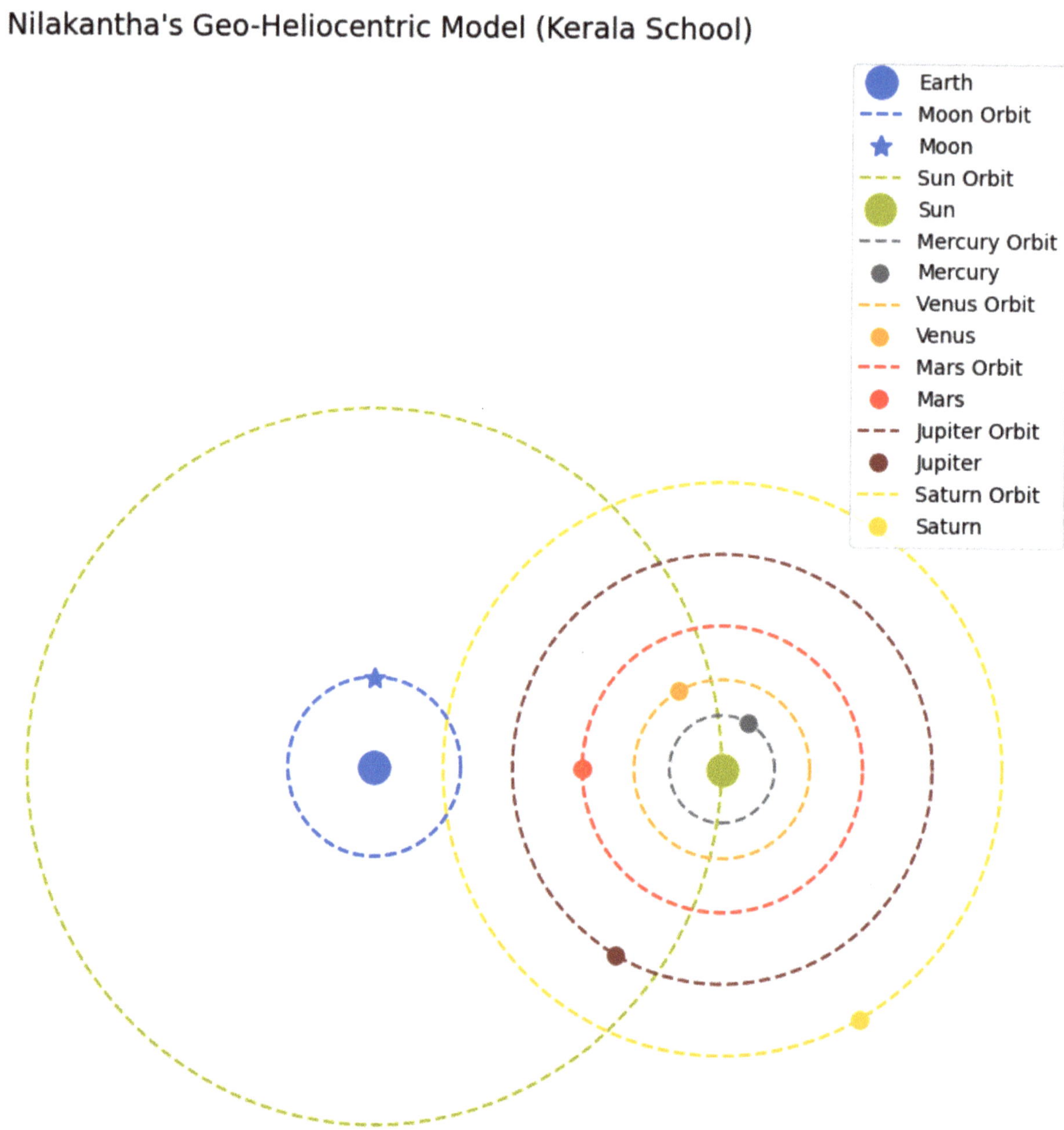

Figure 8.3: Diagram illustrating the geo-heliocentric planetary model proposed by Nīlakaṇṭha Somayājī of the Kerala School (c. 15th c.). In this model, the Moon and Sun orbit the central Earth, while the other five naked-eye planets (Mercury, Venus, Mars, Jupiter, and Saturn) orbit the Sun. This model represents a significant conceptual refinement within the later Aryabhatan tradition, providing a more accurate geometric explanation for certain observed planetary motions.

8.2.4 Continuity Through Computation

The Kerala astronomers explicitly saw themselves as continuing and improving Aryabhata's legacy. They frequently cited the *Aryabhatiya* and wrote commentaries on it. Core computational techniques inherited from the Aryabhatan tradition, such as the ***kuṭṭākāra*** algorithm for solving linear indeterminate equations (Gp.32-33, extensively elaborated by Bhaskara I), remained fundamental tools for calendrical computations (*pañcāṅga*) and precise astronomical predictions.

This sustained focus on computational power and refinement ensured the vitality of Indian mathematical astronomy well into the early modern period.

The work of Mādhava and the Kerala School represents a zenith in pre-modern world mathematics, bridging ancient trigonometry and the beginnings of analysis. By framing their groundbreaking work as commentary and refinement (*vyākhyāna*) upon the Aryabhatan tradition, they quietly extended a scientific lineage Aryabhata had catalyzed nearly a millennium earlier. Their work embodies the same spirit of rigorous inquiry, computational precision, and cosmic exploration that defined the best of the Aryabhatan tradition, showcasing its enduring influence within India.

***

Chapter Nine

The Global Journey: Transmission to the Islamic World and Europe

Seeds Across Continents: The Enduring Legacy of Indian Mathematics and Astronomy Abroad

9.0 Introduction: Seeds Across Continents

While Aryabhata's work and the tradition it spawned flourished within India, its influence did not remain confined to the subcontinent. Key mathematical and astronomical concepts originating from Aryabhata and the broader Indian scientific milieu embarked on a complex journey westward, significantly shaping scientific developments, particularly during the **Islamic Golden Age** (roughly 8th-13th centuries CE), and subsequently contributing to the mathematical foundations of medieval and Renaissance Europe. This transmission occurred through intricate pathways involving trade routes (linking India with Persia and the Mediterranean), diplomatic missions, dedicated translation efforts, and scholarly exchange, demonstrating the interconnectedness of intellectual history across cultures.

Figure 9.1: Conceptual map tracing the westward flow of Indian mathematical and astronomical concepts

9.1 The transmission of knowledge

As Indian astronomy (*jyotiḥśāstra*) developed sophisticated computational methods, this knowledge gradually spread westward, finding fertile ground in the burgeoning intellectual centers of the nascent Islamic world. A crucial vehicle for this early transmission appears to have been astronomical texts and tables brought to the court of the second Abbasid Caliph, **al-Mansur** (reigned 754–775 CE), in Baghdad around 770-773 CE, reportedly by an Indian embassy.

- **The *Zīj al-Sindhind*:** These Indian works, referred to in Arabic sources as the ***Sindhind***, were translated into Arabic by scholars like Muḥammad ibn Ibrahim **al-Fazārī** and Ya'qūb ibn Ṭāriq. The term *Sindhind* itself is widely accepted by scholars as an Arabic adaptation of the Sanskrit word ***Siddhānta*** ("established doctrine" or astronomical treatise), explicitly signaling its Indian origins. While the exact Indian text(s) translated remain debated, many scholars believe they were heavily based on, or directly versions of, **Brahmagupta's *Brāhmasphuṭasiddhānta*** (c. 628 CE). Since Brahmagupta's work itself engaged critically with and built upon Aryabhata's system, the *Sindhind* indirectly incorporated key elements traceable to the Aryabhatan school, including:
    - The decimal place-value system and operations with zero.
    - Trigonometric methods based on the sine function (*jyā*).
    - Specific astronomical parameters and *Yuga*-based time reckoning concepts.
    - Computational algorithms for planetary positions, possibly including methods related to the *kuṭṭākāra*.
- **Integration into Islamic Astronomy:** The *Zīj al-Sindhind* represented one of the earliest systematic efforts to integrate these powerful Indian computational techniques into the Islamic astronomical tradition. This Indian stream flowed alongside, and dynamically interacted with, the geometric planetary models inherited from the Greeks (primarily Ptolemy's *Almagest*, translated as *al-Majisṭī*) and observational practices potentially influenced by Persian and Sasanian traditions. This synthesis created a uniquely powerful and innovative scientific environment in centers like Baghdad.

9.2 Al-Khwārizmī: The Algorithmic Conduit

A pivotal figure in absorbing, synthesizing, and disseminating these Indian methods was **Muḥammad ibn Mūsā al-Khwārizmī** (c. 780–c. 850 CE), a Persian scholar associated with Baghdad's renowned House of Wisdom (*Bayt al-Ḥikmah*). His own astronomical tables (*Zīj*) were heavily influenced by the *Sindhind* tradition, adapting Indian computational methods for use within the Islamic calendar and geographical context. More profoundly for global history, Al-Khwārizmī authored foundational works that introduced key Indian mathematical concepts to the Arabic-speaking world and, through later Latin translations, to Europe:

- **The Hindu-Arabic Numeral System:** His treatise *Kitāb al-Jamʿ wa-l-tafrīq bi-ḥisāb al-Hind* ("The Book of Addition and Subtraction According to the Hindu Calculation"), now surviving mainly through its 12th-century Latin translation, ***Algoritmi de numero Indorum*** ("Al-Khwārizmī on the Hindu Art of Reckoning"), provided a clear, practical exposition of the Indian positional decimal system (digits 1-9) and the crucial concept and computational use of **zero** (*ṣifr*, derived from Sanskrit *śūnya*). This work was instrumental in the system's eventual westward diffusion, and Al-Khwārizmī's own name, Latinized as "***Algoritmi***," became the etymological root of the modern term "**algorithm**," forever linking him to the procedural methods he helped popularize.

- **Algebra:** His most famous work, *Al-Kitāb al-mukhtaṣar fī ḥisāb al-jabr wa'l-muqābala* ("The Compendious Book on Calculation by Completion and Balancing"), systematized the solving of linear and quadratic equations, drawing potentially on Indian, Babylonian, and Greek sources but presenting them in a novel, practical framework. The term "***al-jabr***" from its title gave rise to the word "**algebra**." While distinct from Aryabhata's specific algebraic methods (like *kuṭṭākāra*), Al-Khwarizmi's work established algebra as a distinct mathematical discipline in the Islamic world and later in Europe.

- **Trigonometry:** Al-Khwārizmī's *Zīj* incorporated sine tables derived from Indian methods (***jyā***). The Indian half-chord function (*jyā*) was adopted into Arabic as ***jayb***. This term, through a complex history involving transliteration and later Latin translation (likely by translators like Gerard of Cremona or Robert of Chester working in Spain in the 12th century), was rendered as ***sinus*** (Latin for "bay," "curve," or "fold," possibly due to a misreading of the unvowelled Arabic script *jb* as *jaib* meaning "pocket" or "bosom"). This linguistic journey ultimately gave the world the modern trigonometric term "**sine**." Subsequent Islamic astronomers, such as **al-Battānī** (c. 858–929 CE), further refined sine tables to high accuracy and developed tables for other trigonometric functions (tangent, cotangent, secant, cosecant), building upon the Indian foundation transmitted via works like the *Sindhind* and Al-Khwārizmī's *Zīj*.

Al-Khwārizmī and his contemporaries did not merely act as passive transmitters; they actively synthesized Indian computational techniques with Greek geometric models and algebraic methods, fostering centuries of remarkable mathematical and astronomical advancement within the diverse cultures of the Islamic world.

9.3 Al-Bīrūnī: The Critical Observer and Documenter

Centuries after the initial transmission via the *Sindhind*, another Persian polymath, **Abū Rayḥān al-Bīrūnī** (c. 973–1048 CE), provided an unparalleled window into Indian science. Following extensive travels and residency in regions of present-day Pakistan and Northwest India (accompanying the campaigns of Mahmud of Ghazni) and learning Sanskrit, Al-Bīrūnī authored his monumental *Kitāb Taḥqīq mā li-l-Hind* ("An Inquiry into India," commonly known as ***Alberuni's India***), completed around 1030 CE. This work offers an extraordinarily detailed, critical, and often objective account of Indian sciences, mathematics, philosophy, religion, and society, based on his direct engagement with Sanskrit texts and discussions with Indian scholars (*paṇḍitas*).

- **Discussion of Aryabhata:** Al-Bīrūnī discussed Aryabhata's ideas, including his parameters, *yuga* theory, and the controversial theory of **Earth's rotation**, with remarkable knowledge and insight. He recognized Aryabhata's stature, referring to him as one of "the most famous of their astronomers," and accurately presented the relative motion argument for Earth's rotation.

- **Critique of Rotation:** However, reflecting the dominant scientific consensus of his time, which was heavily influenced by Aristotelian physics and Ptolemaic astronomy, Al-Bīrūnī ultimately **rejected** the theory of Earth's rotation. Like Brahmagupta before him (whose criticisms Al-Bīrūnī also discusses), he pointed to the lack of compelling physical arguments or direct observational proof available to counter the standard objections. He meticulously documented the arguments both for and against but sided with the static Earth model prevalent in the mainstream Greco-Arabic tradition.

- **Preservation of Knowledge:** Despite his critical assessments on certain points, Al-Bīrūnī's meticulous documentation **preserved and transmitted** an immense amount of accurate information about various schools of Indian astronomy and mathematics, including specific details about Aryabhata's parameters, methods (like the sine table), and cosmological ideas. His work remains an invaluable source for understanding the state of Indian science in the 11th century and its historical context.

Al-Bīrūnī's work exemplifies the sophisticated level of cross-cultural scientific dialogue occurring during this period, characterized by both admiration for Indian computational prowess and critical evaluation based on the established scientific frameworks of the Islamic world.

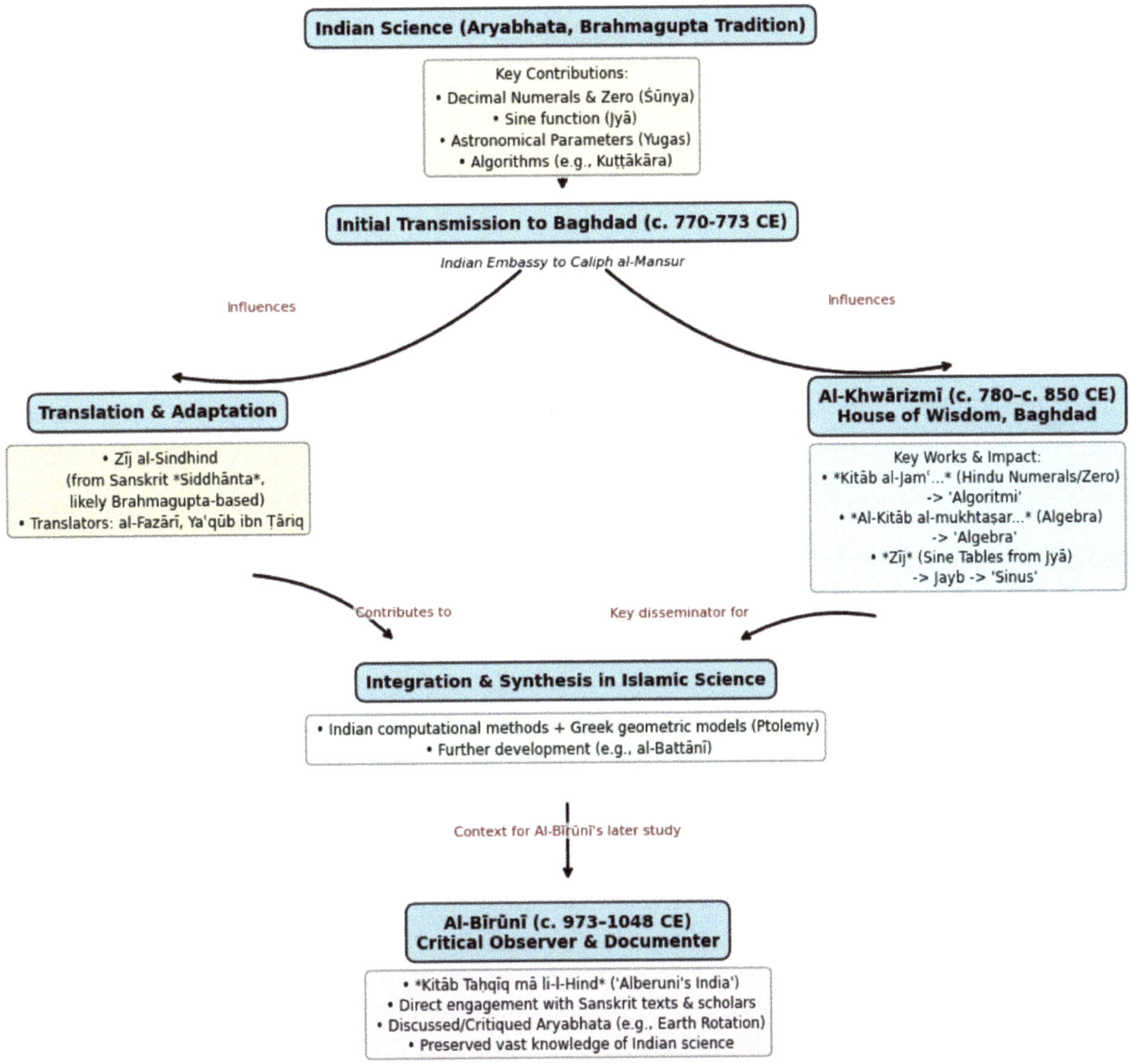

Figure 9.2: Flowchart illustrating Indian science's transmission to the Islamic world: initial transfer (embassies, Zīj al-Sindhind) from Aryabhata/Brahmagupta, Al-Khwārizmī's pivotal dissemination of concepts (decimals, zero, sine, algebra), and Al-Bīrūnī's critical documentation.

9.4 Europe's Numerical Revolution: The Echoes Reach the West

While Indian mathematical and astronomical concepts profoundly shaped Islamic science, a related but distinct transformation, heavily mediated by Arabic scholarship, began slowly unfolding in medieval Europe: the adoption of the **Hindu-Arabic numeral system**. This "numerical revolution," though gradual, fundamentally altered European mathematics, commerce, and eventually provided the computational bedrock for the Scientific Revolution. A key catalyst in this process was Leonardo of Pisa, better known as **Fibonacci** (c. 1170 – c. 1250), and his highly influential book ***Liber Abaci*** ("Book of Calculation" or "Book of the Abacus," first edition 1202 CE, significantly revised 1228 CE).

- **Fibonacci's Role:** Having traveled and studied mathematics with Arab teachers in North Africa (particularly Béjaïa in modern Algeria, a major trading port), Fibonacci recognized the immense practical superiority of the Indian positional decimal system (using digits 0-9 with place value, which he called *novem cifre indorum plus zephirum*) over the cumbersome Roman numerals and the complex abacus-based methods then dominant in European commerce and scholarship. His *Liber Abaci* was not merely a theoretical text but a practical manual that systematically introduced and strongly advocated for this "***modus Indorum***" (method of the Indians) for performing arithmetic operations (addition, subtraction, multiplication, division, fractions) and solving a wide range of practical problems in bookkeeping, interest calculation, currency conversion, and algebra.

- **The Journey of Numerals:** The conceptual origin of this system lies squarely in India. While Aryabhata himself used an alphabetic positional system (Gk.2), demonstrating mastery of the place value principle, the distinct glyphs for digits 1-9 and, crucially, the operational symbol for zero (*śūnya*), evolved within India over the preceding centuries. This fully formed decimal system, clearly articulated with rules for zero by Brahmagupta (BSS XVIII.30-35), traveled westward through Persia and the Arab world. Al-Khwārizmī's work (*De numero Indorum*) was vital in its formal transmission to the Arabic world, and Fibonacci became its most effective champion and popularizer in Latin Europe.

- Gradual Adoption: The adoption of Hindu-Arabic numerals in Europe was not immediate. It faced resistance for centuries from traditionalists accustomed to the abacus and Roman numerals, and sometimes suspicion due to its **association with Arab culture** (or **its origins in the Islamic world**). However, its undeniable efficiency for complex calculations eventually led to its triumph, particularly among merchants (the maestri d'abbaco in Italy) and later within universities and scientific circles during the Renaissance. This streamlined calculation facilitated complex bookkeeping, and provided the essential mathematical language for the burgeoning scientific advancements of the 16th and 17th centuries.

9.5 The Complex Journey of Trigonometry

While the numerals found eventual, widespread acceptance, the sophisticated trigonometry developed in India (***jyā***) and significantly refined in the Islamic world experienced a more fragmented and delayed transmission into the mainstream of European mathematics.

- **Awareness vs. Adoption:** Aryabhata's *jyā* (sine) concept and table (Gp.10, Gp.12), along with the advanced trigonometric knowledge of Islamic scholars (including tables for other functions), were known to European translators working primarily in Spain (Toledo) and Sicily from the 12th century onwards. The term ***sinus*** entered the Latin vocabulary through these translations.

- **Dominance of Chord Method:** However, the underlying computational methods from India (like Aryabhata's difference technique) or the highly advanced infinite series developed later by the Kerala School (which had no known direct transmission pathway to Europe at this time) seem to have been less readily adopted or fully understood initially compared to the revival of Greek geometric methods based on Ptolemy's **chords**. Figures like Georg von Peurbach (1423–1461) and his student **Regiomontanus** (Johannes Müller von Königsberg, 1436–1476) played a key role in reconstructing and promoting Ptolemaic trigonometry based on chords in the 15th century.

- **European Reconstruction:** Consequently, European trigonometry largely had to be rebuilt, initially on these revived geometric foundations. Only later, during the 16th and 17th centuries, did European mathematicians like Viète, Copernicus (in the trigonometric sections of *De Revolutionibus*), Rheticus (who produced extensive sine tables), and ultimately Newton and Euler develop their own powerful analytical framework for trigonometry, fully integrating the sine function and its relatives within the emerging calculus.

Thus, while the concept and the name of the sine function reached Europe from India via Arabic intermediaries, the primary computational engine and theoretical framework for trigonometry in Europe during the early Renaissance were initially reconstructed more directly from Greek sources, before being transformed by later European innovations.

9.6 Global Impact Summary

Fibonacci's introduction of Hindu-Arabic numerals, rooted ultimately in Indian mathematical innovations exemplified and clearly articulated by figures like Aryabhata and Brahmagupta, marked a crucial node in centuries of transcontinental knowledge exchange. While other sophisticated Indian contributions followed more complex or delayed paths into European science, the numerical system itself sparked a transformation that irrevocably shaped the mathematical capabilities of the modern world.

The table below summarizes the transmission and impact of key concepts associated with the Aryabhata tradition:

Concept	Origin in India	Transmission & Development via Islamic World	Impact in Europe	Notes
Positional Numerals	Principle of place value (Aryabhata, Gp.2). Digits 1-9 and placeholder Śūnya (Zero) evolved within the tradition (Brahmagupta explicit).	Adopted & systematized (Al-Khwārizmī, De numero Indorum). Widely used, foundational for calculation (ḥisāb al-hindī).	Introduced & popularized (Fibonacci, Liber Abaci, 1202 CE). Led to gradual adoption (13th-16th c.), revolutionizing arithmetic & commerce.	Basis of modern global numeral system. Transmission primarily via Arabic works.
Zero (Śūnya)	Concept of zero as a placeholder & number with computational rules (Brahmagupta, BSS XVIII). Implicit in Aryabhata's place value.	Essential component of the positional system (Al-Khwārizmī, ṣifr). Used computationally in arithmetic & algebra.	Introduced with numerals (cifra/zephirum). Essential for place value. Met initial resistance but ultimately indispensable for modern math.	Transformed mathematics, enabling algorithms, algebra, and calculus.
Trigonometry (Sine)	Sine function (Jyā) as half-chord length. Table of differences & generative rule (Aryabhata, Gk.10, 12). Infinite series expansions (Kerala School).	Adopted (jayb), tables calculated & refined (al-Battānī), expanded to other functions (tangent, etc.). Central to Islamic astronomy.	Term "sinus" was adopted (via Latin translations, 12th c.). Concept known, but geometric methods (Ptolemy's chords) more influential initially than Indian computational ones.	European trigonometry was rebuilt geometrically (Regiomontanus) then analytically (16th-17th c.). No direct transmission of the Kerala series was known until much later.

Concept	Origin in India	Transmission & Development via Islamic World	Impact in Europe	Notes
Earth's Rotation Theory	Proposed by Aryabhata (Gola.9) as a kinematic explanation via relative motion.	Discussed (Al-Bīrūnī) but largely rejected by mainstream based on physical/Aristotelian arguments & lack of perceived effects.	Idea known but gained little traction in the West until Copernicus & heliocentric context (16th c.) provided a new framework for physical arguments.	Remained a minority view for over a millennium globally. Conceptually significant but limited immediate impact on dominant astronomical models.
Algebraic Methods	Methods for solving linear indeterminate equations (Kuṭṭākāra, Aryabhata Gp.32-33, Brahmagupta). Various solutions for quadratic equations.	Significant developments in algebra as a discipline (Al-jabr, Al-Khwārizmī). Integrated solutions for various equation types. Foundational for Islamic mathematics.	Influenced European algebra via translations (e.g., of Al-Khwārizmī's Al-jabr). Contributed to establishing algebra as a formal discipline in Europe.	Indian methods (like Kuṭṭākāra for specifics) influenced Islamic algebra, which then strongly shaped European developments in the field.

Note: Influence often occurred through multiple stages, intermediaries, and synthesis with other traditions (Greek, Persian). Direct text-to-text transmission was less common than the migration and adaptation of concepts and methods. Some eastward transmission to China also occurred.

***

Chapter Ten

Cosmic Harmony, Mathematics & Ethics

The Soul of the Aryabhatiya: Finding Order in the Universe and Virtue in Knowledge

10.0 Introduction: Beyond Calculation

Aryabhata's *Aryabhatiya* stands as a monument to mathematical ingenuity and astronomical precision. Yet, beyond the intricate calculations, algorithms, and cosmological models, his work, particularly when viewed through the lens of its cultural and philosophical context in Gupta India, reveals deeper dimensions. These concern the perceived nature of the cosmos, the profound role attributed to mathematics in understanding it, and the ethical responsibilities inherent in the pursuit and preservation of scientific knowledge. Aryabhata's scientific endeavors appear embedded within a worldview that perceived the universe not merely as a collection of phenomena, but as an ordered, harmonious system (***ṛta***), accessible and comprehensible through the rigorous application of rational inquiry, epitomized by mathematics (***gaṇita***).

10.1 Cosmic Harmony & Mathematics as Divine Language

Aryabhata's astronomy seems rooted in a conviction, prevalent across various schools of Indian philosophy (including strands of Vedānta, Sāṃkhya, and Nyāya-Vaiśeṣika), that the universe (*jagat* or *brahmāṇḍa*) possesses an inherent structure, rhythm, and rational order (***ṛta*** or *dharma* in a cosmic sense) that can be apprehended through the precise language of mathematics. This perspective intimately linked scientific investigation with broader metaphysical ideas about reality, causality, and cosmic cycles.

10.1.1 The Invocation to Brahman and Mathematical Order

The very opening verse of the *Aryabhatiya* (Gītikāpāda 1 / Chapter 3.1) establishes this connection:

> **"Having paid respectful obeisance to Ka (Brahma/Prajāpati), the One (*ekam*) and the Many (*anekam*), the True Divinity (*satyāṃ devatām*), the Supreme Brahman (*paraṃ brahma*), Aryabhata sets forth three things: mathematics (*gaṇitaṃ*), the reckoning of time (*kālakriyāṃ*), and the sphere (*golaṃ*)."**

Scholars debate the precise intent—whether primarily devotional, merely conventional, or deeply philosophical. However, a compelling interpretation suggests it frames the entire scientific exposition within a worldview where **Brahman** represents the ultimate, unifying reality or principle underlying the diverse phenomena of the cosmos ("one and many"). In this view, **mathematics (** is presented not just as a human tool, but as the essential language or structure through which this unified, rational order manifests and can be understood. Exploring mathematical principles thus becomes a legitimate, perhaps even profound, pathway to comprehending the inherent harmony of the cosmos.

10.1.2 Cyclical Time and Mathematical Symmetry

The concept of vast, recurring cycles of time (*yugas, manvantaras, kalpas*) is central to traditional Indian cosmology, reflecting a view of the universe as eternally rhythmic. Aryabhata's adoption and specific, mathematically precise definition of these cycles served immediate practical purposes for astronomical calculation (ensuring planetary mean motions aligned over long periods). However, it also resonated with this philosophical understanding of eternal cosmic rhythms.

His significant innovation was defining the four sub-yugas (*yugapādas*) within a *mahāyuga* as having **equal durations** (1,080,000 years each), departing from the unequal 4:3:2:1 ratio found in many Purāṇic texts. This move, prioritizing **mathematical symmetry and computational coherence** over strict adherence to potentially inconsistent mythological chronologies, can be seen as an attempt to harmonize the concept of cosmic order (***ṛta***) with rational, calculable cycles. It suggests a belief that the underlying structure of cosmic time is fundamentally mathematical and symmetrical, a view vigorously defended by Bhaskara I based on the requirements of astronomical consistency.

10.1.3 Mathematics as the Language of Cosmic Structure

Within this framework, mathematics transcends its role as a mere descriptive or predictive tool; it becomes the very language through which the fundamental, rational structure of the universe reveals itself. The precise astronomical constants Aryabhata calculated—the specific number of planetary revolutions per *yuga*, the derived length of the sidereal year, the remarkably accurate value of π, the calculated diameter of the Earth—were likely seen not just as successful human measurements or approximations, but as discoveries reflecting the pre-existing, harmonious design embedded within the cosmos. Pursuing mathematics with rigor and precision was thus arguably perceived as a way to engage directly with this underlying reality, to read the "book of nature" written in the language of numbers and geometry.

10.1.4 Integration of Science and Philosophy

Aryabhata's work exemplifies the integration of scientific inquiry with broader philosophical perspectives often characteristic of Indian intellectual history. While he avoids explicit metaphysical speculation, the underlying assumption of a rationally ordered, mathematically describable universe aligns well with philosophical systems like Sāṃkhya (with its interplay of *puruṣa* and *prakṛti*) or certain schools of Vedānta (exploring the relationship between the ultimate Brahman and the manifest world). These philosophical backgrounds provided a context where mapping the observable cosmos mathematically could be understood as exploring the ordered expression of deeper truths or principles. Aryabhata's astronomy, therefore, was likely perceived by him and his intended audience as more than predictive science; it was an exploration of the rational harmony believed to permeate existence, a *darśana* (way of seeing/philosophy) grounded in calculation.

This vision, linking mathematical precision to cosmic order and potentially to ultimate reality, likely contributed significantly to the reverence accorded to Aryabhata's work and its enduring influence within the Indian scientific tradition, inspiring centuries of commentary, refinement, and further discovery by scholars who saw themselves as participating in this ongoing quest to understand the universe through the power of mathematics.

10.2 Knowledge Ethics: Integrity, Intellect, and Preservation

Embedded within Aryabhata's scientific text and the tradition that interpreted it are implicit but strong ethical considerations regarding the pursuit and dissemination of knowledge (*jñāna*). These concern the scholar's responsibility for accuracy, the integrity of methods, the crucial role of discerning intellect in achieving true understanding, and the importance of preserving knowledge faithfully.

10.2.1 Integrity and the Warning Against Distortion

Accuracy and intellectual honesty appear paramount. Astronomical knowledge, particularly concerning timekeeping, calendar construction (*pañcāṅga*), and eclipse prediction, had significant social, agricultural, and ritual importance in ancient India. Errors, whether accidental or deliberate, could undermine not only scientific credibility but also social order and trust. Aryabhata concludes his entire work with a stern warning in the final verse of the Golapāda:

> **"This [work], known from the beginning by the name *Aryabhatiya*..., conforms to self-existent [eternal truth] (*svāyaṃbhuvaṃ*) and is always true. He who creates a counterfeit/distortion (*pratikañcukaṃ*) of it causes the destruction of his own merit (*sukṛta*) and lifespan (*āyuṣoḥ*)."*(Gola.50)***

This powerful injunction against creating a ***pratikañcukaṃ*** (a counterfeit, plagiarism, or deliberate distortion) frames intellectual dishonesty as a grave ethical failing with profound personal consequences (loss of merit and life). It underscores the high value placed on preserving the integrity of foundational scientific knowledge and implicitly charges future scholars with the responsibility of accurate transmission.

Verse Gola.49 also speaks to this theme, contrasting the "supreme jewel of true knowledge" (*sajjñānottamaṃ ratnam*) rescued from the "ocean of true and false knowledge" (*sadasajjñānasamudrāt*), implying an ethical duty to distinguish truth from error.

10.2.2 Svamati: The Authority of Reasoned Intellect

Alongside adherence to correct methods and respect for tradition, Aryabhata highlights the vital role of the scholar's own reasoning faculty. Verse Gola.49 concludes by stating that this true knowledge was rescued "by me using the boat of my own intellect (***svamatināvā***)."

- ***Svamati*** ("own intellect," "own reasoning," "own mind") signifies the importance Aryabhata placed on individual critical intelligence (*buddhi*, *mati*) in the scientific process. While he clearly built upon and respected tradition (invoking Brahman, potentially acknowledging teachers), he asserts that true understanding and progress require the active engagement of the scholar's own rational analysis and verification.
- This implies an ethical duty for the scholar not merely to passively receive or transmit knowledge (*śruti*, received wisdom), but to actively understand, analyze, question, verify through calculation (*gaṇita*) and observation

(*pratyakṣa* where possible), and refine it.

- This potentially positions reasoned inquiry (***svamati***) as a valid, even necessary, means of knowledge (*pramāṇa*) alongside, or perhaps even capable of correcting, traditional authority or previous texts (*śāstra*) within the scientific domain. It reflects the dynamic tension between continuity and innovation characteristic of progressive scholarly traditions. Aryabhata's own bold reforms (equal yugas, Earth rotation) exemplify the application of *svamati* to challenge established norms based on mathematical reasoning.

10.2.3 Ethical Pursuit as a Path to Truth

In the worldview potentially informing Aryabhata, the ethical pursuit of knowledge—conducted with integrity, diligence, meticulousness (*pramāṇa*-based approach), intellectual honesty, and critical self-awareness—was likely seen as more than just a professional standard. It could be viewed as a disciplined path (*sādhana*) toward understanding deeper truths (*satya*) about the ordered cosmos (***ṛta***). This perspective imbued the scholarly endeavor (*śāstra abhyāsa*) with a profound sense of purpose and responsibility, likely contributing to the meticulousness and resilience observed in the long tradition of Indian mathematical astronomy.

10.2.4 Tradition and Innovation: An Ethical Balance

Aryabhata's work itself embodies the ethical balance between respecting inherited knowledge and daring to innovate. He operated within the existing *Siddhāntic* framework, used established terminology, and acknowledged foundational principles (implicitly or explicitly). Yet, he did not hesitate to challenge or reform aspects—like *yuga* lengths, the cause of diurnal motion, specific astronomical parameters, or perhaps even the volume of a sphere—where his own calculations (*gaṇita*) and reasoning (***svamati***) led him to different conclusions.

Verse Gola.49 further hints at this dynamic by suggesting he rescued true knowledge that was submerged in an ocean of both truth and falsehood. This implies that tradition itself is not monolithic or infallible; it requires critical evaluation by the discerning intellect to separate the "jewel of true knowledge" from accumulated errors or outdated concepts. The ethical scholar, therefore, has a responsibility both to preserve the authentic core of the tradition and to refine it through rigorous inquiry.

10.3 Conclusion: The Enduring Chord of Knowledge

Aryabhata's *Aryabhatiya*, viewed beyond its remarkable mathematical and astronomical contributions, implicitly embodies a profound ethical framework for the pursuit of scientific knowledge. It underscores the paramount importance of meticulous accuracy, intellectual honesty, and the faithful preservation of hard-won truths. Simultaneously, it champions the power of individual intellect (***svamati***) to critically engage with tradition, discern truth from falsehood through reasoned analysis and calculation, and innovate where necessary to achieve a more coherent and accurate understanding of the cosmos.

This synthesis of revering tradition while boldly employing independent reason reveals a sophisticated understanding of knowledge. The perceived harmony of the universe, accessible through the language of mathematics, imbued this scientific quest with a deep sense of purpose. The ethical commitment to rigor and integrity, coupled with the intellectual freedom to question and refine, likely contributed significantly to the vitality and longevity of the scientific tradition Aryabhata inaugurated and profoundly shaped. While scientific models inevitably evolve and are superseded, this ethical core—the integrity of the method, the responsibility of the seeker, and the dynamic interplay between tradition and reasoned innovation—remains a timeless foundation for the scientific endeavor across all cultures and ages. The careful transmission of these ideas, often through meticulously copied palm-leaf manuscripts passed down through generations, further underscores the profound value placed on this intellectual and ethical heritage.

***

Chapter Eleven

Conclusion

The Enduring Spark of Aryabhata's Cosmos

As Aryabhata's verses fade from these pages, his vision reverberates across sixteen centuries, a radiant thread connecting his cosmos to ours. Writing this book has been a profound act of rediscovery, revealing a genius—barely in his early twenties when he composed the *Aryabhatiya*—whose intellect transcended the bounds of 5th-century India. How did he achieve such visionary foresight? With the curiosity of youth, observational tools, and a fearless drive to challenge tradition, Aryabhata wove keen observation and groundbreaking mathematical clarity into a universe where reason unveiled cosmic harmony, crafting insights that echo with astonishing modern resonance.

Born in the vibrant intellectual crucible of Gupta India, and his work further illuminated by Bhaskara I's vital interpretations, Aryabhata was more than an astronomer or mathematician; he was a visionary synthesist. His masterpiece reveals a mind adept at abstraction and algorithmic thinking, driven by both practical necessity and the demands of astronomical precision. We have explored the foundations of his universe: from his poetic alphabetical numerical system designed to capture cosmic scales, to his revolutionary claim of Earth's eastward rotation—a bold assertion that sparked centuries of debate and set him apart. He courageously rejected prevailing myths, like demonic entities causing eclipses, instead revealing a cosmos where the divine dwells in the measurable and predictable.

His mathematical arsenal—the decimal place-value system, precise algorithms for roots, a remarkably accurate π, the nascent trigonometry of sine and versine, and the groundbreaking "Pulverizer" algorithm for indeterminate equations—powered a celestial mechanics of unparalleled elegance. This synthesis allowed him to harmonize lunar and solar cycles through precise cosmic time parameters, establish a definitive epoch-based calendar, and develop a sophisticated epicyclic model that predicted planetary motions with remarkable accuracy for his era. He envisioned a spherical Earth at the heart of a three-dimensional celestial sphere, debunking flat-Earth notions with reasoned analogies and shadow-based eclipse explanations.

This journey has recast Aryabhata as a timeless trailblazer. His emphasis on logic and justification, though debated by contemporaries, laid a foundation for future discoveries, from the Kerala School's pre-calculus insights to the global spread of the decimal system and sine function via Islamic scholars. His advanced thinking, driven by a rare fusion of empirical precision and theoretical boldness, anticipated concepts that would shape global science centuries later.

Far from a mere historical artifact, the *Aryabhatiya* endures as a testament to the timeless spirit of inquiry—humanity's quest to understand the cosmos through reason, observation, and the universal language of mathematics. Rediscovering Aryabhata has not only reaffirmed his genius but also rebranded him as a beacon of intellectual courage. His youthful brilliance, ignited in India's Golden Age, still kindles a vision that marries scientific rigor with cosmic wonder, guiding us as we continue to explore the universe's boundless beauty and our place within it.

Chapter Twelve

Bibliography

Primary Sources

- Brahmagupta. *Brāhmasphuṭasiddhānta*. Edited by Sudhakara Dvivedi. Benares: Braj Bhusan Das & Co., 1902.
- Sen, S. N., and A. K. Bag. *The Śulbasūtras: Sources of Indian Mathematics*. New Delhi: Indian National Science Academy, 1983.
- Shukla, Kripa Shankar, and K. V. Sarma, eds. and trans. *Āryabhaṭīya of Āryabhaṭa: Critically Edited with Introduction, English Translation, Notes, Comments and Indexes*. With the commentary (bhāṣya) of Bhāskara I and Somayājī. New Delhi: Indian National Science Academy, 1976.

Secondary Sources

- Al-Biruni, Abu Rayhan Muhammad ibn Ahmad. *Alberuni's India: An Account of the Religion, Philosophy, Literature, Geography, Chronology, Astronomy, Customs, Laws and Astrology of India about A.D. 1030*. Translated by Edward C. Sachau. 2 vols. London: Kegan Paul, Trench, Trübner & Co., 1910. Reprint, Delhi: Motilal Banarsidass, 1993.
- Altekar, A. S. *State and Government in Ancient India*. Delhi: Motilal Banarsidass, 1958.
- Berggren, J. Lennart. *Episodes in the Mathematics of Medieval Islam*. New York: Springer-Verlag, 1986.
- Billard, Roger. *L'Astronomie Indienne: Investigation des Textes Sanskrits et des Données Numériques*. Paris: École Française d'Extrême-Orient, 1971.
- Clark, Walter Eugene, trans. *The Āryabhaṭīya of Āryabhaṭa: An Ancient Indian Work on Mathematics and Astronomy*. Chicago: University of Chicago Press, 1930.
- Datta, Bibhutibhushan, and Avadhesh Narayan Singh. *History of Hindu Mathematics: A Source Book*. Bombay: Asia Publishing House, 1962. Originally published in two parts, Lahore: Motilal Banarsidass, 1935 and 1938.
- Fibonacci, Leonardo (Leonardo Pisano). *Fibonacci's Liber Abaci: A Translation into Modern English of Leonardo Pisano's Book of Calculation*. Translated by Laurence E. Sigler. New York: Springer-Verlag, 2002.

- Hayashi, Takao. *The Bakhshali Manuscript: An Ancient Indian Mathematical Treatise*. Groningen: Egbert Forsten, 1995.
- Ifrah, Georges. *The Universal History of Numbers: From Prehistory to the Invention of the Computer*. Translated by David Bellos, E. F. Harding, Sophie Wood, and Ian Monk. New York: John Wiley & Sons, 2000.
- Joseph, George Gheverghese. *The Crest of the Peacock: Non-European Roots of Mathematics*. 3rd ed. Princeton, NJ: Princeton University Press, 2011.
- Kuppanna Sastri, T. S., and K. V. Sarma. *Collected Papers on Jyotisa*. Tirupati: Kendriya Sanskrit Vidyapeetha, 1989.
- Mookerji, Radha Kumud. *Ancient Indian Education: Brahmanical and Buddhist*. London: Macmillan and Co., 1947. Reprint, Delhi: Motilal Banarsidass, 1989.
- Mookerji, Radha Kumud. *The Gupta Empire*. Bombay: Hind Kitabs, 1947. Reprint, Delhi: Motilal Banarsidass, 1989.
- Ohashi, Yukio. "Astronomical Instruments in Classical Siddhāntas." In *History of Oriental Astronomy: Proceedings of an IAU Colloquium No. 91, New Delhi, India, 13–16 November 1985*, edited by S. M. Razaullah Ansari, 115–127. Cambridge: Cambridge University Press, 1987.
- Pingree, David. *Census of the Exact Sciences in Sanskrit*. Series A, Vols. 1–5. Philadelphia: American Philosophical Society, 1970–1994.
- Pingree, David. *Jyotiḥśāstra: Astral and Mathematical Literature*. A History of Indian Literature, Vol. VI, Fasc. 4. Wiesbaden: Otto Harrassowitz, 1981.
- Plofker, Kim. *Mathematics in India*. Princeton, NJ: Princeton University Press, 2009.
- Ptolemy, Claudius. *Ptolemy's Almagest*. Translated and annotated by G. J. Toomer. Princeton, NJ: Princeton University Press, 1998.
- Sen, S. N., and K. S. Shukla, eds. *History of Astronomy in India*. New Delhi: Indian National Science Academy, 1985.
- SenGupta, P. C. "The Aryabhatiya." *Journal of the Department of Letters, University of Calcutta* 16 (1927): 1–56.
- Singh, Upinder. *A History of Ancient and Early Medieval India: From the Stone Age to the 12th Century*. Delhi: Pearson Longman, 2008.
- Stallings, William. *Cryptography and Network Security: Principles and Practice*. 8th ed. Hoboken, NJ: Pearson, 2019.
- Thapar, Romila. *Early India: From the Origins to AD 1300*. Berkeley: University of California Press, 2002.
- Van der Waerden, B. L. *Geometry and Algebra in Ancient Civilizations*. Berlin: Springer-Verlag, 1983.

Online Resources

- NASA/JPL Horizons System. *Solar System Dynamics, Jet Propulsion Laboratory*. Accessed May 22, 2025. https://ssd.jpl.nasa.gov/horizons/.

Chapter Thirteen

Glossary

A

- **Abda:** Year; often specifically a solar year (ravyabda).
- **Adhimāsa:** Intercalary or extra lunar month; inserted periodically (approx. every 2.7 years) to reconcile lunar and solar calendars (leap month), based on Aryabhata's Yuga parameters (Ch 5.5).
- **Ahan:** Day; often used for the civil day.
- **Ahargaṇa:** "Heap of days"; the count of civil (sāvana) days elapsed since a specific epoch (for Aryabhata, the beginning of the Kali Yuga at Lanka sunrise), fundamental for calculating planetary positions for any given date (Ch 5.7).
- **Ahorātra:** A full day-and-night cycle (24 hours); cosmologically, also used for Brahmā's Day and Night cycle.
- **AI (Artificial Intelligence):** Digital technologies used in the creation process, including OCR, large language models for preliminary translation exploration, literature searching, and organizing information; also used in generating some images (Disclosure, Acknowledgements, Preface).
- **Akṣa:** Terrestrial latitude; an observer's angular distance North or South of the Earth's Equator (Ch 6.7, Gola.14, Gola.31).
- **Akṣajyā:** R·sin(latitude); the sine of the terrestrial latitude (where R is the standard radius, usually 3438 minutes of arc, Gk.10). Calculated from gnomon observations (Ch 6.9, Gola.31).
- **Akṣadṛkkarma:** Latitude visibility correction; a subtle angular correction applied to calculated longitude in positional astronomy (especially near the horizon) to account for the combined effect of celestial latitude and terrestrial latitude (Ch 6.10, Gola.35).
- **Akṣavalana:** Latitude deflection; an angular correction in eclipse calculations accounting for the apparent rotation of the local coordinate system relative to the ecliptic poles due to the observer's latitude (Ch 6.11, Gola.45).
- **Al-jabr:** Arabic term ("completion" or "restoration") from Al-Khwārizmī's treatise title, which gave rise to the word "Algebra" (Ch 9.2).

- **Algoritmi:** Latinization of Al-Khwārizmī's name, root of the modern English word "Algorithm"; reflects the procedural nature of his mathematical methods (Ch 9.2).

- **Al-Khwārizmī, Muḥammad ibn Mūsā:** Persian scholar (c. 780–c. 850 CE) pivotal in transmitting Indian mathematical and astronomical concepts (like decimal numerals, zero, and sine) to the Islamic world and Europe (Ch 9.2).

- **Almagest:** The comprehensive astronomical treatise by Claudius Ptolemy (c. 150 CE), representing the culmination of Greco-Roman astronomy; contrasted with Aryabhata's system (Intro, Ch 7).

- **Amāvāsyā:** New moon; the conjunction of the Sun and Moon, marking the start of the lunar month and the time when solar eclipses are possible (Gola.38).

- **Aṃśa:** Degree; 1/360th of a circle (Kk.2). Synonymous with bhāga. Also used in Gola.4 for "degrees" of elongation from the Sun, potentially referring to time-degrees (kālāṃśa).

- **Anomaly (Kendra):** The angle representing a planet's deviation from uniform mean motion; used in Aryabhata's epicyclic model (and other geocentric models) to calculate corrections (manda and śīghra) for variable speed and synodic phenomena (Ch 5.9, Kk.16).

- **Anuloma:** "With the grain"; Direct or prograde motion, generally eastward along the ecliptic; used to describe the direction of Śīghroccha motion relative to the planet (Kk.21).

- **Apakrama:** Declination; the angular distance of a celestial body North or South of the celestial equator (Ch 6.8, Gola.25). Maximum declination of the Sun is the obliquity of the ecliptic (Gk.9, Gola.1).

- **Apamaṇḍala:** The Ecliptic; the great circle representing the Sun's apparent annual path against the background stars; the fundamental reference plane for planetary longitudes (Ch 6.1, Gola.1).

- **Apogee (Uccha):** A point in a body's orbit (or model) corresponding to its slowest speed or greatest distance from the Earth. Aryabhata uses Mandoccha (slow apogee) for the manda correction and Śīghroccha (fast apogee) for the śīghra correction (Ch 5.3, Gk.4, Kk.4).

- **Ārkṣī Vināḍikā:** Sidereal vināḍikā; a small time unit (1/3600 of a sidereal day, approx. 24 seconds) linked to sidereal time (Kk.2).

- **Arka:** A name for the Sun (Gk.4, Gk.7, Gola.2, Gola.4, Gola.37, Gola.38). See also Ravi, Sūrya.

- **Arkāgrā:** R·sin(amplitude); the sine value used to calculate the Sun's amplitude, its angular distance from the East/West point on the horizon at rising/setting, based on latitude and declination (Ch 6.9, Gola.30).

- **Āsanna:** Approximate, near; used by Aryabhata to describe his value for π, acknowledging it is not an exact value (Ch 4.4, Gp.10).

- **Aśvayuj / Aśvinī:** The first lunar mansion (nakṣatra), located near the beginning point of Aries (0° ecliptic longitude). Used as a reference point for time cycles like Jupiter Years (Ch 5.3, Kk.4).

- **Avamarātra / Kṣayatithi:** Omitted lunar day (tithi); a day dropped from the calendar count (kṣaya) to reconcile the count of lunar tithis with civil days over long periods (Ch 5.5, Kk.6 context).

- **Avarga:** "Non-class"; the second set of consonants (y to h) used in Aryabhata's alphabetical number system (Gk.2), representing multiples of ten (30 to 100).

- **Ayana:** Course or path; refers to the Sun's half-yearly journey north or south relative to the equator (between solstices), causing seasons (Gola.1).

- **Ayanadṛkkarma:** Solstitial/Seasonal visibility correction; a correction applied to calculated longitude in positional astronomy (near horizon) accounting for effects related to the ecliptic's inclination and the season (Ch 6.10, Gola.36).

- **Ayanavalana:** Declination/Seasonal deflection; an angular correction in eclipse calculations related to the ecliptic's tilt relative to the celestial equator, affecting the apparent orientation of contact points (Ch 6.11, Gola.45).

- **Azimuth:** Horizontal direction, measured as an angle along the horizon from a cardinal point (usually North or South). Calculated using spherical trigonometry (Ch 6.9, Gola.29).

B

- **Baḍavāmukha:** "Mare's Mouth"; a mythical location often associated with the South Pole region, integrated into Aryabhata's spherical Earth cosmology (Ch 6.4, Ch 6.6, Gola.6, Gola.12, Gola.16).

- **Bhachakra / Bhagola / Khagola:** The celestial sphere; the conceptual sphere of stars and celestial bodies surrounding the Earth, used as a geometric model for positional astronomy (Ch 6.8, Gola.5).

- **Bhagaṇa:** Revolution; a complete orbit (360°). Aryabhata lists the total number of revolutions completed by planets, Sun, Moon, nodes, and apogees in a Mahāyuga as fundamental parameters (Ch 3.3, Gk.3).

- **Bhāga:** Degree; 1/360th of a circle (Kk.2). Synonymous with aṃśa.

- **Bhāni:** Asterisms or fixed stars; described by Aryabhata as appearing to move westward due to Earth's eastward rotation (Gola.9).

- **Bhaskara I:** Influential 7th-century commentator on Aryabhata's Aryabhatiya. His detailed commentary (bhāṣya) is the earliest surviving and crucial for understanding Aryabhata's work (Intro, Ch 1.2).

- **Bhauma:** A name for Mars (Gk.7, Kk.15). See also Kuja.

- **Bhṛgu:** A name for Venus (Gk.4, Gk.7, Gp.1, Kk.15, Gola.3, Gola.4). See also Śukra.

- **Bhūcchāyā / Kṣiticchāyā:** Earth's shadow; Aryabhata correctly identifies this as the physical cause of lunar eclipses (Ch 6.10, Gola.37).

- **Bhūdivasa:** Earth-day; typically refers to the civil (sāvana) day, measured from sunrise to sunrise (Kk.5, Kk.11). Also used in Gk.3 for Earth's rotation count (sidereal days).

- **Bhūgola:** The Earth-sphere; Aryabhata emphatically asserts the Earth is spherical (Ch 6.3, Gola.5).

- **Bhukti:** Motion, particularly daily motion (angular speed) of a celestial body. Can be mean (madhyama) or true (sphuṭa) (Kk.14, Kk.24).

- **Bhūmi:** The Earth (Kk.15). See also Bhūgola.

- **Bhūvyāsa:** Earth's diameter; given by Aryabhata as 1,050 yojanas (Ch 3.7, Gk.7), a remarkably accurate value for his time. Used in calculations involving the Earth's shadow cone (Gola.39, Gola.40) and horizon dip (Gola.15, Gola.34).

- **Brahma:** The creator deity in Hinduism; invoked by Aryabhata at the beginning of his work (Gk.1, Gp.1), linking scientific inquiry to a sense of cosmic order (Ch 10.1.1). Aryabhata's alphabetical cipher was said by Bhaskara I to come from Brahma's grace (Ch 3.2).

- **Brahman:** The ultimate, impersonal reality in Hindu philosophy; invoked by Aryabhata (Gk.1), representing the ultimate unifying principle underlying the cosmos (Ch 10.1.1).

- **Brāhmapakṣa:** Astronomical school or computational tradition following Brahmagupta's parameters and system, contrasting with the Ārya-pakṣa (Ch 8.1.3).

- **Brāhmasphuṭasiddhānta:** Brahmagupta's major astronomical treatise (c. 628 CE), containing his significant contributions to mathematics and astronomy, as well as his critiques of Aryabhata's system (Ch 8.1).

- **Bṛhaspati:** A name for Jupiter (Gp.1). See also Guru.

- **Budha:** The planet Mercury (Gk.4, Gk.7, Gp.1, Kk.15, Gola.3, Gola.4).

C

- **Calculus Precursors:** Mathematical concepts and techniques developed by the Kerala School (building on the Aryabhatan tradition), such as infinite series expansions for trigonometric functions, which contain the essence of later calculus (Ch 8.2).

- **Calendars:** Systems for organizing time based on celestial cycles (solar, lunar, civil). Indian calendars are typically luni-solar, requiring adjustments like adhimāsa and avamarātra (Ch 5.5).

- **Chandra:** The Moon (Gk.4, Gola.3, Gola.4, Gola.37). See also Indu, Śaśi.

- **Cāndra:** Lunar; refers to time measured by the Moon's phases, specifically the synodic month (New Moon to New Moon) (Ch 5.4, Kk.5).

- **Canva:** An online graphic design platform used for generating some images in the book, potentially involving AI features (Disclosure).

- **Cardinal Points:** North, South, East, West; defined on the observer's horizon (Ch 6.8, Gola.18).

- **Celestial Equator (Viṣuvat):** The great circle on the celestial sphere formed by projecting Earth's equator into space; the primary reference line for declination and right ascension (Ch 6.8, Gola.19, Gola.25).

- **Celestial Latitude (Vikṣepa):** The angular distance of a celestial body North or South of the ecliptic plane. Accounted for in Aryabhata's system, especially for the Moon (Gk.9, Gola.3) and crucial for eclipse prediction.

- **Celestial Longitude:** Angular distance measured eastward along the ecliptic from the vernal equinox (0° Aries); the primary coordinate for tracking planetary positions in Aryabhata's system.

- **Celestial Sphere (Bhachakra, Bhagola, Khagola):** The conceptual sphere of stars and celestial bodies surrounding the Earth, rotating daily in traditional models (Gola.10) or appearing to rotate in Aryabhata's model of Earth's spin (Gola.9); used as a geometric model (Gola.5).
- **Chaitra:** A lunar month (approx. March-April), designated by Aryabhata as the start of the year in the śuklādi calendar system (Kk.11).
- **Chara / Charadala:** Ascensional difference / Half ascensional difference; an angular amount (related to latitude and declination) that quantifies the variation in rising/setting times and daylight duration. Calculated from the kṣitijyā (Gola.26) and used as a time correction (Gola.27).
- **Chāyā:** Shadow; used for the gnomon's shadow (Ch 2.1, Ch 4.5, Gola.29) or the Earth's shadow in eclipses (Ch 6.10, Gola.37).
- **Chnā:** Syllabic code used by Aryabhata (Gk.5) encoding the number 27, representing the number of Mahāyugas elapsed in the current Manvantara at the start of the Kali Yuga.
- **Chord:** Geometric line segment connecting two points on a circle. Used in Greek trigonometry (Ptolemy), conceptually related to the Indian sine (jyā) which is a half-chord (Ch 7.2.2).
- **Chitighana:** "Sum-pile"; term used in formulas for sums of series, specifically the sum of triangular numbers (Gp.20).
- **Civil Day (Sāvana, Bhūdivasa):** The practical day used in calendars, typically measured from sunrise to sunrise for an observer on Earth. This is the unit used for the ahargaṇa (Ch 5.4, Kk.5).
- **Colors (Lunar Eclipse):** The observed changing colors (smoky, black, tawny, copper-red) of the Moon during a total lunar eclipse, meticulously described by Aryabhata (Ch 6.11, Gola.46).
- **Commentaries (Bhāṣya):** Scholarly works interpreting and explaining foundational texts like the Aryabhatiya. Bhaskara I's commentary is the most important for understanding Aryabhata's original meaning (Intro, Ch 1.2).
- **Compound Proportion:** Mathematical method for solving problems involving multiple interdependent proportional relationships (Rule of Five, Rule of Seven, etc.), described by Aryabhata (Ch 4.6, Gp.27, Gp.28).
- **Conjunctions (Yoga):** The apparent alignment of two or more celestial bodies in the sky. The Sun-Moon conjunction is the New Moon (Amāvāsyā) (Kk.5). Planetary conjunctions are calculated based on their relative motions (Ch 5.3, Kk.3).
- **Cosmic Harmony (Ṛta):** The concept of inherent universal order, structure, and rhythm, often linked in Indian thought to mathematical principles and seen as a pathway to deeper truth (Ch 10.1).
- **Cosmological Model (Geocentric):** Aryabhata's model of the cosmos, where the Earth (spherical and rotating) is located at or near the center, and the Sun, Moon, and planets revolve around it (Ch 6.3, Gola.5); distinct from a heliocentric model (like Copernicus') or a purely static geocentric model (like Ptolemy's).
- **Critique of Aryabhata:** The scholarly debates and disagreements sparked by Aryabhata's work, notably Brahmagupta's criticisms regarding Earth's rotation, parameters, and other points (Ch 8.1).

- **Cryptography (RSA):** A modern encryption algorithm whose security relies on mathematical principles (like finding modular inverses using the Extended Euclidean Algorithm) related to Aryabhata's Kuṭṭākāra (Ch 4.7, Ch 9.6, Epilogue).

- **Cube and Cube Root:** Geometric concept of a six-sided solid with equal edges (cube) and the arithmetic operation of multiplying three equal quantities (cubing); algorithm for extracting the cube root (Ch 4.3, Gp.3, Gp.5).

D

- **Daśagītikā:** "Ten Giti verses"; the first section of the Aryabhatiya (actually 13 verses in total), containing the invocation, the alphabetical number system, and core astronomical parameters (Intro, Ch 3).

- **Day (Divasa, Dina):** A fundamental unit of time (Kk.1). See also Civil Day, Sidereal Day, Lunar Day.

- **Decimal Place-Value System:** A numerical system where the value of a digit depends on its position or 'place' (sthāna) relative to the units place, based on powers of ten. Clearly articulated by Aryabhata (Gp.2) and foundational to modern arithmetic; includes the crucial role of zero (śūnya) as a placeholder (Ch 4.2).

- **Declination (Apakrama):** The angular distance of a celestial body North or South of the celestial equator (Ch 6.8, Gola.25). Maximum declination of the Sun is the obliquity of the ecliptic (Gk.9, Gola.1).

- **Dedication:** Section of the book dedicating the work to seekers of truth and the author's family.

- **Deferent:** The main large circle in epicyclic models, upon which the center of the epicycle moves, typically centered on or near the Earth (Ch 7.1.2). See also Kakṣyā.

- **Dhana:** Positive; signifying an additive correction or quantity in calculations (Ch 5.9, Kk.20, Kk.21, Gola.35, Gola.36).

- **Dharma:** Cosmic law, inherent order, duty, righteousness. Used in a broad sense to imply the natural, ordered structure of the universe.

- **Dhūmra:** Smoky, dusky; the appearance of the Moon when entering/leaving the penumbra during a lunar eclipse (Ch 6.11, Gola.46).

- **Diameter (Viṣkambha):** The length of a straight line segment passing through the center of a circle or sphere and terminating at its boundaries. Aryabhata provides diameters for the Earth, Sun, and Moon (Gk.7) and calculates the Earth's shadow diameter (Gola.40).

- **Diophantine Equations (Linear Indeterminate Equations):** Algebraic equations with more variables than equations, for which integer solutions are sought (e.g., $ax + by = c$). Aryabhata's Kuṭṭākāra is an algorithm for solving these (Ch 4.7).

- **Diurnal Motion:** The apparent daily movement of celestial bodies across the sky from East to West, caused by the rotation of the observer's reference frame (Earth's rotation in Aryabhata's model, Gola.9) or the celestial sphere itself (Pravaha wind model, Gola.10).

- **Dṛkkarma:** "Visibility corrections"; subtle angular adjustments (Akṣa- and Ayana-dṛkkarma) applied to calculated longitudes to account for perspective effects due to observer's location and the geometry of the celestial sphere near the horizon (Ch 6.10, Gola.35, Gola.36).

- **Dṛkchāyā:** "Shadow of vision"; likely refers to the dip of the horizon caused by Earth's curvature, making the actual horizon appear slightly lower than the geometric horizon. May also include empirical adjustments for refraction (Ch 6.9, Gola.34).

- **Dṛggatijyā:** R·sin(Parallax angle); the sine value used in calculating the parallax correction (Ch 6.9, Gola.32, Gola.33).

- **Dṛkkṣepajyā:** R·sin(Zenith Distance); equivalent to R·cos(Altitude). Used in parallax calculations (Ch 6.9, Gola.32).

E

- **Earth (Bhūmi, Bhūgola):** The planet Earth (Kk.15). Aryabhata describes it as spherical (Bhūgola, Gola.5), composed of five elements (Gola.6), located at the center of the orbits (Gola.5), and rotating on its axis (Gola.9).

- **Earth: Rotation Theory:** Aryabhata's radical hypothesis that the Earth spins eastward on its axis once daily, causing the apparent westward motion of the heavens (Gk.3, Gola.9); contrasted with the traditional static-Earth view (Gola.10) and Ptolemy's model (Ch 7.1.1).

- **Earth: Shadow (Bhūcchāyā):** See Bhūcchāyā.

- **Earth: Spherical Nature:** The concept that the Earth is round on all sides (sarvato vṛttaḥ, Gola.5, Gola.6), supported by arguments like the Kadamba flower analogy (Gola.7) and implications for time zones (Gola.13) and horizon dip (Gola.15).

- **Eclipses (Grahaṇa):** The obscuration of a celestial body. Aryabhata provides the correct physical explanation (Moon covering Sun, Earth's shadow covering Moon, Gola.37) and detailed computational methods for predicting their circumstances (Ch 6.11).

- **Eclipses: Calculation:** Geometric and trigonometric methods for determining eclipse timing, duration (sthiti, vimarda), and magnitude (grāsa) (Ch 6.11).

- **Eclipses: Colors (Lunar):** The observed changing colors during a lunar eclipse (Gola.46).

- **Eclipses: Conditions:** Requirements for an eclipse to occur, primarily New/Full Moon phase and proximity to a lunar node (pātāsanna) (Gola.38).

- **Ecliptic (Apamaṇḍala):** The great circle on the celestial sphere representing the Sun's apparent annual path, tilted relative to the celestial equator; the primary reference plane for planetary longitudes and latitudes (Ch 6.1, Gola.1).

- **Elements (Five):** The classical elements (earth, water, fire, air, ether) believed to compose the Earth and other entities in traditional philosophy (Gola.6). Aryabhata associates Sun with fire, Moon with water, and Earth with earth, shadow with darkness (Gola.37).

- **Elongation (from Sun):** The angular distance of a celestial body from the Sun in the sky; relevant for visibility

(Gola.4) and the śīghra anomaly calculation.

- **Epoch:** The precise starting date and time for astronomical calculations in a system. For Aryabhata's main system, this is the beginning of the current Kali Yuga (sunrise on Friday, Feb 18, 3102 BCE Julian at Lanka) (Ch 5.7).

- **Epicycles (Pratimaṇḍala):** The smaller circle in a geometric model (like Ptolemy's or Aryabhata's) used to explain the observed irregular motions of planets. The planet moves on the epicycle while the epicycle's center moves on a larger deferent circle. Aryabhata's system calculates corrections (manda, śīghra) derived from this geometric concept (Ch 5.9).

- **Equation of Center:** The correction (mandaphala) applied to a planet's mean longitude to account for the variation in speed caused by the eccentricity of its orbit, modeled using the manda epicycle (Ch 5.9).

- **Equant:** A geometric device used by Ptolemy (but not Aryabhata) in his planetary models to ensure uniform angular motion of the epicycle center from a point offset from the deferent's center (Ch 7.1.2).

- **Equator (Celestial, Viṣuvat):** The great circle on the celestial sphere formed by projecting Earth's equator into space; the primary reference line for declination and right ascension (Ch 6.8, Gola.19, Gola.25).

- **Equator (Terrestrial):** The great circle midway between Earth's geographic poles. Lanka is located on the terrestrial equator in Aryabhata's model (Gola.13).

- **Equinox (Viṣuvat):** The points in time or on the celestial sphere where the Sun crosses the celestial equator. Day and night are of approximately equal length everywhere on Earth at the equinoxes (Ch 6.9, Gola.31).

- **Ethics (of Knowledge):** Moral considerations regarding the pursuit, acquisition, preservation, and dissemination of scientific knowledge, emphasized by Aryabhata in the final verses of his work (Ch 6.12, Ch 10.2).

- **Euclidean Algorithm:** An efficient algorithm for computing the greatest common divisor (GCD) of two integers. The forward reduction phase of Aryabhata's Kuṭṭākāra algorithm is based on this principle (Ch 4.7).

F

- **Fibonacci (Leonardo of Pisa):** Italian mathematician (c. 1170 – c. 1250) who was instrumental in popularizing the Hindu-Arabic numeral system in Europe through his book Liber Abaci (Ch 9.4).

- **Finite Differences:** A mathematical technique involving differences between successive values in a sequence. Aryabhata used a second-order finite difference algorithm to construct or verify his sine difference table (Ch 3.12).

- **Fortnight (Pakṣa):** Half of a lunar month, comprising either the waxing (śukla) or waning (kṛṣṇa) phase. A lunar eclipse can occur at the end of the fortnight (pakṣānte = Full Moon) (Gola.38).

- **Formulas (Summary):** Table in Chapter 6.13 summarizing key mathematical formulas from the Golapāda.

G

- **Ga:** Syllabic code used by Aryabhata (Gk.5) encoding the number 3, representing the number of Yuga quarters elapsed at the start of the Kali Yuga.

- **Gaṇita:** Mathematics, calculation, computation. One of the three main subjects of the Aryabhatiya (Gk.1), forming the second section (Gaṇitapāda) and the essential toolkit for astronomy (Intro, Ch 4).
- **Gaṇitapāda:** "Section on Mathematics"; the second section (33 verses) of the Aryabhatiya, detailing arithmetic, geometry, algebra, and trigonometry (Intro, Ch 4).
- **Geocentric Model:** A model of the cosmos in which the Earth is located at or near the center, and all other celestial bodies (Sun, Moon, planets, stars) revolve around it. Aryabhata's model is geocentric with regard to orbits, but incorporates Earth's rotation (Ch 6.3, Ch 7.1).
- **Geometry:** The branch of mathematics dealing with shapes, spatial relationships, and their properties. A key component of Aryabhata's work (Gaṇitapāda) and crucial for understanding the celestial sphere (Golapāda) and epicyclic models (Ch 4, Ch 6).
- **Ghaṭikā:** Unit of time, 1/60th of a day (24 minutes). Synonymous with Nāḍī. Also refers to the inflow water clock used for measuring these intervals (Ch 2.2, Kk.1).
- **Gnomon (Śaṅku):** A vertical rod or pillar used for measuring time and position by observing the length and direction of its shadow. A fundamental observational instrument in ancient astronomy (Ch 2.1, Ch 4.5, Gola.28, Gola.29, Gola.31).
- **Gola:** Sphere. Used for the celestial sphere (khagola) and the Earth-sphere (bhūgola). Also refers to model spheres (gola yantra) used for visualization and teaching (Ch 6.8, Gola.22). One of the three main subjects of the Aryabhatiya (Gk.1), forming the fourth section (Golapāda).
- **Gola Yantra:** A physical model of the celestial sphere or a part of it (like an armillary sphere), described by Aryabhata as a rotating wooden sphere used to visualize celestial motion synchronized with time (Ch 6.8, Gola.22).
- **Golapāda:** "Section on the Sphere"; the fourth and final section (50 verses) of the Aryabhatiya, detailing the geometry of the celestial sphere, Earth's rotation, and eclipse calculations (Intro, Ch 6).
- **Graha:** "Seizer"; refers to the planets in traditional Indian astronomy, including the Sun and Moon (Gk.3, Gola.2, Kk.3).
- **Grahaṇa:** "Seizing"; an eclipse, specifically solar or lunar (Gola.37).
- **Grāsa:** "Mouthful"; the magnitude of an eclipse, representing the maximum fraction of the eclipsed body's diameter that is covered (Ch 6.11, Gola.44).
- **Gupta Empire:** The historical period in India (c. 320-550 CE) considered a "Golden Age" for art, literature, and science, during which Aryabhata lived and worked near Kusumapura/Pataliputra (Intro, Ch 1.1).
- **Guru:** "Heavy," "Teacher"; the planet Jupiter (Gk.3, Gk.4, Gk.7, Gp.1, Kk.4, Kk.15, Gola.3, Gola.4).

H

- **Half-Correction Rule:** An iterative step in Aryabhata's epicyclic calculation where half of the calculated correction (phala) is added back to the anomaly angle (kendra) to refine its value, leading to greater accuracy (Ch 5.10, Kk.22).
- **Harmony (Cosmic, Ṛta):** The concept of inherent order and regularity in the universe, which Indian thought often perceived as being describable and comprehensible through mathematics (Ch 10.1).
- **Heliacal Rising/Setting:** The first appearance of a star or planet in the morning twilight just before sunrise, or its last appearance in the evening twilight just after sunset. Aryabhata provides visibility thresholds based on elongation from the Sun (Gola.4).
- **Hindu-Arabic Numerals:** The positional decimal system using symbols for digits 1-9 and zero (śūnya), which originated in India and was transmitted to the Islamic world and then to Europe (Ch 9.4).
- **Horizon (Kṣitija):** The boundary line separating the visible sky from the obscured part of the Earth. It is a fundamental reference plane in the observer's local coordinate system (Ch 6.8, Gola.18).
- **Horizon Dip (Dṛkchāyā):** The apparent lowering of the horizon due to the curvature of the Earth, meaning less than half the celestial sphere is visible from a point on the surface (Gola.15, Gola.34).
- **Horāśāstra:** Astrology; while Aryabhata's work is primarily mathematical astronomy, his precise calculations formed the necessary basis for astrological computations (Ch 5.1).
- **Hour Angle (H):** Angular distance of a celestial body East or West of the observer's meridian, measured along the celestial equator. Related to local sidereal time.

I

- **IAST (International Alphabet of Sanskrit Transliteration):** A standard system for transliterating Sanskrit text using Latin letters. Used in the book for Sanskrit terms and verses.
- **Image Generation:** The process of creating visual figures or diagrams, with some in the book created programmatically (Python) or using AI tools (Canva) (Disclosure).
- **Indu:** A name for the Moon (Gola.2, Gola.38). See also Chandra, Śaśi.
- **Infinite Series:** Mathematical series with an infinite number of terms. The Kerala School (building on Aryabhata) developed series expansions for trigonometric functions representing precursors to calculus (Ch 8.2).
- **Intellect (Svamati):** One's own intellect, reasoning, or mind; emphasized by Aryabhata as a crucial tool for navigating existing knowledge, discerning truth, and advancing understanding (Ch 6.12, Gola.49).
- **Interest Calculation:** Rules for calculating simple interest (Gp.25, Gp.31), demonstrating the application of algebra to practical problems.
- **Intercalary Month (Adhimāsa):** See Adhimāsa.
- **Introduction:** The opening section of the book, setting the context and structure of the exploration of the Aryabhatiya.

- **Invocation (Maṅgalācharaṇa):** An auspicious verse or prayer at the beginning of a text, typically paying homage to divine principles or deities; used by Aryabhata (Gk.1, Gp.1).
- **Inverse Rule of Three:** A variation of the Rule of Three applied when quantities are inversely proportional (Ch 4.6, Gp.27).
- **Islamic Golden Age:** A period (roughly 8th-13th centuries CE) of flourishing intellectual and scientific activity in the Islamic world, crucial as an intermediary in the transmission of Indian science to Europe (Ch 9.1).
- **Iteration:** Repeating a calculation process, using the result of one step as the input for the next, to achieve greater accuracy or convergence. Used by Aryabhata in his epicyclic calculations (Ch 5.10).

J

- **Jala Yantra:** Water clock. See Ghaṭikā.
- **Jayb:** The Arabic term for the sine function, adopted from the Sanskrit jyā, which later became the Latin "sinus" (Ch 9.2).
- **Jyeṣṭhadeva:** A prominent scholar of the Kerala School (c. 1500–1575), author of Yuktibhāṣā (Ch 8.2.2).
- **Jupiter (Guru):** The fifth planet (from the Sun), known in Indian astronomy as Guru or Bṛhaspati (Gk.3, Gp.1). Aryabhata provides its revolution count (Gk.3) and defines "Jupiter Years" based on its motion through the zodiac (Kk.4).
- **Jupiter Years (Gurorabdāḥ):** A calendrical cycle associated with Jupiter's motion through the zodiac, defined by Aryabhata as its sidereal revolutions multiplied by 12 (the number of signs) (Ch 5.3, Kk.4).
- **Jyā / Jīvā / Jyārdha:** "Bow-string" / "Half-chord"; Aryabhata's foundational trigonometric function, representing R·sin(θ) (the length of the half-chord subtending an angle θ in a circle of radius R). He provides a table of differences for this function (Gk.10) and a rule for its calculation (Gk.12).
- **Jyotiḥśāstra:** Traditional Indian astronomy and astrology, encompassing both mathematical calculation (gaṇita) and observational practice. Aryabhata's work fits within and builds upon this tradition (Ch 1.1).

K

- **Kadamba Flower Analogy:** An analogy used by Aryabhata (Gola.7) to explain how beings can live all around a spherical Earth, suggesting that "down" is always relative to the center.
- **Kakṣyā:** Orbit, path, or deferent circle of a planet in the geocentric model. Also used for the vast circumference of the celestial sphere (Ch 3.6, Gk.6).
- **Kāla:** Time. One of the three main subjects of the Aryabhatiya (Gk.1), forming the third section (Kālakriyāpāda).
- **Kālakriyāpāda:** "Section on the Reckoning of Time"; the third section (25 verses) of the Aryabhatiya, defining time units, cosmic cycles, and methods for calculating planetary positions based on time elapsed (Intro, Ch 5).

- **Kalā:** Arc-minute; 1/60th of a degree. Also called liptā. Aryabhata establishes a fundamental equivalence between 1 kalā of celestial rotation and 1 prāṇa (breath) of time (Gk.11, Kk.2).

- **Kali Yuga:** The last and current age (yuga) in a Mahāyuga cycle. Its beginning (Feb 18, 3102 BCE Julian) serves as the epoch for Aryabhata's calculations (Ch 3.5, Gk.5).

- **Kalpa:** An immense cosmological time period, a "Day of Brahmā," representing a cycle of cosmic creation/manifestation. Consists of 1008 Mahāyugas according to Aryabhata's parameters (Gk.5), though common tradition gives 1000 (Kk.8). Approx. 4.32 billion years (Ch 3.5, Ch 5.6).

- **Kapila:** Tawny, brownish-red; the appearance of the Moon deep within the Earth's umbra but near its edge during a total lunar eclipse (Ch 6.11, Gola.46).

- **Karṇa:** "Ear" or "Hypotenuse"; in a right triangle, the side opposite the right angle (Ch 4.5, Gp.14, Gp.17). In epicyclic geometry, it often refers to the varying true distance from the Earth to the planet on the epicycle, crucial for refining position calculations (Ch 5.10, Kk.23).

- **Kendra:** Anomaly. See Anomaly.

- **Kerala School:** A highly innovative school of mathematics and astronomy in South India (c. 14th-16th c.), which explicitly built upon and extended the Aryabhatan tradition, notable for developing infinite series (calculus precursors) and refined astronomical models (Ch 8.2).

- **Khagola:** The celestial sphere. See Bhachakra.

- **Khaṇḍa-jyās:** "Sine segments" or "sine differences"; the 24 values listed by Aryabhata in Gk.10, which represent the successive differences between R-sine values (Ch 3.10).

- **Kinematic Simplification:** Explaining observed motion by proposing a simpler, more elegant underlying motion, without necessarily addressing the forces involved; used by Bhaskara I to support Earth's rotation (Ch 1.2).

- **Koṇa:** "Angle," "Corner"; identified with the planet Saturn in Gp.1. See also Śani.

- **Koṭi:** "End," "Peak"; the perpendicular side in a right triangle; trigonometrically, $R \cdot \cos(\theta)$ (Ch 4.5). Also refers to the number ten million (Gp.2).

- **Koṭiphala:** "Perpendicular-result"; the cosine component of the correction calculated from the R-cosine of the anomaly angle, used in refining planetary positions and distances (Ch 5.9, Kk.19).

- **Kṛṣṇa:** Black, dark; the appearance of the Moon during the partial phase of a lunar eclipse as it enters the umbra (Gola.46). Also refers to the waning fortnight (pakṣa).

- **Kṛta Yuga:** The first age in a Mahāyuga cycle. Followed by Tretā, Dvāpara, and Kali (Ch 3.5).

- **Kṣaya:** Decrease, loss; used for omitted lunar days (kṣayatithi or avamarātra) (Ch 5.5).

- **Kṣayavṛddhi:** Decrease and increase; refers to the variation in the length of daylight and night throughout the year due to latitude and the Sun's changing declination (Gola.19).

- **Kṣetra:** Field, area, geometric figure. Also used for space or the celestial sphere (Kk.2, Kk.25).
- **Kṣiticchāyā:** Earth's shadow. See Bhūcchāyā.
- **Kṣitija:** The horizon; the line separating the visible sky from the Earth, the boundary of the observer's local celestial sphere (Ch 6.8, Gola.18).
- **Kṣitijyā:** R·sin(Ascensional Difference); the sine value used to calculate the variation in daylight duration due to latitude and declination (Ch 6.9, Gola.26).
- **Kṣityunnati:** Elevation above the horizon; Altitude (Gola.28). See also Altitude.
- **Kuja:** "Earth-born"; a name for Mars (Gk.3, Gk.4, Gk.7, Gp.1, Kk.15, Gola.3). See also Bhauma.
- **Kudinārdham:** Half of a civil day; approximately 12 hours, the average length of daylight for an observer on Earth away from the poles (Gola.17).
- **Kusumapura:** "City of Flowers"; identified with Pataliputra (modern Patna), Aryabhata's likely place of work when he composed the Aryabhatiya around 499 CE (Intro, Ch 1.1, Gp.1).
- **Kuṭṭākāra:** "Pulverizer" or "Reducer"; Aryabhata's algorithm for finding integer solutions to linear indeterminate equations (aN □ C mod b), a sophisticated algebraic technique crucial for astronomical calculations involving cycles (Ch 4.7, Gp.32-33).

L

- **Lagna:** The Ascendant; the point of the ecliptic rising on the eastern horizon at a given moment. Calculated based on the observer's latitude and local time using unequal rising times (chara) (Ch 6.9, Gola.27 context).
- **Lambaka:** R·cos(latitude); the sine of the co-latitude (90° - latitude). Used in various trigonometric calculations related to latitude (Ch 6.9, Gola.26, Gola.28, Gola.30, Gola.35).
- **Lambana:** Parallax; the apparent shift in an object's position when viewed from different locations (e.g., from the Earth's surface versus its center). Significant for the Moon due to its proximity and requires correction for accurate position and eclipse timing (Ch 6.9, Gola.32, Gola.33).
- **Laṅkā:** Hypothetical location on the Earth's equator at 0° longitude, serving as the prime meridian for Aryabhata's astronomical system and the reference point for his epoch (Ch 5.7, Gola.13).
- **Latitude (Celestial):** See Vikṣepa.
- **Latitude (Terrestrial):** See Akṣa.
- **Leap Month:** See Adhimāsa.
- **Liber Abaci:** Leonardo of Pisa's influential book (1202 CE) that introduced and popularized the Hindu-Arabic numeral system and Indian arithmetic methods in Europe (Ch 9.4).
- **Liptā:** Arc-minute; 1/60th of a degree. Synonymous with kalā (Kk.2).

- **Longitude (Celestial):** Angular distance measured eastward along the ecliptic from the vernal equinox (0° Aries).

- **Longitude (Terrestrial):** Angular distance measured east or west from a prime meridian (Lanka).

M

- **Madhyama:** Mean or average; refers to mean planetary positions or motions calculated without corrections for anomalies (Ch 5.8, Ch 5.9, Kk.10).

- **Mādhava of Sangamagrāma:** Founder of the Kerala School of mathematics (c. 1340 – c. 1425), renowned for developing infinite series expansions for trigonometric functions (Ch 8.2.1).

- **Mahāyuga:** The "Great Age," a fundamental cosmological cycle of 4,320,000 solar years in Aryabhata's system, defined as the period required for the mean positions of the Sun, Moon, and planets (including nodes and apogees) to theoretically return to a conjunction at 0° Aries (Ch 3.3, Gk.3). Subdivided by Aryabhata into four equal Yugapādas (Gk.5).

- **Manda:** "Slow"; the first epicyclic correction, accounting for the equation of center or anomaly related to orbital eccentricity/apsidal line, causing variations in speed (Ch 5.9).

- **Mandakendra:** The angle of the slow anomaly; the angular distance between a planet's mean position and its Mandoccha (Ch 5.9, Kk.16).

- **Mandaphala:** The angular correction calculated from the manda anomaly, added to or subtracted from the mean longitude (Ch 5.9, Kk.20).

- **Mandoccha:** "Slow apogee"; the point in a planet's orbit (or its model) corresponding to its slowest motion or greatest distance (aphelion/apogee). The reference point for the manda correction (Ch 5.9, Kk.16).

- **Maṅgalācharaṇa:** Auspicious invocation at the beginning of a text (Gk.1, Gp.1).

- **Manvantara:** A secondary cosmic time cycle, consisting of 72 Mahāyugas according to Aryabhata's parameters (Gk.5). 14 Manvantaras make up a Kalpa.

- **Mars (Kuja, Bhauma):** The fourth planet (from the Sun) (Gk.3, Gp.1).

- **Mathematics (Gaṇita):** The science of calculation, computation, and quantitative reasoning. One of the three main subjects in the Aryabhatiya, providing the essential tools for astronomy (Intro, Ch 4, Gk.1).

- **Mean Daily Motion (Bhukti):** The average angular distance a celestial body travels in one day, derived from its total revolutions in a Yuga (Ch 5.8).

- **Mean Position (Madhyama Graha):** The average longitude of a celestial body calculated based on its mean motion and the elapsed time (ahargaṇa) since the epoch, before applying corrections for anomalies (Ch 5.8).

- **Meridian (Yāmyottara):** The great circle passing through the celestial poles and the observer's zenith and nadir. Objects reach their highest (or lowest) point in their daily path when crossing the meridian (Ch 6.8, Gola.18).

- **Meru:** Mythical cosmic mountain at the Earth's center in traditional cosmology. Rationalized by Aryabhata to likely represent the Earth's North Pole or the northern termination of the rotational axis in his spherical model (Ch 6.6, Gola.6, Gola.11, Gola.12).

- **Mercury (Budha):** The first planet (from the Sun) (Gk.4, Gp.1).
- **Miśraka-vyavahāra:** Rules concerning mixtures (alligation), a practical application of weighted averages described by Aryabhata (Ch 4.6, Gp.29).
- **Moon (Chandra, Śaśi, Indu):** Earth's natural satellite (Gk.3). Aryabhata provides its parameters, calculates its motions, accounts for its latitude (vikṣepa), and uses its position relative to the Sun/shadow for eclipse calculations (Ch 5, Ch 6.2, Ch 6.11).
- **Mṛjjalaśikhivāyumaya:** Made of earth, water, fire, and air; described as the composition of the Earth-sphere (Gola.5). Ether (ākāśa) is added in Gola.6 for the five elements.

N

- **Nāḍī:** Unit of time, 1/60th of a day (24 minutes). Synonymous with Ghaṭikā. Measured using water clocks (Ch 2.2, Kk.1).
- **Nakṣatra:** Lunar mansion; one of 27 (or 28) divisions of the zodiac used for tracking the Moon's position. Also refers generally to stars or asterisms (Gk.3, Kk.5).
- **Nākṣatra:** Sidereal; refers to time measured relative to the distant fixed stars, particularly the sidereal day (Earth's rotation period) (Ch 5.4, Kk.5).
- **Nīlakaṇṭha Somayājī:** Important astronomer of the Kerala School (c. 1444–1544), author of the Tantrasaṅgraha and a commentary on the Aryabhatiya; proposed a geo-heliocentric model (Ch 8.2.2).
- **Nimagnaṃ:** Submerged, lost, obscured; used by Aryabhata to describe the state of true knowledge before he "rescued" it (Gola.49).
- **Nodes (Pāta):** Points where the orbit of a celestial body (specifically the Moon's orbit) intersects the ecliptic plane. Crucial for determining if an eclipse will occur (Gk.4, Gola.38).
- **North Pole:** The northern termination of Earth's rotational axis, projected onto the celestial sphere as the North Celestial Pole (Ch 6.6). Associated with Meru in Aryabhata's cosmology.
- **Nṛṣi:** A term in Gk.7 related to the definition of the Yojana, often interpreted as referencing human height or a similar standard measure.
- **Number System (Aryabhata's):** The methods used by Aryabhata to represent numbers. Includes his unique alphabetical system (Gk.2) and his articulation of the decimal place-value principle (Gp.2).

O

- **Obliquity of the Ecliptic:** The angle (approx. 24° for Aryabhata, Gk.9) between the plane of the celestial equator and the plane of the ecliptic. Causes seasons (Gola.1) and variations in day/night length.
- **Observation (Pratyakṣa):** The act of directly observing celestial phenomena or making measurements using

instruments (like the gnomon or water clock). Provides the empirical data that grounds astronomical theory and calculation (Ch 2, Ch 6.9).

- **OCR (Optical Character Recognition):** Digital technology used to extract text (Sanskrit in this case) from images or physical documents (Disclosure).
- **Omitted Lunar Day (Avamarātra):** See Avamarātra.
- **Orbit (Kakṣyā):** The curved path followed by a celestial body as it revolves around another (e.g., Moon around Earth, Earth around Sun conceptually, planets around Earth in the geocentric model). Aryabhata defines the scale of these orbits in yojanas (Gk.6, Kk.13).
- **Orientation (Eclipse):** The apparent angle of the contact points (where the eclipsing body touches the eclipsed body) on the disk during an eclipse, calculated using valana corrections (Ch 6.11, Gola.45).
- **Overture (Cosmic):** Used in Chapter 3.0 to describe the Gītikāpāda as the foundational opening section setting the stage for Aryabhata's system.

P

- **Pāda:** Quarter. Used for the four equal divisions (yugapādas) of the Mahāyuga (Gk.5, Kk.7 context). Also used for the four sections of the Aryabhatiya (Intro).
- **Pakṣa:** "Wing"; a fortnight (half a lunar month). Waning (kṛṣṇa) or waxing (śukla) fortnight.
- **Pañcāṅga:** Traditional Indian almanac, a calendar based on five elements (lunar day/tithi, weekday/vāra, lunar mansion/nakṣatra, yoga, karaṇa); relies on precise astronomical calculations (Ch 5.2, Ch 5.5).
- **Parallax (Lambana):** The apparent shift in an object's position when viewed from different locations. Parallax in altitude occurs because the observer is on the Earth's surface, not its center (Ch 6.9, Gola.32, Gola.33). Parallax in longitude affects the timing of phenomena like solar eclipses.
- **Parameśvara:** A scholar of the Kerala School (c. 1370–1460) who contributed to observational astronomy and revised parameters within the Aryabhatan tradition (Ch 8.2.2).
- **Pāta:** "Falling," "Intersection"; the ascending or descending node of a body's orbit (where it crosses the ecliptic). Lunar nodes are crucial for eclipse possibility (Ch 6.2, Gola.3, Gola.38).
- **Pataliputra:** Ancient name for the city where Aryabhata likely worked; identified with Kusumapura (Intro).
- **Pi (π):** The mathematical constant representing the ratio of a circle's circumference to its diameter ($\approx$ 3.1415 9...). Aryabhata provides a remarkably accurate value of 3.1416 (Ch 4.4, Gp.10).
- **Place Value:** The principle in a positional numeral system where the value of a digit is determined by its position within the number (Ch 4.2, Gp.2).
- **Planets (Graha):** Celestial bodies (including Sun and Moon in traditional Indian astronomy) whose positions and motions are calculated and predicted by Aryabhata (Ch 5.8, Ch 5.9).
- **Poles (Celestial/Terrestrial):** The northern and southern endpoints of the axis of rotation. Earth's poles are

the geographic poles; their projection onto the celestial sphere defines the celestial poles, around which daily apparent motion occurs (Ch 6.8, Gola.18, Gola.24).

- **Positional Astronomy:** The branch of astronomy concerned with the positions and motions of celestial bodies on the celestial sphere, often using coordinate systems (e.g., equatorial, horizontal, ecliptic) (Ch 6.8).
- **Prāṇa:** "Breath"; a traditional Indian unit of time, standardized within Aryabhata's system such that 6 prāṇas equal 1 vināḍikā (approx. 4 modern seconds per prāṇa). Aryabhata establishes a crucial 1:1 equivalence between 1 prāṇa of time and 1 kalā (arc-minute) of celestial rotation (Gk.11, Kk.2).
- **Pratikañchuka:** "Counter-jacket"; a counterfeit, plagiarism, or deliberate distortion of a text or knowledge. Aryabhata issues a stern warning against creating one of his Aryabhatiya (Ch 6.12, Gola.50).
- **Pratiloma:** "Against the grain"; Retrograde or backward motion. Used to describe the westward apparent motion of the Śīghroccha for outer planets or the direction of motion relative to a reference point (Gk.4, Kk.21).
- **Pratimaṇḍala:** Epicycle; the smaller circle in epicyclic models (Ch 5.9).
- **Pratyakṣa:** Direct perception, observation; one of the valid means of knowledge (pramāṇa) in Indian philosophy, complementing theoretical calculation (Ch 10.2.2).
- **Pravaha Vāyu:** The "Provector Wind"; a cosmic wind invoked in traditional Indian cosmology as the cause of the daily rotation of the heavens around a static Earth. Presented by Aryabhata (Gola.10) as an alternative to his Earth rotation theory.
- **Precession of the Equinoxes:** The slow westward drift of the vernal equinox along the ecliptic, causing the sidereal year to be slightly longer than the tropical year. While a complex phenomenon, its effect on long-term parameters was known or accounted for in Indian astronomy systems (Ch 5.12).
- **Preface:** The author's introductory remarks on the book's motivation, scope, and approach.
- **Prime Meridian (Laṅkā):** The zero-longitude line on Earth (passing through Lanka in Aryabhata's system) used as a reference for terrestrial and celestial longitudes and time zones (Ch 5.7, Gola.13).
- **Prime Vertical (Samamaṇḍala):** The great circle on the celestial sphere passing through the observer's Zenith, Nadir, and the exact East and West points on the horizon (Ch 6.8, Gola.20).
- **Proportion (Rule of Three, Trairāśika):** A fundamental mathematical method for solving problems involving direct proportionality: if A yields B, then C yields X, where X = (C × B) / A. An indispensable tool used throughout the Aryabhatiya for scaling parameters and converting units (Ch 4.6, Gp.26).
- **Ptolemy, Claudius:** Influential Greco-Egyptian astronomer (c. 150 CE), author of the Almagest; his geocentric system and mathematical methods (chords, equants) are compared and contrasted with Aryabhata's (Ch 7).
- **Pulverizer Algorithm:** See Kuṭṭākāra.
- **Purāṇas:** Traditional Hindu texts containing cosmologies, genealogies, and narratives. Some Purāṇic chronological ideas (like unequal yuga lengths) differed from the mathematically consistent framework used by Aryabhata (Ch 1.1, Ch 3.5, Ch 5.6).

- **Puruṣa:** Man, person; sometimes used as a unit of measure (e.g., human height) for defining the Yojana (Ch 3.7, Gk.7).

- **Pythagorean Theorem:** The geometric theorem stating that in a right triangle, the square of the hypotenuse is equal to the sum of the squares of the other two sides ($a^2 + b^2 = c^2$). Explicitly stated by Aryabhata (Gp.17) and used in various calculations (gnomon shadows, eclipse durations) (Ch 4.5, Ch 6.9, Ch 6.11).

Q

- **Quadratic Equations:** Algebraic equations in which the highest power of the variable is 2 (e.g., $ax^2 + bx + c = 0$). Aryabhata's methods for finding terms in series or solving for unknowns in interest problems imply the ability to solve these (Ch 4.6, Gp.21, Gp.24, Gp.25).

- **Quarters (Yuga Pādas):** The four equal divisions (Kṛta, Tretā, Dvāpara, Kali) of a Mahāyuga (each lasting 1,080,000 years) in Aryabhata's mathematically consistent system (Ch 3.5, Gk.5).

R

- **Rāhu:** Mythological demon head believed in some traditions to cause eclipses by swallowing the Sun or Moon. Aryabhata explicitly rejects this myth in favor of a physical shadow-based explanation (Ch 6.10, Gola.37).

- **Rāśi:** "Heap," "Quantity"; a Zodiac Sign, representing 1/12th of the ecliptic (30°) (Kk.2).

- **Ravi:** A name for the Sun (Gk.3, Gp.1, Kk.3, Kk.5, Kk.12, Gola.2, Gola.4, Gola.17, Gola.47). See also Arka, Sūrya.

- **Reasoning (Yukti):** Rational argument, logical justification, or proof. Emphasized by Bhaskara I in defending Aryabhata's points (Ch 1.2) and by the Kerala School (Yuktibhāṣā) (Ch 8.2.2).

- **Relative Motion:** The perception of motion from a moving frame of reference. Aryabhata used the analogy of a man in a boat to explain Earth's rotation as causing the apparent motion of stationary stars (Ch 6.5, Gola.9).

- **Relativity (of Perspective/Time):** The concept that perceived direction (like "down") or the duration of time cycles (like day/night) is relative to the observer's position and motion on a spherical Earth (Ch 6.6, Ch 6.7, Gola.12, Gola.17).

- **Retrograde Motion (Pratiloma):** The apparent temporary backward (westward) movement of a planet against the backdrop of stars, caused by the combined motions of the Earth and the planet in their orbits. Accounted for by the śīghra correction in Aryabhata's model (Ch 5.9).

- **Revolutions (Bhagaṇa):** See Bhagaṇa.

- **Ṛta:** Cosmic order, truth, natural law. The concept of an inherent, rational structure to the universe that can be understood through science and mathematics (Ch 10.1).

- **Ṛṇa:** Debt, negative; signifying a subtractive correction or quantity in calculations (Ch 5.9, Kk.21, Gola.35, Gola.36).

- **Romaka:** Location 180° longitude from Lanka in Aryabhata's equatorial cities model, possibly representing

Rome or Alexandria conceptually (Gola.13).

- **Roots (Square, Cube):** Algorithms for extracting the square root (Gp.4) and cube root (Gp.5) of numbers, essential for many calculations.
- **Rotation (Earth's Axial):** See Earth: Rotation Theory.
- **RSA Cryptography:** A widely used modern encryption algorithm whose security relies on mathematical principles related to finding modular inverses, a specific application of the Extended Euclidean Algorithm which is the basis for Aryabhata's Kuṭṭākāra (Ch 4.7, Ch 9.6).
- **Rule of Three (Trairāśika):** See Trairāśika.

S

- **Sakṛṣṇatāmra:** Dark coppery red; the characteristic color of the Moon deep within the Earth's umbra during a total lunar eclipse ("blood moon") (Ch 6.11, Gola.46).
- **Samamaṇḍala:** The Prime Vertical; the great circle passing through the observer's Zenith, Nadir, East, and West points (Ch 6.8, Gola.20).
- **Sampradāya:** Tradition, lineage of teaching; the established schools of thought or scientific transmission in Indian intellectual history.
- **Sāṃkhya:** One of the classical schools of Indian philosophy, with concepts (like Prakriti and Purusha) that touch upon the nature of reality and cosmic evolution, potentially forming part of the philosophical backdrop for scientific inquiry (Ch 5.1, Ch 10.1.4).
- **Śaṅku:** A gnomon (vertical staff for measuring shadows). See Gnomon.
- **Śani:** The planet Saturn (Gk.3, Gk.4, Gk.7, Gp.1, Kk.3, Kk.15, Gola.3, Gola.4). See also Koṇa.
- **Śara:** "Arrow"; the versine function, R(1-cosθ), representing the height of a circular segment measured from the midpoint of the arc to the midpoint of the chord (Ch 4.4, Gp.9, Gp.12).
- **Śaśi:** "Hare-marked"; a name for the Moon (Gk.3, Gp.1, Kk.3, Kk.5, Kk.15, Gola.37, Gola.46). See also Chandra, Indu.
- **Śāstra:** Science, scripture, authoritative text; refers generally to bodies of systematized knowledge.
- **Saturn (Śani, Koṇa):** The sixth planet (from the Sun) (Gk.3, Gp.1).
- **Satya:** Truth, reality; the ultimate aim of philosophical and scientific inquiry (Ch 10.2.3).
- **Sāvana:** Civil; refers to time measured by civil days, typically sunrise to sunrise. The ahargaṇa is a count of sāvana days (Ch 5.4, Kk.5).
- **Saura:** Solar; refers to time measured by the Sun's motion, specifically the solar year (Ch 5.4, Kk.5).
- **Seasons:** Caused by the tilt of the Earth's axis (or the ecliptic relative to the celestial equator) as the Earth orbits

the Sun. Aryabhata describes this geometry (Gola.1).

- **Sexagesimal:** Base-60 number system. Influence from Mesopotamia seen in the division of the day and circle into 60 units and subunits (Nāḍī, Vināḍikā, Degree, Minute, Second) (Ch 5.2).
- **Shadow:** See Chāyā, Bhūcchāyā.
- **Siddhānta:** "Established doctrine"; a type of comprehensive Indian astronomical treatise, typically blending mathematical astronomy with cosmological ideas. Aryabhata's work can be seen as a tantra (manual) within this broader tradition (Ch 1.1).
- **Sidereal:** Relative to the distant fixed stars (Nākṣatra). Used for rotation periods (sidereal day) and orbital periods measured against the background star field (Ch 5.4).
- **Śīghra:** "Fast"; the second epicyclic correction, related to synodic phenomena (like conjunctions, oppositions, and apparent retrograde motion) and the variable elongation from the Sun (Ch 5.9).
- **Śīghrakendra:** The angle of the fast anomaly; the angular distance between the position (often the once-corrected position) of a planet and its Śīghroccha (Ch 5.9, Kk.16).
- **Śīghraphala:** The angular correction calculated from the śīghra anomaly, applied to the (usually once-corrected) longitude to find the true position (Ch 5.9, Kk.21).
- **Śīghroccha:** "Fast apogee"; a reference point (often linked to the mean Sun or its position) for the śīghra correction (Ch 5.9, Kk.16).
- **Sindhind:** Arabic term for Indian astronomical works (Siddhāntas) transmitted to Baghdad (c. 8th century CE), particularly those related to or derived from Brahmagupta's school, carrying Indian mathematical and astronomical concepts westward (Ch 9.1).
- **Sine (Indian):** See Jyā.
- **Sinus:** Latin term for the sine function, derived via the Arabic term jayb (from Sanskrit jyā), entering European mathematics through translations of Arabic works (Ch 9.2).
- **Solar:** Relating to the Sun (Saura). Refers to time measured by the Sun's apparent motion, especially the solar year (Ch 5.4).
- **Sphere:** See Gola, Celestial Sphere, Earth-Sphere.
- **Spherical Trigonometry:** The branch of mathematics dealing with the relationships between angles and sides of triangles on the surface of a sphere. Implicitly used throughout the Golapāda for calculations on the celestial sphere (Ch 6.9).
- **Sphuṭa:** "Clear," "True"; refers to the corrected, true position or motion of a planet after applying epicyclic corrections, representing its observed location (Ch 5.9, Kk.20, Kk.21).
- **Sphuṭīkaraṇa:** The process of calculating the true position (sphuṭa) from the mean position (madhyama) by applying corrections.

- **Square and Square Root:** Geometric concept of a four-sided figure with equal sides (square) and multiplying two equal quantities (squaring); algorithm for extracting the square root of a number (Ch 4.3, Gp.3, Gp.4).
- **Stars:** Distant celestial bodies (Bhāni, Nakṣatra). Treated as fixed points in Aryabhata's model, whose apparent daily motion is caused by Earth's rotation (Gola.9).
- **Sthāna:** Place, position; fundamental concept in the decimal place-value system where the value of a digit depends on its position (Ch 4.2, Gp.2).
- **Sthiti / Sthityardha:** (Half) Duration of the entire eclipse event, including the partial phases (from first to last contact). Calculated using the sthityardha radius (Gola.41, Gola.42).
- **Śukla:** White, bright; the waxing fortnight of the lunar month (from New Moon to Full Moon).
- **Śuklādi:** Convention where the lunar month begins on the day after the New Moon conjunction (śukla pakṣa). Used by Aryabhata (Kk.11).
- **Śukra:** A name for Venus (Gk.4, Gk.7, Gp.1, Kk.15, Gola.3, Gola.4). See also Bhṛgu.
- **Śulba Sūtras:** Ancient Indian texts (predating Aryabhata) containing geometric rules and methods for constructing Vedic altars, demonstrating early geometric knowledge (Ch 4.1).
- **Śūnya:** Empty, void; the concept and (later) symbol for zero. Crucial for the operational power of the decimal place-value system (Ch 4.2).
- **Sun (Ravi, Sūrya, Arka):** The star at the center of our solar system. In Aryabhata's geocentric model, it orbits the Earth (Kk.15); its apparent path (ecliptic, Gola.1) and eclipses (Gola.37) are calculated.
- **Sūrya Siddhānta:** An influential, traditional Indian astronomical treatise (extant version is later than Aryabhata), representing a different school of astronomical parameters and methods compared to Aryabhata's (Ch 1.1).
- **Svamati:** One's own intellect, reasoning, or mind; emphasized by Aryabhata as a valid and necessary means of knowledge for discerning truth and advancing science, complementing tradition (Ch 6.12, Gola.49, Ch 10.2.2).
- **Synodic Period:** The period required for a celestial body (especially the Moon or a planet) to return to the same position relative to the Sun as seen from Earth (e.g., synodic month = New Moon to New Moon).

T

- **Tantrasaṅgraha:** A major astronomical treatise by Nīlakaṇṭha Somayājī of the Kerala School (c. 1500), which expanded upon and updated Aryabhata's system (Ch 8.2.2).
- **Tatparā:** "Beyond that"; a very small unit of angular measure, 1/60th of an arcsecond (vikalā) (Kk.2).
- **Taylor Series:** Infinite series representing functions as a sum of terms involving their derivatives at a single point. Equivalents for trigonometric functions were developed by the Kerala School (Ch 8.2.1).
- **Tithi:** A lunar day, defined as the time taken for the Moon's elongation from the Sun to increase by 12°. Approximately 30 tithis make up a synodic month. The difference between the count of tithis and civil days

necessitates omitted days (avamarātra) (Ch 5.5, Kk.1).

- **Trairāśika:** The Rule of Three. See Proportion.
- **Tradition (Sampradāya):** The body of inherited knowledge, texts (śāstra), and practices passed down through generations of scholars. Aryabhata built upon and engaged critically with existing traditions (Ch 1.1, Ch 10.2.4).
- **Transmission (Global):** The process by which Indian scientific concepts (numerals, zero, sine, algorithms) were carried westward, primarily via the Islamic world (Sindhind, Al-Khwārizmī), influencing mathematics and astronomy in Europe (Fibonacci) (Ch 9).
- **Tretā Yuga:** The second age in a Mahāyuga cycle, between Kṛta and Dvāpara (Ch 3.5).
- **Trijyā:** "Three-radius" (possibly a misnomer or referring to the third quadrant sine); the standard radius used in Aryabhata's sine table calculations, equal to 3438 minutes of arc (Gk.10). Used in various trigonometric formulas (Ch 6.9).
- **Trigonometry:** The branch of mathematics dealing with the relationships between angles and sides of triangles, particularly using functions like sine, cosine, and tangent. Aryabhata developed a sophisticated sine-based system (Jyā) (Ch 3.10, Ch 4.4, Ch 5.9, Ch 6.9).
- **True Position/Motion (Sphuṭa):** The corrected, observed position or motion of a planet after applying epicyclic corrections (manda, śīghra) to its mean position/motion. The ultimate goal of planetary calculation (Ch 5.9, Kk.20, Kk.21).

U

- **Uccha:** "High"; refers generally to an apogee or related reference point for calculating anomalies (manda or śīghra) (Kk.4, Kk.16). See also Apogee, Mandoccha, Śīghroccha.
- **Ujjayinī:** Important ancient Indian city (modern Ujjain) in central India, a major center for astronomy. Aryabhata calculates its latitude (approx. 22.5° N, Gola.14), likely based on local gnomon observations.
- **Umbra:** The darkest, central part of a shadow where the light source is completely blocked. Earth's umbra is the focus of lunar eclipse calculations (Gola.39, Gola.40).
- **Unequal Rising Times:** The phenomenon where different zodiac signs (or 30° segments of the ecliptic) take different amounts of time to rise above the horizon at a given latitude (except at the equator), due to the tilt of the ecliptic. Calculated using the chara correction (Gola.27).
- **Unmaṇḍala:** The Equatorial Horizon or Six o'clock line; the great circle on the celestial sphere passing through the observer's East and West points and the celestial poles (Gola.19).
- **Utkramajyā:** Versed sine; R(1-cosθ). Mathematically equivalent to Śara. Used in calculations involving the ecliptic's inclination (Gola.36).

V

- **Valana:** "Deflection," "Inclination"; angular corrections (Akṣa- and Ayana-valana) applied in eclipse calculations to determine the precise orientation of the contact points on the eclipsed body's disk from the observer's perspective (Ch 6.11, Gola.45).

- **Varga:** "Class," "Group," "Square"; refers to the first set of consonants (k to m) in Aryabhata's alphabetical number system (Gk.2), representing digits 1-25. Also means the square of a number (Gp.3).

- **Varṣa:** Year (Kk.1, Kk.7).

- **Vedānta:** One of the classical schools of Hindu philosophy; along with others, it contributed to the philosophical backdrop of Indian science, emphasizing the exploration of reality (Ch 10.1.4).

- **Vedic Period:** Ancient period in India associated with the Vedas and early rituals, some aspects of which (like altar construction geometry in the Śulba Sūtras) influenced later mathematics (Ch 4.1).

- **Venus (Bhṛgu, Śukra):** The second planet (from the Sun) (Gk.4, Gp.1).

- **Versed Sine:** See Śara, Utkramajyā.

- **Vikalā / Viliptā:** Arcsecond; 1/60th of an arc-minute (kalā) (Kk.2).

- **Vikṣepa:** "Scattering," "Deviation"; Celestial latitude, the angular distance of a body North or South of the ecliptic plane (Gk.9, Gola.3).

- **Vilomā:** Against the grain, reverse; Retrograde motion. Used to describe the motion of the lunar nodes (Gk.4).

- **Vimarda / Vimardārdha:** "Crushing," "Rubbing"; (Half) Duration of the total or annular phase of an eclipse, calculated using the vimardārdha radius (Gola.41, Gola.43).

- **Vimardārdha Radius:** Half the absolute difference of the diameters of the eclipsing and eclipsed bodies; determines the geometric limit for totality or annularity (Gola.41).

- **Vināḍikā / Vighaṭikā:** Unit of time, 1/60th of a nāḍī (approx. 24 seconds) (Kk.1, Kk.2). 6 prāṇas = 1 vināḍikā.

- **Viparyaya:** Inversion, reversal; used in the Inverse Rule of Three (Gp.27).

- **Viṣkambha:** Diameter. See Diameter.

- **Visibility:** The condition under which a celestial body can be seen, affected by factors like elongation from the Sun (Gola.4) or the magnitude of an eclipse (Gola.48). Positional corrections (dṛkkarma) are also related to visibility near the horizon (Ch 6.10).

- **Viṣuvat:** "Equally divided"; refers to both the Celestial Equator (the line) and the Equinox (the points in time or on the ecliptic where the Sun crosses the equator) (Gola.19, Gola.25, Gola.30, Gola.31).

- **Volume calculation:** Methods for calculating the volume of geometric solids like cubes (Gp.3), prisms/pyramids (Gp.6), and spheres (Gp.7). Note: Aryabhata's sphere volume formula is uniquely interpreted and appears geometrically inaccurate compared to modern formula.

- **Vṛnda:** Sanskrit term for a billion (10^9) in Aryabhata's list of powers of ten (Gp.2).

- **Vṛtta:** Circle; used for the horizon circle (Gola.18), the ecliptic or other celestial circles (Gola.19, Gola.25), or an epicycle (Vṛttam, Kk.17).

W

- **Water Clock (Ghaṭikā Yantra):** A device (typically inflow type) used to measure time by the flow of water, defining units like the nāḍī/ghaṭikā (Ch 2.2, Kk.1).

Y

- **Yajña:** Vedic ritual or sacrifice, often requiring precise astronomical timing for auspiciousness (Ch 5.1).
- **Yamakoṭi:** Mythical city located 90° East of Lanka on the equator in Aryabhata's model, used in illustrating time zones (Gola.13).
- **Yojana:** A unit of distance used in ancient India. Its exact modern equivalent is debated (estimated as 5-9 miles or 8-15 km). Used by Aryabhata to define the scale of the cosmos and the diameters of celestial bodies (Ch 3.6, Ch 3.7, Gk.6, Gk.7).
- **Yoga:** Conjunction; the joining or alignment of celestial bodies (Kk.3, Kk.5). Also used more generally for combination or calculation steps (Gola.31).
- **Yuga:** Age or Era; a long cosmological time period. Refers to the Mahāyuga and its subdivisions (Kṛta, Tretā, Dvāpara, Kali) (Ch 3.5).
- **Yuga: Equal vs Unequal division:** Aryabhata's system divides the Mahāyuga into four equal yugapādas, contrasting with the traditional unequal (4:3:2:1) Purāṇic division (Ch 3.5, Ch 5.6, Ch 8.1.2).
- **Yugapāda:** A quarter of a Mahāyuga (1,080,000 years) in Aryabhata's system (Gk.5, Kk.7).
- **Yukti:** Reasoning, rationale, argument, proof. Emphasized in scholarly discourse, particularly in texts like Yuktibhāṣā, for justifying mathematical theorems (Ch 1.2, Ch 8.2.2).
- **Yuktibhāṣā:** "Rationale-Language" or "Exposition of Proofs"; an important text by Jyeṣṭhadeva of the Kerala School (c. 1500-1575) providing detailed rationales for the school's mathematical discoveries (Ch 8.2.2).

Z

- **Zenith:** The point on the celestial sphere directly above the observer (Ch 6.8, Gola.18).
- **Zenith Distance:** The angular distance of a celestial body from the observer's zenith. Complementary to altitude (ZD = 90° - Alt) (Ch 6.9).
- **Zero (Śūnya):** The concept of a null value or emptiness and the placeholder digit in the decimal system. Crucial for the positional notation articulated by Aryabhata (Gp.2) and its computational power (Ch 4.2). Its computational rules were further developed by Brahmagupta (Ch 9.4).
- **Zīj:** Islamic astronomical handbook, typically containing tables for calculating celestial positions and related phenomena, along with explanatory text. Zīj al-Sindhind refers to those based on Indian sources (Ch 9.1).

- **Zodiac:** The belt of 12 constellations (rāśis) along the ecliptic, used as a reference frame for celestial longitude (Ch 6.1).

Chapter Fourteen

Index

Note: References are to Chapter. Verse number for the four pādas (e.g., Gk.3, Gp.10, Kk.5, Gola.9) or to Chapter. Section number for contextual chapters (e.g., Ch 1.2, Ch 7.1). 'Intro' refers to the main book Introduction, 'Pref' to Preface.

H

I

J

K

N

O

P

R

S

T

U

V

Z

***

Chapter Fifteen

Audio Content

Addtional resources for learning

The author provides this audio content free of charge. Please note that hosting platforms may have their own access fees or subscription models, which are determined by their respective terms and conditions and are separate from this offering.

AUDIO CONTENT TO HELP WITH LEARNING (WHOLE BOOK COVERED)

APPLE

YOUTUBE

CASTBOX

AMAZON MUSIC

SPOTIFY

IHEART

AMAZON CA

AUDIBLE

www.ingramcontent.com/pod-product-compliance
Lightning Source LLC
LaVergne TN
LVHW070831220926
848486LV00030B/267

* 9 7 8 1 0 6 9 4 8 8 7 2 5 *